Mathematical Statistics

Old School

John I. Marden
Department of Statistics
University of Illinois at Urbana-Champaign

Typeset using the memoir package (Madsen and Wilson, 2015) with LATEX (Lamport, 1994).

Preface

My idea of mathematical statistics encompasses three main areas: The mathematics needed as a basis for work in statistics; the mathematical methods for carrying out statistical inference; and the theoretical approaches for analyzing the efficacy of various procedures. This book is conveniently divided into three parts roughly corresponding to those areas.

Part I introduces distribution theory, covering the basic probability distributions and their properties. Here we see distribution functions, densities, moment generating functions, transformations, the multivariate normal distribution, joint marginal and conditional distributions, Bayes theorem, and convergence in probability and distribution.

Part II is the core of the book, focussing on inference, mostly estimation and hypothesis testing, but also confidence intervals and model selection. The emphasis is on frequentist procedures, partly because they take more explanation, but Bayesian inference is fairly well represented as well. Topics include exponential family and linear regression models; likelihood methods in estimation, testing, and model selection; and bootstrap and randomization techniques.

Part III considers statistical decision theory, which evaluates the efficacy of procedures. In earlier years this material would have been considered the essence of mathematical statistics — UMVUEs, the CRLB, UMP tests, invariance, admissibility, minimaxity. Since much of this material deals with small sample sizes, few parameters, very specific models, and super-precise comparisons of procedures, it may now seem somewhat quaint. Certainly, as statistical investigations become increasingly complex, there being a single optimal procedure, or a simply described set of admissible procedures, is highly unlikely. But the discipline of stating clearly what the statistical goals are, what types of procedures are under consideration, and how one evaluates the procedures, is key to preserving statistics as a coherent intellectual area, rather than just a handy collection of computational techniques.

This material was developed over the last thirty years teaching various configurations of mathematical statistics and decision theory courses. It is currently used as a main text in a one-semester course aimed at master's students in statistics and in a two-semester course aimed at Ph. D. students in statistics. Both courses assume a prerequisite of a rigorous mathematical statistics course at the level of Hogg, McKean, and Craig (2013), though the Ph. D. students are generally expected to have learned the material at a higher level of mathematical sophistication.

Much of Part I constitutes a review of the material in Hogg, et. al., hence does not need to be covered in detail, though the material on conditional distributions, the multivariate normal, and mapping and the Δ-method in asymptotics (Chapters 6, 7, and 9) may need extra emphasis. The masters-level course covers a good chunk of Part II, particularly Chapters 10 through 16. It would leave out the more technical sections on likelihood asymptotics (Sections 14.4 through 14.7), and possibly the material on regularization and least absolute deviations in linear regression (Sections 12.5 and 12.6). It would also not touch Part III. The Ph. D.-level course can proceed more quickly through Part I, then cover Part II reasonably comprehensively. The most typical topics in Part III to cover are the optimality results in testing and estimation (Chapters 19 and 21), and general statistical decision theory up through the James-Stein estimator and randomized procedures (Sections 20.1 through 20.7). The last section of Chapter 20 and the whole of Chapter 22 deal with necessary conditions for admissibility, which would be covered only if wishing to go deeper into statistical decision theory.

The mathematical level of the course is a bit higher than that of Hogg et al. (2013) and in the same ballpark as texts like the mathematical statistics books Bickel and Doksum (2007), Casella and Berger (2002), and Knight (1999), the testing/estimation duo Lehmann and Romano (2005) and Lehmann and Casella (2003), and the more decision-theoretic treatments Ferguson (1967) and Berger (1993). A solid background in calculus and linear algebra is necessary, and real analysis is a plus. The later decision-theoretic material needs some set theory and topology. By restricting primarily to densities with respect to either Lebesgue measure or counting measure, I have managed to avoid too much explicit measure theory, though there are places where "with probability one" statements are unavoidable. Billingsley (1995) is a good resource for further study in measure theoretic probability.

Notation for variables and parameters mostly follows the conventions that capital letters represent random quantities, and lowercase represent specific values and constants; bold letters indicate vectors or matrices, while non-bolded ones are scalars; and Latin letters represent observed variables and constants, with Greek letters representing parameters. There are exceptions, such as using the Latin "p" is as a parameter, and functions will usually be non-bold, even when the output is multidimensional.

This book would not exist if I didn't think I understood the material well enough to teach it. To the extent I do, thanks go to my professors at the University of Chicago, especially Raj Bahadur, Michael Perlman, and Michael Wichura.

Contents

Part I

Distribution Theory

Chapter 1

Distributions and Densities

1.1 Introduction

This chapter kicks off Part I, in which we present the basic probability concepts needed for studying and developing statistical procedures. We introduce probability distributions, transformations, and asymptotics. Part II covers the core ideas and methods of statistical inference, including frequentist and Bayesian approaches to estimation, testing, and model selection. It is the main focus of the book. Part III tackles the more esoteric part of mathematical statistics: decision theory. The main goal is to evaluate inference procedures, to determine which do a good job. Optimality, admissibility, and minimaxity are the main topics.

1.2 Probability

We quickly review the basic definition of a probability distribution. Starting with the very general, suppose X is a random object. It could be a single variable, a vector, a matrix, or something more complicated, e.g., a function, infinite sequence, or image. The **space** of X is $\mathcal{X}$, the set of possible values X can take on. A probability distribution on X, or on $\mathcal{X}$, is a function P that assigns a value in $[0,1]$ to subsets of $\mathcal{X}$. For "any" subset $A \subset \mathcal{X}$, $P[A]$ is the probability $X \in A$. It can also be written $P[X \in A]$. (The quotes on "any" are to point out that technically, only subsets in a "sigma field" of subsets of $\mathcal{X}$ are allowed. We will gloss over that restriction, not because it is unimportant, but because for our purposes we do not get into too much trouble doing so.)

In order for P to be a probability distribution, it has to satisfy two axioms:

1. $P[\mathcal{X}] = 1$;

2. If $A_1, A_2, \ldots$ are disjoint ($A_i \cap A_j = \varnothing$ for $i \neq j$), then

$$P[\cup_{i=1}^{\infty} A_i] = \sum_{i=1}^{\infty} P[A_i]. \tag{1.1}$$

The second axiom means to refer to finite unions as well as infinite ones. Using these axioms, along with the restriction that $0 \leq P[A] \leq 1$, all the usual properties of probabilities can be derived. Some such follow.

Complement. The complement of a set A is $A^C = \mathcal{X} - A$, that is, everything that is not in A (but in $\mathcal{X}$). Clearly, A and A^C are disjoint, and their union is everything:

$$A \cap A^C = \varnothing, \quad A \cup A^C = \mathcal{X}, \tag{1.2}$$

so,

$$1 = P[\mathcal{X}] = P[A \cup A^C] = P[A] + P[A^C], \tag{1.3}$$

which means

$$P[A^C] = 1 - P[A]. \tag{1.4}$$

That is, the probability the object does not land in A is 1 minus the probability that it does land in A.

Empty set. $P[\varnothing] = 0$, because the empty set is the complement of $\mathcal{X}$, which has probability 1.

Union of two (nondisjoint) sets. If A and B are not disjoint, then it is not necessarily true that $P[A \cup B] = P[A] + P[B]$. But $A \cup B$ can be separated into two disjoint sets: the set A and the part of B not in A, which is $[B \cap A^c]$. Then

$$P[A \cup B] = P[A] + P[B \cap A^c]. \tag{1.5}$$

Now $B = (B \cap A) \cup (B \cap A^c)$, and $(B \cap A)$ and $(B \cap A^c)$ are disjoint, so

$$P[B] = P[B \cap A] + P[B \cap A^c] \Rightarrow P[B \cap A^c] = P[B] - P[A \cap B]. \tag{1.6}$$

Then stick that formula into (1.5), so that

$$P[A \cup B] = P[A] + P[B] - P[A \cap B]. \tag{1.7}$$

The above definition doesn't help much in specifying a probability distribution. In principle, one would have to give the probability of every possible subset, but luckily there are simplifications.

We will deal primarily with random variables and finite collections of random variables. A random variable has space $\mathcal{X} \subset \mathbb{R}$, the real line. A collection of p random variables has space $\mathcal{X} \subset \mathbb{R}^p$, the p-dimensional Euclidean space. The elements are usually arranged in some convenient way, such as in a vector (row or column), matrix, multidimensional array, or triangular array. Mostly, we will have them arranged as a row vector $\mathbf{X} = (X_1, \ldots, X_n)$ or a column vector.

Some common ways to specify the probabilities of a collection of p random variables include

1. Distribution functions (Section 1.3);

2. Densities (Sections 1.4 and 1.5);

3. Moment generating functions (or characteristic functions) (Section 2.5);

4. Representations.

Distribution functions and characteristic functions always exist, moment generating functions do not. The densities we will deal with are only those with respect to Lebesgue measure or counting measure, or combinations of the two — which means for us, densities do not always exist. By "representation" we mean the random variables are expressed as a function of some other random variables. Section 1.6.2 contains a simple example.

1.3   Distribution functions

The **distribution function** for **X** is the function $F : \mathbb{R}^p \to [0,1]$ given by

$$F(\mathbf{x}) = F(x_1,\ldots,x_p) = P[X_1 \leq x_1,\ldots,X_p \leq x_p]. \tag{1.8}$$

Note that F is defined on all of $\mathbb{R}^p$, not just the space $\mathcal{X}$. In principal, given F, one can figure out the probability of all subsets $A \subset \mathcal{X}$ (although no one would try), which means F uniquely identifies P, and *vice versa*. If F is a continuous function, then **X** is termed continuous. Generally, we will indicate random variable with capital letters, and the values they can take on with lowercase letters.

For a single random variable X, the distribution function is $F(x) = P[X \leq x]$ for $x \in \mathbb{R}$. This function satisfies the following properties:

1. $F(x)$ is nondecreasing in x;

2. $\lim_{x \to -\infty} F(x) = 0$;

3. $\lim_{x \to \infty} F(x) = 1$;

4. For any x, $\lim_{y \downarrow x} F(y) = F(x)$.

The fourth property is that F is continuous from the right. It need not be continuous. For example, suppose X is the number of heads (i.e., 0 or 1) in one flip of a fair coin, so that $\mathcal{X} = \{0,1\}$, and $P[X = 0] = P[X = 1] = 1/2$. Then

$$F(x) = \begin{cases} 0 & \text{if} & x < 0 \\ 1/2 & \text{if} & 0 \leq x < 1 \\ 1 & \text{if} & x \geq 1 \end{cases}. \tag{1.9}$$

Now $F(1) = 1$, and if $y \uparrow 1$, say, $y = 1 - 1/m$, then for $m = 1,2,3,\ldots$, $F(1 - 1/m) = 1/2$, which approaches $1/2$, not 1. On the other hand, if $y \downarrow 1$, say $y = 1 + 1/m$, then $F(1 + 1/m) = 1$, which does approach 1.

A jump in F at a point x means that $P[X = x] > 0$; in fact, the probability is the height of the jump. If F is continuous at x, then $P[X = x] = 0$. Figure 1.1 shows a distribution function with jumps at 1 and 6, which means that the probability X equals either of those points is positive, the probability being the height of the gaps (which are $1/4$ in this plot). Otherwise, the function is continuous, hence no other single value has positive probability. Note also the flat part between 1 and 4, which means that $P[1 < X \leq 4] = 0$.

Not only do all distribution functions for random variables satisfy those four properties, but any function F that satisfies those four is a legitimate distribution function. Similar results hold for finite collections of random variables:

1. $F(x_1,\ldots,x_p)$ is nondecreasing in each x_i, holding the others fixed;

2. $\lim_{x_i \to -\infty} F(x_1, x_2,\ldots,x_p) = 0$ for any of the x_i's;

3. $\lim_{x \to \infty} F(x, x,\ldots,x) = 1$;

4. For any $(x_1,\ldots,x_p)$, $\lim_{y_1 \downarrow x_1,\ldots,y_p \downarrow x_p} F(y_1,\ldots,y_p) = F(x_1,\ldots,x_p)$.

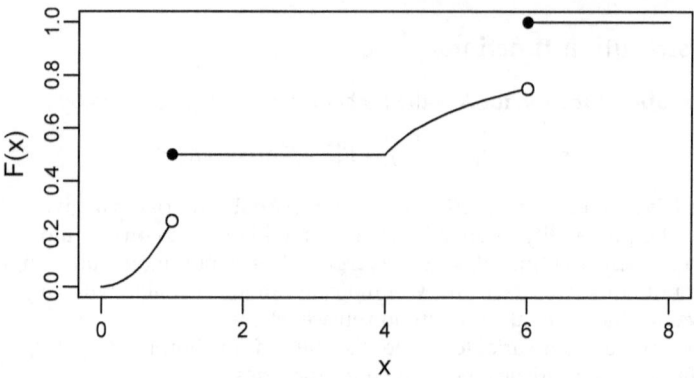

Figure 1.1: A distribution function.

1.4   PDFs: Probability density functions

A **density with respect to Lebesgue measure on** $\mathbb{R}^p$, which we simplify to "pdf" for "probability density function," is a function $f : \mathcal{X} \to [0, \infty)$ such that for any subset $A \subset \mathcal{X}$,

$$P[A] = \int \int \cdots \int_A f(x_1, x_2, \ldots, x_p) dx_1 dx_2 \ldots dx_p. \tag{1.10}$$

If $\mathbf{X}$ has a pdf, then it is continuous. In fact, its distribution function is differentiable, f being the derivative of F:

$$f(x_1, \ldots, x_p) = \frac{\partial}{\partial x_1} \cdots \frac{\partial}{\partial x_p} F(x_1, \ldots, x_p). \tag{1.11}$$

There are continuous distributions that do not have pdfs, as in Section 1.6.2. Any pdf has to satisfy the following two properties:

1. $f(x_1, \ldots, x_p) \geq 0$ for all $(x_1, \ldots, x_p) \in \mathcal{X}$;

2. $\int \int \cdots \int_{\mathcal{X}} f(x_1, x_2, \ldots, x_p) dx_1 dx_2 \ldots dx_p = 1$.

It is also true that any function f satisfying those two conditions is a pdf of a legitimate probability distribution. Table 1.1 contains some famous univariate (so that $p = 1$ and $\mathcal{X} \subset \mathbb{R}$) distributions with their pdfs. For later convenience, the means and variances (see Section 2.2) are included. The Γ in the table is the gamma function, defined by

$$\Gamma(\alpha) = \int_0^\infty x^{\alpha-1} e^{-x} dx \quad \text{for } \alpha > 0. \tag{1.12}$$

There are many more important univariate densities, such as the F and noncentral versions of the t, χ^2, and F. The most famous multivariate distribution is the multivariate normal. We will look at that one in Chapter 7. The next section presents a simple bivariate distribution.

Name	Space $\mathcal{X}$	pdf $f(x)$	Mean	Variance
Normal : $N(\mu,\sigma^2)$ $\mu \in \mathbb{R}, \sigma^2 > 0$	$\mathbb{R}$	$\frac{1}{\sqrt{2\pi}\,\sigma}e^{-(x-\mu)^2/(2\sigma^2)}$	μ	σ^2
Uniform(a,b) $a < b$	(a,b)	$\frac{1}{b-a}$	$\frac{a+b}{2}$	$\frac{(b-a)^2}{12}$
Exponential(λ) $\lambda > 0$	$(0,\infty)$	$\lambda e^{-\lambda x}$	$\frac{1}{\lambda}$	$\frac{1}{\lambda^2}$
Gamma(α,λ) $\alpha > 0, \lambda > 0$	$(0,\infty)$	$\frac{\lambda^\alpha}{\Gamma(\alpha)}e^{-\lambda x}x^{\alpha-1}$	$\frac{\alpha}{\lambda}$	$\frac{\alpha}{\lambda^2}$
Beta(α,β) $\alpha > 0, \beta > 0$	$(0,1)$	$\frac{\Gamma(\alpha+\beta)}{\Gamma(\alpha)\Gamma(\beta)}x^{\alpha-1}(1-x)^{\beta-1}$	$\frac{\alpha}{\alpha+\beta}$	$\frac{\alpha\beta}{(\alpha+\beta)^2(\alpha+\beta+1)}$
Cauchy	$\mathbb{R}$	$\frac{1}{\pi}\frac{1}{1+x^2}$	*	*
Laplace	$\mathbb{R}$	$\frac{1}{2}e^{-|x|}$	0	2
Logistic	$\mathbb{R}$	$\frac{e^x}{(1+e^x)^2}$	0	$\frac{\pi^2}{3}$
Chi-square : χ^2_ν $\nu = 1,2,\ldots$	$(0,\infty)$	$\frac{1}{\Gamma(\nu/2)2^{\nu/2}}x^{\nu/2-1}e^{-x/2}$	ν	2ν
Student's t_ν $\nu = 1,2,\ldots$	$\mathbb{R}$	$\frac{\Gamma((\nu+1)/2)}{\Gamma(\nu/2)\sqrt{\nu\pi}}\left(1+\frac{t^2}{\nu}\right)^{-\frac{\nu+1}{2}}$	0 if $\nu \geq 2$	$\frac{\nu}{\nu-2}$ if $\nu \geq 3$

$* = $ Doesn't exist

Table 1.1: Some common probability density functions.

1.4.1 A bivariate pdf

Suppose (X,Y) has space

$$W = \{(x,y)\,|\,0 < x < 1,\ 0 < y < 1\} \tag{1.13}$$

and pdf

$$f(x,y) = c(x+y). \tag{1.14}$$

The constant c is whatever it needs to be so that the pdf integrates to 1, i.e.,

$$1 = c\int_0^1\int_0^1 (x+y)dydx = c\int_0^1 (x+\tfrac{1}{2})dx = c(\tfrac{1}{2}+\tfrac{1}{2}) = c. \tag{1.15}$$

So the pdf is simply $f(x,y) = x + y$. Some values of the distribution function are

$$F(0,0) = 0;$$

$$F(\tfrac{1}{2}, \tfrac{1}{4}) = \int_0^{\frac{1}{2}} \int_0^{\frac{1}{4}} (x+y)dydx = \int_0^{\frac{1}{2}} (\tfrac{1}{4}x + \tfrac{1}{32})dx = \frac{1}{32} + \frac{1}{32} = \frac{1}{16};$$

$$F(\tfrac{1}{2}, 2) = \int_0^{\frac{1}{2}} \int_0^{1} (x+y)dydx = \int_0^{\frac{1}{2}} (x + \tfrac{1}{2})dx = \frac{1}{8} + \frac{1}{4} = \frac{3}{8};$$

$$F(2,1) = 1. \tag{1.16}$$

Other probabilities:

$$\begin{aligned}
P[X+Y \le \tfrac{1}{2}] &= \int_0^{\frac{1}{2}} \int_0^{\frac{1}{2}-x} (x+y)dydx \\
&= \int_0^{\frac{1}{2}} (x\,(\tfrac{1}{2} - x) + \tfrac{1}{2}\,(\tfrac{1}{2} - x)^2)dx \\
&= \int_0^{\frac{1}{2}} (\tfrac{1}{8} - \tfrac{1}{2}x^2)dx \\
&= \frac{1}{24}, \tag{1.17}
\end{aligned}$$

and for $0 < y < 1$,

$$P[Y \le y] = \int_0^1 \int_0^y (x+w)dwdx = \int_0^1 (xy + \tfrac{1}{2}y^2)dx = \frac{y}{2} + \frac{y^2}{2} = \frac{1}{2}\,y(1+y), \tag{1.18}$$

which is the distribution function of Y, at least for $0 < y < 1$. The pdf for Y is then found by differentiating:

$$f_Y(y) = F_Y'(y) = y + \frac{1}{2} \quad \text{for } 0 < y < 1. \tag{1.19}$$

1.5 PMFs: Probability mass functions

A **discrete** random variable is one for which $\mathcal{X}$ is a countable (which includes finite) set. Its probability can be given by its **probability mass function**, which we will call "pmf," $f : \mathcal{X} \to [0,1]$ given by

$$P[\{(x_1, \ldots, x_p)\}] = P[\mathbf{X} = \mathbf{x}] = f(\mathbf{x}) = f(x_1, \ldots, x_p), \tag{1.20}$$

where $\mathbf{x} = (x_1, \ldots, x_p)$. The pmf gives the probabilities of the individual points. (Measure-theoretically, the pmf is the density *with respect to counting measure* on $\mathcal{X}$.) The probability of any subset A is the sum of the probabilities of the individual points in A. Table 1.2 contains some popular univariate discrete distributions.

The distribution function of a discrete random variable is a pure jump function, that is, it is flat except for jumps of height $f(x)$ at x for each $x \in \mathcal{X}$. See Figure 2.4 on page 33 for an example. The most famous multivariate discrete distribution is the multinomial, which we look at in Section 2.5.3.

Name	Space $\mathcal{X}$	pmf $f(x)$	Mean	Variance
Bernoulli(p) $0 < p < 1$	$\{0, 1\}$	$p^x(1-p)^{1-x}$	p	$p(1-p)$
Binomial(n, p) $n = 1, 2, \ldots; 0 < p < 1$	$\{0, 1, \ldots, n\}$	$\binom{n}{x} p^x(1-p)^{n-x}$	np	$np(1-p)$
Poisson(λ) $\lambda > 0$	$\{0, 1, 2, \ldots\}$	$e^{-\lambda}\frac{\lambda^x}{x!}$	λ	λ
Discrete Uniform(a, b) a, b integers, $a < b$	$\{a, a+1, \ldots, b\}$	$\frac{1}{b-a+1}$	$\frac{a+b}{2}$	$\frac{(b-a+1)^2-1}{12}$
Geometric(p) $0 < p < 1$	$\{0, 1, 2, \ldots\}$	$p(1-p)^x$	$\frac{1-p}{p}$	$\frac{1-p}{p^2}$
Negative Binomial(K, p) $K = 1, 2, \ldots; 0 < p < 1$	$\{0, 1, 2, \ldots\}$	$\binom{x+K-1}{K-1}p^K(1-p)^x$	$K\frac{1-p}{p}$	$K\frac{1-p}{p^2}$

Table 1.2: Some common probability mass functions.

1.6 Distributions without pdfs or pmfs

Distributions need not be either discrete or have a density with respect to Lebesgue measure. We present here some simple examples.

1.6.1 Late start

Consider waiting for a train to leave. It will not leave early, but very well may leave late. There is a positive probability, say 10%, it will leave exactly on time. If it does not leave on time, there is a continuous distribution for how late it leaves. Thus it is not totally discrete, but not continuous, either, so it has neither a pdf nor pmf. It does have a distribution function, because everything does. A possible one is

$$F(x) = \begin{cases} 0 & \text{if } x < 0 \\ 0.1 & \text{if } x = 0 \\ 1 - 0.9\exp(-x/100) & \text{if } x > 0 \end{cases}, \tag{1.21}$$

where x is the number of minutes late. Figure 1.2 sketches this F.

Is this a legitimate distribution function? It is easy to see it is nondecreasing, once one notes that $1 - 0.9\exp(-x/100) > 0.1$ if $x > 0$. The limits are ok as $x \to \pm\infty$. It is also continuous from the right, where the only tricky spot is $\lim_{x\downarrow 0} F(x)$, which goes to $1 - 0.9\exp(0) = 0.1 = F(0)$, so it checks.

One can then find the probabilities of various late times, e.g., it has no chance of leaving early, 10% chance of leaving exactly on time, $F(60) = 1 - 0.9\exp(-60/100) \approx$ 0.506 chance of being at most one hour late, $F(300) = 1 - 0.9\exp(-300/100) \approx 0.955$ chance of being at most five hours late, etc. (Sort of like Amtrak.)

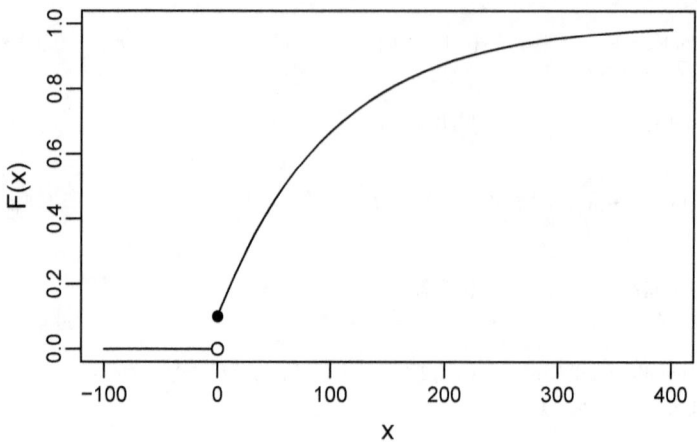

Figure 1.2: The distribution function for a late start.

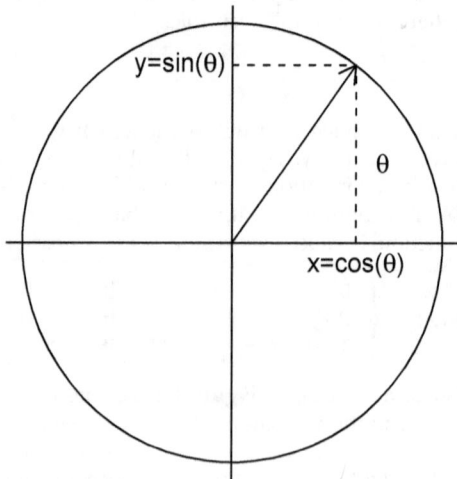

Figure 1.3: Illustration of the spinner.

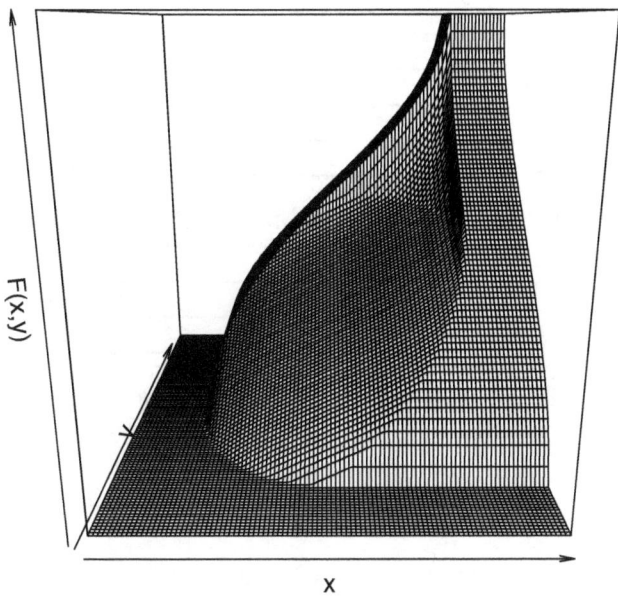

Figure 1.4: The sketch of the distribution function in (1.22) for the spinner example.

1.6.2 Spinner

Imagine a spinner whose pointer is one unit in length. It is spun so that it is equally likely to be pointing in any direction. The random quantity is the (x, y) location of the end of the pointer, so that $\mathcal{X} = \{(x, y) \in \mathbb{R}^2 \mid x^2 + y^2 = 1\}$, the circle with radius 1. The distribution of (X, Y) is not discrete because the point can land anywhere on the circle. On the other hand, it does not have a density with respect to Lebesgue measure on $\mathbb{R}^2$ because the integral over the circle is the volume above the circle under the pdf, that volume being 0.

But there is a distribution function. The $F(x, y)$ is the arc length of the part(s) of the circle that has x-coordinate less than or equal to x and y-coordinate less than or equal to y, divided by total arc length (which is 2π):

$$F(x, y) = \frac{\text{arc length}(\{(u, v) \mid u^2 + v^2 = 1, u \le x, v \le y\})}{2\pi}. \tag{1.22}$$

Figure 1.4 has a sketch of F.

Fortunately, there is an easier way to describe the distribution. For any point (x, y) on the circle, one can find the angle with the x-axis of the line connecting $(0, 0)$ and (x, y), $\theta = \text{Angle}(x, y)$, so that $x = \cos(\theta)$ and $y = \sin(\theta)$. See Figure 1.3. For uniqueness' sake, take $\theta \in [0, 2\pi)$. Then (x, y) being uniform on the circle implies that θ is uniform from 0 to 2π. Then the distribution of (X, Y) can be described via

$$(X, Y) = (\cos(\Theta), \sin(\Theta)), \quad \text{where } \Theta \sim \text{Uniform}[0, 2\pi). \tag{1.23}$$

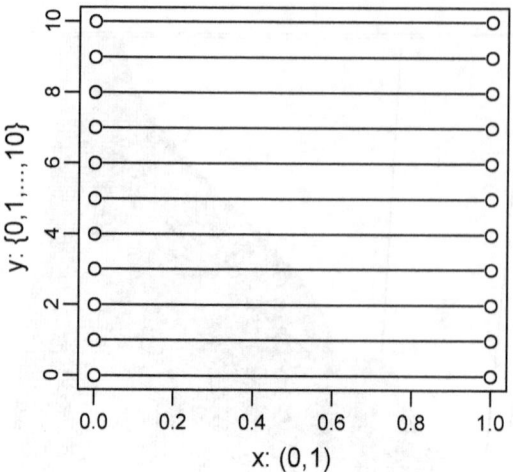

Figure 1.5: The sketch of the space $\mathcal{W}$ in (1.24).

Such a description is called a *representation*, in that we are representing one set of random variables as a function of another set (which in this case is just the one Θ).

1.6.3 Mixed-type densities

Imagine now a two-stage process, where one first chooses a coin out of an infinite collection of coins, then flips the coin $n = 10$ times. The coins have different probabilities of heads x, so that over the population of coins, $X \sim \text{Uniform}(0,1)$. Let Y be the number of heads among the 10 flips. Then what is random is the pair (X, Y). This vector is neither discrete nor continuous: X is continuous and Y is discrete. The space is a union of 11 ($= n + 1$) line segments,

$$\mathcal{W} = \{(x,0) \,|\, 0 < x < 1\} \cup \{(x,1) \,|\, 0 < x < 1\} \cup \cdots \cup \{(x,10) \,|\, 0 < x < 1\}. \quad (1.24)$$

See Figure 1.5.

The density can still be given as $f(x,y)$, but now the x part is discrete, and the y part is continuous. Then the probability of any set involves summing over the x and integrating over the y. E.g.,

$$P[X < 1/2 \ \& \ Y \ge 5] = \int_0^{1/2} \sum_{y=5}^{10} f(x,y), \quad (1.25)$$

where

$$f(x,y) = \binom{10}{y} x^y (1-x)^{10-y}. \quad (1.26)$$

This idea can be extended to any number of random variables, some discrete and some continuous.

1.7 Exercises

Exercise 1.7.1. Suppose the random variable X is always equal to the constant c. That is, the space is {c}, and $P[X = c] = 1$. (a) Let $F(x)$ be the distribution function of X. What are the values of $F(x)$ for (i) $x < c$, (ii) $x = c$, and (iii) $x > c$? (b) Now let f be the pmf of X. What are the values of $f(x)$ for (i) $x < c$, (ii) $x = c$, and (iii) $x > c$?

Exercise 1.7.2. Suppose (X, Y) is a random vector with space $\{(1,2), (2,1)\}$, where $P[(X, Y) = (1,2)] = P[(X, Y) = (2,1)] = 1/2$. Fill in the table with the values of $F(x, y)$:

$y \downarrow; x \rightarrow$	0	1	2	3
3				
2				
1				
0				

Exercise 1.7.3. Suppose (X, Y) is a continuous two-dimensional random vector with space $\{(x, y) \mid 0 < x < 1, 0 < y < 1, x + y < 1\}$. (a) Which of the following is the best sketch of the space?

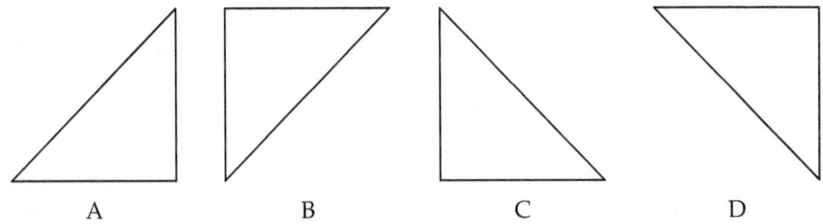

A   B   C   D

(b) The density is $f(x, y) = c$ for (x, y) in the space. What is c? (c) Find the following values: (i) $F(0.1, 0.2)$, (ii) $F(0.8, 1)$, (iii) $F(0.8, 1.5)$, (iv) $F(0.7, 0.8)$.

Exercise 1.7.4. Continue with the distribution in Exercise 1.7.3, but focus on just X. (a) What is the space of X? (b) For x in that space, what is the distribution function $F_X(x)$? (c) For x in that space, $F_X(x) = F(x, y)$ for what values of y in the range $[0, 1]$? (d) For x in that space, what is the pdf $f_X(x)$?

Exercise 1.7.5. Now take (X, Y) with space $\{(x, y) \mid 0 < x < y < 1\}$. (a) Of the spaces depicted in Exercise 1.7.3, which is the best sketch of the space in this case? (b) Suppose (X, Y) has pdf $f(x, y) = 2$, for (x, y) in the space. Let $W = Y/X$. What is the space of W? (c) Find the distribution function of W, $F_W(w)$, for w in the space. [Hint: Note that $F_W(w) = P[W \le w] = P[Y \le wX]$. The set in the probability is then a triangle, for which the area can be found.] (d) Find the pdf of W, $f_W(w)$, for w in the space.

Exercise 1.7.6. Suppose (X_1, X_2) is uniformly distributed over the unit square, that is, the space is $\{(x_1, x_2) \mid 0 < x_1 < 1, 0 < x_2 < 1\}$, and the pdf is $f(x_1, x_2) = 1$ for (x_1, x_2) in the space. Let $Y = X_1 + X_2$. (a) What is the space of Y? (b) Find the distribution function $F_Y(y)$ of Y. [Hint: Draw the picture of the space of (X_1, X_2), and sketch the region for which $x_1 + x_2 \le y$, as in the figures:

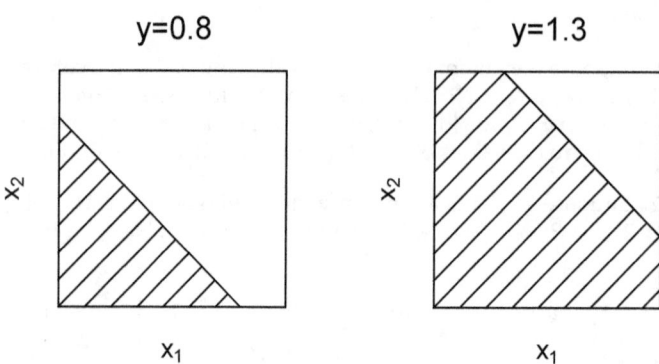

Then find the area of that region. Do it separately for $y < 1$ and $y \geq 1$.] (c) Show that the pdf of Y is $f_Y(y) = y$ if $y \in (0,1)$ and $f_Y(y) = 2 - y$ if $y \in [1,2)$. Sketch the pdf. It has a **tent distribution**.

Exercise 1.7.7. Suppose $X \sim \text{Uniform}(0,1)$, and let $Y = |X - 1/4|$. (a) What is the space of Y? (b) Find the distribution function of Y. [Specify it in pieces: $y < 0$, $0 < y < a$, $a < y < b$, $y > b$. What are a and b?] (c) Find the pdf of Y.

Exercise 1.7.8. Set $X = \cos(\Theta)$ and $Y = \sin(\Theta)$, where $\Theta \sim \text{Uniform}(0,2\pi)$. (a) What is the space $\mathcal{X}$ of X? (b) For $x \in \mathcal{X}$, find $F(x) = P[X \leq x]$. [Hint: Figure out which θ's correspond to $X \leq x$. The answer should have a $\cos^{-1}$ in it.] (c) Find the pdf of X. (d) Is the pdf of Y the same as that of X?

Exercise 1.7.9. Suppose $U \sim \text{Uniform}(0,1)$, and $(X,Y) = (U, 1 - U)$. Let $F(x,y)$ be the distribution function of (X,Y). (a) Find and sketch the space of (X,Y). (b) For which values of (x,y) is $F(x,y) = 1$? (c) For which values of (x,y) is $F(x,y) = 0$? (d) Find $F(3/4,3/4)$, $F(3/2,3/4)$, and $F(3/4,7/8)$.

Exercise 1.7.10. (a) Use the definition of the gamma function in (1.12) to help show that

$$\int_0^\infty x^{\alpha-1}e^{-\lambda x} = \lambda^{-\alpha}\Gamma(\alpha) \qquad (1.27)$$

for $\alpha > 0$ and $\lambda > 0$, thus justifying the constant in the gamma pdf in Table 1.1. (b) Use integration by parts to show that $\Gamma(\alpha + 1) = \alpha\Gamma(\alpha)$ for $\alpha > 0$. (c) Show that $\Gamma(1) = 1$, hence with part (b), $\Gamma(n) = (n-1)!$ for positive integer n.

Exercise 1.7.11. The gamma distribution given in Table 1.1 has two parameters: α is the **shape** and λ is the **rate**. (Alternatively, the second parameter may be given by $\beta = 1/\lambda$, which is called the **scale** parameter.) (a) Sketch the pdfs for shape parameters $\alpha = .5, .8, 1, 2$, and 5, with $\lambda = 1$. What do you notice? What is qualitatively different about the behavior of the pdfs near $x = 0$ depending on whether $\alpha < 1$, $\alpha = 1$, or $\alpha > 1$? (b) Now fix $\alpha = 1$ and sketch the pdfs for $\lambda = .5, 1$, and 5. What do you notice about the shapes? (c) Fix $\alpha = 5$, and explore the pdfs for different rates.

Exercise 1.7.12. (a) The Exponential(λ) distribution is a special case of the gamma. What are the corresponding parameters? (b) The χ_ν^2 is a special case of the gamma. What are corresponding parameters. (c) The Uniform(0,1) is a special case of the beta. What are the corresponding parameters? (d) The Cauchy is a special case of Student's t_ν. For which ν?

Exercise 1.7.13. Let $Z \sim N(0,1)$, and let $W = Z^2$. What is the space of W? (a) Write down the distribution function of W as an integral over the pdf of Z. (b) Show that the pdf of W is

$$g(w) = \frac{1}{\sqrt{2\pi w}}\, e^{-w/2}. \tag{1.28}$$

[Hint: Differentiate the distribution function from part (a). Recall that

$$\frac{d}{dw} \int_{a(w)}^{b(w)} f(z)dz = f(b(w))b'(w) - f(a(w))a'(w). \;] \tag{1.29}$$

(c) The distribution of W is χ^2_ν (see Table 1.1) for which ν? (d) The distribution of W is a special case of a gamma. What are the parameters? (e) Show that by matching (1.28) with the gamma or chi-square density, we have that $\Gamma(1/2) = \sqrt{\pi}$.

Exercise 1.7.14. Now suppose $Z \sim N(\mu,1)$, and let $W = X^2$, which is called **noncentral chi-square** on one degree of freedom. (Section 7.5.3 treats noncentral chi-squares more generally.) (a) What is the space of W? (b) Show that the pdf of W is

$$g_\mu(w) = g(w)\, e^{-\frac{1}{2}\mu^2}\, \frac{e^{\mu\sqrt{w}} + e^{-\mu\sqrt{w}}}{2}, \tag{1.30}$$

where g is the pdf in (1.28). Note that the last fraction is $\cosh(\mu\sqrt{w})$.

Exercise 1.7.15. The logistic distribution has space $\mathbb{R}$ and pdf $f(x) = e^x(1+e^x)^{-2}$ as in Table 1.1. (a) Show that the pdf is symmetric about 0, i.e., $f(x) = f(-x)$ for all x. (b) Show that the distribution function is $F(x) = e^x/(1+e^x)$. (c) Let $U \sim \text{Uniform}(0,1)$. Thinking of u as a probability of some event. The odds of that event are $u/(1-u)$, and the log odds or **logit** is $\text{logit}(u) = \log(u/(1-u))$. Show that $X = \text{logit}(U) \sim \text{Logistic}$, which may explain where the name came from. [Hint: Find $F_X(x) = P[\log(U/(1-U)) \leq x]$, and show that equals the distribution function in part (b).]

Chapter 2

Expected Values, Moments, and Quantiles

2.1 Definition of expected value

The distribution function F contains all there is to know about the distribution of a random vector, but it is often difficult to take in all at once. Quantities that summarize aspects of the distribution are often helpful, including moments (means and variances, e.g.) and quantiles, which are discussed in this chapter. Moments are special cases of expected values.

We start by defining expected value in the pdf and pmf cases. There are many $\mathbf{X}$'s that have neither a pmf nor pdf, but even in those cases we can often find the expected value.

Definition 2.1. *Expected value. Suppose $\mathbf{X}$ has pdf f, and $g : \mathcal{X} \to \mathbb{R}$. If*

$$\int \cdots \int_{\mathcal{X}} |g(x_1, \ldots, x_p)| f(x_1, \ldots, x_p) dx_1 \ldots dx_p < \infty, \tag{2.1}$$

then the expected value of $g(\mathbf{X})$, $E[g(\mathbf{X})]$, exists and

$$E[g(\mathbf{X})] = \int \cdots \int_{\mathcal{X}} g(x_1, \ldots, x_p) f(x_1, \ldots, x_p) dx_1 \ldots dx_p. \tag{2.2}$$

If $\mathbf{X}$ has pmf f, and

$$\sum \cdots \sum_{(x_1, \ldots, x_p) \in \mathcal{X}} |g(x_1, \ldots, x_p)| f(x_1, \ldots, x_p) < \infty, \tag{2.3}$$

then the expected value of $g(\mathbf{X})$, $E[g(\mathbf{X})]$, exists and

$$E[g(\mathbf{X})] = \sum \cdots \sum_{(x_1, \ldots, x_p) \in \mathcal{X}} g(x_1, \ldots, x_p) f(x_1, \ldots, x_p). \tag{2.4}$$

The requirement (2.1) or (2.3) that the absolute value of the function must have a finite integral/sum is there to eliminate ambiguous situations. For example, consider the Cauchy distribution with pdf $f(x) = 1/(\pi(1 + x^2))$ and space $\mathbb{R}$, and take $g(x) = x$, so we wish to find $E[X]$. Consider

$$\int_{-\infty}^{\infty} |x| f(x) dx = \int_{-\infty}^{\infty} \frac{1}{\pi} \frac{|x|}{1 + x^2} dx = 2 \int_{0}^{\infty} \frac{1}{\pi} \frac{x}{1 + x^2} dx. \tag{2.5}$$

For large $|x|$, the integrand is on the order of $1/|x|$, which does not have a finite integral. More precisely, it is not hard to show that

$$\frac{x}{1+x^2} > \frac{1}{2x} \quad \text{for } x > 1. \tag{2.6}$$

Thus

$$\int_{-\infty}^{\infty} \frac{1}{\pi} \frac{|x|}{1+x^2} dx > \int_{1}^{\infty} \frac{1}{\pi} \frac{1}{x} \, dx = \frac{1}{\pi} \log(x) \, |_{1}^{\infty} = \frac{1}{\pi} \log(\infty) = \infty. \tag{2.7}$$

In this case we say that "the expected value of the Cauchy does not exist." By the symmetry of the density, it would be natural to expect the expected value to be 0. But what we have is

$$E[X] = \int_{-\infty}^{0} xf(x)dx + \int_{0}^{\infty} xf(x)dx = -\infty + \infty = \text{Undefined}. \tag{2.8}$$

That is, we cannot do the integral, so the expected value is not defined.

One could allow $+\infty$ and $-\infty$ to be legitimate values of the expected value, e.g., say that $E[X^2] = +\infty$ for the Cauchy, as long as the value is unambiguous. We are not allowing that possibility formally, but informally will on occasion act as though we do.

Expected values cohere in the proper way, that is, if $\mathbf{Y}$ is a random vector that is a function of $\mathbf{X}$, say $\mathbf{Y} = h(\mathbf{X})$, then for a function g of $\mathbf{Y}$,

$$E[g(\mathbf{Y})] = E[g(h(\mathbf{X}))], \tag{2.9}$$

if the latter exists. This property helps in finding the expected values when representations are used. For example, in the spinner case (1.23),

$$E[X] = E[\cos(\Theta)] = \frac{1}{2\pi} \int_{0}^{2\pi} \cos(\theta)d\theta = 0, \tag{2.10}$$

where the first expected value has X as the random variable, for which we do not have a pdf, and the second expected value has Θ as the random variable, for which we do have a pdf (the Uniform$[0, 2\pi)$).

One important feature of expected values is their linearity, which follows by the linearity of integrals and sums:

Lemma 2.2. *For any random variables X, Y, and constant c,*

$$E[cX] = cE[X] \quad \text{and} \quad E[X + Y] = E[X] + E[Y], \tag{2.11}$$

if the expected values exist.

The lemma can be used to show more involved linearities, e.g.,

$$E[aX + bY + cZ + d] = aE[X] + bE[Y] + cE[Z] + d \tag{2.12}$$

(since $E[d] = d$ for a constant d), and

$$E[g(\mathbf{X}) + h(\mathbf{X})] = E[g(\mathbf{X})] + E[h(\mathbf{X})]. \tag{2.13}$$

Warning. Be aware that for non-linear functions, the expected value of a function is **NOT** the function of the expected value, i.e.,

$$E[g(X)] \neq g(E[X]) \tag{2.14}$$

unless $g(x)$ is linear, or you are lucky. For example,

$$E[X^2] \neq E[X]^2, \tag{2.15}$$

unless X is a constant. (Which is fortunate, because otherwise all variances would be 0. See (2.20) below.)

2.1.1 Indicator functions

An indicator function is one that takes on only the values 0 and 1. It is usually given as $I_A(\mathbf{x})$ or $I[\mathbf{x} \in A]$, or simply $I[A]$, for a subset $A \subset \mathcal{X}$, where A contains the values for which the function is 1:

$$I_A(\mathbf{x}) = I[\mathbf{x} \in A] = I[A] = \begin{cases} 1 & \text{if} \quad \mathbf{x} \in A \\ 0 & \text{if} \quad \mathbf{x} \notin A \end{cases}. \tag{2.16}$$

These functions give alternative expressions for probabilities in terms of expected values as in

$$E[I_A[\mathbf{X}]] = 1 \times P[\mathbf{X} \in A] + 0 \times P[\mathbf{X} \notin A] = P[A]. \tag{2.17}$$

2.2 Means, variances, and covariances

Means, variances, and covariances are particular expected values. For a random variable, the mean is just its expected value:

$$\text{The } \textbf{mean} \text{ of } X = E[X] \quad (\text{often denoted } \mu). \tag{2.18}$$

(From now on, we will usually suppress the phrase "if it exists" when writing expected values, but think of it to yourself when reading "E.") The variance is the expected value of the deviation from the mean, squared:

$$\text{The } \textbf{variance} \text{ of } X = Var[X] = E[(X - E[X])^2] \quad (\text{often denoted } \sigma^2). \tag{2.19}$$

The **standard deviation** is the square root of the variance. It is often a nicer quantity because it is in the same units as X, and measures the "typical" size of the deviation of X from its mean.

A very useful formula for finding variances is

$$Var[X] = E[X^2] - E[X]^2, \tag{2.20}$$

which can be seen, letting $\mu = E[X]$, as follows:

$$E[(X - \mu)^2] = E[X^2 - 2X\mu + \mu^2] = E[X^2] - 2E[X]\mu + \mu^2 = E[X^2] - \mu^2. \tag{2.21}$$

With two random variables, (X, Y), say, there is in addition the covariance:

$$\text{The } \textbf{covariance} \text{ of } X \text{ and } Y = Cov[X, Y] = E[(X - E[X])(Y - E[Y])]. \tag{2.22}$$

The covariance measures a type of relationship between X and Y. Notice that the expectand is positive when X and Y are both greater than or both less than their respective means, and negative when one is greater and one less. Thus if X and Y tend to go up or down together, the covariance will be positive, while if when one goes up the other goes down, the covariance will be negative. Note also that it is symmetric, $Cov[X, Y] = Cov[Y, X]$, and $Cov[X, X] = Var[X]$.

As for the variance in (2.20), we have the formula

$$Cov[X, Y] = E[XY] - E[X]E[Y]. \tag{2.23}$$

The correlation coefficient is a normalization of the covariance, which is generally easier to interpret:

The **correlation coefficient** of X and $Y = Corr[X, Y] = \dfrac{Cov[X, Y]}{\sqrt{Var[X]Var[Y]}}$ (2.24)

if $Var[X] > 0$ and $Var[Y] > 0$. This is a unitless quantity that measures the linear relationship of X and Y. It is bounded by -1 and $+1$. To verify this fact, we first need the following.

Lemma 2.3. *Cauchy-Schwarz. For random variables (U, V),*

$$E[UV]^2 \leq E[U^2]E[V^2], \tag{2.25}$$

with equality if and only if

$$U = 0 \text{ or } V = \beta U \text{ with probability } 1, \tag{2.26}$$

for $\beta = E[UV]/E[U^2]$.

Here, the phrase "with probability 1" means $P[U = 0] = 1$ or $P[V = \beta U] = 1$.

Proof. The lemma is easy to see if U is always 0, because then $E[UV] = E[U^2] = 0$. Suppose it is not, so that $E[U^2] > 0$. Consider

$$E[(V - bU)^2] = E[V^2 - 2bUV + b^2U^2] = E[V^2] - 2bE[UV] + b^2E[U^2]. \tag{2.27}$$

Because the expectand on the left is nonnegative for any b, so is its expected value. In particular, it is nonnegative for the b that minimizes the expected value, which is easy to find:

$$\frac{\partial}{\partial b} E[(V - bU)^2] = -2E[UV] + 2bE[U^2], \tag{2.28}$$

and setting that to 0 yields $b = \beta$ where $\beta = E[UV]/E[U^2]$. Then

$$E[V^2] - 2\beta E[UV] + \beta^2 E[U^2] = E[V^2] - 2\frac{E[UV]}{E[U^2]}E[UV] + \left(\frac{E[UV]}{E[U^2]}\right)^2 E[U^2]$$

$$= E[V^2] - \frac{E[UV]^2}{E[U^2]} \geq 0, \tag{2.29}$$

from which (2.25) follows.

There is equality in (2.25) if and only if there is equality in (2.29), which means that $E[(V - \beta U)^2] = 0$. Because the expectand is nonnegative, its expected value can be 0 if and only if it is 0, i.e.,

$$(V - \beta U)^2 = 0 \text{ with probability 1.} \tag{2.30}$$

But that equation implies the second part of (2.26), proving the lemma. $\qquad \square$

For variables (X, Y), apply the lemma with $U = X - E[X]$ and $V = Y - E[Y]$:

$$E[(X - E[X])(Y - E[Y])]^2 \leq E[(X - E[X])^2]E[(Y - E[Y])^2]$$
$$\Longleftrightarrow \quad Cov[X, Y]^2 \leq Var[X]Var[Y]. \tag{2.31}$$

Thus from (2.24), if the variances are positive and finite,

$$-1 \leq Corr[X, Y] \leq 1. \tag{2.32}$$

Furthermore, if there is an equality in (2.31), then either X is a constant, or

$$Y - E[Y] = b(X - E[X]) \Leftrightarrow Y = \alpha + \beta X, \tag{2.33}$$

where

$$\beta = \frac{Cov[X, Y]}{Var[X]} \text{ and } \alpha = E[Y] - \beta E[X]. \tag{2.34}$$

In this case,

$$Corr[X, Y] = \begin{cases} 1 & \text{if } \beta > 0 \\ -1 & \text{if } \beta < 0 \end{cases}. \tag{2.35}$$

Thus the correlation coefficient measures the linearity of the relationship between X and Y, $+1$ meaning perfectly positively linearly related, -1 meaning perfectly negatively linearly related.

2.2.1 Uniform on a triangle

Suppose (X, Y) has pdf $f(x, y) = 2$ for $(x, y) \in W = \{(x, y) \mid 0 < x < y < 1\}$, which is the upper-left triangle of the unit square, as in Figure 2.1.

One would expect the correlation to be positive, since the large y's tend to go with larger x's, but the correlation would not be $+1$, because the space is not contained in a straight line. To find the correlation, we need to perform some integrals:

$$E[X] = \int_0^1 \int_0^y x2dxdy = \int_0^1 y^2dy = \frac{1}{3}, \quad E[Y] = \int_0^1 \int_0^y y2dxdy = 2\int_0^1 y^2dy = \frac{2}{3},$$

$$E[X^2] = \int_0^1 \int_0^y x^2 2dxdy = \int_0^1 \frac{2y^3}{3}dy = \frac{1}{6}, \quad E[Y^2] = \int_0^1 \int_0^y y^2 2dxdy = 2\int_0^1 y^3dy = \frac{1}{2},$$

$$\text{and } E[XY] = \int_0^1 \int_0^y xy2dxdy = \int_0^1 y^3dy = \frac{1}{4}. \tag{2.36}$$

Then

$$Var[X] = \frac{1}{6} - \frac{1}{3^2} = \frac{1}{18}, \quad Var[Y] = \frac{1}{2} - \left(\frac{2}{3}\right)^2 = \frac{1}{18}, \quad Cov[X, Y] = \frac{1}{4} - \frac{1}{3} \times \frac{2}{3} = \frac{1}{36}, \tag{2.37}$$

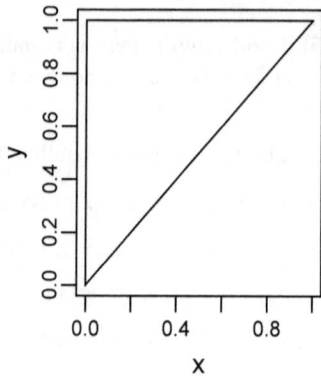

Figure 2.1: The space $\mathcal{W} = \{(x, y) \mid 0 < x < y < 1\}$.

and, finally,

$$Corr[X, Y] = \frac{1/36}{\sqrt{(1/18)(1/18)}} = \frac{1}{2}. \tag{2.38}$$

This value does seem plausible: positive but not too close to 1.

2.2.2   Variance of linear combinations & affine transformations

A **linear combination** of the variables, $X_1, \ldots, X_p$, is a function of the form

$$b_1 X_1 + \cdots + b_p X_p, \tag{2.39}$$

for constants $b_1, \ldots, b_p$. An **affine transformation** just adds a constant:

$$a + b_1 X_1 + \cdots + b_p X_p. \tag{2.40}$$

Thus they are almost the same, and if you want to add the (constant) variable $X_0 \equiv 1$, you can think of an affine transformation as a linear combination, as one does when setting up a linear regression model with intercept. Here we find formulas for the variance of an affine transformation.

Start with $a + bX$:

$$\begin{aligned}
Var[a + bX] &= E[(a + bX - E[a + bX])^2] \\
&= E[(a + bX - a - bE[X])^2] \\
&= E[b^2(X - E[X])^2] \\
&= b^2 E[(X - E[X])^2] \\
&= b^2 Var[X]. \tag{2.41}
\end{aligned}$$

The constant a goes away (it does not contribute to the variability), and the constant b is squared. For a linear combination of two variables, the variance involves the two variances, as well as the covariance:

$$
\begin{aligned}
Var[a + b_1 X_1 + b_2 X_2] &= E[(a + b_1 X_1 + b_2 X_2 - E[a + b_1 X_1 + b_2 X_2])^2] \\
&= E[(b_1(X_1 - E[X_1]) + b_2(X_2 - E[X_2]))^2] \\
&= b_1^2 E[(X_1 - E[X_1])^2] + 2b_1 b_2 E[(X_1 - E[X_1])(X_2 - E[X_2])] \\
&\qquad + b_2^2 E[(X_2 - E[X_2])^2] \\
&= b_1^2 Var[X_1] + b_2^2 Var[X_2] + 2b_1 b_2 Cov[X_1, X_2].
\end{aligned}
\tag{2.42}
$$

With p variables, we have

$$
Var\left[a + \sum_{i=1}^{p} b_i X_i\right] = \sum_{i=1}^{p} b_i^2 Var[X_i] + 2\sum\sum_{1 \le i < j \le p} b_i b_j Cov[X_i, X_j].
\tag{2.43}
$$

Covariances between two linear combinations work similarly. That is,

$$
Cov\left[a + \sum_{i=1}^{p} b_i X_i, c + \sum_{i=1}^{q} d_i Y_i\right] = \sum_{i=1}^{p} \sum_{j=1}^{q} b_i d_j Cov[X_i, Y_j].
\tag{2.44}
$$

These formulas can be made simpler using matrix and vector notation, which we do in the next section.

2.3 Vectors and matrices

The mean of a vector or matrix of random variables is the corresponding vector or matrix of means. That is, if $\mathbf{X}$ is an $n \times 1$ column vector, $\mathbf{X} = (X_1, \dots, X_n)'$ (the prime means transpose), then

$$
E[\mathbf{X}] = \begin{pmatrix} E[X_1] \\ \vdots \\ E[X_n] \end{pmatrix}.
\tag{2.45}
$$

If $\mathbf{X}$ is a row vector, $1 \times p$, then $E[\mathbf{X}] = (E[X_1], \dots, E[X_p])$. More generally, if $\mathbf{X}$ is an $n \times p$ matrix, then so is its mean:

$$
E[\mathbf{X}] = E\left[\begin{pmatrix} X_{11} & X_{12} & \cdots & X_{1p} \\ X_{21} & X_{22} & \cdots & X_{2p} \\ \vdots & \vdots & \ddots & \vdots \\ X_{n1} & X_{n2} & \cdots & X_{np} \end{pmatrix}\right] = \begin{pmatrix} E[X_{11}] & E[X_{12}] & \cdots & E[X_{1p}] \\ E[X_{21}] & E[X_{22}] & \cdots & E[X_{2p}] \\ \vdots & \vdots & \ddots & \vdots \\ E[X_{n1}] & E[X_{n2}] & \cdots & E[X_{np}] \end{pmatrix}.
\tag{2.46}
$$

The linearity in Lemma 2.2 holds for linear/affine transformations of vectors and matrices as well. If $\mathbf{X}$ is $n \times 1$, then for fixed $m \times n$ matrix $\mathbf{B}$ and $m \times 1$ vector $\mathbf{a}$,

$$
E[\mathbf{a} + \mathbf{B}\mathbf{X}] = \mathbf{a} + \mathbf{B}E[\mathbf{X}],
\tag{2.47}
$$

and if $\mathbf{X}$ is $n \times p$, for matrices $\mathbf{A}$ $(m \times q)$, $\mathbf{B}$ $(m \times n)$ and $\mathbf{C}$ $(p \times q)$,

$$
\mathbf{A} + E[\mathbf{B}\mathbf{X}\mathbf{C}] = \mathbf{A} + \mathbf{B}E[\mathbf{X}]\mathbf{C}.
\tag{2.48}
$$

These formulas can be proved by writing out the individual elements, and noting that each is a linear combination of the random variables.

A $1 \times p$ vector $\mathbf{X}$ yields p variances, the $Var[X_i]$'s, but also the $\binom{p}{2}$ covariances, the $Cov[X_i, X_j]$'s. These are usually conveniently arranged in a $p \times p$ matrix, the **covariance matrix**:

$$\mathbf{\Sigma} = Cov[\mathbf{X}] = \begin{pmatrix} Var[X_1] & Cov[X_1, X_2] & \cdots & Cov[X_1, X_p] \\ Cov[X_2, X_1] & Var[X_2] & \cdots & Cov[X_2, X_p] \\ \vdots & \vdots & \ddots & \vdots \\ Cov[X_p, X_1] & Cov[X_p, X_2] & \cdots & Var[X_p] \end{pmatrix}. \qquad (2.49)$$

The same matrix will work for $\mathbf{X}$ and $\mathbf{X}'$, that is, a column vector or a row vector. This matrix is symmetric, i.e., $\mathbf{\Sigma}' = \mathbf{\Sigma}$. (The covariance matrix of a matrix $\mathbf{X}$ of random variables is typically defined by first changing the matrix $\mathbf{X}$ into a long vector, then defining the $Cov[\mathbf{X}]$ to be the covariance of that vector.) A compact way to define the covariance is

$$Cov[\mathbf{X}] = \begin{cases} E[(\mathbf{X} - E[\mathbf{X}])(\mathbf{X} - E[\mathbf{X}])'] & \text{if } \mathbf{X} \text{ is a column vector} \\ E[(\mathbf{X} - E[\mathbf{X}])'(\mathbf{X} - E[\mathbf{X}])] & \text{if } \mathbf{X} \text{ is a row vector} \end{cases} . \qquad (2.50)$$

A convenient, and important to remember, formula for the covariance of an affine transformation follows.

Lemma 2.4. *For fixed* $\mathbf{a}$ *and* $\mathbf{B}$, *where* $\mathbf{X}$ *is a column vector,*

$$Cov[\mathbf{a} + \mathbf{BX}] = \mathbf{B}Cov[\mathbf{X}]\mathbf{B}'. \qquad (2.51)$$

This equation is an example of a "sandwich" formula, with the $\mathbf{B}$'s as the bread. It is not hard to show that similarly, for $\mathbf{X}$ being a row vector,

$$Cov[\mathbf{a} + \mathbf{X}\mathbf{B}'] = \mathbf{B}Cov[\mathbf{X}]\mathbf{B}'. \qquad (2.52)$$

Note that this lemma is a matrix version of (2.41).

Proof.

$$\begin{aligned} Cov[\mathbf{a} + \mathbf{BX}] &= Cov[(\mathbf{a} + \mathbf{BX} - E[\mathbf{a} + \mathbf{BX}])(\mathbf{a} + \mathbf{BX} - E[\mathbf{a} + \mathbf{BX}])'] \quad \text{by (2.50)} \\ &= Cov[\mathbf{B}(\mathbf{X} - E[\mathbf{X}])(\mathbf{B}(\mathbf{X} - E[\mathbf{X}]))'] \\ &= Cov[\mathbf{B}(\mathbf{X} - E[\mathbf{X}])(\mathbf{X} - E[\mathbf{X}])'\mathbf{B}'] \\ &= \mathbf{B}Cov[(\mathbf{X} - E[\mathbf{X}])(\mathbf{X} - E[\mathbf{X}])']\mathbf{B}' \quad \text{by (2.48)} \\ &= \mathbf{B}Cov[\mathbf{X}]\mathbf{B}' \quad \text{again by (2.50)}. \end{aligned} \qquad (2.53)$$

$\square$

This lemma leads to a simple formula for the variance of $a + b_1 X_1 + \cdots + b_p X_p$:

$$Var[a + b_1 X_1 + \cdots + b_p X_p] = \mathbf{b}Cov[\mathbf{X}]\mathbf{b}', \qquad (2.54)$$

because we can write $a + b_1 X_1 + \cdots + b_p X_p = a + \mathbf{X}\mathbf{b}'$ for $\mathbf{b} = (b_1, \ldots, b_p)$. Thus $\mathbf{a} = a$ and $\mathbf{B} = \mathbf{b}$ in (2.52). Compare this formula to (2.43).

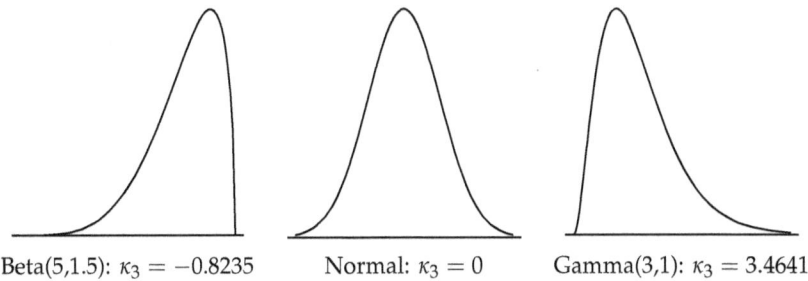

Beta(5,1.5): $\kappa_3 = -0.8235$ Normal: $\kappa_3 = 0$ Gamma(3,1): $\kappa_3 = 3.4641$

Figure 2.2: Some pdfs illustrating skewness.

2.4 Moments

The mean, variance, and covariance are special cases of what are called **moments**. Moments of a random variable provide summaries of its distribution. The kth **raw moment** of a random variable is the expected value of its kth power, where $k = 1, 2, \ldots$. The kth **central moment** is the expected value of the kth power of its deviation from the mean μ, at least for $k > 1$:

$$k^{th} \text{ raw moment} = \mu'_k = E[X^k], \quad k = 1, 2, \ldots;$$
$$k^{th} \text{ central moment} = \mu_k = E[(X - \mu)^k], \quad k = 2, 3, \ldots. \tag{2.55}$$

Thus $\mu'_1 = \mu = E[X]$, $\mu'_2 = E[X^2]$, and $\mu_2 = \sigma^2 = Var[X] = \mu'_2 - \mu'^2_1$. It is not hard, but a bit tedious, to figure out the kth central moment from the first k raw moments, and *vice versa*. It is not uncommon for given moments not to exist. In particular, if the kth moment does not exist, then neither does any higher moment.

The first two moments measure the center and spread of the distribution. The third central moment is generally a measure of **skewness**, where symmetric distributions have 0 skewness, a heavier tail to the right than to the left would have a positive skewness, and a heavier tail to the left would have a negative skewness. Usually it is normalized so that it is not dependent on the variance:

$$\text{Skewness} = \kappa_3 = \frac{\mu_3}{\sigma^3}. \tag{2.56}$$

See Figure 2.2, where the plots show negative, zero, and positive skewness, respectively.

The fourth central moment is a measure of **kurtosis**. It, too, is normalized:

$$\text{Kurtosis} = \kappa_4 = \frac{\mu_4}{\sigma^4} - 3. \tag{2.57}$$

The normal distribution has $\mu_4/\sigma^4 = 3$, so that subtracted "3" in (2.57) means the kurtosis of a normal is 0. It is not particularly easy to figure out what kurtosis means in general, but for nice unimodal densities, it measures "boxiness." A negative kurtosis indicates a density more boxy than the normal, such as the uniform. A positive

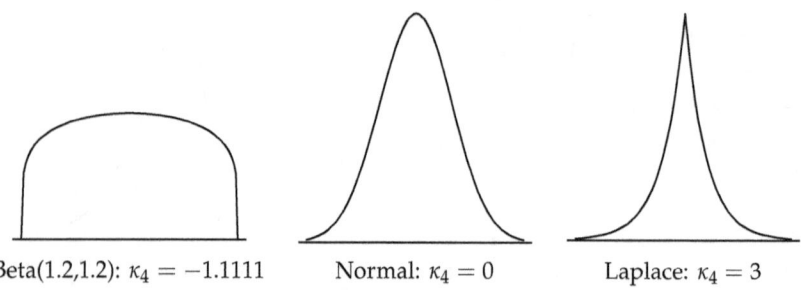

Beta(1.2,1.2): $\kappa_4 = -1.1111$ Normal: $\kappa_4 = 0$ Laplace: $\kappa_4 = 3$

Figure 2.3: Some symmetric pdfs illustrating kurtosis.

kurtosis indicates a pointy middle and heavy tails, such as the Laplace. Figure 2.3 compares some symmetric pdfs, going from boxy to normal to pointy.

The first several moments of a random variable do not characterize it. That is, two different distributions could have the same first, second, and third moments. Even if they agree on all moments, and all moments are finite, the two distributions might not be the same, though that's rare. See Exercise 2.7.20. The next section (Section 2.5) presents the moment generating function, which does determine the distribution under conditions.

Multivariate distributions have the regular moments for the individual component random variables, but also have **mixed moments**. For a p-variate random variable $(X_1, \ldots, X_p)$, mixed moments are expected values of products of powers of the X_i's. So for $\mathbf{k} = (k_1, \ldots, k_p)$, the kth raw mixed moment is $E[\prod X_i^{k_i}]$, and the kth central moment is $E[\prod (X_i - \mu_i)^{k_i}]$, assuming these expected values exist. Thus for two variables, the $(1,1)^{th}$ central moment is the covariance.

2.5 Moment and cumulant generating functions

The **moment generating function (mgf** for short) is a meta-moment in a way, since it can be used to find all the moments of $\mathbf{X}$. If $\mathbf{X}$ is $p \times 1$, it is a function from $\mathbb{R}^p \to [0, \infty]$ given by

$$M_{\mathbf{X}}(\mathbf{t}) = E\left[e^{t_1 X_1 + \cdots + t_p X_p}\right] = E[e^{\mathbf{t} \cdot \mathbf{X}}] \qquad (2.58)$$

for $\mathbf{t} = (t_1, \ldots, t_p)$. (For p-dimensional vectors $\mathbf{a}$ and $\mathbf{b}$, $\mathbf{a} \cdot \mathbf{b} = a_1 b_1 + \cdots + a_p b_p$ is called their **dot product**. Its definition does not depend on the type of vectors, row or column, just that they have the same number of elements.) The mgf does not always exist, that is, often the integral or sum defining the expected value diverges. An infinite mgf for some values of $\mathbf{t}$ is ok, as long as it is finite for $\mathbf{t}$ in a neighborhood of $\mathbf{0}_p$, in which case the mgf uniquely determines the distribution of $\mathbf{X}$.

Theorem 2.5. Uniqueness of mgf. *If for some $\epsilon > 0$,*

$$M_{\mathbf{X}}(\mathbf{t}) < \infty \text{ and } M_{\mathbf{X}}(\mathbf{t}) = M_{\mathbf{Y}}(\mathbf{t}) \text{ for all } \mathbf{t} \text{ such that } \|\mathbf{t}\| \leq \epsilon, \qquad (2.59)$$

then $\mathbf{X}$ and $\mathbf{Y}$ have the same distribution.

If one knows complex variables, the **characteristic function** is superior because it always exists. It is defined as $\phi_\mathbf{X}(\mathbf{t}) = E[\exp(i\mathbf{t} \cdot \mathbf{X})]$, and also uniquely defines the distribution. In fact, most proofs of Theorem 2.5 first show the uniqueness of characteristic functions, then argue that the conditions of the theorem guarantee that the mgf $M(\mathbf{t})$ can be extended to an analytic function of complex $\mathbf{t}$, which for imaginary $\mathbf{t}$ yields the characteristic function. Billingsley (1995) is a good reference for the proofs of the uniquenesses of mgfs (his Section 30) and characteristic functions (his Theorem 26.2).

The uniqueness in Theorem 2.5 is the most useful property of mgfs, but they can also be handy for generating (mixed) moments.

Lemma 2.6. *Suppose* $\mathbf{X}$ *has mgf such that for some* $\epsilon > 0$,

$$M_\mathbf{X}(\mathbf{t}) < \infty \text{ for all } \mathbf{t} \text{ such that } \|\mathbf{t}\| \leq \epsilon. \tag{2.60}$$

Then for any nonnegative integers $k_1, \ldots, k_p$,

$$E[X_1^{k_1} X_2^{k_2} \cdots X_p^{k_p}] = \frac{\partial^{k_1 + \cdots + k_p}}{\partial t_1^{k_1} \cdots \partial t_p^{k_p}} M_\mathbf{X}(\mathbf{t}) \Big|_{\mathbf{t} = 0_p}, \tag{2.61}$$

which is finite.

Notice that this lemma implies that all mixed moments are finite under the condition (2.60). The basic idea is straightforward. Assuming the derivatives and expectation can be interchanged,

$$\frac{\partial^{k_1 + \cdots + k_p}}{\partial t_1^{k_1} \cdots \partial t_p^{k_p}} E[e^{\mathbf{t} \cdot \mathbf{X}}]\Big|_{\mathbf{t} = 0_p} = E[\frac{\partial^{k_1 + \cdots + k_p}}{\partial t_1^{k_1} \cdots \partial t_p^{k_p}} e^{\mathbf{t} \cdot \mathbf{X}}\Big|_{\mathbf{t} = 0_p}]$$

$$= E[X_1^{k_1} X_2^{k_2} \cdots X_p^{k_p}]. \tag{2.62}$$

But justifying that interchange requires some careful analysis. If interested, Section 2.5.4 provides the details when $p = 1$.

Specializing to a random variable X, the mgf is

$$M_X(t) = E[e^{tX}]. \tag{2.63}$$

If it exists for t in a neighborhood of 0, then all moments of X exist, and

$$\frac{\partial^k}{\partial t^k} M_X(t)\Big|_{t=0} = E[X^k]. \tag{2.64}$$

The **cumulant generating function** is the log of the moment generating function,

$$c_\mathbf{X}(\mathbf{t}) = \log(M_\mathbf{X}(\mathbf{t})). \tag{2.65}$$

It generates the **cumulants**, which are defined by what the cumulant generating function generates, i.e., for a random variable, the kth cumulant is

$$\gamma_k = \frac{\partial^k}{\partial t^k} c_X(t)\Big|_{t=0}. \tag{2.66}$$

Mixed cumulants for multivariate $\mathbf{X}$ are found by taking mixed partial derivatives, analogous to (2.61).

Cumulants are often easier to work with than moments. The first four are

$$\gamma_1 = E[X] = \mu_1 = \mu,$$
$$\gamma_2 = Var[X] = \mu_2 = \sigma^2,$$
$$\gamma_3 = E[(X - E[X])^3] = \mu_3, \quad \text{and}$$
$$\gamma_4 = E[(X - E[X])^4] - 3\,Var[X]^2 = \mu_4 - 3\mu_2^2 = \mu_4 - 3\sigma^4. \tag{2.67}$$

The skewness (2.56) and kurtosis (2.57) are then simple functions of the cumulants:

$$\text{Skewness}[X] = \kappa_3 = \frac{\gamma_3}{\sigma^3} \quad \text{and} \quad \text{Kurtosis}[X] = \kappa_4 = \frac{\gamma_4}{\sigma^4}. \tag{2.68}$$

2.5.1 Normal distribution

A $Z \sim N(0,1)$ is called a **standard normal**. Its mgf is

$$M_Z(t) = E[e^{tZ}] = \frac{1}{\sqrt{2\pi}} \int_{-\infty}^{\infty} e^{tz} e^{-\frac{1}{2}z^2}\,dz = \frac{1}{\sqrt{2\pi}} \int_{-\infty}^{\infty} e^{-\frac{1}{2}(z^2 - 2tz)}\,dz. \tag{2.69}$$

In the exponent, complete the square with respect to the z: $z^2 - 2tz = (z - t)^2 - t^2$. Then

$$M_Z(t) = e^{\frac{1}{2}t^2} \int_{-\infty}^{\infty} \frac{1}{\sqrt{2\pi}} e^{-\frac{1}{2}(z-t)^2}\,dz = e^{-\frac{1}{2}t^2}. \tag{2.70}$$

The second equality holds because the integrand in the middle expression is the pdf of a $N(t,1)$, which means the integral is 1.

The cumulant generating function is then a simple quadratic:

$$c_Z(t) = \frac{t^2}{2}, \tag{2.71}$$

and it is easy to see that

$$c_Z'(0) = 0, \quad c''(0) = 1, \quad c'''(t) = 0. \tag{2.72}$$

Thus the mean is 0 and variance is 1 (not surprisingly), and all other cumulants are 0. In particular, the skewness and kurtosis are both 0.

It is a little messier, but the same technique shows that if $X \sim N(\mu, \sigma^2)$,

$$M_X(t) = e^{\mu t + \sigma^2 t^2/2}. \tag{2.73}$$

2.5.2 Gamma distribution

The gamma distribution has two parameters: $\alpha > 0$ is the *shape* parameter, and $\lambda > 0$ is the *rate* parameter. Its space is $\mathcal{X} = (0, \infty)$, and as in Table 1.1 on page 7 its pdf is

$$f(x \,|\, \alpha, \lambda) = \frac{\lambda^\alpha}{\Gamma(\alpha)}\, x^{\alpha-1} e^{-\lambda x}, \quad x \in (0, \infty). \tag{2.74}$$

If $\alpha = 1$, then this distribution is the Exponential(λ) in Table 1.1.

The mgf is

$$
M_X(t) = E[e^{tX}] = \frac{\lambda^\alpha}{\Gamma(\alpha)} \int_0^\infty e^{tx} x^{\alpha-1} e^{-\lambda x}
$$

$$
= \frac{\lambda^\alpha}{\Gamma(\alpha)} \int_0^\infty x^{\alpha-1} e^{-(\lambda-t)x}. \tag{2.75}
$$

That integral needs $(\lambda - t) > 0$ to be finite, so we need $t < \lambda$, which means the mgf is finite for a neighborhood of zero, since $\lambda > 0$. Now the integral at the end of (2.75) looks like the gamma density but with $\lambda - t$ in place of λ. Thus that integral equals the inverse of the constant in the Gamma($\alpha, \lambda - t$), so that

$$
E[e^{tX}] = \frac{\lambda^\alpha}{\Gamma(\alpha)} \frac{\Gamma(\alpha)}{(\lambda - t)^\alpha}
$$

$$
= \left(\frac{\lambda}{\lambda - t} \right)^\alpha, \quad t < \lambda. \tag{2.76}
$$

We will use the cumulant generating function $c_X(t) = \log(M_X(t))$ to obtain the mean and variance, because it is slightly easier. Thus

$$
c_X'(t) = \frac{\partial}{\partial t} \, \alpha(\log(\lambda) - \log(\lambda - t)) = \frac{\alpha}{\lambda - t} \implies E[X] = c_X'(0) = \frac{\alpha}{\lambda}, \tag{2.77}
$$

and

$$
c_X''(t) = \frac{\partial^2}{\partial t^2} \, \alpha(\log(\lambda) - \log(\lambda - t)) = \frac{\alpha}{(\lambda - t)^2} \implies Var[X] = c_X''(0) = \frac{\alpha}{\lambda^2}. \tag{2.78}
$$

In general, the kth cumulant (2.66) is

$$
\gamma_k = (k-1)! \, \frac{\alpha}{\lambda^k}, \tag{2.79}
$$

and in particular

$$
\text{Skewness}[X] = \frac{2\alpha/\lambda^3}{\alpha^{3/2}/\lambda^3} = \frac{2}{\sqrt{\alpha}} \quad \text{and} \quad \text{Kurtosis}[X] = \frac{6\alpha/\lambda^4}{\alpha^2/\lambda^4} = \frac{6}{\alpha}. \tag{2.80}
$$

Thus the skewness and kurtosis depends on just the shape parameter α. Also, they are positive, but tend to 0 as α increases.

2.5.3 Binomial and multinomial distributions

A Bernoulli trial is an event that has just two possible outcomes, often called "success" and "failure." For example, flipping a coin once is a trial, and one might declare that heads is a success. In many medical studies, a single person's outcome is often a success or failure. Such a random variable Z has space $\{0,1\}$, where 1 denotes success and 0 failure. The distribution is completely specified by the probability of a success, denoted $p : p = P[Z = 1]$.

The binomial is a model for counting the number of successes in n trials, e.g., the number of heads in ten flips of a coin, where the trials are independent (formally defined in Section 3.3) and have the same probability p of success. As in Table 1.2 on page 9,

$$X \sim \text{Binomial}(n, p) \implies f_X(x) = \binom{n}{x} p^x (1-p)^{n-x}, \quad x \in \mathcal{X} = \{0, 1, \ldots, n\}. \quad (2.81)$$

The fact that this pmf sums to 1 relies on the **binomial theorem**:

$$(a + b)^n = \sum_{x=0}^{n} \binom{n}{x} a^x b^{n-x}, \quad (2.82)$$

with $a = p$ and $b = 1 - p$. This theorem also helps in finding the mgf:

$$\begin{aligned} M_X(t) = E[e^{tX}] &= \sum_{x=0}^{n} e^{tx} f_X(x) \\ &= \sum_{x=0}^{n} e^{tx} \binom{n}{x} p^x (1-p)^{n-x} \\ &= \sum_{x=0}^{n} \binom{n}{x} (pe^t)^x (1-p)^{n-x} \\ &= (pe^t + 1 - p)^n. \quad (2.83) \end{aligned}$$

It is finite for all $t \in \mathbb{R}$, as is the case for any bounded random variable.

Now $c_X(t) = \log(M_X(t)) = n \log(pe^t + 1 - p)$ is the cumulant generating function. The first two cumulants are

$$E[X] = c_X'(0) = n \frac{pe^t}{pe^t + 1 - p} \bigg|_{t=0} = np, \quad (2.84)$$

and

$$\begin{aligned} Var[X] &= c_X''(0) \\ &= n \left(\frac{pe^t}{pe^t + 1 - p} - \frac{(pe^t)^2}{(pe^t + 1 - p)^2} \right) \bigg|_{t=0} \\ &= n(p - p^2) = np(1 - p). \quad (2.85) \end{aligned}$$

(In Section 4.3.5 we will exhibit an easier approach.)

The **multinomial** distribution also models the results of n trials, but here there are K possible categories for each trial. E.g., one may roll a die n times, and see whether it is a one, two, ..., or six (so $K = 6$); or one may randomly choose n people, each of whom is then classified as *short*, *medium*, or *tall* (so $K = 3$). As for the binomial, the trials are assumed independent, and the probability of an individual trial coming up in category k is p_k, so that $p_1 + \cdots + p_K = 1$. The random vector is $\mathbf{X} = (X_1, \ldots, X_K)$, where X_k is the number of observations from category k. Letting $\mathbf{p} = (p_1, \ldots, p_K)$, we have

$$\mathbf{X} \sim \text{Multinomial}(n, \mathbf{p}) \implies f_{\mathbf{X}}(\mathbf{x}) = \binom{n}{\mathbf{x}} p_1^{x_1} \cdots p_K^{x_K}, \quad \mathbf{x} \in \mathcal{X}, \quad (2.86)$$

where the space consists of all possible ways K nonnegative integers can sum to n:

$$\mathcal{X} = \{\mathbf{X} \in \mathbb{R}^K \mid x_k \in \{0,\ldots,n\} \text{ for each } k, \text{ and } x_1 + \cdots + x_K = n\}, \tag{2.87}$$

and for $\mathbf{x} \in \mathcal{X}$,

$$\binom{n}{\mathbf{x}} = \left(\frac{n!}{x_1! \cdots x_K!}\right). \tag{2.88}$$

This pmf is related to the **multinomial theorem**:

$$(a_1 + \cdots + a_K)^n = \sum_{\mathbf{x} \in \mathcal{X}} \binom{n}{\mathbf{x}} a_1^{x_1} \cdots a_K^{x_K}. \tag{2.89}$$

Note that the binomial is a special case of the multinomial with $K = 2$:

$$X \sim \text{Binomial}(n,p) \implies (X, n-X) \sim \text{Multinomial}(n, (p, 1-p)). \tag{2.90}$$

Now for the mgf. It is a function of $\mathbf{t} = (t_1, \ldots, t_K)$:

$$\begin{aligned}
M_{\mathbf{X}}(\mathbf{t}) = E[e^{\mathbf{t} \cdot \mathbf{X}}] &= \sum_{\mathbf{x} \in \mathcal{X}} e^{\mathbf{t} \cdot \mathbf{X}} f_{\mathbf{X}}(\mathbf{x}) \\
&= \sum_{\mathbf{x} \in \mathcal{X}} e^{t_1 x_1} \cdots e^{t_K x_K} \binom{n}{\mathbf{x}} p_1^{x_1} \cdots p_K^{x_K} \\
&= \sum_{\mathbf{x} \in \mathcal{X}} \binom{n}{\mathbf{x}} (p_1 e^{t_1})^{x_1} \cdots (p_K e^{t_K})^{x_K} \\
&= (p_1 e^{t_1} + \cdots + p_K e^{t_K})^n < \infty \text{ for all } \mathbf{t} \in \mathbb{R}^K. \tag{2.91}
\end{aligned}$$

The mean and variance of each X_k can be found much as for the binomial. We find that

$$E[X_k] = np_K \text{ and } Var[X_k] = np_k(1 - p_k). \tag{2.92}$$

(In fact, these results are not surprising since the individual X_k are binomial.) For the covariance between X_1 and X_2, we first find

$$\begin{aligned}
E[X_1 X_2] &= \frac{\partial^2}{\partial t_1 \partial t_2} M_{\mathbf{X}}(\mathbf{t})\Big|_{t=0_K} \\
&= n(n-1)(p_1 e^{t_1} + \cdots + p_K e^{t_K})^{n-2} p_1 e^{t_1} p_2 e^{t_2}\Big|_{t=0_K} \\
&= n(n-1) p_1 p_2. \tag{2.93}
\end{aligned}$$

(The cumulant generating function works as well.) Thus

$$Cov[X_1, X_2] = n(n-1) p_1 p_2 - (np_1)(np_2) = -np_1 p_2. \tag{2.94}$$

Similarly, $Cov[X_k, X_l] = -np_k p_l$ if $k \neq l$. It does make sense for the covariance to be negative, since the more there are in category 1, the fewer are available for category 2.

2.5.4   Proof of the moment generating lemma

Here we prove Lemma 2.6 when $p = 1$. The main mathematical challenge is proving that we can interchange derivatives and expected values. We will use the **dominated convergence theorem** from real analysis and measure theory. See, e.g., Theorem 16.4 in Billingsley (1995). Suppose $g_n(\mathbf{x}), n = 0, 1, 2, \cdots$, and $g(\mathbf{x})$ are functions such that $\lim_{n\to\infty} g_n(\mathbf{x}) = g(\mathbf{x})$ for each $\mathbf{x}$. The theorem states that if there is a function $h(\mathbf{x})$ such that $|g_n(\mathbf{x})| \leq h(\mathbf{x})$ for all n, and $E[h(\mathbf{X})] < \infty$, then

$$\lim_{n\to\infty} E[g_n(\mathbf{X})] = E[g(\mathbf{X})]. \tag{2.95}$$

The assumption in Lemma 2.6 is that for some $\epsilon > 0$, the random variable X has $M(t) < \infty$ for $|t| \leq \epsilon$. We show that for $|t| < \epsilon$,

$$M^{(k)}(t) \equiv \frac{\partial^k}{\partial t^k} M(t) = E[X^k e^{tX}], \quad \text{and} \quad E[|X|^k e^{tX}] < \infty, \quad k = 0, 1, 2, \ldots. \tag{2.96}$$

The lemma follows by setting $t = 0$. Exercise 2.7.22(a) in fact proves a somewhat stronger result than the above inequality:

$$E[|X|^k e^{|sX|}] < \infty, \quad |s| < \epsilon. \tag{2.97}$$

The $k = 0^{th}$ derivative is just the function itself, so that (2.96) for $k = 0$ is $M(t) = E[\exp(tX)] < \infty$, which is what we have assumed. Now assume (2.96) holds for $k = 0, \ldots, m$, and consider $k = m + 1$. Since $|t| < \epsilon$, we can take $\epsilon' = (\epsilon - |t|)/2 > 0$, so that $|t| + \epsilon' < \epsilon$. Then by (2.96),

$$\frac{M^{(m)}(t + \delta) - M^{(m)}(t)}{\delta} = E\left[\frac{X^m e^{(t+\delta)X} - X^m e^{tX}}{\delta}\right]$$

$$= E\left[X^m e^{tX} \frac{e^{\delta X} - 1}{\delta}\right] \quad \text{for } 0 < |\delta| \leq \epsilon'. \tag{2.98}$$

Here we apply the dominated convergence theorem to the term in the last expectation, where $g_n(x) = x^m \exp(tx)(\exp(\delta_n x) - 1)/\delta_n$, with $\delta_n = \epsilon'/n \to 0$. Exercise 2.7.22(b) helps to show that

$$|g_n(x)| \leq |x|^m e^{|tx|} \frac{e^{\epsilon'|x|} - 1}{\epsilon'} \equiv h(x). \tag{2.99}$$

Now (2.97) applied with $k = m$, and $s = |t|$ and $s = |t| + \epsilon'$, shows that $E[h(X)] < \infty$. Hence the dominated convergence theorem implies that (2.95) holds, meaning we can take $\delta \to 0$ on both sides of (2.98). The left-hand side is the $(m + 1)^{st}$ derivative of M, and in the expected value $(\exp(\delta x) - 1)/\delta \to x$. That is,

$$M^{(m+1)}(t) = E[X^{m+1} e^{tX}]. \tag{2.100}$$

Then induction, along with (2.97), proves (2.96).

The proof for general p runs along the same lines. The induction step is performed on multiple indices, one for each k_i in the mixed-moment in (2.61).

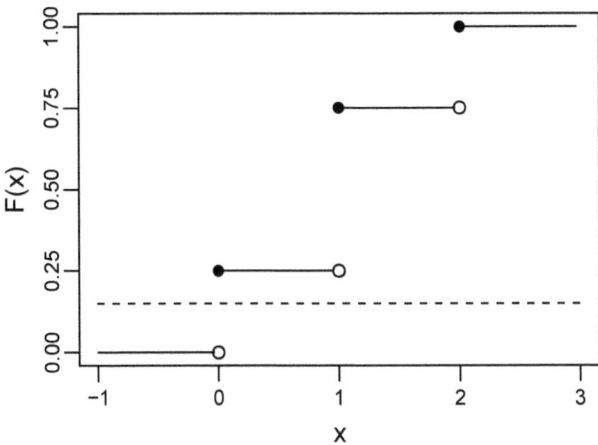

Figure 2.4: The distribution function for a Binomial(2,1/2). The dotted line is where $F(x) = 0.15$.

2.6 Quantiles

A positional measure for a random variable is one that gives the value that is in a certain relation to the rest of the values. For example, the 0.25^{th} quantile is the value such that the random variable is below the value 25% of the time, and above it 75% of the time. The median is the $(1/2)^{th}$ quantile. Ideally, for $q \in [0,1]$, the qth quantile is the value η_q such that $F(\eta_q) = q$, where F is the distribution function. That is, $\eta_q = F^{-1}(q)$. Unfortunately, F does not have an inverse for all q unless it is strictly increasing, which leaves out all discrete random variables. Even in the continuous case, the inverse might not be unique, e.g., there may be a flat spot in F. For example, consider the pdf $f(x) = 1/2$ for $x \in (0,1) \cup (2,3)$. Then any number x between 1 and 2 has $F(x) = 1/2$, so that there is no unique median. Thus the definition is a bit more involved.

Definition 2.7. *For $q \in (0,1)$, a qth quantile of the random variable X is any value η_q such that*

$$P[X \leq \eta_q] \geq q \ \text{ and } \ P[X \geq \eta_q] \geq 1 - q. \tag{2.101}$$

With this definition, there is at least one quantile for each q for any distribution, but there is no guarantee of uniqueness without some additional assumptions.

As mentioned above, if the distribution function is strictly increasing in x for all $x \in \mathcal{X}$, where the space $\mathcal{X}$ is a (possibly infinite) interval, then $\eta_q = F^{-1}(q)$ uniquely. For example, if X is Exponential(1), then $F(x) = 1 - e^{-x}$ for $x > 0$, so that $\eta_q = -\log(1-q)$ for $q \in (0,1)$.

By contrast, consider $X \sim$ Binomial(2,1/2), whose distribution function is given in Figure 2.4. At $x = 0$, $P[X \leq 0] = 0.25$ and $P[X \geq 0] = 1$. Thus 0 is a quantile for any $q \in (0, 0.25]$. The horizontal dotted line in the graph is where $F(x) = 0.15$. It never hits the distribution function, but it passes through the gap at $x = 0$, hence its quantile is 0. But $q = 0.25 = F(x)$ hits an entire interval of points between 0 and 1. Thus any of those values is its quantile, i.e., $\eta_{0.25}$. The complete set of quantiles for $q \in (0, 1)$ is

$$\eta_q = \begin{cases} 0 & \text{if } q \in (0, 0.25) \\ [0,1] & \text{if } q = 0.25 \\ 1 & \text{if } q \in (0.25, 0.75) \\ [1,2] & \text{if } q = 0.75 \\ 2 & \text{if } q \in (0.75, 1) \end{cases} . \qquad (2.102)$$

2.7 Exercises

Exercise 2.7.1. (a) Let $X \sim \text{Beta}(\alpha, \beta)$. Find $E[X(1 - X)]$. (Give the answer in terms of a rational polynomial in α, β.) (b) Find $E[X^a(1 - X)^b]$ for nonnegative integers a and b.

Exercise 2.7.2. The Geometric(p) distribution is a discrete distribution with space being the nonnegative integers. It has pmf $f(x) = p(1 - p)^x$, for parameter $p \in (0, 1)$. If one is flipping a coin with $p = P[\text{Heads}]$, then X is the number of tails before the first head, assuming independent flips. (a) Find the moment generating function, $M(t)$, of X. For what t is it finite? (b) Find $E[X]$ and $Var[X]$.

Exercise 2.7.3. Prove (2.23), i.e., that $Cov[X, Y] = E[XY] - E[X]E[Y]$ if the expected values exist.

Exercise 2.7.4. Suppose $Y_1, \ldots, Y_n$ are uncorrelated random variables with the same mean μ and same variance σ^2. Let $\mathbf{Y} = (Y_1, \ldots, Y_n)'$. (a) Write down $E[\mathbf{Y}]$ and $Cov[\mathbf{Y}]$. (b) For an $n \times 1$ vector $\mathbf{a}$, show that $\mathbf{a}'\mathbf{Y}$ has mean $\mu \sum a_i$ and variance $\sigma^2 \|\mathbf{a}\|^2$, where "$\|\mathbf{a}\|$" is the **norm** of the vector $\mathbf{a}$:

$$\|\mathbf{a}\| = \sqrt{a_1^2 + \cdots + a_n^2}. \qquad (2.103)$$

Exercise 2.7.5. Suppose $\mathbf{X}$ is a 3×1 vector with covariance matrix $\sigma^2 \mathbf{I}_3$. (a) Find the matrix $\mathbf{A}$ so that $\mathbf{AX}$ is the vector of deviations, i.e.,

$$\mathbf{D} = \mathbf{AX} = \begin{pmatrix} X_1 - \overline{X} \\ X_2 - \overline{X} \\ X_3 - \overline{X} \end{pmatrix}. \qquad (2.104)$$

(b) Find the $\mathbf{B}$ for which $Cov[\mathbf{D}] = \sigma^2 \mathbf{B}$. How does it compare to $\mathbf{A}$? (c) What is the correlation between two elements of $\mathbf{D}$? (d) Let $\mathbf{c}$ be the 1×3 vector such that $\mathbf{c}\mathbf{X} = \overline{X}$. What is $\mathbf{c}$? Find $\mathbf{cc}' = \|\mathbf{c}\|^2$, $Var[\mathbf{c}\mathbf{X}]$, and $\mathbf{cA}$. (e) Find

$$Cov\left[\begin{pmatrix} \mathbf{c} \\ \mathbf{A} \end{pmatrix} \mathbf{X} \right]. \qquad (2.105)$$

From that matrix (it should be 4×4), read off the covariance of $\overline{X}$ with the deviations.

Exercise 2.7.6. Here, $\mathbf{Y}$ is an $n \times 1$ vector with $Cov[\mathbf{Y}] = \sigma^2 \mathbf{I}_n$. Also, $E[Y_i] = \beta x_i$, $i = 1, \ldots, n$, where $\mathbf{x} = (x_1, \ldots, x_n)'$ is a fixed set of constants (not all zero), and β is a parameter. This model is simple linear regression with an intercept of zero. Let $U = \mathbf{x}'\mathbf{Y}$. (a) Find $E[U]$ and $Var[U]$. (b) Find the constant c so that $E[U/c] = \beta$. (Then U/c is an unbiased estimator of β.) What is $Var[U/c]$?

Exercise 2.7.7. Suppose $\mathbf{X} \sim \text{Multinomial}(n, \mathbf{p})$, where $\mathbf{X}$ and $\mathbf{p}$ are $1 \times K$. Show that

$$Cov[\mathbf{X}] = n(\text{diag}(\mathbf{p}) - \mathbf{p}'\mathbf{p}), \tag{2.106}$$

where $\text{diag}(\mathbf{p})$ is the $K \times K$ diagonal matrix

$$\text{diag}(\mathbf{p}) = \begin{pmatrix} p_1 & 0 & 0 & \cdots & 0 \\ 0 & p_2 & 0 & \cdots & 0 \\ 0 & 0 & p_3 & \cdots & 0 \\ \vdots & \vdots & \vdots & \ddots & \vdots \\ 0 & 0 & 0 & \cdots & p_K \end{pmatrix}. \tag{2.107}$$

[Hint: You can use the results in (2.92) and (2.94).]

Exercise 2.7.8. Suppose X and Y are random variables with $Var[X] = \sigma_X^2$, $Var[Y] = \sigma_Y^2$ and $Cov[X, Y] = \eta$. (a) Find $Cov[aX + bY, cX + dY]$ directly (i.e., using (2.44)). (b) Now using the matrix manipulations, find the covariance matrix for

$$\begin{pmatrix} a & b \\ c & d \end{pmatrix} \begin{pmatrix} X \\ Y \end{pmatrix}. \tag{2.108}$$

Does the covariance term in the resulting matrix equal the answer in part (a)?

Exercise 2.7.9. Suppose

$$Cov[(X, Y)] = \begin{pmatrix} 1 & 2 \\ 2 & 5 \end{pmatrix}. \tag{2.109}$$

(a) Find the constant b so that X and $Y - bX$ are uncorrelated. (b) For that b, what is $Var[Y - bX]$?

Exercise 2.7.10. Suppose $\mathbf{X}$ is $p \times 1$ and $\mathbf{Y}$ is $q \times 1$. Then $Cov[\mathbf{X}, \mathbf{Y}]$ is defined to be the $p \times q$ matrix with elements $Cov[X_i, Y_j]$:

$$Cov[\mathbf{X}, \mathbf{Y}] = \begin{pmatrix} Cov[X_1, Y_1] & Cov[X_1, Y_2] & \cdots & Cov[X_1, Y_q] \\ Cov[X_2, Y_1] & Cov[X_2, Y_2] & \cdots & Cov[X_2, Y_q] \\ \vdots & \vdots & \ddots & \vdots \\ Cov[X_p, Y_1] & Cov[X_p, Y_2] & \cdots & Cov[X_p, Y_q] \end{pmatrix}. \tag{2.110}$$

(a) Show that $Cov[\mathbf{X}, \mathbf{Y}] = E[(\mathbf{X} - E[\mathbf{X}])(\mathbf{Y} - E[\mathbf{Y}])'] = E[\mathbf{XY}'] - E[\mathbf{X}]E[\mathbf{Y}]'$. (b) Suppose $\mathbf{A}$ is $r \times p$ and $\mathbf{B}$ is $s \times q$. Show that $Cov[\mathbf{AX}, \mathbf{BY}] = \mathbf{A}Cov[\mathbf{X}, \mathbf{Y}]\mathbf{B}'$.

Exercise 2.7.11. Let $Z \sim N(0, 1)$, and $W = Z^2$, so $W \sim \chi_1^2$ as in Exercise 1.7.13. (a) Find the moment generating function $M_W(t)$ of W by integrating over the pdf of Z, i.e., find $\int e^{tz^2} f_Z(z) dz$. For which values of t is $M_W(t)$ finite? (b) From (2.76), the moment generating function of a Gamma(α, λ) random variable is $\lambda^\alpha / (\lambda - t)^\alpha$ when $t < \lambda$. For what values of α and λ does this mgf equal that of W? Is that result as it should be, i.e, the mgf of χ_1^2?

Exercise 2.7.12. As in Table 1.1 (page 7), the Laplace distribution (also known as the double exponential) has space $(-\infty, \infty)$ and pdf $f(x) = (1/2)e^{-|x|}$. (a) Show that the Laplace has mgf $M(t) = 1/(1 - t^2)$. [Break the integral into two parts, according to the sign of x.] (b) For which t is the mgf finite?

Exercise 2.7.13. Continue with $X \sim$ Laplace as in Exercise 2.7.12. (a) Show that for k even, $E[X^k] = \Gamma(k+1) = k!$ [Hint: It is easiest to do the integral directly, noting that by symmetry it is twice the integral over $(0, \infty)$.] (b) Use part (a) to show that $Var[X] = 2$ and $Kurtosis[X] = 3$.

Exercise 2.7.14. Suppose $(X, Y) = (\cos(\Theta), \sin(\Theta))$, where $\Theta \sim$ Uniform$(0, 2\pi)$. (a) Find $E[X], E[X^2], E[X^3], E[X^4]$, and $Var[X]$, Skewness$[X]$, and Kurtosis$[X]$. (b) Find $Cov[X, Y]$ and $Corr[X, Y]$. (c) Find $E[X^2 + Y^2]$ and $Var[X^2 + Y^2]$.

Exercise 2.7.15. (a) Show that the mgf of the Poisson(λ) is $e^{\lambda(e^t - 1)}$. (b) Find the kth cumulant of the Poisson(λ) as a function of λ and k.

Exercise 2.7.16. (a) Fill in the skewness and kurtosis for the indicated distributions (if they exist). The "$\cos(\Theta)$" is the X from Exercise 2.7.14.

Distribution	Skewness	Kurtosis
Normal(0,1)		
Uniform(0,1)		
Exponential(1)		
Laplace		
Cauchy		
$\cos(\Theta)$		
Poisson(1/2)		
Poisson(20)		

(b) Which of the given distributions with zero skewness is most "boxy," according to the above table? (c) Which of the given distributions with zero skewness has the most "pointy-middled/fat-tailed," according to the above table? (d) Which of the given distributions is most like the normal (other than the normal), according to the above table? Which is least like the normal? (Ignore the distributions whose skewness and/or kurtosis does not exist.)

Exercise 2.7.17. The logistic distribution has space $\mathbb{R}$ and pdf $f(x) = e^x(1 + e^x)^{-2}$ as in Table 1.1. Show that the qth quantile is $\eta_q = \log(q/(1-q))$, which is logit(q).

Exercise 2.7.18. This exercises uses $X \sim$ Logistic. (a) Exercise 1.7.15 shows that X can be represented as $X = \log(U/(1-U))$ where $U \sim$ Uniform$(0, 1)$. Show that the mgf of X is

$$M_X[e^{tX}] = E[e^{t \log(U/(1-U))}] = \Gamma(1 + t)\Gamma(1 - t). \tag{2.111}$$

For which values of t is that equation valid? [Hint: Write the integrand as a product of powers of u and $1 - u$, and notice that it looks like the beta pdf without the constant.] (b) The **digamma function** is defined to be $\psi(\alpha) = d \log(\Gamma(\alpha))/d\alpha$. The **trigamma function** is its derivative, $\psi'(\alpha)$. Show that the variance of the logistic is $\pi^2/3$. You can use the fact that $\psi'(1) = \pi^2/6$. (c) Show that

$$Var[X] = 2 \int_0^\infty x^2 \frac{e^{-x}}{(1 + e^{-x})^2} = 4\eta(2), \quad \text{where } \eta(s) = \sum_{k=1}^\infty \frac{(-1)^{k-1}}{k^s}. \tag{2.112}$$

The function η is the **Dirichlet eta function**, and $\eta(2) = \pi^2/12$. [Hint: For the first equality in (2.112), use the fact that the pdf of the logistic is symmetric about 0, i.e., $f(x) = f(-x)$. For the second equality, use the expansion $(1-z)^{-2} = \sum_{k=1}^{\infty} k z^{k-1}$ for $|z| < 1$, then integrate each term over x, noting that each term has something like a gamma pdf.]

Exercise 2.7.19. If $X \sim N(\mu, \sigma^2)$, then $Y = \exp(X)$ has a **lognormal distribution**. (a) Show that the kth raw moment of Y is $\exp(k\mu + k^2\sigma^2/2)$. [Hint: Note that $E[Y^k] = M_X(k)$, where M_X is the mgf of X.] (b) Show that for $t > 0$, the mgf of Y is infinite. Thus the conditions for Lemma 2.6 do not hold, but the moments are finite anyway. [Hint: Write $M_Y(t) = E[\exp(t\exp(X))] = c\int(\exp(t\exp(x) - (x-\mu)^2/(2\sigma^2))dx$. Then show that for $t > 0$, $t\exp(x)/((x-\mu)^2/(2\sigma^2)) \to \infty$ as $x \to \infty$, which means there is some x_0 such that the exponent in the integral is greater than 0 for $x > x_0$. Thus $M_Y(t) > c\int_{x_0}^{\infty} 1dx = \infty$.]

Exercise 2.7.20. Suppose Z has pmf $p_Z(z) = c\exp(-z^2/2)$ for $z = 0, \pm1, \pm2, \ldots$. That is, the space of Z is $\mathbb{Z}$, the set of all integers. Here, $c = 1/\sum_{z\in\mathbb{Z}}\exp(-z^2/2)$. Let $W = \exp(Z)$. (a) Show that $E[W^k] = \exp(k^2/2)$. [Hint: Write $E[W^k] = E[\exp(kZ)] = c\sum_{z\in\mathbb{Z}}\exp(kz - k^2/2)$. Then complete the square in the exponent wrt k, and change the summation to that over $z - k$.] (b) Show that W has the same raw moments as the lognormal Y in Exercise 2.7.19 when $\mu = 0$ and $\sigma^2 = 1$. Do W and Y have the same distribution? (See Durrett (2010) for this W and an extension.)

Exercise 2.7.21. Suppose the random variable X has mgf $M(t)$ that is finite for $|t| \leq \epsilon$ for some $\epsilon > 0$. This exercise shows that all moments of X are finite. (a) Show that $E[\exp(t|X|)] < \infty$ for $|t| \leq \epsilon$. [Hint: Note that for such t, $M(t)$ and $M(-t)$ are both finite, and $\exp(t|X|) < \exp(tX) + \exp(-tX)$ for any t and X. Then take expected values of both sides of that inequality.] (b) Write $\exp(t|X|)$ in its series expansion $(\exp(a) = \sum_{k=0}^{\infty} a^k/k!)$, and show that if $t > 0$, for any integer k, $|X|^k \leq \exp(t|X|)k!/t^k$. Argue that then $E[|X|^k] < \infty$.

Exercise 2.7.22. Continue with the setup in Exercise 2.7.21. Here we prove some facts needed for the proof of Lemma 2.6 in Section 2.5.4. (a) Fix $|t| < \epsilon$, and show that there exists a $\delta \in (0, \epsilon)$ such that $|t + \delta| < \epsilon$. Thus $M(t + \delta) < \infty$. Write $M(t + \delta) = E[\exp(t|X|)\exp(\delta|X|)]$. Expand $\exp(\delta|x|)$ as in Exercise 2.7.21(b) to show that for any integer k, $|X|^k\exp(t|X|) \leq \exp((t + \delta)|X|)k!/\delta^k$. Argue that therefore $E[|X|^k\exp(t|X|)] < \infty$. (b) Suppose $\delta \in (0, \epsilon')$. Show that

$$\left|\frac{e^{\delta x} - 1}{\delta}\right| \leq \frac{e^{\epsilon'|x|} - 1}{\epsilon'}. \tag{2.113}$$

[Hint: Expand the exponential again to obtain $(\exp(\delta x) - 1)/\delta = \sum_{k=1}^{\infty} \delta^{k-1}x^k/k!$. Then take absolute values, noting that in the sum, all the terms satisfy $\delta^{k-1}|x|^k \leq \epsilon'^{k-1}|x|^k$. Finally, reverse the expansion step.]

Exercise 2.7.23. Verify the quantiles of the Binomial$(2, 1/2)$ given in (2.102) for $q \in (0.25, 1)$.

Exercise 2.7.24. Suppose X has the "late start" distribution function as in (1.21) and Figure 1.2 on page 10, where $F(x) = 0$ if $x < 0$, $F(0) = 1/10$, and $F(x) = 1 - (9/10)e^{-x/100}$ if $x > 0$. Find the quantiles for all $q \in (0, 1)$.

Exercise 2.7.25. Imagine wishing to guess the value of a random variable X before you see it. (a) If you guess m and the value of X turns out to be x, you lose $(x - m)^2$ dollars. What value of m will minimize your expected loss? Show that $m = E[X]$ minimizes $E[(X - m)^2]$ over m, assuming that $Var[X] < \infty$. [Hint: Write the expected loss as $E[X^2] - 2mE[X] + m^2$, then differentiate wrt m and set to 0.] (b) What is the minimum value? (c) Suppose instead you lose $|x - m|$, which has relatively smaller penalties for large errors than does squared error loss. Assume that X has a continuous distribution with pdf f and finite mean. Show that $E[|X - m|]$ is minimized by m being any median of X. [Hint: Write the expected value as

$$E[|X - m|] = \int_{-\infty}^{m} |x - m| f(x) dx + \int_{m}^{\infty} |x - m| f(x) dx$$

$$= -\int_{-\infty}^{m} (x - m) f(x) dx + \int_{m}^{\infty} (x - m) f(x) dx, \qquad (2.114)$$

then differentiate and set to 0. Use the fact that $P[X = m] = 0$.] The minimum value here is called the mean absolute deviation from the median. (d) Now suppose the penalty is different depending on whether your guess is too small or too large. That is, for some $q \in (0, 1)$, you lose $q|x - m|$ if $x > m$, and $(1 - q)|x - m|$ if $x < m$. Show that the expected value of this loss is minimized by m being any q^{th} quantile of X.

Exercise 2.7.26. The **interquartile range** of a distribution is defined to be the difference between the two quartiles, that is, it is IQR $= \eta_{0.75} - \eta_{0.25}$ (at least if the quartiles are unique). Find the interquartile range for a $N(\mu, \sigma^2)$ random variable.

Chapter 3

Marginal Distributions and Independence

3.1 Marginal distributions

Given the distribution of a vector of random variables, it is possible in principle to find the distribution of any individual component of the vector, or any subset of components. To illustrate, consider the distribution of the scores (Assignment, Exams) for a statistics class, where each variable has values "Lo" and "Hi":

Assignments	Exams Lo	Hi	Marginal of Assignments	
Lo	0.3178	0.2336	0.5514	(3.1)
Hi	0.1028	0.3458	0.4486	
Marginal of Exams	0.4206	0.5794	1	

Thus about 32% of the students did low on both assignments and exams, and about 35% did high on both. But notice it is also easy to figure out the percentages of people who did low or high on the individual scores, e.g.,

$$P[\text{Assigment} = \text{Lo}] = 0.5514 \text{ and (hence) } P[\text{Assigment} = \text{Hi}] = 0.4486. \quad (3.2)$$

These numbers are in the margins of the table (3.1), hence the distribution of assignments alone, and of exams alone, are called **marginal** distributions. The distribution of (Assignments, Exams) together is called the **joint** distribution.

More generally, given the joint distribution of (the big vector) $(\mathbf{X}, \mathbf{Y})$, one can find the marginal distribution of the vector $\mathbf{X}$, and the marginal distribution of the vector $\mathbf{Y}$. (We don't have to take consecutive components of the vector, e.g., given $(X_1, X_2, \ldots, X_5)$, we could be interested in the marginal distribution of (X_1, X_3, X_4), say.)

Actually, the words *joint* and *marginal* can be dropped. The joint distribution of $(\mathbf{X}, \mathbf{Y})$ is just the distribution of $(\mathbf{X}, \mathbf{Y})$; the marginal distribution of $\mathbf{X}$ is just the distribution of $\mathbf{X}$, and the same for $\mathbf{Y}$. The extra verbiage can be helpful, though, when dealing with different types of distributions in the same breath.

Before showing how to find the marginal distributions from the joint, we should deal with the spaces. Let $\mathcal{W}$ be the joint space of $(\mathbf{X}, \mathbf{Y})$, and $\mathcal{X}$ and $\mathcal{Y}$ be the marginal spaces of $\mathbf{X}$ and $\mathbf{Y}$, respectively. Then

$$\mathcal{X} = \{\mathbf{x} \mid (\mathbf{x}, \mathbf{y}) \in \mathcal{W} \text{ for some } \mathbf{y}\} \text{ and } \mathcal{Y} = \{\mathbf{y} \mid (\mathbf{x}, \mathbf{y}) \in \mathcal{W} \text{ for some } \mathbf{x}\}. \quad (3.3)$$

For example, consider the joint space $\mathcal{W} = \{(x,y) \,|\, 0 < x < y < 1\}$, sketched in Figure 2.1 on page 22. The marginal spaces $\mathcal{X}$ and $\mathcal{Y}$ are then both $(0,1)$.

There are various approaches to finding the marginal distributions from the joint. First, suppose $F(\mathbf{x}, \mathbf{y})$ is the distribution function for $(\mathbf{X}, \mathbf{Y})$ jointly, and $F_{\mathbf{X}}(\mathbf{x})$ is that for $\mathbf{x}$ marginally. Then (assuming $\mathbf{x}$ is $p \times 1$ and $\mathbf{y}$ is $q \times 1$),

$$
\begin{aligned}
F_{\mathbf{X}}(\mathbf{x}) &= P[X_1 \le x_1, \ldots, X_p \le x_p] \\
&= P[X_1 \le x_1, \ldots, X_p \le x_p, Y_1 \le \infty, \ldots, Y_q \le \infty] \\
&= F(x_1, \ldots, x_p, \infty, \ldots, \infty).
\end{aligned}
\tag{3.4}
$$

That is, you put ∞ in for the variables you are not interested in, because they are certainly less than infinity.

The mgf is equally easy. Suppose $M(\mathbf{t}, \mathbf{s})$ is the mgf for $(\mathbf{X}, \mathbf{Y})$ jointly, so that

$$
M(\mathbf{t}, \mathbf{s}) = E[e^{\mathbf{t} \cdot \mathbf{X} + \mathbf{s} \cdot \mathbf{Y}}].
\tag{3.5}
$$

To eliminate the dependence on $\mathbf{Y}$, we now set $\mathbf{s}$ to zero, that is, the mgf of $\mathbf{X}$ alone is

$$
M_{\mathbf{X}}(\mathbf{t}) = E[e^{\mathbf{t} \cdot \mathbf{X}}] = E[e^{\mathbf{t} \cdot \mathbf{X} + \mathbf{0}_q \cdot \mathbf{Y}}] = M(\mathbf{t}, \mathbf{0}_q).
\tag{3.6}
$$

3.1.1 Multinomial distribution

Given $\mathbf{X} \sim \text{Multinomial}(n, \mathbf{p})$ as in (2.86), one may wish to find the marginal distribution of a single component, e.g., X_1. It should be binomial, because now for each trial a success is that the observation is in the first category. To show this fact, we find the mgf of X_1 by setting $t_2 = \cdots = t_K = 0$ in (2.91):

$$
M_{X_1}(t) = M_{\mathbf{X}}((t, 0, \ldots, 0)) = (p_1 e^t + p_2 + \cdots + p_K)^n = (p_1 e^t + 1 - p_1)^n,
\tag{3.7}
$$

which is indeed the mgf of a binomial as in (2.83). Specifically, $X_1 \sim \text{Binomial}(n, p_1)$.

3.2 Marginal densities

More challenging, but also more useful, is to find the marginal density from the joint density, assuming it exists. Suppose the joint distribution of the two random variables, (X, Y), has pmf $f(x, y)$, and space $\mathcal{W}$. Then X has a pmf, $f_X(x)$, as well. To find it in terms of f, write

$$
\begin{aligned}
f_X(x) &= P[X = x \text{ (and } Y \text{ can be anything)}] \\
&= \sum_{y \,|\, (x,y) \in \mathcal{W}} P[X = x, Y = y] \\
&= \sum_{y \,|\, (x,y) \in \mathcal{W}} f(x, y).
\end{aligned}
\tag{3.8}
$$

That is, you add up all the $f(x, y)$ for that value of x, as in the table (3.1). The same procedure works if $\mathbf{X}$ and $\mathbf{Y}$ are vectors. The set of $\mathbf{y}$'s we are summing over we will call the **conditional space** of $\mathbf{Y}$ given $\mathbf{X} = \mathbf{x}$, and denote it by $\mathcal{Y}_{\mathbf{x}}$:

$$
\mathcal{Y}_{\mathbf{x}} = \{\mathbf{y} \in \mathcal{Y} \,|\, (\mathbf{x}, \mathbf{y}) \in \mathcal{W}\}.
\tag{3.9}
$$

With $\mathcal{W} = \{(x,y) \mid 0 < x < y < 1\}$, for any $x \in (0,1)$, y ranges from x to 1, hence

$$\mathcal{Y}_x = (x,1). \tag{3.10}$$

In the coin example in Section 1.6.3, for any probability of heads x, the range of Y is the same (see Figure 1.5 on page 12), so that $\mathcal{Y}_x = \{0,1,\ldots,n\}$ for any $x \in (0,1)$.

To summarize, in the general discrete case, we have

$$f_\mathbf{x}(\mathbf{x}) = \sum_{y \in \mathcal{Y}_x} f(\mathbf{x},\mathbf{y}), \quad \mathbf{x} \in \mathcal{X}. \tag{3.11}$$

3.2.1 Ranks

The National Opinion Research Center Amalgam Survey of 1972 asked people to rank three types of areas in which to live: City over 50,000, Suburb (within 30 miles of a City), and Country (everywhere else). The table (3.12) shows the results (Duncan and Brody, 1982), with respondents categorized by their current residence.

Ranking (City, Suburb, Country)	Residence City	Suburb	Country	Total	
$(1,2,3)$	210	22	10	242	
$(1,3,2)$	23	4	1	28	
$(2,1,3)$	111	45	14	170	(3.12)
$(2,3,1)$	8	4	0	12	
$(3,1,2)$	204	299	125	628	
$(3,2,1)$	81	126	152	359	
Total	637	500	302	1439	

That is, a ranking of $(1,2,3)$ means that person ranks living in the city best, suburbs next, and country last. There were 242 people in the sample with that ranking, 210 of whom live in the city (so they should be happy), 22 of whom live in the suburbs, and just 10 of whom live in the country.

The random vector here is (X,Y,Z), say, where X represents the rank of city, Y that of suburb, and Z that of country. The space consists of the six permutations of 1, 2, and 3:

$$\mathcal{W} = \{(1,2,3),(1,3,2),(2,1,3),(2,3,1),(3,1,2),(3,2,1)\}, \tag{3.13}$$

as in the first column of the table. Suppose the total column is our population, so that there are 1439 people all together, and we randomly choose a person from this population. Then the (joint) distribution of the person's ranking (X,Y,Z) is given by

$$f(x,y,z) = P[(X,Y,Z) = (x,y,z)] = \begin{cases} 242/1439 & \text{if } (x,y,z) = (1,2,3) \\ 28/1439 & \text{if } (x,y,z) = (1,3,2) \\ 170/1439 & \text{if } (x,y,z) = (2,1,3) \\ 12/1439 & \text{if } (x,y,z) = (2,3,1) \\ 628/1439 & \text{if } (x,y,z) = (3,1,2) \\ 359/1439 & \text{if } (x,y,z) = (3,2,1) \end{cases}. \tag{3.14}$$

This distribution could use some summarizing, e.g., what are the marginal distributions of X, Y, and Z? For each ranking $x = 1,2,3$, we have to add over the possible rankings of Y and Z, so that

$$f_X(1) = f(1,2,3) + f(1,3,2) = \frac{28 + 242}{1439} = 0.1876;$$

$$f_X(2) = f(2,1,3) + f(2,3,1) = \frac{170 + 12}{1439} = 0.1265;$$

$$f_X(3) = f(3,1,2) + f(3,2,1) = \frac{628 + 359}{1439} = 0.6859. \tag{3.15}$$

Thus city is ranked third over 2/3 of the time. The marginal rankings of suburb and country can be obtained similarly.   □

3.2.2 PDFs

Again for two variables, suppose now that the pdf is $f(x,y)$. We know that the distribution function is related to the pdf via

$$F(x,y) = \int_{(-\infty,x]\cap \mathcal{X}} \int_{(-\infty,y]\cap \mathcal{Y}_u} f(u,v)dvdu. \tag{3.16}$$

From (3.4), to obtain the distribution function, we set $y = \infty$, which means in the inside integral, we can remove the "$(-\infty, y]$" part:

$$F_X(x) = \int_{(-\infty,x]\cap \mathcal{X}} \int_{\mathcal{Y}_u} f(u,v)dvdu. \tag{3.17}$$

Then the pdf of X is found by taking the derivative with respect to x for $x \in \mathcal{X}$, which here just means stripping away the outer integral and setting $u = x$ (and $v = y$, if we wish):

$$f_X(x) = \frac{\partial}{\partial x} F_X(x) = \int_{\mathcal{Y}_x} f(x,y)dy. \tag{3.18}$$

Thus instead of summing over the y as in (3.11), we integrate. This procedure is often called "integrating out y."

Consider the example in Section 2.2.1, where $f(x,y) = 2$ for $0 < x < y < 1$. From (3.10), for $x \in (0,1)$, we have $\mathcal{Y}_x = (x,1)$, hence

$$f_X(x) = \int_{\mathcal{Y}_x} f(x,y)dy = \int_x^1 2dy = 2(1 - x). \tag{3.19}$$

With vectors, the process is the same, just embolden the variables:

$$f_{\mathbf{x}}(\mathbf{x}) = \int_{\mathcal{Y}_{\mathbf{x}}} f(\mathbf{x}, \mathbf{y})d\mathbf{y}. \tag{3.20}$$

3.3 Independence

Much of statistics is geared towards evaluation of relationships between variables: Does smoking cause cancer? Do cell phones? What factors explain the rise in asthma? The absence of a relationship, **independence**, is also important. Two sets A and B are independent if $P[A \cap B] = P[A] \times P[B]$. The definition for random variables is similar:

Definition 3.1. *Suppose* $(\mathbf{X}, \mathbf{Y})$ *has joint distribution P, and marginal spaces* $\mathcal{X}$ *and* $\mathcal{Y}$, *respectively. Then* $\mathbf{X}$ *and* $\mathbf{Y}$ *are* **independent** *if*

$$P[\mathbf{X} \in A \text{ and } \mathbf{Y} \in B] = P[\mathbf{X} \in A] \times P[\mathbf{Y} \in B] \text{ for all } A \subset \mathcal{X} \text{ and } B \subset \mathcal{Y}. \qquad (3.21)$$

Also, if $(\mathbf{X}^{(1)}, \dots, \mathbf{X}^{(K)})$ *has distribution P, and the vector* $\mathbf{X}^{(k)}$ *has space* $\mathcal{X}_k$, *then* $\mathbf{X}^{(1)}, \dots,$ $\mathbf{X}^{(K)}$ *are (mutually)* **independent** *if*

$$P[\mathbf{X}^{(1)} \in A_1, \dots, \mathbf{X}^{(K)} \in A_K] = P[\mathbf{X}^{(1)} \in A_1] \times \cdots \times P[\mathbf{X}^{(K)} \in A_K]$$
$$\text{for all } A_1 \subset \mathcal{X}_1, \dots, A_K \subset \mathcal{X}_K. \quad (3.22)$$

The basic idea in independence is that what happens with one variable does not affect what happens with another. There are a number of useful equivalences for independence of $\mathbf{X}$ and $\mathbf{Y}$. (Those for mutual independence of K vectors hold similarly.)

- **Distribution functions**: $\mathbf{X}$ and $\mathbf{Y}$ are independent if and only if

$$F(\mathbf{x}, \mathbf{y}) = F_{\mathbf{X}}(\mathbf{x}) \times F_{\mathbf{Y}}(\mathbf{y}) \text{ for all } \mathbf{x} \in \mathbb{R}^p, \mathbf{y} \in \mathbb{R}^q. \qquad (3.23)$$

- **Expected values of products of functions**: $\mathbf{X}$ and $\mathbf{Y}$ are independent if and only if

$$E[g(\mathbf{X})h(\mathbf{Y})] = E[g(\mathbf{X})] \times E[h(\mathbf{Y})] \qquad (3.24)$$

  for all functions $g : \mathcal{X} \to \mathbb{R}$ and $h : \mathcal{Y} \to \mathbb{R}$ whose expected values exist.

- **MGFs**: Suppose the marginal mgfs of $\mathbf{X}$ and $\mathbf{Y}$ are finite for $\mathbf{t}$ and $\mathbf{s}$ in neighborhoods of zero (respectively in $\mathbb{R}^p$ and $\mathbb{R}^q$). Then $\mathbf{X}$ and $\mathbf{Y}$ are independent if and only if

$$M(\mathbf{t}, \mathbf{s}) = M_{\mathbf{X}}(\mathbf{t})M_{\mathbf{Y}}(\mathbf{s}) \qquad (3.25)$$

  for all $(\mathbf{t}, \mathbf{s})$ in a neighborhood of zero in $\mathbb{R}^{p+q}$.

The second item can be used to show that independent random variables are uncorrelated, because as in (2.23), $Cov[X, Y] = E[XY] - E[X]E[Y]$, and (3.24) shows that $E[XY] = E[X]E[Y]$ if X and Y are independent. Be aware that the implication does *not* go the other way, that is, X and Y can have correlation 0 and still not be independent. For example, suppose $\mathcal{W} = \{(0, 1), (0, -1), (1, 0), (-1, 0)\}$, and $P[(X, Y) = (x, y)] = 1/4$ for each $(x, y) \in \mathcal{W}$. Then it is not hard to show that $E[X] = E[Y] = 0$, and that $E[XY] = 0$ (in fact, $XY = 0$ always), hence $Cov[X, Y] = 0$. But X and Y are not independent, e.g., take $A = \{0\}$ and $B = \{0\}$. Then

$$P[X = 0 \text{ and } Y = 0] = 0 \neq P[X = 0]P[Y = 0] = \frac{1}{2} \times \frac{1}{2}. \qquad (3.26)$$

3.3.1 Independent exponentials

Suppose U and V are independent Exponential(1)'s. The mgf of an Exponential(1) is $1/(1-t)$ for $t < 1$. See (2.76), which gives the mgf of Gamma(α, λ) as $(\lambda/(\lambda - t))^{\alpha}$ for $t < \lambda$. Thus the mgf of (U, V) is

$$M_{(U,V)}(t_1, t_2) = M_U(t_1)M_V(t_2) = \frac{1}{1-t_1} \frac{1}{1-t_2}, \quad t_1 < 1, t_2 < 1. \qquad (3.27)$$

Now let
$$X = U + V \quad \text{and} \quad Y = U - V. \tag{3.28}$$

Are X and Y independent? What are their marginal distributions? We can start by looking at the mgf:

$$
\begin{aligned}
M_{(X,Y)}(s_1, s_2) &= E[e^{s_1 X + s_2 Y}] \\
&= E[e^{s_1(U+V) + s_2(U-V)}] \\
&= E[e^{(s_1+s_2)U + (s_1-s_2)V}] \\
&= M_{(U,V)}(s_1 + s_2, s_1 - s_2) \\
&= \frac{1}{1 - s_1 - s_2} \frac{1}{1 - s_1 + s_2}.
\end{aligned} \tag{3.29}
$$

This mgf is finite if $s_1 + s_2 < 1$ and $s_1 - s_2 < 1$, which is a neighborhood of $(0,0)$. If X and Y are independent, then this mgf must factor into the two individual mgfs. It does not appear to factor. More formally, the marginal mgfs are

$$M_X(s_1) = M_{(X,Y)}(s_1, 0) = \frac{1}{(1 - s_1)^2}, \tag{3.30}$$

and

$$M_Y(s_2) = M_{(X,Y)}(0, s_2) = \frac{1}{(1 - s_2)(1 + s_2)} = \frac{1}{1 - s_2^2}. \tag{3.31}$$

Note that the first one is that of a Gamma$(2,1)$, so that $X \sim$ Gamma$(2,1)$. The one for Y may not be recognizable, but it turns out to be the mgf of a Laplace, as in Exercise 2.7.12. Notice that

$$M_{(X,Y)}(s_1, s_2) \neq M_X(s_1)M_Y(s_2), \tag{3.32}$$

hence X and Y are *not* independent. They are uncorrelated, however:

$$
\begin{aligned}
Cov[X, Y] = Cov[U + V, U - V] &= Var[U] - Var[V] - Cov[U, V] + Cov[V, U] \\
&= 1 - 1 = 0.
\end{aligned} \tag{3.33}
$$

3.3.2 Spaces and densities

Suppose (X, Y) is discrete, with pmf $f(x,y) > 0$ for $(x,y) \in \mathcal{W}$, and f_X and f_Y are the marginal pmfs of X and Y, respectively. Then applying (3.21) to the singleton sets $\{x\}$ and $\{y\}$ for $x \in \mathcal{X}$ and $y \in \mathcal{Y}$ shows that

$$P[X = x, Y = y] = P[X = x] \times P[Y = y], \tag{3.34}$$

which translates to

$$f(x,y) = f_X(x)f_Y(y). \tag{3.35}$$

In particular, this equation shows that if $x \in \mathcal{X}$ and $y \in \mathcal{Y}$, then $f(x,y) > 0$, hence $(x,y) \in \mathcal{W}$. That is, if X and Y are independent, then

$$\mathcal{W} = \mathcal{X} \times \mathcal{Y}, \tag{3.36}$$

the "rectangle" created from the marginal spaces.

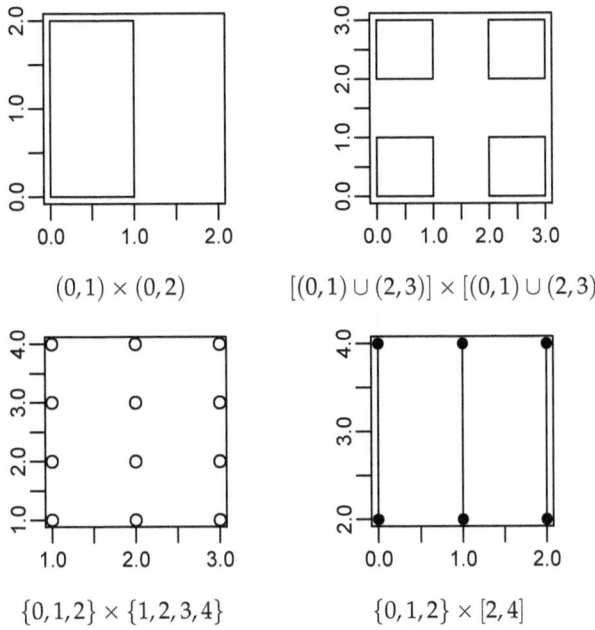

Figure 3.1: Some Cartesian products, i.e., rectangles.

Formally, given sets $A \subset \mathbb{R}^p$ and $B \subset \mathbb{R}^q$, the **Cartesian product** or **rectangle** $A \times B$ is defined to be

$$A \times B = \{(\mathbf{x}, \mathbf{y}) \mid \mathbf{x} \in A \text{ and } \mathbf{y} \in B\} \subset \mathbb{R}^{p+q}. \tag{3.37}$$

The set may not be a rectangle in the usual sense, although it will be if $p = q = 1$ and A and B are both intervals. Figure 3.1 has some examples. Of course, $\mathbb{R}^{p+q}$ is a rectangle itself, being $\mathbb{R}^p \times \mathbb{R}^q$. The result (3.36) holds in general.

Lemma 3.2. *If* $\mathbf{X}$ *and* $\mathbf{Y}$ *are independent, then the spaces can be taken so that (3.36) holds.*

This lemma implies that if the joint space can not be a rectangle, then $\mathbf{X}$ and $\mathbf{Y}$ are not independent. Consider the example in Section 2.2.1, where $W = \{(x, y) \mid 0 < x < y < 1\}$, a triangle. If we take a square below that triangle, such as $(0.8, 1) \times (0.8, 1)$, then

$$P[(X, Y) \in (0.8, 1) \times (0.8, 1)] = 0 \text{ but } P[X \in (0.8, 1)]P[Y \in (0.8, 1)] > 0, \tag{3.38}$$

so that X and Y are not independent. The result extends to more variables. In Section 3.2.1 on ranks, the marginal spaces are

$$\mathcal{X} = \mathcal{Y} = \mathcal{Z} = \{1, 2, 3\}, \tag{3.39}$$

but

$$\mathcal{W} \neq \mathcal{X} \times \mathcal{Y} \times \mathcal{Z} = \{(1, 1, 1), (1, 1, 2), (1, 1, 3), \ldots, (3, 3, 3)\}, \tag{3.40}$$

in particular because $\mathcal{W}$ has only 6 elements, while the product space has $3^3 = 27$.

The factorization in (3.34) is necessary and sufficient for independence in the discrete and continuous cases, or mixed-type densities as in Section 1.6.3.

Lemma 3.3. *Suppose* $\mathbf{X}$ *has marginal density* $f_{\mathbf{X}}(\mathbf{x})$, *and* $\mathbf{Y}$ *has marginal density* $f_{\mathbf{Y}}(\mathbf{y})$. *Then* $\mathbf{X}$ *and* $\mathbf{Y}$ *are independent if and only if the distribution of* $(\mathbf{X}, \mathbf{Y})$ *can be given by density*

$$f(\mathbf{x}, \mathbf{y}) = f_{\mathbf{X}}(\mathbf{x}) f_{\mathbf{Y}}(\mathbf{y}) \tag{3.41}$$

and space

$$\mathcal{W} = \mathcal{X} \times \mathcal{Y}. \tag{3.42}$$

In the lemma, we say "can be given" since in the continuous case, we can change the densities or spaces on sets of probability zero without changing the distribution. We can simplify a little, that is, as long as the space and joint pdf factor, we have independence.

Lemma 3.4. *Suppose* $(\mathbf{X}, \mathbf{Y})$ *has joint density* $f(\mathbf{x}, \mathbf{y})$. *Then* $\mathbf{X}$ *and* $\mathbf{Y}$ *are independent if and only if the density can be written as*

$$f(\mathbf{x}, \mathbf{y}) = g(\mathbf{x}) h(\mathbf{y}) \tag{3.43}$$

for some functions g and h, and

$$\mathcal{W} = \mathcal{X} \times \mathcal{Y}. \tag{3.44}$$

This lemma is not presuming the g and h are actual densities, although they certainly could be.

3.3.3 IID

A special case of independence has the vectors with the exact same distribution, as well as being independent. That is, $\mathbf{X}^{(1)}, \ldots, \mathbf{X}^{(K)}$ are independent, and all have the same marginal distribution. We say the vectors are **iid**, meaning "independent and identically distributed." This type of distribution often is used to model random samples, where n individuals are chosen from a (virtually) infinite population, and p variables are recorded on each. Then $K = n$ and the $\mathbf{X}^{(i)}$'s are $p \times 1$ vectors. If the marginal density is $f_{\mathbf{X}}$ for each $\mathbf{X}^{(i)}$, with marginal space $\mathcal{X}$, then the joint density of the entire sample is

$$f(\mathbf{x}^{(1)}, \ldots, \mathbf{x}^{(n)}) = f_{\mathbf{X}}(\mathbf{x}^{(1)}) \cdots f_{\mathbf{X}}(\mathbf{x}^{(n)}) \tag{3.45}$$

with space

$$\mathcal{W} = \mathcal{X} \times \cdots \times \mathcal{X} = \mathcal{X}^n. \tag{3.46}$$

3.4 Exercises

Exercise 3.4.1. Let (X, Y, Z) be the ranking variables as in (3.14). (a) Find the marginal distributions of Y and Z. What is the most popular rank of suburb? Of country? (b) Find the marginal space and marginal pmf of (X, Y). How does the space differ from that of (X, Y, Z)?

Exercise 3.4.2. Suppose U and V are iid with finite variance, and let $X = U + V$ and $Y = U - V$, as in (3.28). (a) Show that X and Y are uncorrelated. (b) Suppose U and V both have space $(0, \infty)$. Without knowing the pdfs, what can you say about the independence of X and Y? [Hint: What is the space of (X, Y)?]

Exercise 3.4.3. Suppose U and V are iid $N(0,1)$, and let $X = U + V$ and $Y = U - V$ again. (a) Find the mgf of the joint (U, V), $M_{(U,V)}(t_1, t_2)$. (b) Find the mgf of (X, Y), $M_{(X,Y)}(s_1, s_2)$. Show that it factors into the mgfs of X and Y. (c) What are the marginal distributions of X and Y? [Hint: See (2.73)].

Exercise 3.4.4. Suppose (X, Y) is uniform over the unit disk, so that the space is $\mathcal{W} = \{(x, y) \mid x^2 + y^2 < 1\}$ and $f(x, y) = 1/\pi$ for $(x, y) \in \mathcal{W}$. (a) What are the (marginal) spaces of X and Y? Are X and Y independent? (b) For $x \in \mathcal{X}$ (the marginal space of X), what is $\mathcal{Y}_x$ (the conditional space of Y given $X = x$)? (c) Find the (marginal) pdf of X.

Exercise 3.4.5. Let Z_1, Z_2, and Z_3 be independent, each with space $\{-1, +1\}$, and $P[Z_i = -1] = P[Z_i = +1] = 1/2$. Set

$$X_1 = Z_1 Z_3 \quad \text{and} \quad X_2 = Z_2 Z_3. \tag{3.47}$$

(a) What is the space of (X_1, X_2)? (b) Are X_1 and X_2 independent? (c) Now let $X_3 = Z_1 Z_2$. Are X_1 and X_3 independent? (d) Are X_2 and X_3 independent? (e) What is the space of (X_1, X_2, X_3)? (f) Are X_1, X_2 and X_3 mututally independent? (g) Now let $U = X_1 X_2 X_3$? What is the space of U?

Exercise 3.4.6. For each given pdf $f(x, y)$ and space $\mathcal{W}$ for (X, Y), answer true or false to the three statements: (i) X and Y are independent; (ii) The space of (X, Y) is a rectangle; (iii) $Cov[X, Y] = 0$.

	$f(x, y)$	$\mathcal{W}$
(a)	$c_1 x y$	$\{(x, y) \mid 0 < x < 1, 0 < y < 1\}$
(b)	$c_2 x y$	$\{(x, y) \mid 0 < x < 1, 0 < y < 1, x + y < 1\}$
(c)	$c_3(x + y)$	$\{(x, y) \mid 0 < x < 1, 0 < y < 1\}$

$$\tag{3.48}$$

The c_i's are constants.

Exercise 3.4.7. Suppose $\Theta \sim \text{Uniform}(0, 2\pi)$, and define $X = \cos(\Theta), Y = \sin(\Theta)$. Also, set $R = \sqrt{X^2 + Y^2}$. True or false? (a) X and Y are independent. (b) The space of (X, Y) is a rectangle. (c) $Cov[X, Y] = 0$. (d) R and Θ are independent. (e) The space of (R, Θ) is a rectangle. (f) $Cov[R, \Theta] - 0$.

Chapter 4

Transformations: DFs and MGFs

A major task of mathematical statistics is finding, or approximating, the distributions of random variables that are functions of other random variables. Important examples are estimators, hypothesis tests, and predictors. This chapter and the next will address finding exact distributions. There are many approaches, and which one to use may not always be obvious. Chapters 8 and 9 consider large-sample approximations. The following sections run through a number of possibilities, though the granddaddy of them all, using Jacobians, has its own Chapter 5.

4.1 Adding up the possibilities

If $\mathbf{X}$ is discrete, then any function $\mathbf{Y} = g(\mathbf{X})$ will also be discrete, hence its pmf can be found by adding up all the probabilities that correspond a given $\mathbf{y}$:

$$f_{\mathbf{Y}}(\mathbf{y}) = P[\mathbf{Y} = \mathbf{y}] = P[g(\mathbf{X}) = \mathbf{y}] = \sum_{\mathbf{x} \mid g(\mathbf{x}) = \mathbf{y}} f_{\mathbf{X}}(\mathbf{x}), \quad \mathbf{y} \in \mathcal{Y}. \tag{4.1}$$

Of course, that final summation may or not be easy to find.

One situation in which it is easy is when g is a one-to-one and onto function from $\mathcal{X}$ to $\mathcal{Y}$, so that there exists an inverse function,

$$g^{-1} : \mathcal{Y} \to \mathcal{X}; \quad g(g^{-1}(y)) = y \text{ and } g^{-1}(g(x)) = x. \tag{4.2}$$

Then, with $f_{\mathbf{X}}$ being the pmf of $\mathbf{X}$,

$$f_{\mathbf{Y}}(\mathbf{y}) = P[g(\mathbf{X}) = \mathbf{y}] = P[\mathbf{X} = g^{-1}(\mathbf{y})] = f_{\mathbf{X}}(g^{-1}(\mathbf{y})). \tag{4.3}$$

For example, if $X \sim \text{Poisson}(\lambda)$, and $Y = X^2$, then $g(x) = x^2$, hence $g^{-1}(y) = \sqrt{y}$ for $y \in \mathcal{Y} = \{0, 1, 4, 9, \ldots\}$. The pmf of Y is then

$$f_Y(y) = f_X(\sqrt{y}) = e^{-\lambda} \frac{\lambda^{\sqrt{y}}}{\sqrt{y}!}, \quad y \in \mathcal{Y}. \tag{4.4}$$

Notice that it is important to have the spaces correct. E.g., this g is not one-to-one if the space is $\mathbb{R}$, and the "$\sqrt{y}$" makes sense only if y is the square of a nonnegative integer.

We consider some more examples.

4.1.1 Sum of discrete uniforms

Suppose $\mathbf{X} = (X_1, X_2)$, where X_1 and X_2 are independent, and

$$X_1 \sim \text{Discrete Uniform}(0,1) \quad \text{and} \quad X_2 \sim \text{Discrete Uniform}(0,2). \qquad (4.5)$$

Note that $X_1 \sim \text{Bernoulli}(1/2)$. We are after the distribution of $Y = X_1 + X_2$. The space of Y can be seen to be $\mathcal{Y} = \{0,1,2,3\}$. This function is not one-to-one, e.g., there are two $\mathbf{x}$'s that sum to 1: $(0,1)$ and $(1,0)$. This is a small enough example that we can just write out all the possibilities:

$$f_Y(0) = P[X_1 + X_2 = 0] = P[\mathbf{X} = (0,0)] = \frac{1}{2} \times \frac{1}{3} = \frac{1}{6};$$

$$f_Y(1) = P[X_1 + X_2 = 1] = P[\mathbf{X} = (0,1) \text{ or } \mathbf{X} = (1,0)] = \frac{1}{2} \times \frac{1}{3} + \frac{1}{2} \times \frac{1}{3} = \frac{2}{6};$$

$$f_Y(2) = P[X_1 + X_2 = 2] = P[\mathbf{X} = (0,2) \text{ or } \mathbf{X} = (1,1)] = \frac{1}{2} \times \frac{1}{3} + \frac{1}{2} \times \frac{1}{3} = \frac{2}{6};$$

$$f_Y(3) = P[X_1 + X_2 = 3] = P[\mathbf{X} = (1,2)] = \frac{1}{2} \times \frac{1}{3} = \frac{1}{6}. \qquad (4.6)$$

4.1.2 Convolutions for discrete variables

Here we generalize the previous example a bit by assuming X_1 has pmf f_1 and space $\mathcal{X}_1 = \{0,1,\ldots,a\}$, and X_2 has pmf f_2 and space $\mathcal{X}_2 = \{0,1,\ldots,b\}$. Both a and b are positive integers, or either could be $+\infty$. Then Y has space

$$\mathcal{Y} = \{0,1,\ldots,a+b\}. \qquad (4.7)$$

To find $f_Y(y)$, we need to sum up the probabilities of all (x_1, x_2)'s for which $x_1 + x_2 = y$. These pairs can be written $(x_1, y - x_1)$, and require $x_1 \in \mathcal{X}_1$ as well as $y - x_1 \in \mathcal{X}_2$. That is, for fixed $y \in \mathcal{Y}$,

$$0 \le x_1 \le a \text{ and } 0 \le y - x_1 \le b \Longrightarrow \max\{0, y-b\} \le x_1 \le \min\{a, y\}. \qquad (4.8)$$

For example, with $a = 3$ and $b = 3$, the following table shows which x_1's correspond to each y:

$x_1 \downarrow$; $x_2 \rightarrow$	0	1	2	3
0	0	1	2	3
1	1	2	3	4
2	2	3	4	5
3	3	4	5	6

$$(4.9)$$

Each value of y appears along a diagonal, so that

$$
\begin{aligned}
y = 0 &\Rightarrow x_1 = 0; \\
y = 1 &\Rightarrow x_1 = 0, 1; \\
y = 2 &\Rightarrow x_1 = 0, 1, 2; \\
y = 3 &\Rightarrow x_1 = 0, 1, 2, 3; \\
y = 4 &\Rightarrow x_1 = 1, 2, 3; \\
y = 5 &\Rightarrow x_1 = 2, 3; \\
y = 6 &\Rightarrow x_1 = 3.
\end{aligned}
\qquad (4.10)
$$

Thus in general, for $y \in \mathcal{Y}$,

$$f_Y(y) = P[X_1 + X_2 = y]$$

$$= \sum_{x_1 = \max\{0, y-b\}}^{\min\{a, y\}} P[X_1 = x_1, X_2 = y - x_1]$$

$$= \sum_{x_1 = \max\{0, y-b\}}^{\min\{a, y\}} f_1(x_1) f_2(y - x_1). \tag{4.11}$$

This formula is called the **convolution** of f_1 and f_2. In general, the convolution of two random variables is the distribution of the sum.

To illustrate, suppose X_1 has the pmf of the Y in (4.6), and X_2 is Discrete Uniform(0,3), so that $a = b = 3$ and

$$f_1(0) = f_1(3) = \frac{1}{6}, \ f_1(1) = f_1(2) = \frac{1}{3} \text{ and } f_2(x_2) = \frac{1}{4}, \ x_2 = 0, 1, 2, 3. \tag{4.12}$$

Then $\mathcal{Y} = \{0, 1, \ldots, 6\}$, and

$$f_Y(0) = \sum_{x_1=0}^{0} f_1(x_1) f_2(0 - x_1) = \frac{1}{24};$$

$$f_Y(1) = \sum_{x_1=0}^{1} f_1(x_1) f_2(1 - x_1) = \frac{1}{24} + \frac{1}{12} = \frac{3}{24};$$

$$f_Y(2) = \sum_{x_1=0}^{2} f_1(x_1) f_2(2 - x_1) = \frac{1}{24} + \frac{1}{12} + \frac{1}{12} = \frac{5}{24};$$

$$f_Y(3) = \sum_{x_1=0}^{3} f_1(x_1) f_2(3 - x_1) = \frac{1}{24} + \frac{1}{12} + \frac{1}{12} + \frac{1}{24} = \frac{6}{24};$$

$$f_Y(4) = \sum_{x_1=1}^{3} f_1(x_1) f_2(4 - x_1) = \frac{1}{12} + \frac{1}{12} + \frac{1}{24} = \frac{5}{24};$$

$$f_Y(5) = \sum_{x_1=2}^{3} f_1(x_1) f_2(5 - x_1) = \frac{1}{12} + \frac{1}{24} = \frac{3}{24};$$

$$f_Y(6) = \sum_{x_1=3}^{3} f_1(x_1) f_2(6 - x_1) = \frac{1}{24}. \tag{4.13}$$

Check that the $f_Y(y)$'s do sum to 1.

4.1.3 Sum of two Poissons

An example for which $a = b = \infty$ has X_1 and X_2 independent Poissons, with parameters λ_1 and λ_2, respectively. Then $Y = X_1 + X_2$ has space $\mathcal{Y} = \{0, 1, \cdots\}$, the same as the spaces of the X_i's. In this case, for fixed y, $x_1 = 0, \ldots, y$, hence

$$
\begin{aligned}
f_Y(y) &= \sum_{x_1=0}^{y} f_1(x_1) f_2(y - x_1) \\
&= \sum_{x_1=0}^{y} e^{-\lambda_1} \frac{\lambda_1^{x_1}}{x_1!} e^{-\lambda_2} \frac{\lambda_2^{y-x_1}}{(y-x_1)!} \\
&= e^{-\lambda_1 - \lambda_2} \sum_{x_1=0}^{y} \frac{1}{x_1!(y-x_1)!} \lambda_1^{x_1} \lambda_2^{y-x_1} \\
&= e^{-\lambda_1 - \lambda_2} \frac{1}{y!} \sum_{x_1=0}^{y} \frac{y!}{x_1!(y-x_1)!} \lambda_1^{x_1} \lambda_2^{y-x_1} \\
&= e^{-\lambda_1 - \lambda_2} \frac{1}{y!} (\lambda_1 + \lambda_2)^y, \quad\quad\quad\quad\quad (4.14)
\end{aligned}
$$

the last step using the binomial theorem in (2.82). But that last expression is the Poisson pmf, i.e.,

$$
Y \sim \text{Poisson}(\lambda_1 + \lambda_2). \quad\quad\quad\quad\quad (4.15)
$$

This fact can be proven also using mgfs. See Section 4.3.

4.2   Distribution functions

Suppose the $p \times 1$ vector $\mathbf{X}$ has distribution function $F_{\mathbf{X}}$, and $Y = g(\mathbf{X})$ for some function

$$
g : \mathcal{X} \longrightarrow \mathcal{Y} \subset \mathbb{R}. \quad\quad\quad\quad\quad (4.16)
$$

Then the distribution function F_Y of Y is

$$
F_Y(y) = P[Y \le y] = P[g(\mathbf{X}) \le y], \quad y \in \mathbb{R}. \quad\quad\quad\quad\quad (4.17)
$$

The final probability is in principle obtainable from the distribution of $\mathbf{X}$, which solves the problem. If Y has a pdf, we can then find that by differentiating, as we did in (1.17). Exercises 1.7.13 and 1.7.14 already gave previews of this approach for the χ_1^2 distribution, as did Exercise 1.7.15 for the logistic.

4.2.1   Convolutions for continuous random variables

Suppose (X_1, X_2) has pdf $f(x_1, x_2)$. For initial simplicity, we will take the space to be the entire plane, $\mathcal{X} = \mathbb{R}^2$, noting that f could be 0 over wide swaths of the space. Let $Y = X_1 + X_2$, so it has space $\mathcal{Y} = \mathbb{R}$. Its distribution function is

$$
F_Y(y) = P[X_1 + X_2 \le y] = P[X_2 \le y - X_1] = \int_{-\infty}^{\infty} \int_{-\infty}^{y-x_1} f(x_1, x_2) dx_2 dx_1. \quad (4.18)
$$

The pdf is found by differentiating with respect to y, which replaces the inner integral with the integrand evaluated at $x_2 = y - x_1$:

$$
f_Y(y) = F_Y'(y) = \int_{-\infty}^{\infty} f(x_1, y - x_1) dx_1. \quad\quad\quad\quad\quad (4.19)
$$

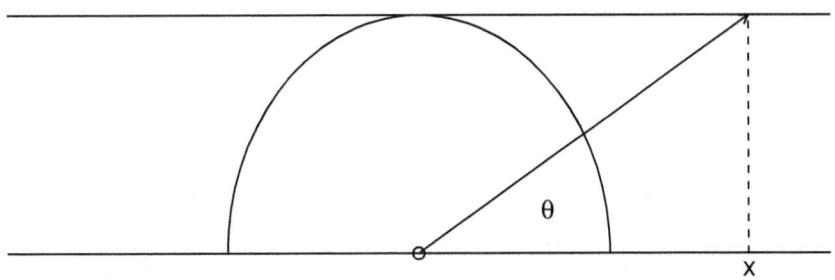

Figure 4.1: An arrow is shot at an angle of θ, which is chosen randomly between 0 and π. Where it hits the line one unit high is the value x.

This convolution formula is the analog of (4.11) in the discrete case.

When evaluating that integral, we must be careful of when f is 0. For example, if X_1 and X_2 are iid Uniform(0,1)'s, we would integrate $f(x_1, w - x_1) = 1$ over just $x_1 \in (0, w)$ if $w \in (0, 1)$, or over just $x_1 \in (w - 1, 1)$ if $w \in [1, 2)$. In fact, Exercise 1.7.6 did basically this procedure to find the tent distribution.

As another illustration, suppose X_1 and X_2 are iid Exponential(λ), so that $\mathcal{Y} = (0, \infty)$. Then for fixed $y > 0$, the x_1 runs from 0 to y. Thus

$$f_Y(y) = \int_0^y \lambda^2 e^{-\lambda x_1} e^{-\lambda(y - x_1)} dx_1 = \lambda^2 y e^{-\lambda y}, \tag{4.20}$$

which is a Gamma(2, λ). This is a special case of the sum of independent gammas, as we will see in Section 5.3.

4.2.2 Uniform $\rightarrow$ Cauchy

Imagine an arrow shot into the air from the origin (the point $(0,0)$) at an angle θ to the ground (x-axis). It goes straight (no gravity or friction) to hit the ceiling one unit high ($y = 1$). The horizontal distance between the origin and where it hits the ceiling is x. See Figure 4.1. If θ is chosen uniformly from 0 to π, what is the density of X?

In the figure, look at the right triangle formed by the origin, the point where the arrow hits the ceiling, and the drop from that point to the x-axis. That is, the triangle that connects the points $(0, 0), (x, 1)$, and $(x, 0)$. The cotangent of θ is the base over the height of that triangle, which is x over one (the length of the dotted line segment): $x = \cot(\theta)$. Note that smaller values of θ correspond to larger values of x. Thus the distribution function of X is

$$F_X(x) = P[X \le x] = P[\cot(\Theta) \le x] = P[\Theta \ge \operatorname{arccot}(x)] = 1 - \frac{1}{\pi} \operatorname{arccot}(x), \tag{4.21}$$

the final equality following since Θ has pdf $1/\pi$ for $\theta \in (0, \pi)$. The pdf of X is found by differentiating, so all we need is to remember or derive or Google the derivative of the inverse cotangent, which is $-1/(1 + x^2)$. Thus

$$f_X(x) = \frac{1}{\pi} \frac{1}{1 + x^2}, \tag{4.22}$$

the Cauchy pdf from Table 1.1 on page 7.

4.2.3 Probability transform

Generating random numbers on a computer is a common activity in statistics, e.g., for inference using techniques such as the bootstrap and Markov chain Monte Carlo, for randomizing subjects to treatments, and for assessing the performance of various procedures. It it easy to generate independent Uniform$(0,1)$ random variables — Easy in the sense that many people have worked very hard for many years to develop good methods that are now available in most statistical software. Actually, the numbers are not truly random, but rather pseudo-random, because there is a deterministic algorithm producing them. But you could also question whether randomness exists in the real world anyway. Even flipping a coin is deterministic, if you know all the physics in the flip. (At least classical physics. Quantum physics is beyond me.)

One usually is not satisfied with uniforms, but rather has normals or gammas, or something more complex, in mind. There are many clever ways to create the desired random variables from uniforms (see for example the Box-Muller transformation in Section 5.4.4), but the most basic uses the inverse distribution function. We suppose $U \sim$ Uniform$(0,1)$, and wish to generate an X that has given distribution function F. Assume that F is continuous, and strictly increasing for $x \in \mathcal{X}$, so that the quantiles $F^{-1}(u)$ are well-defined for every $u \in (0,1)$. Consider the random variable

$$W = F^{-1}(U). \tag{4.23}$$

Then its distribution function is

$$F_W(w) = P[W \le w] = P[F^{-1}(U) \le w] = P[U \le F(w)] = F(w), \tag{4.24}$$

where the last step follows because $U \sim$ Uniform$(0,1)$ and $0 \le F(w) \le 1$. But that equation means that W has the desired distribution function, hence to generate an X, we generate a U and take $X = F^{-1}(U)$.

To illustrate, suppose we wish $X \sim$ Exponential(λ). The distribution function F of X is zero for $x \le 0$, and

$$F(x) = \int_0^x \lambda e^{-\lambda w} dw = 1 - e^{-\lambda x} \text{ for } x > 0. \tag{4.25}$$

Thus for $u \in (0,1)$,

$$u = F(x) \Longrightarrow x = F^{-1}(u) = -\log(1-u)/\lambda. \tag{4.26}$$

One limitation of this method is that F^{-1} is not always computationally simple. Even in the normal case there is no closed form expression. In addition, this method does not work directly for generating multivariate $\mathbf{X}$, because then F is not invertible.

The approach works for non-continuous distributions as well, but care must be taken because u may fall in a gap. For example, suppose X is Bernoulli$(1/2)$. Then for $u = 1/2$, we have $x = 0$, since $F(0) = 1/2$, but no other value of $u \in (0,1)$ has an x with $F(x) = u$. The fix is to define a substitute for the inverse, F^-, where $F^-(u) = 0$ for $0 < u < 1/2$, and $F^-(u) = 1$ for $1/2 < u < 1$. Thus half the time $F^-(U)$ is 0, and half the time it is 1. More generally, if u is in a gap, set $F^-(u)$ to the value of x for that gap. Mathematically, we define for $0 < u < 1$,

$$F^-(u) = \min\{x \mid F(x) \geq u\}, \tag{4.27}$$

which will yield $X =^{\mathcal{D}} F^-(U)$ for continuous and noncontinuous F.

The process can be reversed, which is useful in hypothesis testing, to obtain p-values. That is, suppose X has continuous strictly increasing (on $\mathcal{X}$) distribution function F, and let

$$U = F(X). \tag{4.28}$$

Then the distribution function of U is

$$F_U(u) = P[F(X) \leq u] = P[X \leq F^{-1}(u)] = F(F^{-1}(u)) = u, \tag{4.29}$$

which is the distribution function of a Uniform$(0, 1)$, i.e.,

$$F(X) \sim \text{Uniform}(0, 1). \tag{4.30}$$

When F is not continuous, $F(X)$ is stochastically larger than a uniform:

$$P[F(X) \leq u] \leq u, \ \ u \in (0, 1). \tag{4.31}$$

That is, it is at least as likely as a uniform to be larger than u. See Exercise 4.4.16. (We look at stochastic ordering again in Definition 18.1 on page 306.)

4.2.4 Location-scale families

One approach to modeling univariate data is to assume a particular shape of the distribution, then let the mean and variance vary, as in the normal. More generally, since the mean or variance may not exist, we use the terms "location" and "scale." We define such families next.

Definition 4.1. *Let Z have distribution function F on $\mathbb{R}$. Then for $\mu \in \mathbb{R}$ and $\sigma > 0$, $X = \mu + \sigma Z$ has the **location-scale family** distribution based on F with location parameter μ and scale parameter σ.*

The distribution function for X in the definition is then

$$F(x \mid \mu, \sigma) = F\left(\frac{x - \mu}{\sigma}\right). \tag{4.32}$$

If Z has pdf f, then by differentiation we have that the pdf of X is

$$f(x \mid \mu, \sigma) = \frac{1}{\sigma} f\left(\frac{x - \mu}{\sigma}\right). \tag{4.33}$$

If Z has the moment generating function $M(t)$, then that for X is

$$M(t \mid \mu, \sigma) = e^{t\mu} M(t\sigma). \tag{4.34}$$

The normal distribution is the most famous location-scale family. Let $Z \sim N(0,1)$, the standard normal distribution, and set $X = \mu + \sigma Z$. It is immediate that $E[X] = \mu$ and $Var[X] = \sigma^2$. The pdf of X is then

$$\phi(x \mid \mu, \sigma^2) = \frac{1}{\sigma} \phi\left(\frac{x-\mu}{\sigma}\right), \text{ where } \phi(z) = \frac{1}{\sqrt{2\pi}} e^{-\frac{1}{2}z^2},$$

$$= \frac{1}{\sqrt{2\pi}\sigma} e^{-\frac{1}{2\sigma}(x-\mu)^2}, \tag{4.35}$$

which is the $N(\mu, \sigma^2)$ density given in Table 1.1. The mgf of Z is $M(t) = \exp(t^2/2)$ from (2.70), hence X has mgf

$$M(t \mid \mu, \sigma^2) = e^{t\mu + \frac{1}{2}t^2\sigma^2}, \tag{4.36}$$

as we saw in (2.73).

Other popular location-scale families are based on the uniform, Laplace, Cauchy, and logistic. You primarily see location-scale families for continuous distributions, but they can be defined for discrete distribution. Also, if σ is fixed (at $\sigma = 1$, say), then the family is a location family, and if μ is fixed (at $\mu = 0$, say), it is a scale family. There are multivariate versions of location-scale families as well, such as the multivariate normal in Chapter 7.

4.3   Moment generating functions

Instead of finding the distribution function of Y, one could try to find the mgf. By uniqueness, Theorem 2.5 on page 26, if you recognize the mgf of Y as being for a particular distribution, then you know Y has that distribution.

4.3.1   Uniform $\to$ Exponential

Suppose $X \sim \text{Uniform}(0,1)$, and $Y = -\log(X)$. Then the mgf of Y is

$$M_Y(t) = E[e^{tY}]$$

$$= E[e^{-t\log(X)}]$$

$$= \int_0^1 e^{-t\log(x)} dx$$

$$= \int_0^1 x^{-t} dx$$

$$= \frac{1}{-t+1} x^{-t+1} \Big|_{x=0}^{1}$$

$$= \frac{1}{1-t} \text{ if } t < 1 \text{ (and } +\infty \text{ if not)}. \tag{4.37}$$

This mgf is that of the gamma in (2.76), where $\alpha = \lambda = 1$, which means it is also Exponential(1). Thus the pdf of Y is

$$f_Y(y) = e^{-y} \text{ for } y \in \mathcal{Y} = (0, \infty). \tag{4.38}$$

4.3.2 Sum of independent gammas

The mgf approach is especially useful for the sums of independent random variables (convolutions). For example, suppose $X_1, \ldots, X_K$ are independent, with $X_k \sim$ Gamma(α_k, λ). (They all have the same rate, but are allowed different shapes.) Let $Y = X_1 + \cdots + X_K$. Then its mgf is

$$
\begin{aligned}
M_Y(t) &= E[e^{tY}] \\
&= E[e^{t(X_1 + \cdots + X_K)}] \\
&= E[e^{tX_1} \cdots e^{tX_K}] \\
&= E[e^{tX_1}] \times \cdots \times E[e^{tX_K}] \quad \text{by independence of the } X_k\text{'s} \\
&= M_1(t) \times \cdots \times M_K(t) \quad \text{where } M_k(t) \text{ is the mgf of } X_k \\
&= \left(\frac{\lambda}{\lambda - t}\right)^{\alpha_1} \cdots \left(\frac{\lambda}{\lambda - t}\right)^{\alpha_K} \quad \text{for } t < \lambda, \text{ by (2.76)} \\
&= \left(\frac{\lambda}{\lambda - t}\right)^{\alpha_1 + \cdots + \alpha_K}.
\end{aligned}
\tag{4.39}
$$

But then this mgf is that of Gamma$(\alpha_1 + \cdots + \alpha_K, \lambda)$. Thus

$$
X_1 + \cdots + X_K \sim \text{Gamma}(\alpha_1 + \cdots + \alpha_K, \lambda).
\tag{4.40}
$$

What if the λ's are not equal? Then it is still easy to find the mgf, but it would not be the Gamma mgf, or anything else we have seen so far.

4.3.3 Linear combinations of independent normals

Suppose $X_1, \ldots, X_K$ are independent, $X_k \sim N(\mu_k, \sigma_k^2)$. Consider the affine transformation

$$
Y = a + b_1 X_1 + \cdots + b_K X_K.
\tag{4.41}
$$

It is straightforward, from Section 2.2.2, to see that

$$
E[Y] = a + b_1 \mu_1 + \cdots + b_K \mu_K \quad \text{and} \quad Var[Y] = b_1^2 \sigma_1^2 + \cdots + b_K^2 \sigma_K^2,
\tag{4.42}
$$

since independence implies that all the covariances are 0. But those equations do not give the entire distribution of Y. We need the mgf:

$$
\begin{aligned}
M_Y(t) &= E[e^{tY}] \\
&= E[e^{t(a + b_1 X_1 + \cdots + b_K X_K)}] \\
&= e^{at} E[e^{(tb_1)X_1}] \times \cdots \times E[e^{(tb_K)X_K}] \quad \text{by independence} \\
&= e^{at} M(tb_1 \mid \mu_1, \sigma_1^2) \times \cdots \times M(tb_K \mid \mu_K, \sigma_K^2) \\
&= e^{at} e^{tb_1 \mu_1 + \frac{1}{2} t^2 b_1^2 \sigma_1^2} \cdots e^{tb_K \mu_K + \frac{1}{2} t^2 b_K^2 \sigma_K^2} \\
&= e^{t(a + b_1 \mu_1 + \cdots + b_K \mu_K) + \frac{1}{2} t^2 (b_1^2 \sigma_1^2 + \cdots + b_K^2 \sigma_K^2)} \\
&= M(t \mid a + b_1 \mu_1 + \cdots + b_K \mu_K, b_1^2 \sigma_1^2 + \cdots + b_K^2 \sigma_K^2),
\end{aligned}
\tag{4.43}
$$

by (4.36). Notice that in going from the third to fourth step, we have changed the variable for the mgfs from t to the $b_k t$'s, which is legitimate. The final mgf in (4.43) is indeed the mgf of a normal, with the appropriate mean and variance, i.e.,

$$Y \sim N(a + b_1 \mu_1 + \cdots + b_K \mu_K, b_1^2 \sigma_1^2 + \cdots + b_K^2 \sigma_K^2). \tag{4.44}$$

If the X_k's are iid $N(\mu, \sigma^2)$, then (4.44) can be used to show that

$$\sum X_k \sim N(K\mu, K\sigma^2) \quad \text{and} \quad \overline{X} \sim N(\mu, \sigma^2/K). \tag{4.45}$$

4.3.4   Normalized means

The central limit theorem, which we present in Section 8.6, is central to statistics as it justifies using the normal distribution in certain non-normal situations. Here we assume we have $X_1, \ldots, X_n$ iid with any distribution, as long as the mgf $M(t)$ of X_i exists for t in a neighborhood of 0. The normalized mean is

$$W_n = \sqrt{n}\, \frac{\overline{X} - \mu}{\sigma}, \tag{4.46}$$

where $\mu = E[X_i]$ and $\sigma^2 = Var[X_i]$, so that $E[W_n] = 0$ and $Var[W_n] = 1$. If the X_i are normal, then $W_n \sim N(0, 1)$. The central limit theorem implies that even if the X_i's are not normal, W_n is "approximately" normal if n is "large." We will look at the cumulant generating function of W_n, and compare it to the normal's.

First, the mgf of W_n:

$$\begin{aligned}
M_n(t) &= E[e^{tW_n}] \\
&= E[e^{t\sqrt{n}(\overline{X} - \mu)/\sigma}] \\
&= e^{-t\sqrt{n}\mu/\sigma}\, E[e^{(t/(\sqrt{n}\sigma))\sum X_i}] \\
&= e^{-t\sqrt{n}\mu/\sigma} \prod E[e^{(t/(\sqrt{n}\sigma))X_i}] \\
&= e^{-t\sqrt{n}\mu/\sigma}\, M(t/(\sqrt{n}\sigma))^n. \tag{4.47}
\end{aligned}$$

Letting $c(t)$ be the cumulant generating function of X_i, and $c_n(t)$ be that of W_n, we have that

$$c_n(t) = -\frac{t\sqrt{n}\mu}{\sigma} + nc\left(\frac{t}{\sqrt{n}\sigma}\right). \tag{4.48}$$

We know that $c_n'(0)$ is the mean, which is 0 in this case, and $c_n''(0)$ is the variance, which here is 1. For higher derivatives, the first term in (4.48) vanishes, and each derivative brings out another $1/(\sqrt{n}\sigma)$ from $c(t)$, hence the kth cumulant of W_n is

$$c_n^{(k)}(0) = nc^{(k)}(0)\frac{1}{(\sqrt{n}\sigma)^k}. \tag{4.49}$$

Letting γ_k be the kth cumulant of X_i, and $\gamma_{n,k}$ be that of W_n,

$$\gamma_{n,k} = \frac{1}{n^{k/2-1}}\frac{1}{\sigma^k}\,\gamma_k, \quad k = 3, 4, \ldots. \tag{4.50}$$

For the normal, all cumulants are 0 after the first two. Thus the closer the $\gamma_{n,k}$'s are to 0, the closer the distribution of W_n is to $N(0,1)$ in some sense. There are then two factors: The larger n, the closer these cumulants are to 0 for $k > 2$. Also, the smaller γ_k/σ^k in absolute value, the closer to 0.

For example, if the X_i's are Exponential(1), or Gamma(1,1), then from (2.79) the cumulants are $\gamma_k = (k-1)!$, and $\gamma_{n,k} = (k-1)!/n^{k/2-1}$. The Poisson($\lambda$) has cumulant generating function $c(t) = \lambda(e^t - 1)$, so all its derivatives are λe^t, hence $\gamma_k = \lambda$, and

$$\gamma_{n,k} = \frac{1}{n^{k/2-1}} \frac{1}{\sigma^{k/2}} \gamma_k = \frac{1}{n^{k/2-1}} \frac{1}{\lambda^{k/2}} \lambda = \frac{1}{(n\lambda)^{k/2-1}}. \tag{4.51}$$

The Laplace has pdf $(1/2)e^{-|x|}$ for $x \in \mathbb{R}$. Its cumulants are

$$\gamma_k = \begin{cases} 0 & \text{if } k \text{ is odd} \\ 2(k-1)! & \text{if } k \text{ is even} \end{cases}. \tag{4.52}$$

Thus the variance is 2, and

$$\gamma_{n,k} = \gamma_k = \begin{cases} 0 & \text{if } k \text{ is odd} \\ (k-1)!/(2n)^{k/2-1} & \text{if } k \text{ is even} \end{cases}. \tag{4.53}$$

Here is a small table with some values of $\gamma_{k,n}$:

$\gamma_{n,k}$		Normal	Exponential	Poisson			Laplace	
k	n			$\lambda = 1/10$	$\lambda = 1$	$\lambda = 10$		
3	1	0	2.000	3.162	1.000	0.316	0.000	
	10	0	0.632	1.000	0.316	0.100	0.000	
	100	0	0.200	0.316	0.100	0.032	0.000	
4	1	0	6.000	10.000	1.000	0.100	3.000	(4.54)
	10	0	0.600	1.000	0.100	0.010	0.300	
	100	0	0.060	0.100	0.010	0.001	0.030	
5	1	0	24.000	31.623	1.000	0.032	0.000	
	10	0	0.759	1.000	0.032	0.001	0.000	
	100	0	0.024	0.032	0.001	0.000	0.000	

For each distribution, as n increases, the cumulants do decrease. Also, the exponential is closer to normal that the Poisson(1/10), but the Poisson(1) is closer than the exponential, and the Poisson(10) is even closer. The Laplace is symmetric, so its odd cumulants are 0, automatically making it relatively close to normal. Its kurtosis is a bit worse than the Poisson(1), however.

4.3.5 Bernoulli and binomial

The distribution of a random variable that takes on just the values 0 and 1 is completely specified by giving the probability it is 1. Such a variable is called **Bernoulli**.

Definition 4.2. *If Z has space $\{0,1\}$, then it is Bernoulli with parameter $p = P[Z = 1]$, written*

$$Z \sim \text{Bernoulli}(p). \tag{4.55}$$

The pmf can then be written

$$f(z) = p^z (1-p)^{1-z}. \tag{4.56}$$

Note that Bernoulli(p) = Binomial$(1, p)$. Rather than defining the binomial through its pmf, as in Table 1.2 on page 9, a better alternative is to base the definition on Bernoullis.

Definition 4.3. *If $Z_1, \ldots, Z_n$ are iid Bernoulli(p), then $X = Z_1 + \cdots + Z_n$ is binomial with parameters n and p, written*

$$X \sim \text{Binomial}(n, p). \tag{4.57}$$

The binomial counts the number of successes in n independent trials, where the Bernoullis represent the individual trials, with a "1" indicating success. The moments, and mgf, of a Bernoulli are easy to find. In fact, because $Z^k = Z$ for $k = 1, 2, \ldots,$

$$E[Z] = E[Z^2] = 0 \times (1-p) + 1 \times p = p \Rightarrow Var[Z] = p - p^2 = p(1-p), \tag{4.58}$$

and

$$M_Z(t) = E[e^{tZ}] = e^0 (1-p) + e^t p = pe^t + 1 - p. \tag{4.59}$$

Thus for $X \sim \text{Binomial}(n, p)$,

$$E[X] = nE[Z_i] = np, Var[X] = nVar[Z_i] = np(1-p), \tag{4.60}$$

and

$$M_X(t) = E[e^{t(Z_1 + \cdots + Z_n)}] = E[e^{tZ_1}] \cdots E[e^{tZ_n}] = M_Z(t)^n = (pe^t + 1 - p)^n. \tag{4.61}$$

This mgf is the same as we found in (2.83), meaning we really are defining the same binomial.

4.4   Exercises

Exercise 4.4.1. Suppose (X, Y) has space $\{1, 2, 3\} \times \{1, 2, 3\}$ and pmf $f(x, y) = (x + y)/c$ for some constant c. (a) Are X and Y independent? (b) What is the constant c? (c) Let $W = X + Y$. Find $f_W(w)$, the pmf of W.

Exercise 4.4.2. If $\mathbf{x}$ and $\mathbf{y}$ are vectors of length n, then Kendall's distance between $\mathbf{x}$ and $\mathbf{y}$ measures the extent to which they have a positive relationship in the sense that larger values of $\mathbf{x}$ go with larger values of $\mathbf{y}$. It is defined by

$$d(\mathbf{x}, \mathbf{y}) = \sum\sum_{1 \leq i < j \leq n} I[(x_i - x_j)(y_i - y_j) < 0]. \tag{4.62}$$

The idea is to plot the points, and draw a line segment between each pair of points (x_i, y_i) and (x_j, y_j). If any segment has a negative slope, then the x's and y's for that pair go in the wrong direction. (Look ahead to Figure 18.1 on page 308 for an illustration.) Kendall's distance then counts the number of such pairs. If $d(\mathbf{x}, \mathbf{y}) = 0$, then the plot shows a nondecreasing pattern. Worst is if $d(\mathbf{x}, \mathbf{y}) = n(n-1)/2$, since then all pairs go in the wrong direction.

Suppose $\mathbf{Y}$ is uniformly distributed over the permutations of the integers 1, 2, 3, that is, Y has space

$$\mathcal{Y} = \{(1,2,3), (1,3,2), (2,1,3), (2,3,1), (3,1,2), (3,2,1)\}, \qquad (4.63)$$

and $P(\mathbf{Y} = \mathbf{y}) = 1/6$ for each $\mathbf{y} \in \mathcal{Y}$. (a) Find the pmf of $U = d((1,2,3), \mathbf{Y})$. [Write out the $d((1,2,3), \; y)$'s for the $\mathbf{y}$'s in $\mathcal{Y}$.] (b) Kendall's τ is defined to be $T = 1 - 4U/n(n-1)$. It normalizes Kendall's distance so that it acts like a correlation coefficient, i.e., $T = -1$ if $\mathbf{x}$ and $\mathbf{y}$ have an exact negative relationship, and $T = 1$ if they have an exact positive relationship. In this problem, $n = 3$. Find the space of T, and its pmf.

Exercise 4.4.3. Continue with the $\mathbf{Y}$ and U from Exercise 4.4.2, where $n = 3$. We can write

$$U = d((1,2,3), \mathbf{y}) = \sum_{i=1}^{2} \sum_{j=i+1}^{3} I[-(y_i - y_j) < 0] = \sum_{i=1}^{2} \sum_{j=i+1}^{3} I[y_i > y_j]. \qquad (4.64)$$

Then

$$U = U_1 + U_2, \qquad (4.65)$$

where $U_i = \sum_{j=i+1}^{3} I[y_i > y_j]$. (a) Find the space and pmf of (U_1, U_2). Are U_1 and U_2 independent? (b) What are the marginal distributions of U_1 and U_2?

Exercise 4.4.4. Suppose U and V are independent with the same Geometric(p) distribution. (So they have space the nonnegative integers and pmf $f(u) = p(1-p)^u$.) Let $T = U + V$. (a) What is the space of T? (b) Find the pmf of T using convolutions as in (4.11). [The answer should be $(1+t)p^2(1-p)^t$. This is a negative binomial distribution, as we will see in Exercise 4.4.15.]

Exercise 4.4.5. Suppose U_1 is Discrete Uniform$(0, 1)$ and U_2 is Discrete Uniform$(0, 2)$, where U_1 and U_2 are independent. (a) Find the moment generating function, $M_{U_1}(t)$. (b) Find the moment generating function, $M_{U_2}(t)$. (c) Now let $U = U_1 + U_2$. Find the moment generating function of U, $M_U(t)$, by multiplying the two individual mgfs. (d) Use the mgf to find the pmf of U. (It is given in (4.6).) Does this U have the same distribution as that in Exercises 4.4.2 and 4.4.3?

Exercise 4.4.6. Define the 3×1 random vector $\mathbf{Z}$ to be Trinoulli(p_1, p_2, p_3) if it has space $\mathcal{Z} = \{(1,0,0), (0,1,0), (0,0,1)\}$ and pmf

$$f_{\mathbf{Z}}(1,0,0) = p_1, \quad f_{\mathbf{Z}}(0,1,0) = p_2, \quad f_{\mathbf{Z}}(0,0,1) = p_3, \qquad (4.66)$$

where the $p_i > 0$ and $p_1 + p_2 + p_3 = 1$. (a) Find the mgf of $\mathbf{Z}$, $M_{\mathbf{Z}}(t_1, t_2, t_3)$. (b) If $\mathbf{Z}_1, \ldots, \mathbf{Z}_n$ are iid Trinoulli(p_1, p_2, p_3), then $\mathbf{X} = \mathbf{Z}_1 + \cdots + \mathbf{Z}_n$ is Trinomial$(n, (p_1, p_2, p_3))$. What is the space $\mathcal{X}$ of $\mathbf{X}$? (c) What is the mgf of $\mathbf{X}$? This mgf is that of which distribution that we have seen before?

Exercise 4.4.7. Let $\mathbf{X} = (X_1, X_2, X_3, X_4) \sim$ Multinomial$(n, (p_1, p_2, p_3, p_4))$. Find the mgf of $\mathbf{Y} = (X_1, X_2, X_3 + X_4)$. What is the distribution of $\mathbf{Y}$?

Exercise 4.4.8. Suppose (X_1, X_2) has space $(a, b) \times (c, d)$ and pdf $f(x_1, x_2)$. Show that the limits of integration for the convolution (4.19) $\int f(x_1, y - x_1) dx_1$ are $\max\{a, y - d\} < x_1 < \min\{b, y - c\}$.

Exercise 4.4.9. Let X_1 and X_2 be iid $N(0,1)$'s, and set $Y = X_1 + X_2$. Using the convolution formula (4.19), show that $Y \sim N(0,2)$. [Hint: In the exponent of the integrand, complete the square with respect to x_1, then note you are integrating over what looks like a $N(0,1/2)$ pdf.]

Exercise 4.4.10. Suppose Z has the distribution function $F(z)$, pdf $f(z)$, and mgf $M(t)$. Let $X = \mu + \sigma Z$, where $\mu \in \mathbb{R}$ and $\sigma > 0$. Thus we have a location-scale family, as in Definition 4.1 on page 55. (a) Show that the distribution function of X is $F((x - \mu)/\sigma)$, as in (4.32). (b) Show that the pdf of X is $(1/\sigma)f((x - \mu)/\sigma)$, as in (4.33). (c) Show that the mgf of X is $\exp(t\mu)M(t\sigma)$, as in (4.34).

Exercise 4.4.11. Show that the mgf of a $N(\mu, \sigma^2)$ is $\exp(t\mu + (1/2)t^2\sigma^2)$, as in (4.36). For what values of t is the mgf finite?

Exercise 4.4.12. Consider the location-scale family based on Z. Suppose Z has finite skewness κ_3 in (2.56) and kurtosis κ_4 in (2.57). Show that $X = \mu + \sigma Z$ has the same skewness and kurtosis as Z as long as $\sigma > 0$. [Hint: Show that $E[(X - \mu)^k] = \sigma^k E[Z^k]$.]

Exercise 4.4.13. In Exercise 2.7.12, we found that the Laplace has mfg $1/(1 - t^2)$ for $|t| > 1$. Suppose U and V are independent Exponential(1), and let $Y = U - V$. Find the mgf of Y. What is the distribution of Y? [Hint: Look back at Section 3.3.1.]

Exercise 4.4.14. Suppose $X_1, \ldots, X_K$ are iid $N(\mu, \sigma^2)$. Use (4.44) to show that (4.45) holds, i.e., that $\sum X_k \sim N(K\mu, K\sigma^2)$ and $\overline{X} \sim N(\mu, \sigma^2/K)$.

Exercise 4.4.15. Consider a coin with probability of heads being p, and flip it a number of times independently, until you see a heads. Let Z be the number of tails flipped before the first head. Then $Z \sim \text{Geometric}(p)$. Exercise 2.7.2 showed that the mgf of Z is $M_Z(t) = p(1 - e^t(1-p))^{-1}$ for t's such that it is finite. Suppose $Z_1, \ldots, Z_K$ are iid Geometric(p), and $Y = Z_1 + \cdots + Z_K$. Then Y is the number of tails before the Kth head. It is called Negative Binomial(K, p). (a) What is the mgf of Y, $M_Y(t)$? (b) For $\alpha > 0$ and $0 < x < 1$, define

$$_1F_0(\alpha; -; x) = \sum_{y=0}^{\infty} \frac{\Gamma(y + \alpha)}{\Gamma(\alpha)} \frac{x^y}{y!}. \tag{4.67}$$

It can be shown using a Taylor series that $_1F_0(\alpha; -; x) = (1 - x)^{-\alpha}$. (Such functions arise again in Exercise 7.8.21. Exercise 2.7.18(c) used $_1F_0(2; -; x)$.) We can then write $M_Y(t) = p^c {}_1F_0(\alpha; -; x)$ for what c, α and x? (c) Table 1.2 gives the pmf of Y to be

$$f_Y(y) = \binom{y + K - 1}{K - 1} p^K (1 - p)^y. \tag{4.68}$$

Find the mgf of this f_Y to verify that it is the pmf for the negative binomial as defined in this exercise. [Hint: Write the binomial coefficient out, and replace two of the factorials with Γ functions.] (d) Show that for $K > 1$,

$$\delta(y) \equiv \frac{K - 1}{y + K - 1} \tag{4.69}$$

has $E[\delta(Y)] = p$, so is an unbiased estimator of p. [Hint: Write down the $\sum \delta(y) f_Y(y)$, then factor out a p and note that you have the pmf of another negative binomial.]

Exercise 4.4.16. Let F be the distribution function of the random variable X, which may not be continuous. Take $u \in (0, 1)$. The goal is to show (4.31), that $P[F(X) \le u] \le u$. Let $x^* = F^-(u) = \min\{x \mid F(x) \ge u\}$ as in (4.27). (a) Suppose $F(x)$ is continuous at $x = x^*$. Show that $F(x^*) = u$, and $F(x) \le u$ if and only if $x \le x^*$. [It helps to draw a picture of the distribution function.] Thus $P[F(X) \le u] = P[X \le x^*] = F(x^*) = u$. (b) Suppose $F(x)$ is *not* continuous at $x = x^*$. Show that $P[X \ge x^*] \ge 1 - u$, and $F(x) \le u$ if and only if $x < x^*$. Thus $P[F(X) \le u] = P[X < x^*] = 1 - P[X \ge x^*] \le 1 - (1 - u) = u$.

Exercise 4.4.17. Let (X, Y) be uniform over the unit disk: The space is $\mathcal{W} = \{(x, y) \mid x^2 + y^2 < 1\}$, and $f(x, y) = 1/\pi$ for $(x, y) \in \mathcal{W}$. Let $U = X^2$. (a) Show that the distribution function of U can be written as

$$F_U(u) = \frac{4}{\pi} \int_0^{\sqrt{u}} \sqrt{1 - x^2}\, dx \qquad (4.70)$$

for $u \in (0, 1)$. (b) Find the pdf of U. It should be a Beta(α, β). What are α and β?

Exercise 4.4.18. Suppose $X_1, X_2, \ldots, X_n$ are independent, all with the same continuous distribution function $F(x)$ and space $\mathcal{X}$. Let Y be their maximum:

$$Y = \max\{X_1, X_2, \ldots, X_n\}, \qquad (4.71)$$

and $F_Y(y)$ be the distribution function of Y. (a) What is the space of Y? (b) Explain why $Y \le y$ if and only if $X_1 \le y$ & $X_2 \le y$ & $\cdots$ & $X_n \le y$. (c) Explain why $P[Y \le y] = P[X_1 \le y] \times P[X_2 \le y] \times \cdots \times P[X_n \le y]$. (d) Thus we can write $F_Y(y) = u^a$ for some u and a. What are u and a?

For the rest of this exercise, suppose the X_i's above are Uniform$(0, 1)$ (and still independent). (e) For $0 < x < 1$, what is $F(x)$? (f) What is $F_Y(y)$, in terms of y and n? (g) What is $f_Y(y)$, the pdf of Y? This is the pdf of what distribution? (Give the name and the parameters.)

Exercise 4.4.19. In the following questions, X and Y are independent. Yes or no? (a) If $X \sim$ Gamma(α, λ) and $Y \sim$ Gamma(β, λ), where $\alpha \ne \beta$, is $X + Y$ gamma? (b) If $X \sim$ Gamma(α, λ) and $Y \sim$ Gamma(α, δ), where $\lambda \ne \delta$, is $X + Y$ gamma? (c) If $X \sim$ Poisson(λ) and $Y \sim$ Poisson(δ), is $X + Y$ Poisson? (d) If X and Y are Exponential(λ), is $X + Y$ exponential? (e) If $X \sim$ Binomial(n, p) and $Y \sim$ Binomial(n, q), where $p \ne q$, is $X + Y$ binomial? (f) If $X \sim$ Binomial(n, p) and $Y \sim$ Binomial(m, p), where $n \ne m$, is $X + Y$ binomial? (g) If X and Y are Laplace, is $X + Y$ Laplace?

Chapter 5

Transformations: Jacobians

In Section 4.1, we saw that finding the pmfs of transformed variables when we start with discrete variables is possible by summing the appropriate probabilities. In the one-to-one case, it is even easier. That is, if $g(\mathbf{x})$ is one-to-one, then the pmf of $\mathbf{Y} = g(\mathbf{X})$ is $f_{\mathbf{Y}}(\mathbf{y}) = f_{\mathbf{X}}(g^{-1}(\mathbf{y}))$. In the continuous case, it is not as straightforward. The problem is that for a pdf, $f_{\mathbf{X}}(\mathbf{x})$ is not $P[\mathbf{X} = \mathbf{x}]$, which is 0. Rather, for a small area $\mathcal{A} \subset \mathcal{X}$ that contains $\mathbf{x}$,

$$P[\mathbf{X} \in \mathcal{A}] \approx f_{\mathbf{X}}(\mathbf{x}) \times \text{Area}(\mathcal{A}). \tag{5.1}$$

Now suppose $g : \mathcal{X} \to \mathcal{Y}$ is one-to-one and onto. Then for $\mathbf{y} \in \mathcal{B} \subset \mathcal{Y}$,

$$P[\mathbf{Y} \in \mathcal{B}] \approx f_{\mathbf{Y}}(\mathbf{y}) \times \text{Area}(\mathcal{B}). \tag{5.2}$$

Since

$$P[\mathbf{Y} \in \mathcal{B}] = P[\mathbf{X} \in g^{-1}(\mathcal{B})] \approx f_{\mathbf{X}}(g^{-1}(\mathbf{y})) \times \text{Area}(g^{-1}(\mathcal{B})), \tag{5.3}$$

where

$$g^{-1}(\mathcal{B}) = \{\mathbf{x} \in \mathcal{X} \mid g(\mathbf{x}) \in \mathcal{B}\}, \tag{5.4}$$

we find that

$$f_{\mathbf{Y}}(\mathbf{y}) \approx f_{\mathbf{X}}(g^{-1}(\mathbf{y})) \times \frac{\text{Area}(g^{-1}(\mathcal{B}))}{\text{Area}(\mathcal{B})}. \tag{5.5}$$

Compare this equation to (4.3). For continuous distributions, we need to take care of the transformation of areas as well as the transformation of $\mathbf{Y}$ itself. The actual pdf of $\mathbf{Y}$ is found by shrinking the $\mathcal{B}$ in (5.5) down to the point $\mathbf{y}$.

5.1 One dimension

When X and Y are random variables, the ratio of areas in (5.5) is easy to find. Because g and g^{-1} are one-to-one, they must be either strictly increasing or strictly decreasing. For $y \in \mathcal{Y}$, take ϵ small enough that $\mathcal{B}_\epsilon \equiv [y, y + \epsilon) \in \mathcal{Y}$. Then, since the area of an interval is just its length,

$$\frac{\text{Area}(g^{-1}(\mathcal{B}_\epsilon))}{\text{Area}(\mathcal{B}_\epsilon)} = \frac{|g^{-1}(y+\epsilon) - g^{-1}(y)|}{\epsilon}$$

$$\rightarrow \left| \frac{\partial}{\partial y} g^{-1}(y) \right| \quad \text{as } \epsilon \rightarrow 0. \tag{5.6}$$

(The absolute value is there in case g is decreasing.) That derivative is called the **Jacobian** of the transformation g^{-1}. That is,

$$f_Y(y) = f_X(g^{-1}(y)) \, |J_{g^{-1}}(y)|, \quad \text{where } J_{g^{-1}}(y) = \frac{\partial}{\partial y} g^{-1}(y). \tag{5.7}$$

This approach needs a couple of assumptions: y is in the interior of $\mathcal{Y}$, and the derivative of $g^{-1}(y)$ exists.

Reprising Example 4.3.1, where $X \sim \text{Uniform}(0,1)$ and $Y = -\log(X)$, we have $\mathcal{Y} = (0, \infty)$, and $g^{-1}(y) = e^{-y}$. Then

$$f_X(e^{-y}) = 1 \quad \text{and} \quad J_{g^{-1}}(y) = \frac{\partial}{\partial y} e^{-y} = -e^{-y} \Longrightarrow f_Y(y) = 1 \times |-e^{-y}| = e^{-y}, \tag{5.8}$$

which is indeed the answer from (4.38).

5.2 General case

For the general case (for vectors), we need to figure out the ratio of the volumes, which is again given by the Jacobian, but for vectors.

Definition 5.1. *Suppose $g : \mathcal{X} \rightarrow \mathcal{Y}$ is one-to-one and onto, where both $\mathcal{X}$ and $\mathcal{Y}$ are open subsets of $\mathbb{R}^p$, and all the first partial derivative of $g^{-1}(\mathbf{y})$ exist. Then the **Jacobian** of the transformation g^{-1} is defined to be*

$$J_{g^{-1}} : \mathcal{Y} \rightarrow \mathbb{R},$$

$$J_{g^{-1}}(\mathbf{y}) = \begin{vmatrix} \frac{\partial}{\partial y_1} g_1^{-1}(\mathbf{y}) & \frac{\partial}{\partial y_2} g_1^{-1}(\mathbf{y}) & \cdots & \frac{\partial}{\partial y_p} g_1^{-1}(\mathbf{y}) \\ \frac{\partial}{\partial y_1} g_2^{-1}(\mathbf{y}) & \frac{\partial}{\partial y_2} g_2^{-1}(\mathbf{y}) & \cdots & \frac{\partial}{\partial y_p} g_2^{-1}(\mathbf{y}) \\ \vdots & \vdots & \ddots & \vdots \\ \frac{\partial}{\partial y_1} g_p^{-1}(\mathbf{y}) & \frac{\partial}{\partial y_2} g_p^{-1}(\mathbf{y}) & \cdots & \frac{\partial}{\partial y_p} g_p^{-1}(\mathbf{y}) \end{vmatrix}, \tag{5.9}$$

where $g^{-1}(\mathbf{y}) = (g_1^{-1}(\mathbf{y}), \ldots, g_p^{-1}(\mathbf{y}))$, and here the "$|\cdot|$" represents the determinant.

The next theorem is from advanced calculus.

Theorem 5.2. *Suppose the conditions in Definition 5.1 hold, and $J_{g^{-1}}(\mathbf{y})$ is continuous and non-zero for $\mathbf{y} \in \mathcal{Y}$. Then*

$$f_Y(\mathbf{y}) = f_X(\mathbf{x}) \times |J_{g^{-1}}(\mathbf{y})|. \tag{5.10}$$

Here the "$|\cdot|$" represents the absolute value.

If you think of g^{-1} as $\mathbf{x}$, you can remember the formula as

$$f_Y(\mathbf{y}) = f_X(\mathbf{x}) \times \left| \frac{d\mathbf{x}}{d\mathbf{y}} \right|. \tag{5.11}$$

5.3 Gamma, beta, and Dirichlet distributions

Suppose X_1 and X_2 are independent, with

$$X_1 \sim \text{Gamma}(\alpha, \lambda) \quad \text{and} \quad X_2 \sim \text{Gamma}(\beta, \lambda), \tag{5.12}$$

so that they have the same rate but possibly different shapes. We are interested in

$$Y_1 = \frac{X_1}{X_1 + X_2}. \tag{5.13}$$

This variable arises, e.g., in linear regression, where R^2 has that distribution under certain conditions. The function taking (x_1, x_2) to y_1 is not one-to-one. To fix that up, we introduce another variable, Y_2, so that the function to the pair (y_1, y_2) is one-to-one. Then to find the pdf of Y_1, we integrate out Y_2. We will take

$$Y_2 = X_1 + X_2. \tag{5.14}$$

Then

$$\mathcal{X} = (0, \infty) \times (0, \infty) \quad \text{and} \quad \mathcal{Y} = (0, 1) \times (0, \infty). \tag{5.15}$$

To find g^{-1}, solve the equation $\mathbf{y} = g(\mathbf{x})$ for $\mathbf{x}$:

$$y_1 = \frac{x_1}{x_1 + x_2}, \quad y_2 = x_1 + x_2 \implies x_1 = y_1 y_2, \quad x_2 = y_2 - x_1$$

$$\implies x_1 = y_1 y_2, \quad x_2 = y_2(1 - y_1), \tag{5.16}$$

hence

$$g^{-1}(y_1, y_2) = (y_1 y_2, y_2(1 - y_1)). \tag{5.17}$$

Using this inverse, we can see that indeed the function is onto the $\mathcal{Y}$ in (5.15), since any $y_1 \in (0, 1)$ and $y_2 \in (0, \infty)$ will yield, via (5.17), x_1 and x_2 in $(0, \infty)$. For the Jacobian:

$$
J_{g^{-1}}(\mathbf{y}) = \begin{vmatrix} \frac{\partial}{\partial y_1} y_1 y_2 & \frac{\partial}{\partial y_2} y_1 y_2 \\ \frac{\partial}{\partial y_1} y_2(1 - y_1) & \frac{\partial}{\partial y_2} y_2(1 - y_1) \end{vmatrix}
$$

$$
= \begin{vmatrix} y_2 & y_1 \\ -y_2 & 1 - y_1 \end{vmatrix}
$$

$$
= y_2(1 - y_1) + y_1 y_2
$$

$$
= y_2. \tag{5.18}
$$

Now because the X_i's are independent gammas, their pdf is

$$f_X(x_1, x_2) = \frac{\lambda^\alpha}{\Gamma(\alpha)} x_1^{\alpha-1} e^{-\lambda x_1} \frac{\lambda^\beta}{\Gamma(\beta)} x_2^{\beta-1} e^{-\lambda x_2}. \tag{5.19}$$

Then the pdf of **Y** is

$$
\begin{aligned}
f_{\mathbf{Y}}(\mathbf{y}) &= f_{\mathbf{X}}(g^{-1}(\mathbf{y})) \times |J_{g^{-1}}(\mathbf{y})| \\
&= f_{\mathbf{X}}(y_1 y_2, y_2(1 - y_1))) \times |y_2| \\
&= \frac{\lambda^\alpha}{\Gamma(\alpha)} (y_1 y_2)^{\alpha - 1} e^{-\lambda y_1 y_2} \frac{\lambda^\beta}{\Gamma(\beta)} (y_2(1 - y_1))^{\beta - 1} e^{-\lambda y_2(1 - y_1)} \times |y_2| \\
&= \frac{\lambda^{\alpha + \beta}}{\Gamma(\alpha)\Gamma(\beta)} y_1^{\alpha - 1}(1 - y_1)^{\beta - 1} y_2^{\alpha + \beta - 1} e^{-\lambda y_2}.
\end{aligned}
\tag{5.20}
$$

To find the pdf of Y_1, we can integrate out y_2:

$$
f_{Y_1}(y_1) = \int_0^\infty f_{\mathbf{Y}}(y_1, y_2) dy_2.
\tag{5.21}
$$

That certainly is a fine approach. But in this case, note that the joint pdf in (5.20) can be factored into a function of just y_1, and a function of just y_2. That fact, coupled with the fact that the space is a rectangle, means that by Lemma 3.4 on page 46 we automatically have that Y_1 and Y_2 are independent.

If we look at the y_2 part of $f_{\mathbf{Y}}(\mathbf{y})$ in (5.20), we see that it looks like a Gamma$(\alpha + \beta, \lambda)$ pdf. We just need to multiply & divide by the appropriate constant:

$$
\begin{aligned}
f_{\mathbf{Y}}(\mathbf{y}) &= \left[\frac{\lambda^{\alpha + \beta}}{\Gamma(\alpha)\Gamma(\beta)} y_1^{\alpha - 1}(1 - y_1)^{\beta - 1} \frac{\Gamma(\alpha + \beta)}{\lambda^{\alpha + \beta}} \right] \left[\frac{\lambda^{\alpha + \beta}}{\Gamma(\alpha + \beta)} y_2^{\alpha + \beta - 1} e^{-\lambda y_2} \right] \\
&= \left[\frac{\Gamma(\alpha + \beta)}{\Gamma(\alpha)\Gamma(\beta)} y_1^{\alpha - 1}(1 - y_1)^{\beta - 1} \right] \left[\frac{\lambda^{\alpha + \beta}}{\Gamma(\alpha + \beta)} y_2^{\alpha + \beta - 1} e^{-\lambda y_2} \right].
\end{aligned}
\tag{5.22}
$$

So in the first line, we have separated the gamma pdf from the rest, and in the second line simplified the y_1 part of the density. Thus that part is the pdf of Y_1, which from Table 1.1 on page 7 is the Beta(α, β) pdf. Note that we have surreptitiously also proven that

$$
\int_0^1 y_1^{\alpha - 1}(1 - y_1)^{\beta - 1} dy_1 = \frac{\Gamma(\alpha)\Gamma(\beta)}{\Gamma(\alpha + \beta)} \equiv B(\alpha, \beta),
\tag{5.23}
$$

which is the **beta function**.

To summarize, we have shown that $X_1/(X_1 + X_2)$ and $X_1 + X_2$ are independent (even though they do not really look independent), and that

$$
\frac{X_1}{X_1 + X_2} \sim \text{Beta}(\alpha, \beta) \quad \text{and} \quad X_1 + X_2 \sim \text{Gamma}(\alpha + \beta, \lambda).
\tag{5.24}
$$

That last fact we already knew, from Example 4.3.2. Also, notice that the beta variable does not depend on the rate λ.

5.3.1   Dirichlet distribution

The Dirichlet distribution is a multivariate version of the beta. We start with the independent random variables $X_1, \ldots, X_K$, $K > 1$, where

$$
X_k \sim \text{Gamma}(\alpha_k, 1).
\tag{5.25}
$$

Then the $K - 1$ vector $\mathbf{Y}$ defined via

$$Y_k = \frac{X_k}{X_1 + \cdots + X_K}, \quad k = 1, \ldots, K - 1, \tag{5.26}$$

has a Dirichlet distribution, written

$$\mathbf{Y} \sim \text{Dirichlet}(\alpha_1, \ldots, \alpha_K). \tag{5.27}$$

There is also a Y_K, which may come in handy, but because $Y_1 + \cdots + Y_K = 1$, it is redundant. Also, as in the beta, the definition is the same if the X_k's are $\text{Gamma}(\alpha_k, \lambda)$. The representation (5.26) makes it easy to find the marginals, i.e.,

$$Y_k \sim \text{Beta}(\alpha_k, \alpha_1 + \cdots + \alpha_{k-1} + \alpha_{k+1} + \cdots + \alpha_K), \tag{5.28}$$

hence the marginal means and variances. Sums of the Y_k's are also beta, e.g., if $K = 4$, then

$$Y_1 + Y_3 = \frac{X_1 + X_3}{X_1 + X_2 + X_3 + X_4} \sim \text{Beta}(\alpha_1 + \alpha_3, \alpha_2 + \alpha_4). \tag{5.29}$$

The space of $\mathbf{Y}$ is

$$\mathcal{Y} = \{ \mathbf{y} \in \mathbb{R}^{K-1} \mid 0 < y_k < 1, k = 1, \ldots, K - 1, \text{ and } y_1 + \cdots + y_{K-1} < 1 \}. \tag{5.30}$$

To find the pdf of $\mathbf{Y}$, we need a one-to-one transformation from $\mathbf{X}$, so we need to append another function of $\mathbf{X}$ to the $\mathbf{Y}$. The easiest choice is

$$W = X_1 + \ldots + X_K, \tag{5.31}$$

so that $g(\mathbf{x}) = (\mathbf{y}, w)$. Then g^{-1} is given by

$$x_1 = wy_1;$$
$$x_2 = wy_2;$$
$$\vdots$$
$$x_{K-1} = wy_{K-1};$$
$$x_K = w(1 - y_1 - \ldots - y_{K-1}). \tag{5.32}$$

It can be shown that the determinant of the Jacobian is w^{K-1}. Exercise 5.6.3 illustrates the calculations for $K = 4$. The joint pdf of $(\mathbf{Y}, W)$ is

$$f_{(\mathbf{Y}, W)}(\mathbf{y}, w) = \left[\prod_{k=1}^{K} \frac{1}{\Gamma(\alpha_K)} \right] \left[\prod_{k=1}^{K-1} y_k^{\alpha_k - 1} \right] (1 - y_1 - \cdots - y_{K-1})^{\alpha_K - 1} w^{\alpha_1 + \cdots + \alpha_K - 1} e^{-w}. \tag{5.33}$$

The joint space of $(\mathbf{Y}, W)$ is $\mathcal{Y} \times (0, \infty)$, together with the factorization of the density in (5.33), means that Lemma 3.4 can be used to show that $\mathbf{Y}$ and W are independent. In addition, $W \sim \text{Gamma}(\alpha_1 + \cdots + \alpha_K, 1)$, which can be seen either by looking at the w-part in (5.33), or by noting that it is the sum of independent gammas with the same rate parameter. Either way, we then have that the constant that goes with the w-part is $1/\Gamma(\alpha_1 + \cdots + \alpha_K)$, hence the pdf of $\mathbf{Y}$ must be

$$f_{\mathbf{Y}}(\mathbf{y}) = \frac{\Gamma(\alpha_1 + \cdots + \alpha_K)}{\Gamma(\alpha_1) \cdots \Gamma(\alpha_K)} y_1^{\alpha_1 - 1} \cdots y_{K-1}^{\alpha_{K-1} - 1} (1 - y_1 - \cdots - y_{K-1})^{\alpha_K - 1}. \tag{5.34}$$

If $K = 2$, then this is a beta pdf. Thus the Dirichlet is indeed an extension of the beta, and

$$\text{Beta}(\alpha, \beta) = \text{Dirichlet}(\alpha, \beta). \tag{5.35}$$

5.4   Affine transformations

Suppose $\mathbf{X}$ is $1 \times p$, and for given $1 \times p$ vector $\mathbf{a}$ and $p \times p$ matrix $\mathbf{B}$, let

$$\mathbf{Y} = g(\mathbf{X}) = \mathbf{a} + \mathbf{X}\mathbf{B}'. \tag{5.36}$$

In order for this transformation to be one-to-one, we need that $\mathbf{B}$ is invertible, which we will assume, so that

$$\mathbf{x} = g^{-1}(\mathbf{y}) = (\mathbf{y} - \mathbf{a})(\mathbf{B}')^{-1}. \tag{5.37}$$

For a matrix $\mathbf{C}$ and vector $\mathbf{z}$, with $\mathbf{w} = \mathbf{z}\mathbf{C}'$, it is not hard to see that

$$\frac{\partial w_i}{\partial z_j} = c_{ij}, \tag{5.38}$$

the ij^{th} element of $\mathbf{C}$. Thus from (5.37), since the "$\mathbf{a}$" part is a constant, the Jacobian is

$$J_{g^{-1}}(\mathbf{y}) = |\mathbf{B}^{-1}| = |\mathbf{B}|^{-1}. \tag{5.39}$$

The invertibility of $\mathbf{B}$ ensures that the absolute value of this Jacobian is between 0 and ∞.

5.4.1   Bivariate normal distribution

We apply this last result to the normal, where X_1 and X_2 are iid $N(0,1)$, $\mathbf{a}$ is a 1×2 vector, and $\mathbf{B}$ is a 2×2 invertible matrix. Then $\mathbf{Y}$ is given as in (5.36), with $p = 2$. The mean and covariance matrix for $\mathbf{X}$ are

$$E[\mathbf{X}] = \mathbf{0}_2 \quad \text{and} \quad Cov[\mathbf{X}] = \mathbf{I}_2, \tag{5.40}$$

so that from (2.47) and (2.53),

$$\boldsymbol{\mu} \equiv E[\mathbf{Y}] = \mathbf{a} \quad \text{and} \quad \boldsymbol{\Sigma} \equiv Cov[\mathbf{Y}] = \mathbf{B}Cov[\mathbf{X}]\mathbf{B}' = \mathbf{B}\mathbf{B}'. \tag{5.41}$$

The space of $\mathbf{X}$, hence of $\mathbf{Y}$, is $\mathbb{R}^2$. To find the pdf of $\mathbf{Y}$, we start with that of $\mathbf{X}$:

$$f_{\mathbf{X}}(\mathbf{x}) = \prod_{i=1}^{2} \frac{1}{\sqrt{2\pi}} e^{-\frac{1}{2}x_i^2} = \frac{1}{2\pi} e^{-\frac{1}{2}\mathbf{x}\mathbf{x}'}. \tag{5.42}$$

Then

$$f_{\mathbf{Y}}(\mathbf{y}) = f_{\mathbf{X}}((\mathbf{y} - \mathbf{a})(\mathbf{B}')^{-1}) \, \text{abs}(|\mathbf{B}|^{-1})$$
$$= \frac{1}{2\pi} e^{-\frac{1}{2} \, (\mathbf{y}-\mathbf{a})(\mathbf{B}')^{-1}((\mathbf{y}-\mathbf{a})(\mathbf{B}')^{-1})'} \, \text{abs}(|\mathbf{B}|)^{-1}. \tag{5.43}$$

Using (5.41), we can write

$$(\mathbf{y} - \mathbf{a})(\mathbf{B}')^{-1}((\mathbf{y} - \mathbf{a})(\mathbf{B}')^{-1})' = (\mathbf{y} - \boldsymbol{\mu})(\mathbf{B}\mathbf{B}')^{-1}(\mathbf{y} - \boldsymbol{\mu})' = (\mathbf{y} - \boldsymbol{\mu})\boldsymbol{\Sigma}^{-1}(\mathbf{y} - \boldsymbol{\mu})', \tag{5.44}$$

and using properties of determinants,

$$\text{abs}(|\mathbf{B}|) = \sqrt{|\mathbf{B}||\mathbf{B}'|} = \sqrt{|\mathbf{B}\mathbf{B}'|} = \sqrt{|\boldsymbol{\Sigma}|}, \tag{5.45}$$

hence the pdf of $\mathbf{Y}$ can be given as a function of the mean and covariance matrix:

$$f_{\mathbf{Y}}(\mathbf{y}) = \frac{1}{2\pi} \frac{1}{\sqrt{|\Sigma|}} e^{-\frac{1}{2}(\mathbf{y}-\boldsymbol{\mu})\Sigma^{-1}(\mathbf{y}-\boldsymbol{\mu})'}. \tag{5.46}$$

This $\mathbf{Y}$ is **bivariate normal**, with mean $\boldsymbol{\mu}$ and covariance matrix Σ, written very much in the same way as the regular normal,

$$\mathbf{Y} \sim N(\boldsymbol{\mu}, \Sigma). \tag{5.47}$$

In particular, the $\mathbf{X}$ here is

$$\mathbf{X} \sim N(\mathbf{0}_2, \mathbf{I}_2). \tag{5.48}$$

The mgf of a bivariate normal is not hard to find given that of the $\mathbf{X}$. Because X_1 and X_2 are iid $N(0,1)$, their mgf is

$$M_{\mathbf{X}}(\mathbf{t}) = e^{\frac{1}{2}t_1^2} e^{\frac{1}{2}t_2^2} = e^{\frac{1}{2}\mathbf{tt}'}, \tag{5.49}$$

where here $\mathbf{t}$ is 1×2. Then with $\mathbf{Y} = \mathbf{a} + \mathbf{XB}'$,

$$\begin{aligned}
M_{\mathbf{Y}}(\mathbf{t}) &= E[e^{\mathbf{Yt}'}] \\
&= E[e^{(\mathbf{a}+\mathbf{XB}')\mathbf{t}'}] \\
&= e^{\mathbf{at}'} E[e^{\mathbf{X}(\mathbf{tB})'}] \\
&= e^{\mathbf{at}'} M_{\mathbf{X}}(\mathbf{tB}) \\
&= e^{\mathbf{at}'} e^{\frac{1}{2}\mathbf{tB}(\mathbf{tB})'} \\
&= e^{\mathbf{at}'} e^{\frac{1}{2}\mathbf{tBB}'\mathbf{t}'} \\
&= e^{\boldsymbol{\mu}\mathbf{t}' + \frac{1}{2}\mathbf{t}\Sigma\mathbf{t}'}, \tag{5.50}
\end{aligned}$$

because $\mathbf{a} = \boldsymbol{\mu}$ and $\mathbf{BB}' = \Sigma$ by (5.41). Compare this mgf to that of the regular normal, in (4.36).

In Chapter 7 we will deal with the general p-dimensional multivariate normal, proceeding exactly the same way as above, except the matrices and vectors have more elements.

5.4.2   Orthogonal transformations and polar coordinates

A $p \times p$ **orthogonal matrix** is a matrix Γ such that

$$\Gamma'\Gamma = \Gamma\Gamma' = \mathbf{I}_p. \tag{5.51}$$

An orthogonal matrix has orthonormal columns (and orthonormal rows), that is, with $\Gamma = (\gamma_1, \ldots, \gamma_p)$,

$$\|\gamma_i\| = 1 \text{ and } \gamma_i'\gamma_j = 0 \text{ if } i \neq j. \tag{5.52}$$

An orthogonal transformation of $\mathbf{X}$ is then

$$\mathbf{Y} = \Gamma\mathbf{X} \tag{5.53}$$

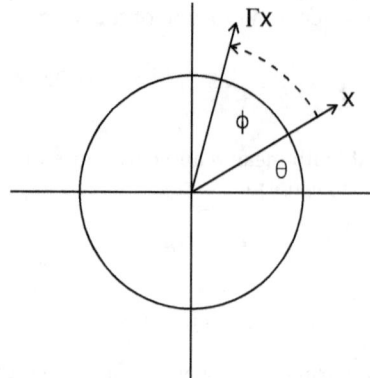

Figure 5.1: The vector $\mathbf{x} = 1.5(\cos(\theta), \sin(\theta))'$ is rotated ϕ radians by the orthogonal matrix $\boldsymbol{\Gamma}$.

for some orthogonal matrix $\boldsymbol{\Gamma}$. This transformation rotates $\mathbf{Y}$ about zero, that is, the length of $\mathbf{Y}$ is the same as that of $\mathbf{X}$, because

$$\|\boldsymbol{\Gamma}\mathbf{z}\|^2 = \mathbf{z}'\boldsymbol{\Gamma}'\boldsymbol{\Gamma}\mathbf{z} = \mathbf{z}'\mathbf{z} = \|\mathbf{z}\|^2, \tag{5.54}$$

but its orientation is different. The Jacobian is ± 1:

$$|\boldsymbol{\Gamma}'\boldsymbol{\Gamma}| = |\mathbf{I}_p| = 1 \implies |\boldsymbol{\Gamma}|^2 = 1. \tag{5.55}$$

When $p = 2$, the orthogonal matrices can be parametrized by the angle. More specifically, the set of all 2×2 orthogonal matrices equals

$$\left\{ \begin{pmatrix} \cos(\phi) & \sin(\phi) \\ \sin(\phi) & -\cos(\phi) \end{pmatrix} \mid \phi \in [0, 2\pi) \right\} \cup \left\{ \begin{pmatrix} \cos(\phi) & -\sin(\phi) \\ \sin(\phi) & \cos(\phi) \end{pmatrix} \mid \phi \in [0, 2\pi) \right\}. \tag{5.56}$$

Note that the first set of matrices in (5.56) has determinant -1, and the second set has determinant 1.

To see the effect on the 2×1 vector $\mathbf{x} = (x_1, x_2)'$, first find the polar coordinates for x_1 and x_2:

$$x_1 = r\cos(\theta) \text{ and } x_2 = r\sin(\theta), \text{ where } r \geq 0 \text{ and } \theta \in [0, 2\pi). \tag{5.57}$$

Then taking an orthogonal matrix $\boldsymbol{\Gamma}$ from the second set in (5.56), which are called rotations, we have

$$\boldsymbol{\Gamma}\mathbf{x} = \begin{pmatrix} \cos(\phi) & -\sin(\phi) \\ \sin(\phi) & \cos(\phi) \end{pmatrix} r \begin{pmatrix} \cos(\theta) \\ \sin(\theta) \end{pmatrix} = r \begin{pmatrix} \cos(\theta + \phi) \\ \sin(\theta + \phi) \end{pmatrix}. \tag{5.58}$$

Thus if $\mathbf{x}$ is at an angle of θ to the x-axis, then $\boldsymbol{\Gamma}$ rotates the vector by the angle ϕ, keeping the length of the vector constant. Figure 5.1 illustrates the rotation of $\mathbf{x}$ with polar coordinates $r = 1.5$ and $\theta = \pi/6$. The orthogonal matrix in (5.58) uses $\phi = \pi/4$. As ϕ goes from 0 to 2π, the vector $\boldsymbol{\Gamma}\mathbf{x}$ makes a complete rotation about the origin.

The first set of matrices in (5.56) are called reflections. They first flip the sign of x_2, then rotate by the angle ϕ.

5.4.3 Spherically symmetric pdfs

A $p \times 1$ random vector $\mathbf{X}$ has a **spherically symmetric distribution** if for any orthogonal $\mathbf{\Gamma}$,

$$\mathbf{X} =^{\mathcal{D}} \mathbf{\Gamma X}, \tag{5.59}$$

meaning $\mathbf{X}$ and $\mathbf{\Gamma X}$ have the same distribution. Suppose $\mathbf{X}$ is spherically symmetric and has a pdf $f_{\mathbf{X}}(\mathbf{x})$. This density can be taken to be spherically symmetric as well, that is,

$$f_{\mathbf{X}}(\mathbf{x}) = f_{\mathbf{X}}(\mathbf{\Gamma x}) \text{ for all } \mathbf{x} \in \mathcal{X} \text{ and orthogonal } \mathbf{\Gamma}. \tag{5.60}$$

We will look more at the $p = 2$ case. Exercise 5.6.10 shows that (5.60) implies that there is a function $h(r)$ for $r \geq 0$ such that

$$f_{\mathbf{X}}(\mathbf{x}) = h(\|\mathbf{x}\|). \tag{5.61}$$

For example, suppose X_1 and X_2 are independent $N(0,1)$'s. Then from Table 1.1, the pdf of $\mathbf{X} = (X_1, X_2)'$ is

$$f_{\mathbf{X}}(\mathbf{x}) = \frac{1}{\sqrt{2\pi}} e^{-\frac{1}{2}x_1^2} \times \frac{1}{\sqrt{2\pi}} e^{-\frac{1}{2}x_2^2} = \frac{1}{2\pi} e^{-\frac{1}{2}(x_1^2 + x_2^2)}, \tag{5.62}$$

hence we can take

$$h(r) = \frac{1}{2\pi} e^{-\frac{1}{2}r^2} \tag{5.63}$$

in (5.61).

Consider the distribution of the polar coordinates $(R, \Theta) = g(X_1, X_2)$ where

$$g(x_1, x_2) = (\|\mathbf{x}\|, \text{Angle}(x_1, x_2)). \tag{5.64}$$

The Angle(x_1, x_2) is taken to be in $[0, 2\pi)$, and is basically the $\arctan(x_2/x_1)$, except that that is not uniquely defined, e.g., if $x_1 = x_2 = 1$, $\arctan(1/1) = \pi/4$ or $5\pi/4$. What we really mean is the unique value θ in $[0, 2\pi)$ for which (5.57) holds. A glitch is that θ is not uniquely defined when $(x_1, x_2) = (0,0)$, since then $r = 0$ and any θ would work. So we assume that $\mathcal{X}$ does not contain $(0,0)$, hence $r \in (0, \infty)$. This requirement does not hurt anything because we are dealing with continuous random variables.

The g^{-1} is already given in (5.57), hence the Jacobian is

$$J_{g^{-1}}(r, \theta) = \begin{vmatrix} \frac{\partial}{\partial r} r \cos(\theta) & \frac{\partial}{\partial \theta} r \cos(\theta) \\ \frac{\partial}{\partial r} r \sin(\theta) & \frac{\partial}{\partial \theta} r \sin(\theta) \end{vmatrix}$$

$$= \begin{vmatrix} \cos(\theta) & -r \sin(\theta) \\ \sin(\theta) & r \cos(\theta) \end{vmatrix}$$

$$= r(\cos(\theta)^2 + \sin(\theta)^2)$$

$$= r. \tag{5.65}$$

You may recall this result from calculus, i.e.,

$$dx_1 dx_2 = r \, dr \, d\theta. \tag{5.66}$$

Then by (5.61), the pdf of (R, Θ) is

$$f_{(R,\Theta)}(r, \theta) = h(r) r. \tag{5.67}$$

Exercise 5.6.11 shows that the space is a rectangle, $(0, \infty) \times [0, 2\pi)$. This pdf can be written as a product of a function of just r, i.e., $h(r)r$, and a function of just θ, i.e., the function "1". Thus R and Θ must be independent, by Lemma 3.4. Since the pdf of Θ is constant, it must be uniform, i.e.,

$$\theta \sim \text{Uniform}[0, 2\pi), \tag{5.68}$$

which has pdf $1/(2\pi)$. Then from (5.67),

$$f_R(r) = h(r)r = [2\pi h(r)r] \left[\frac{1}{2\pi}\right]. \tag{5.69}$$

Applying this formula to the normal example in (5.63), we have that R has pdf

$$f_R(r) = 2\pi h(r)r = r\, e^{-\frac{1}{2}r^2}. \tag{5.70}$$

5.4.4   Box-Muller transformation

The **Box-Muller transformation** (Box and Muller, 1958) is an approach to generating two random normals from two random uniforms that reverses the above polar coordinate procedure. That is, suppose U_1 and U_2 are independent Uniform(0,1). Then we can generate Θ by setting

$$\Theta = 2\pi U_1, \tag{5.71}$$

and R by setting

$$R = F_R^{-1}(U_2), \tag{5.72}$$

where F_R is the distribution function for the pdf in (5.70):

$$F_R(r) = \int_0^r w\, e^{-\frac{1}{2}w^2}\, dw = -e^{\frac{1}{2}w^2}\Big|_{w=0}^r = 1 - e^{-\frac{1}{2}r^2}. \tag{5.73}$$

Inverting $u_2 = F_R(r)$ yields

$$r = F_R^{-1}(u_2) = \sqrt{-2\log(1 - u_2)}. \tag{5.74}$$

Thus, as in (5.57), we set

$$X_1 = \sqrt{-2\log(1 - U_2)}\ \cos(2\pi U_1) \quad \text{and} \quad X_2 = \sqrt{-2\log(1 - U_2)}\ \sin(2\pi U_1), \tag{5.75}$$

which are then independent $N(0,1)$'s. (Usually one sees U_2 in place of the $1 - U_2$ in the logs, but either way is fine because both are Uniform(0,1).)

5.5 Order statistics

The order statistics for a sample $\{x_1, \ldots, x_n\}$ are the observations placed in order from smallest to largest. They are usually designated with indices "(i)," so that the order statistics are

$$x_{(1)}, x_{(2)}, \ldots, x_{(n)}, \tag{5.76}$$

where

$$x_{(1)} = \text{smallest of } \{x_1, \ldots, x_n\} = \min\{x_1, \ldots, x_n\};$$
$$x_{(2)} = \text{second smallest of} \{x_1, \ldots, x_n\};$$

$$\vdots$$

$$x_{(n)} = \text{largest of } \{x_1, \ldots, x_n\} = \max\{x_1, \ldots, x_n\}. \tag{5.77}$$

For example, if the sample is $\{3.4, 2.5, 1.7, 5.2\}$ then the order statistics are $1.7, 2.5, 3.4, 5.2$. If two observations have the same value, then that value appears twice, i.e., the order statistics for $\{3.4, 1.7, 1.7, 5.2\}$ are $1.7, 1.7, 3.4, 5.2$.

These statistics are useful as descriptive statistics, and in nonparametric inference. For example, estimates of the median of a distribution are often based on order statistics, such as the median, or the trimean, which is a linear combination of the two quartiles and the median.

We will deal with $X_1, \ldots, X_n$ being iid, each with distribution function F, pdf f, and space $\mathcal{X}$, so that the space of $\mathbf{X} = (X_1, \ldots, X_n)$ is $\mathcal{X}^n$. Then let

$$\mathbf{Y} = (X_{(1)}, \ldots, X_{(n)}). \tag{5.78}$$

We will assume that the X_i's are distinct, that is, no two have the same value. This assumption is fine in the continuous case, because the probability that two observations are equal is zero. In the discrete case, there may indeed be ties, and the analysis becomes more difficult. The space of $\mathbf{Y}$ is

$$\mathcal{Y} = \{(y_1, \ldots, y_n) \in \mathcal{X}^n \mid y_1 < y_2 < \cdots < y_n\}. \tag{5.79}$$

To find the pdf, start with $\mathbf{y} \in \mathcal{Y}$, and let $\delta > 0$ be small enough so that the intervals $(y_1, y_1 + \delta), (y_2, y_2 + \delta), \ldots, (y_n, y_n + \delta)$ are disjoint. (So take δ less than the all the gaps $y_{i+1} - y_i$.) Then

$$P[y_1 < Y_1 < y_1 + \delta, \ldots, y_n < Y_n < y_n + \delta]$$
$$= P[y_1 < X_{(1)} < y_1 + \delta, \ldots, y_n < X_{(n)} < y_n + \delta]. \tag{5.80}$$

Now the event in the latter probability occurs when any permutation of the X_i's has one component in the first interval, one in the second, etc. E.g., if $n = 3$,

$$P[y_1 < X_{(1)} < y_1 + \delta, y_2 < X_{(2)} < y_2 + \delta, y_3 < X_{(3)} < y_3 + \delta]$$
$$= P[y_1 < X_1 < y_1 + \delta, y_2 < X_2 < y_2 + \delta, y_3 < X_3 < y_3 + \delta]$$
$$+ P[y_1 < X_1 < y_1 + \delta, y_2 < X_3 < y_2 + \delta, y_3 < X_2 < y_3 + \delta]$$
$$+ P[y_1 < X_2 < y_1 + \delta, y_2 < X_1 < y_2 + \delta, y_3 < X_3 < y_3 + \delta]$$
$$+ P[y_1 < X_2 < y_1 + \delta, y_2 < X_3 < y_2 + \delta, y_3 < X_1 < y_3 + \delta]$$
$$+ P[y_1 < X_3 < y_1 + \delta, y_2 < X_1 < y_2 + \delta, y_3 < X_2 < y_3 + \delta]$$
$$+ P[y_1 < X_3 < y_1 + \delta, y_2 < X_2 < y_2 + \delta, y_3 < X_1 < y_3 + \delta]$$
$$= 6 \, P[y_1 < X_1 < y_1 + \delta, y_2 < X_2 < y_2 + \delta, y_3 < X_3 < y_3 + \delta]. \tag{5.81}$$

The last equation follows because the X_i's are iid, hence the six individual probabilities are equal. In general, the number of permutations is $n!$. Thus, we can write

$$P[y_1 < Y_1 < y_1 + \delta, \dots, y_n < Y_n < y_n + \delta] = n! P[y_1 < X_1 < y_1 + \delta,$$
$$\dots, y_n < X_n < y_n + \delta]$$
$$= n! \prod_{i=1}^{n} [F(y_i + \delta) - F(y_i)]. \qquad (5.82)$$

Dividing by δ^n then letting $\delta \to 0$ yields the joint density, which is

$$f_{\mathbf{Y}}(\mathbf{y}) = n! \prod_{i=1}^{n} f(y_i), \quad \mathbf{y} \in \mathcal{Y}. \qquad (5.83)$$

Marginal distributions of individual order statistics, or sets of them, can be obtained by integrating out the ones that are not desired. The process can be a bit tricky, and one must be careful with the spaces. Instead, we will present a representation that leads to the marginals as well as other quantities.

We start with the $U_1, \dots, U_n$ being iid Uniform$(0,1)$, so that the pdf (5.83) of the order statistics $\mathbf{Y} = (U_{(1)}, \dots, U_{(n)})$ is simply $n!$. Consider the first order statistic together with the gaps between consecutive order statistics:

$$G_1 = U_{(1)}, \quad G_2 = U_{(2)} - U_{(1)}, \dots, \quad G_n = U_{(n)} - U_{(n-1)}. \qquad (5.84)$$

These G_i's are all positive, and they sum to $U_{(n)}$, which has range $(0,1)$. Thus the space of $\mathbf{G} = (G_1, \dots, G_n)$ is

$$\mathcal{G} = \{\mathbf{g} \in \mathbb{R}^n \mid 0 < g_i < 1, \ i = 1, \dots, n \ \& \ g_1 + \cdots + g_n < 1\}. \qquad (5.85)$$

The inverse function to (5.84) is

$$u_{(1)} = g_1, \ u_{(2)} = g_1 + g_2, \dots, u_{(n)} = g_1 + \cdots + g_n. \qquad (5.86)$$

Note that this is a linear function of $\mathbf{G}$:

$$\mathbf{Y} = \mathbf{G}\mathbf{A}', \quad \mathbf{A}' = \begin{pmatrix} 1 & 1 & 1 & \cdots & 1 \\ 0 & 1 & 1 & \cdots & 1 \\ 0 & 0 & 1 & \cdots & 1 \\ \vdots & \vdots & \vdots & \ddots & \vdots \\ 0 & 0 & 0 & \cdots & 1 \end{pmatrix}, \qquad (5.87)$$

and $|\mathbf{A}| = 1$. Thus the Jacobian is 1, and the pdf of $\mathbf{G}$ is also $n!$:

$$f_{\mathbf{G}}(\mathbf{g}) = n!, \quad \mathbf{g} \in \mathcal{G}. \qquad (5.88)$$

This pdf is quite simple on its own, but note that it is a special case of the Dirichlet in (5.34) with $K = n + 1$ and all $\alpha_k = 1$. Thus any order statistic is a beta, because it is the sum of the first few gaps, i.e.,

$$U_{(k)} = g_1 + \cdots + g_k \sim \text{Beta}(k, n - k + 1), \qquad (5.89)$$

analogous to (5.29). In particular, if n is odd, then the median of the observations is $U_{(n+1)/2}$, hence

$$\text{Median}\{U_1, \ldots, U_n\} \sim \text{Beta}\left(\frac{n+1}{2}, \frac{n+1}{2}\right), \qquad (5.90)$$

and using Table 1.1 to find the mean and variance of a beta,

$$E[\text{Median}\{U_1, \ldots, U_n\}] = \frac{1}{2} \qquad (5.91)$$

and

$$\text{Var}[\text{Median}\{U_1, \ldots, U_n\}] = \frac{((n+1)/2)(n+1)/2}{(n+1)^2(n+2)} = \frac{1}{4(n+2)}. \qquad (5.92)$$

To obtain the pdf of the order statistics of non-uniforms, we can use the probability transform approach as in Section 4.2.3. That is, suppose the X_i's are iid with (strictly increasing) distribution function F and pdf f. We then have that $\mathbf{X}$ has the same distribution as $(F^{-1}(U_1), \ldots, F^{-1}(U_n))$, where the U_i's are iid Uniform$(0, 1)$. Because F, hence F^{-1}, is increasing, the order statistics for the X_i's match those of the U_i's, that is,

$$(X_{(1)}, \ldots, X_{(n)}) =^{\mathcal{D}} (F^{-1}(U_{(1)}), \ldots, F^{-1}(U_{(n)})). \qquad (5.93)$$

Thus for any particular k, $X_{(k)} =^{\mathcal{D}} F^{-1}(U_{(k)})$. We know that $U_{(k)} \sim \text{Beta}(k, n-k+1)$, hence can find the distribution of $X_{(k)}$ using the transformation with $h(u) = F^{-1}$. Thus $h^{-1} = F$, i.e.,

$$U_{(k)} = h^{-1}(X_{(k)}) = F(X_{(k)}) \Rightarrow J_{h^{-1}}(x) = F'(x) = f(x). \qquad (5.94)$$

The pdf of $X_{(k)}$ is then

$$f_{X_{(k)}}(x) = f_{U_{(k)}}(F(x))f(x) = \frac{\Gamma(n+1)}{\Gamma(k)\Gamma(n-k+1)} F(x)^{k-1}(1 - F(x))^{n-k} f(x). \qquad (5.95)$$

For most F and k, the pdf in (5.95) is not particularly easy to deal with analytically. In Section 9.2, we introduce the Δ-method, which can be used to approximate the distribution of order statistics for large n.

5.6   Exercises

Exercise 5.6.1. Show that if $X \sim \text{Gamma}(\alpha, \lambda)$, then $cX \sim \text{Gamma}(\alpha, \lambda/c)$.

Exercise 5.6.2. Suppose $X \sim \text{Beta}(\alpha, \beta)$, and let $Y = X/(1-X)$ (so $g(x) = x/(1-x)$.) (a) What is the space of Y? (b) Find $g^{-1}(y)$. (c) Find the Jacobian of $g^{-1}(y)$. (d) Find the pdf of Y.

Exercise 5.6.3. Suppose X_1, X_2, X_3, X_4 are random variables, and define the function $(Y_1, Y_2, Y_3, W) = g(X_1, X_2, X_3, X_4)$ by

$$Y_1 = \frac{X_1}{X_1 + X_2 + X_3 + X_4}, \quad Y_2 = \frac{X_2}{X_1 + X_2 + X_3 + X_4}, \quad Y_3 = \frac{X_3}{X_1 + X_2 + X_3 + X_4}, \qquad (5.96)$$

and $W = X_1 + X_2 + X_3 + X_4$, as in (5.26) and (5.31) with $K = 4$. (a) Find the inverse function $g^{-1}(y_1, y_2, y_3, w)$. (b) Find the Jacobian of $g^{-1}(y_1, y_2, y_3, w)$. (c) Show that the determinant of the Jacobian of $g^{-1}(y_1, y_2, y_3, w)$ is w^3. [Hint: The determinant of a matrix does not change if one of the rows is added to another row. In the matrix of derivatives, adding each of the first three rows to the last row can simplify the determinant calculation.]

Exercise 5.6.4. Suppose $\mathbf{X} = (X_1, X_2, \ldots, X_K)$, where the X_k's are independent and $X_k \sim \text{Gamma}(\alpha_k, 1)$, as in (5.25), for $K > 1$. Define $\mathbf{Y} = (Y_1, \ldots, Y_{K-1})$ by $Y_k = X_k / (X_1 + \cdots + X_K)$ for $k = 1, \ldots, K-1$, and $W = X_1 + \cdots + X_K$, so that $(\mathbf{Y}, W) = g(\mathbf{X})$ as in (5.26) and (5.31). Thus $\mathbf{Y} \sim \text{Dirichlet}(\alpha_1, \ldots, \alpha_K)$. (a) Write down the pdf of $\mathbf{X}$. (b) Show that the joint space of $(\mathbf{Y}, W)$ is $\mathcal{Y} \times (0, \infty)$, where $\mathcal{Y}$ is given in (5.30). (You can take as given that the space of $\mathbf{Y}$ is $\mathcal{Y}$ and the space of W is $(0, \infty)$.) (c) Show that the pdf of $(Y_1, \ldots, Y_{K-1}, W)$ is given in (5.33).

Exercise 5.6.5. Suppose U_1 and U_2 are independent Uniform(0,1)'s, and let $(Y_1, Y_2) = g(U_1, U_2)$ be defined by $Y_1 = U_1 + U_2, Y_2 = U_1 - U_2$. (a) Find $g^{-1}(y_1, y_2)$ and the absolute value of the Jacobian, $|J_{g^{-1}}(y_1, y_2)|$. (b) What is the pdf of (Y_1, Y_2)? (c) Sketch the joint space of (Y_1, Y_2). Is it a rectangle? (d) Find the marginal spaces of Y_1 and Y_2. (e) Find the conditional space of Y_2 given $Y_1 = y_1$ for y_1 in the marginal space. [Hint: Do it separately for $y_1 < 1$ and $y_1 \geq 1$.] (f) What is the marginal pdf of Y_1? (We found this tent distribution in Exercise 1.7.6 using distribution functions.)

Exercise 5.6.6. Suppose $Z \sim N(0,1)$ and $U \sim \text{Uniform}(0,1)$, and Z and U are independent. Let $(X, Y) = g(Z, U)$, given by

$$X = \frac{Z}{U} \text{ and } Y = U. \qquad (5.97)$$

The X has the **slash** distribution. (a) Is the space of (X, Y) a rectangle? (b) What is the space of X? (c) What is the space of Y? (d) What is the expected value of X? (e) Find the inverse function $g^{-1}(x, y)$. (f) Find the Jacobian of g^{-1}. (g) Find the pdf of (X, Y). (h) What is the conditional space of Y given $X = x$, $\mathcal{Y}_x$? (i) Find the pdf of X.

Exercise 5.6.7. Suppose $\mathbf{X} = (X_1, X_2)$ is bivariate normal with mean $\boldsymbol{\mu}$ and covariance matrix Σ, where

$$\boldsymbol{\mu} = (\mu_1, \mu_2) \text{ and } \Sigma = \begin{pmatrix} \sigma_{11} & \sigma_{12} \\ \sigma_{21} & \sigma_{22} \end{pmatrix}. \qquad (5.98)$$

Assume that Σ is invertible. (a) Write down the pdf. Find the a, b, c for which

$$(\mathbf{x} - \boldsymbol{\mu}) \Sigma^{-1} (\mathbf{x} - \boldsymbol{\mu})' = a(x_1 - \mu_1)^2 + b(x_1 - \mu_1)(x_2 - \mu_2) + c(x_2 - \mu_2)^2. \qquad (5.99)$$

(b) If $\sigma_{12} \neq 0$, are X_1 and X_2 independent? (c) Suppose $\sigma_{12} = 0$ (so $\sigma_{21} = 0$, too). Show explicitly that the pdf factors into a function of x_1 times a function of x_2. Are X_1 and X_2 independent? (d) Still with $\sigma_{12} = 0$, what are the marginal distributions of X_1 and X_2?

Exercise 5.6.8. Suppose that $(Y_1, \ldots, Y_{K-1}) \sim \text{Dirichlet}(\alpha_1, \ldots, \alpha_K)$, where $K \geq 5$. What is the distribution of (W_1, W_2), where

$$W_1 = \frac{Y_1}{Y_1 + Y_2 + Y_3 + Y_4}, \text{ and } W_2 = \frac{Y_2 + Y_3}{Y_1 + Y_2 + Y_3 + Y_4}? \qquad (5.100)$$

Justify your answer. [Hint: It is easier to work directly with the gammas defining the Y_i's, rather than using pdfs.]

Exercise 5.6.9. Let $\mathbf{G} \sim \text{Dirichlet}(\alpha_1, \ldots, \alpha_K)$ (so that $\mathbf{G} = (G_1, \ldots, G_{K-1})$). Set $\alpha_+ = \alpha_1 + \cdots + \alpha_K$. (a) Show that $E[G_k] = \alpha_k / \alpha_+$ and $Var[G_k] = \alpha_k(\alpha_+ - \alpha_k)/(\alpha_+^2(\alpha_+ + 1))$. [Hint: Use the beta representation in (5.28), and get the mean and variance from Table 1.1 on page 7.] (b) Find $E[G_k G_l]$ for $k \neq l$. (c) Show that $Cov[G_k, G_l] = -\alpha_k \alpha_l / (\alpha_+^2(\alpha_+ + 1))$ for $k \neq l$. (d) Show that

$$Cov[\mathbf{G}] = \frac{1}{\alpha_+(\alpha_+ + 1)} \left(\mathbf{D}(\boldsymbol{\alpha}) - \frac{1}{\alpha_+} \boldsymbol{\alpha}' \boldsymbol{\alpha} \right), \tag{5.101}$$

where $\boldsymbol{\alpha} = (\alpha_1, \ldots, \alpha_{K-1})$ and $\mathbf{D}(\boldsymbol{\alpha})$ is the $(K-1) \times (K-1)$ diagonal matrix with the α_i's on the diagonal:

$$\mathbf{D}(\boldsymbol{\alpha}) = \begin{pmatrix} \alpha_1 & 0 & \cdots & 0 \\ 0 & \alpha_2 & \cdots & 0 \\ \vdots & \vdots & \ddots & \vdots \\ 0 & 0 & \cdots & \alpha_{K-1} \end{pmatrix}. \tag{5.102}$$

Exercise 5.6.10. Suppose $\mathbf{x}$ is 2×1 and $f_{\mathbf{X}}(\mathbf{x}) = f_{\mathbf{X}}(\Gamma \mathbf{x})$ for any $\mathbf{x} \in \mathbb{R}^2$ and orthogonal 2×2 matrix Γ. (a) Write $\mathbf{x}$ in terms of its polar coordinates as in (5.57), i.e., $x_1 = \|x\| \cos(\theta)$ and $x_2 = \|x\| \sin(\theta)$. Define Γ using an angle ϕ as in (5.58). For what ϕ is $\Gamma \mathbf{x} = (\|x\|, 0)'$? (b) Show that there exists a function $h(r)$ for $r \geq 0$ such that $f_{\mathbf{X}}(\mathbf{x}) = h(\|x\|)$. [Hint: Let $h(r) = f_{\mathbf{X}}((r, 0)')$.]

Exercise 5.6.11. Suppose $\mathbf{x} \in \mathbb{R}^2 - \{\mathbf{0}_2\}$ (the two-dimensional plane without the origin), and let $(r, \theta) = g(\mathbf{x})$ be the polar coordinate transformation, as in (5.64). Show that g defines a one-to-one correspondence between $\mathbb{R}^2 - \{\mathbf{0}_2\}$ and the rectangle $(0, \infty) \times [0, 2\pi)$.

Exercise 5.6.12. Suppose $Y_1, \ldots, Y_n$ are independent Exponential(1)'s, and $Y_{(1)}, \ldots, Y_{(n)}$ are their order statistics. (a) Write down the space and pdf of the single order statistic $Y_{(k)}$. (b) What is the pdf of $Y_{(1)}$? What is the name of the distribution?

Exercise 5.6.13. The Gumbel(μ) distribution has space $(-\infty, \infty)$ and distribution function

$$F_\mu(x) = e^{-e^{-(x-\mu)}}. \tag{5.103}$$

(a) Is this a legitimate distribution function? Why or why not? (b) Suppose $X_1, \ldots, X_n$ are independent and identically distributed Gumbel(μ) random variables, and $Y = X_{(n)}$, their maximum. What is the distribution function of Y? (Use the method in Exercise 4.4.18.) (c) This Y is also distributed as a Gumbel. What is the value of the parameter?

The remaining exercises are based on $U_1, \ldots, U_n$ independent Uniform(0,1)'s, where $\mathbf{Y} = (U_{(1)}, \ldots, U_{(n)})$ is the vector of their order statistics. Following (5.86), the order statistics are be represented by $U_{(i)} = G_1 + \cdots + G_i$, where $\mathbf{G} = (G_1, \ldots, G_n) \sim \text{Dirichlet}(\alpha_1, \alpha_2, \ldots, \alpha_{n+1})$, and all the α_i's equal 1.

Exercise 5.6.14. This exercise finds the covariance matrix of $\mathbf{Y}$. (a) Show that

$$Cov[\mathbf{G}] = \frac{1}{(n+1)(n+2)} \left(\mathbf{I}_n - \frac{1}{n+1} \mathbf{1}_n \mathbf{1}_n' \right), \tag{5.104}$$

where $\mathbf{I}_n$ is the $n \times n$ identity matrix, and $\mathbf{1}_n$ is the $n \times 1$ vector of 1's. [Hint: This covariance is a special case of that in Exercise 5.6.9(d).] (b) Show that

$$Cov[\mathbf{Y}] = \frac{1}{(n+1)(n+2)} \left(\mathbf{B} - \frac{1}{n+1}\mathbf{cc}' \right), \tag{5.105}$$

where $\mathbf{B}$ is the $n \times n$ matrix with ij^{th} element $b_{ij} = \min\{i, j\}$, and $\mathbf{c} = (1, 2, \ldots, n)'$. [Hint: Use $\mathbf{Y} = \mathbf{GA}'$ for the $\mathbf{A}$ in (5.87), and show that $\mathbf{B} = \mathbf{AA}'$ and $\mathbf{c} = \mathbf{A1}_n$.] (c) From (5.105), obtain

$$Var[U_{(i)}] = \frac{i(n+1-i)}{(n+1)^2(n+2)} \text{ and } Cov[U_{(i)}, U_{(j)}] = \frac{i(n+1-j)}{(n+1)^2(n+2)} \text{ if } i < j. \tag{5.106}$$

Exercise 5.6.15. Suppose n is odd, so that the sample median is well-defined, and consider the three statistics: the sample mean $\overline{U}$, the sample median U_{med}, and the sample **midrange** U_{mr}, defined to be the midpoint of the minimum and maximum, $U_{mr} = (U_{(1)} + U_{(n)})/2$. The variance of U_{med} is given in (5.92). (a) Find $Var[U_i]$, and $Var[\overline{U}]$. (b) Find $Var[U_{(1)}]$, $Var[U_{(n)}]$, and $Cov[U_{(1)}, U_{(n)}]$. [Use (5.106).] What is $Var[U_{mr}]$? (c) When $n = 1$, the three statistics above are all the same. For $n > 1$ (but still odd), which has the lowest variance, and which has the highest variance? (d) As $n \to \infty$, what is the limit of $Var[U_{med}]/Var[\overline{U}]$? (e) As $n \to \infty$, what is the limit of $Var[U_{mr}]/Var[\overline{U}]$?

Exercise 5.6.16. This exercise finds the joint pdf of $(U_{(1)}, U_{(n)})$, the minimum and maximum of the U_i's. Let $V_1 = G_1$ and $V_2 = G_2 + \cdots + G_n$, so that $(U_{(1)}, U_{(n)}) = g(V_1, V_2) = (V_1, V_1 + V_2)$. (a) What is the space of $(U_{(1)}, U_{(n)})$? (b) The distribution of (V_1, V_2) is Dirichlet. What are the parameters? Write down the pdf of (V_1, V_2). (c) Find the inverse function g^{-1} and the absolute value of the determinant of the Jacobian of g^{-1}. (d) Find the pdf of $(U_{(1)}, U_{(n)})$.

Exercise 5.6.17. Suppose $X \sim \text{Binomial}(n, p)$. This exercise relates the distribution function of the binomial to the beta distribution. (a) Show that for $p \in (0, 1)$, $P[U_{(k+1)} > p] = P[X \le k]$. [Hint: The $(k+1)^{st}$ order statistic is greater than p if and only if how many of the U_i's are less than p? What is that probability?] (b) Conclude that if F is the distribution function of X, then $F(k) = P[\text{Beta}(k+1, n-k) > p]$. This formula is used in the R function pbinom.

Chapter 6

Conditional Distributions

6.1 Introduction

A two-stage process is described in Section 1.6.3, which will appear again in Section 6.4.1, where one first randomly chooses a coin from a population of coins, then flips it independently $n = 10$ times. There are two random variables in this experiment: X, the probability of heads for the chosen coin, and Y, the total number of heads among the n flips. It is given that X is equally likely to be any number between 0 and 1, i.e.,

$$X \sim \text{Uniform}(0,1). \tag{6.1}$$

Also, once the coin is chosen, Y is binomial. If the chosen coin has $X = x$, then we say that *the conditional distribution of Y given X = x* is Binomial(n, x), written

$$Y \mid X = x \sim \text{Binomial}(n, x). \tag{6.2}$$

Together the equations (6.1) and (6.2) describe the distribution of (X, Y).

A couple of other distributions may be of interest. First, what is the marginal, sometimes referred to in this context as *unconditional*, distribution of Y? It is *not* binomial. It is the distribution arising from the entire two-stage procedure, not that arising given a particular coin. The space is $\mathcal{Y} = \{0, 1, \ldots, n\}$, as for the binomial, but the pmf is different:

$$f_Y(y) = P[Y = y] \neq P[Y = y \mid X = x]. \tag{6.3}$$

(That last expression is pronounced *the probability that Y = y given X = x*.)

Also, one might wish to interchange the roles of X and Y, and ask for the conditional distribution of X given $Y = y$ for some y. This distribution is of particular interest in Bayesian inference, as follows. One chooses a coin as before, and then wishes to know its x. It is flipped ten times, and the number of heads observed, y, is used to guess what the x is. More precisely, one then finds the conditional distribution of X given $Y = y$:

$$X \mid Y = y \sim \text{??} \tag{6.4}$$

In Bayesian parlance, the marginal distribution of X in (6.1) is the **prior** distribution of X, because it is your best guess before seeing the data, and the conditional distribution in (6.4) is the **posterior** distribution, determined after you have seen the data.

These ideas extend to random vectors $(\mathbf{X}, \mathbf{Y})$. There are five distributions we consider, three of which we have seen before:

- **Joint.** The joint distribution of $(\mathbf{X}, \mathbf{Y})$ is the distribution of $\mathbf{X}$ and $\mathbf{Y}$ taken together.

- **Marginal.** The two marginal distributions: that of $\mathbf{X}$ alone and that of $\mathbf{Y}$ alone.

- **Conditional.** The two conditional distributions: that of $\mathbf{Y}$ given $\mathbf{X} = \mathbf{x}$, and that of $\mathbf{X}$ given $\mathbf{Y} = \mathbf{y}$.

The next section shows how to find the joint distribution from a conditional and marginal. Further sections look at finding the marginals and reverse conditional, the latter using Bayes theorem. We end with independence, $\mathbf{Y}$ being independent of $\mathbf{X}$ if the conditional distribution of $\mathbf{Y}$ given $\mathbf{X} = \mathbf{x}$ does not depend on $\mathbf{x}$.

6.2   Examples of conditional distributions

When considering the conditional distribution of $\mathbf{Y}$ given $\mathbf{X} = \mathbf{x}$, it may or may not be that the randomness of $\mathbf{X}$ is of interest, depending on the situation. In addition, there is no need for $\mathbf{Y}$ and $\mathbf{X}$ to be of the same type, e.g., in the coin example, X is continuous and Y is discrete. Next we look at some additional examples.

6.2.1   Simple linear regression

The relationship of one variable to another is central to many statistical investigations. The simplest is a linear relationship,

$$Y = \alpha + \beta X + E, \tag{6.5}$$

Here, α and β are fixed, Y is the "dependent" variable, and X is the "explanatory" or "independent" variable. The E is error, needed because one does not expect the variables to be exactly linearly related. Examples include X = Height and Y = Weight, or X = Dosage of a drug and Y some measure of health (cholesterol level, e.g.). The X could be a continuous variable, or an indicator function, e.g., be 0 or 1 according to the sex of the subject.

The normal linear regression model specifies that

$$Y \mid X = x \sim N(\alpha + \beta x, \sigma_e^2). \tag{6.6}$$

In particular,

$$E[Y \mid X = x] = \alpha + \beta x \quad \text{and} \quad Var[Y \mid X = x] = \sigma_e^2. \tag{6.7}$$

(Other models take (6.7) but do not assume normality, or allow $Var[Y \mid X = x]$ to depend on x.) It may be that X is fixed by the experimenter, for example, the dosage x might be preset; or it may be that the X is truly random, e.g., the height of a randomly chosen person would be random. Often, this randomness of X is ignored, and analysis proceeds conditional on $X = x$. Other times, the randomness of X is also incorporated into the analysis, e.g., one might have the marginal

$$X \sim N(\mu_X, \sigma_X^2). \tag{6.8}$$

Chapter 12 goes into much more detail on linear regression models.

6.2.2 Mixture models

The population may consist of a finite or countable number of distinct subpopulations, e.g., in assessing consumer ratings of cookies, there may be a subpopulation of people who like sweetness, and one with those who do not. With K subpopulations, X takes on the values $\{1,\ldots,K\}$. Note that we could have $K = \infty$. These values are indices, not necessarily meaning to convey any ordering. For a normal mixture, the model is

$$Y \mid X = k \sim N(\mu_k, \sigma_k^2). \tag{6.9}$$

Generally there are no restrictions on the μ_k's, but the σ_k^2's may be assumed equal. Also, K may or may not be known. The marginal distribution for X may be unrestricted, i.e.,

$$f_X(k) = p_k, \quad k = 1,\ldots,K, \tag{6.10}$$

where the p_k's are positive and sum to 1, or it may have a specific pmf.

6.2.3 Hierarchical models

Many experiments involve first randomly choosing a number of subjects from a population, then measuring a number of random variables on the chosen subjects. For example, one might randomly choose n third-grade classes from a city, then within each class administer a test to m randomly chosen students. Let X_i be the overall ability of class i, and Y_i the average performance on the test of the students chosen from class i. Then a possible hierarchical model is

$$X_1,\ldots,X_n \text{ are iid} \sim N(\mu,\sigma^2), \text{ and}$$

$$Y_1,\ldots,Y_n \mid X_1 = x_1,\ldots,X_n = x_n \text{ are independent} \sim (N(x_1,\tau^2),\ldots,N(x_n,\tau^2)). \tag{6.11}$$

Here, μ and σ^2 are the mean and variance for the entire population of class means, while x_i is the mean for class i. Interest may center on the overall mean, so the city can obtain funding from the state, as well as for the individual classes chosen, so these classes can get special treats from the local school board.

6.2.4 Bayesian models

A statistical model typically depends on an unknown parameter vector θ, and the objective is to estimate the parameters, or some function of them, or test hypotheses about them. The Bayesian approach treats the data X and the parameter Θ as both being random, hence having a joint distribution. The *frequentist* approach considers the parameters to be fixed but unknown. Both approaches use a model for X, which is a set of distributions indexed by the parameter, e.g., the X_i's are iid $N(\mu,\sigma^2)$, where $\theta = (\mu,\sigma^2)$. The Bayesian approach considers that model *conditional on* $\Theta = \theta$, and would write

$$X_1,\ldots,X_n \mid \Theta = \theta \, (= (\mu,\sigma^2)) \sim \text{ iid } N(\mu,\sigma^2). \tag{6.12}$$

Here, the capital Θ is the random vector, and the lower case θ is the particular value. Then to fully specify the model, the distribution of Θ must be given,

$$\Theta \sim \pi, \tag{6.13}$$

for some prior distribution π. Once the data is x obtained, inference is based on the posterior distribution of $\Theta \mid X = x$.

6.3   Conditional & marginal → Joint

We start with the conditional distribution of $\mathbf{Y}$ given $\mathbf{X} = \mathbf{x}$, and the marginal distribution of $\mathbf{X}$, and find the joint distribution of $(\mathbf{X}, \mathbf{Y})$. Let $\mathcal{X}$ be the (marginal) space of $\mathbf{X}$, and for each $\mathbf{x} \in \mathcal{X}$, let $\mathcal{Y}_\mathbf{x}$ be the conditional space of $\mathbf{Y}$ given $\mathbf{X} = \mathbf{x}$, that is, the space for the distribution of $\mathbf{Y} \mid \mathbf{X} = \mathbf{x}$. Then the (joint) space of $(\mathbf{X}, \mathbf{Y})$ is

$$\mathcal{W} = \{(\mathbf{x}, \mathbf{y}) \mid \mathbf{x} \in \mathcal{X} \ \& \ \mathbf{y} \in \mathcal{Y}_\mathbf{x}\}. \tag{6.14}$$

(In the coin example, and $\mathcal{Y}_x = \{0, 1, \ldots, n\}$, so that in this case the conditional space of Y does not depend on x.)

Now for a function $g(\mathbf{x}, \mathbf{y})$, the conditional expectation of $g(\mathbf{X}, \mathbf{Y})$ given $\mathbf{X} = \mathbf{x}$ is denoted

$$e_g(\mathbf{x}) = E[g(\mathbf{X}, \mathbf{Y}) \mid \mathbf{X} = \mathbf{x}], \tag{6.15}$$

and is defined to be the expected value of the function $g(\mathbf{X}, \mathbf{Y})$ where $\mathbf{X}$ is fixed at $\mathbf{x}$ and $\mathbf{Y}$ has the conditional distribution $\mathbf{Y} \mid \mathbf{X} = \mathbf{x}$. If this conditional distribution has a pdf, say $f_{\mathbf{Y}\mid\mathbf{X}}(\mathbf{y} \mid \mathbf{x})$, then

$$e_g(\mathbf{x}) = \int_{\mathcal{Y}_\mathbf{x}} g(\mathbf{x}, \mathbf{y}) f_{\mathbf{Y}\mid\mathbf{X}}(\mathbf{y} \mid \mathbf{x}) d\mathbf{y}. \tag{6.16}$$

If $f_{\mathbf{X}\mid\mathbf{Y}}$ is a pmf, then we have the summation instead of the integral. It is important to realize that this conditional expectation is a function of $\mathbf{x}$. In the coin example (6.2), with $g(x, y) = y$, the conditional expected number of heads given the chosen coin has parameter x is

$$e_g(x) = E[Y \mid X = x] = E[\text{Binomial}(n, x)] = nx. \tag{6.17}$$

The key to describing the joint distribution is to define the unconditional expected value of g.

Definition 6.1. *Given the conditional distribution of Y given X, and the marginal distribution of X, the joint distribution of (X, Y) is that distribution for which*

$$E[g(\mathbf{X}, \mathbf{Y})] = E[e_g(\mathbf{X})] \tag{6.18}$$

for any function g with finite expected value.

So, continuing the coin example, with $g(x, y) = y$, since marginally $X \sim$ Uniform(0,1),

$$E[Y] = E[e_g(X)] = E[nX] = n \ E[\text{Uniform}(0, 1)] = \frac{n}{2}. \tag{6.19}$$

Notice that this unconditional expected value does not depend on x, while the conditional expected value does, which is as it should be.

This definition yields the joint distribution P on $\mathcal{W}$ by looking at indicator functions g, as in Section 2.1.1. That is, take $A \subset \mathcal{W}$, so that with P being the joint probability distribution for $(\mathbf{X}, \mathbf{Y})$,

$$P[A] = E[I_A(\mathbf{X}, \mathbf{Y})], \tag{6.20}$$

where $I_A(\mathbf{x}, \mathbf{y})$ is the indicator of the set A as in (2.16), so equals 1 if $(\mathbf{x}, \mathbf{y}) \in A$ and 0 if not. Then the definition says that

$$P[A] = E[e_{I_A}(\mathbf{X})] \quad \text{where} \ e_{I_A}(\mathbf{x}) = E[I_A(\mathbf{X}, \mathbf{Y}) \mid \mathbf{X} = \mathbf{x}]. \tag{6.21}$$

We should check that this P is in fact a legitimate probability distribution, and in turn yields the correct expected values. The latter result is proven in measure theory. That it is a probability measure as in (1.1) is not hard to show using that $I_W \equiv 1$, and if A_i's are disjoint,

$$I_{\cup A_i}(\mathbf{x}, \mathbf{y}) = \sum I_{A_i}(\mathbf{x}, \mathbf{y}). \tag{6.22}$$

The $e_{I_A}(\mathbf{x})$ is the conditional probability of A given $\mathbf{X} = \mathbf{x}$, written

$$P[A \mid \mathbf{X} = \mathbf{x}] = P[(\mathbf{X}, \mathbf{Y}) \in A \mid \mathbf{X} = \mathbf{x}] = E[I_A(\mathbf{X}, \mathbf{Y}) \mid \mathbf{X} = \mathbf{x}]. \tag{6.23}$$

6.3.1 Joint densities

If the conditional and marginal distributions have densities, then it is easy to find the joint density. Suppose X has marginal pdf $f_X(\mathbf{x})$ and the conditional pdf of $\mathbf{Y} \mid \mathbf{X} = \mathbf{x}$ is $f_{\mathbf{Y} \mid \mathbf{X}}(\mathbf{y} \mid \mathbf{x})$. Then for any function g with finite expectation, we can take $e_g(\mathbf{x})$ as in (6.16), and write

$$
\begin{aligned}
E[g(\mathbf{X}, \mathbf{Y})] &= E[g(\mathbf{X})] \\
&= \int_{\mathcal{X}} e_g(\mathbf{x}) f_X(\mathbf{x}) d\mathbf{x} \\
&= \int_{\mathcal{X}} \int_{\mathcal{Y}_\mathbf{x}} g(\mathbf{x}, \mathbf{y}) f_{\mathbf{Y} \mid \mathbf{X}}(\mathbf{y} \mid \mathbf{x}) d\mathbf{y} f_X(\mathbf{x}) d\mathbf{x} \\
&= \int_W g(\mathbf{x}, \mathbf{y}) f_{\mathbf{Y} \mid \mathbf{X}}(\mathbf{y} \mid \mathbf{x}) f_X(\mathbf{x}) d\mathbf{x} d\mathbf{y}.
\end{aligned}
\tag{6.24}
$$

The last step is to emphasize that the double integral is indeed integrating over the whole space of $(\mathbf{X}, \mathbf{Y})$. Looking at that last expression, we see that by taking

$$f(\mathbf{x}, \mathbf{y}) = f_{\mathbf{Y} \mid \mathbf{X}}(\mathbf{y} \mid \mathbf{x}) f_X(\mathbf{x}), \tag{6.25}$$

we have that

$$E[g(\mathbf{X}, \mathbf{Y})] = \int_W g(\mathbf{x}, \mathbf{y}) f(\mathbf{x}, \mathbf{y}) d\mathbf{x} d\mathbf{y}. \tag{6.26}$$

Since that equation works for any g with finite expectation, Definition 6.1 implies that $f(\mathbf{x}, \mathbf{y})$ is the joint pdf of $(\mathbf{X}, \mathbf{Y})$. If either or both of the original densities are pmfs, the analysis goes through the same way, with summations in place of integrals.

Equation (6.25) should not be especially surprising. It is analogous to the general definition of conditional probability for sets A and B:

$$P[A \cap B] = P[A \mid B] \times P[B]. \tag{6.27}$$

6.4 Marginal distributions

There is no special trick in obtaining the marginal of $\mathbf{Y}$ given the conditional $\mathbf{Y} \mid \mathbf{X} = \mathbf{x}$ and the marginal of $\mathbf{X}$; just find the joint and integrate out $\mathbf{x}$. Thus

$$f_{\mathbf{Y}}(\mathbf{y}) = \int_{\mathcal{X}_\mathbf{y}} f(\mathbf{x}, \mathbf{y}) d\mathbf{x} = \int_{\mathcal{X}_\mathbf{y}} f_{\mathbf{Y} \mid \mathbf{X}}(\mathbf{y} \mid \mathbf{x}) f_X(\mathbf{x}) d\mathbf{x}. \tag{6.28}$$

6.4.1   Coins and the beta-binomial distribution

In the coin example of (6.1) and (6.2),

$$
\begin{aligned}
f_Y(y) &= \int_0^1 \binom{n}{y} x^y (1-x)^{n-y} dx \\
&= \binom{n}{y} \frac{\Gamma(y+1)\Gamma(n-y+1)}{\Gamma(n+2)} \\
&= \frac{n!}{y!(n-y)!} \frac{y!(n-y)!}{(n+1)!} \\
&= \frac{1}{n+1}, \quad y = 0, 1, \ldots, n,
\end{aligned}
\tag{6.29}
$$

which is the Discrete Uniform$(0,n)$. Note that this distribution is not a binomial, and does not depend on x (it better not!). In fact, this Y is a special case of the following.

Definition 6.2. *Suppose*

$$
Y \mid X = x \sim \text{Binomial}(n, x) \quad \text{and} \quad X \sim \text{Beta}(\alpha, \beta).
\tag{6.30}
$$

Then the marginal distribution of Y is beta-binomial with parameters α, β and n, written

$$
Y \sim \text{Beta-Binomial}(\alpha, \beta, n).
\tag{6.31}
$$

When $\alpha = \beta = 1$, the X is uniform. Otherwise, as above, we can find the marginal pmf to be

$$
f_Y(y) = \frac{\Gamma(\alpha+\beta)}{\Gamma(\alpha)\Gamma(\beta)} \binom{n}{y} \frac{\Gamma(y+\alpha)\Gamma(n-y+\beta)}{\Gamma(n+\alpha+\beta)}, \quad y = 0, 1, \ldots, n.
\tag{6.32}
$$

6.4.2   Simple normal linear model

For another example, take the linear model in (6.6) and (6.8),

$$
Y \mid X = x \sim N(\alpha + \beta x, \sigma_e^2) \quad \text{and} \quad X \sim N(\mu_X, \sigma_X^2).
\tag{6.33}
$$

We could write out the joint pdf and integrate, but instead we will find the mgf, which we can do in steps because it is an expected value. That is, with $g(y) = e^{ty}$,

$$
M_Y(t) = E[e^{tY}] = E[e_g(X)], \quad \text{where} \quad e_g(x) = E[e^{tY} \mid X = x].
\tag{6.34}
$$

We know the mgf of a normal from (4.36), which $Y \mid X = x$ is, hence

$$
e_g(x) = M_{N(\alpha+\beta x, \sigma_e^2)}(t) = e^{(\alpha+\beta x)t + \sigma_e^2 \frac{t^2}{2}}.
\tag{6.35}
$$

The expected value of $e_g(X)$ can also be written as a normal mgf:

$$M_Y(t) = E[e_g(X)] = E[e^{(\alpha+\beta X)t + \sigma_e^2 \frac{t^2}{2}}]$$
$$= e^{\alpha t + \sigma_e^2 \frac{t^2}{2}} E[e^{(\beta t)X}]$$
$$= e^{\alpha t + \sigma_e^2 \frac{t^2}{2}} M_{N(\mu_X, \sigma_X^2)}(\beta t)$$
$$= e^{\alpha t + \sigma_e^2 \frac{t^2}{2}} e^{\beta t \mu_X + \sigma_X^2 \frac{(\beta t)^2}{2}}$$
$$= e^{(\alpha + \beta \mu_X)t + (\sigma_e^2 + \sigma_X^2 \beta^2)\frac{t^2}{2}}$$
$$= \text{mgf of } N(\alpha + \beta \mu_X, \sigma_e^2 + \sigma_X^2 \beta^2). \qquad (6.36)$$

That is, marginally,

$$Y \sim N(\alpha + \beta \mu_X, \sigma_e^2 + \sigma_X^2 \beta^2). \qquad (6.37)$$

6.4.3 Marginal mean and variance

We know that the marginal expected value of $g(X, Y)$ is just the expected value of the conditional expected value:

$$E[g(X, Y)] = E[e_g(X)], \quad \text{where } e_g(x) = E[g(X, Y) \mid X = x]. \qquad (6.38)$$

The marginal variance is not quite the expected value of the conditional variance. First, we will write down the marginal and conditional variances, using the formula $Var[W] = E[W^2] - E[W]^2$ on both:

$$\sigma_g^2 \equiv Var[g(X, Y)] = E[g(X, Y)^2] - E[g(X, Y)]^2 \text{ and}$$
$$v_g(x) \equiv Var[g(X, Y) \mid X = x] = E[g(X, Y)^2 \mid X = x] - E[g(X, Y) \mid X = x]^2. \qquad (6.39)$$

Next, use the conditional expected value result on g^2:

$$E[g(X, Y)^2] = E[e_{g^2}(X)], \quad \text{where } e_{g^2}(x) = E[g(X, Y)^2 \mid X = x]. \qquad (6.40)$$

Now use (6.38) and (6.40) in (6.39):

$$\sigma_g^2 = E[e_{g^2}(X)] - E[e_g(X)]^2 \text{ and}$$
$$v_g(x) = e_{g^2}(x) - e_g(x)^2. \qquad (6.41)$$

Taking expected value over both sides of the second equation and rearranging shows that

$$E[e_{g^2}(X)] = E[v_g(X)] + E[e_g(X)^2], \qquad (6.42)$$

hence

$$\sigma_g^2 = E[v_g(X)] + E[e_g(X)^2] - E[e_g(X)]^2$$
$$= E[v_g(X)] + Var[e_g(X)]. \qquad (6.43)$$

To summarize:

1. The unconditional expected value is the expected value of the conditional expected value.

2. The unconditional variance is the expected value of the conditional variance plus the variance of the conditional expected value.

The second sentence is very analogous to what happens in regression, where the total sum-of-squares equals the regression sum-of-squares plus the residual sum of squares.

These are handy results. For example, in the beta-binomial (Definition 6.2), finding the mean and variance using the pdf (6.32) can be challenging. But using the conditional approach is much easier. Because conditionally Y is Binomial(n, x),

$$e_Y(x) = nx \quad \text{and} \quad v_Y(x) = nx(1 - x). \tag{6.44}$$

Then because $X \sim \text{Beta}(\alpha, \beta)$,

$$E[Y] = nE[X] = n\frac{\alpha}{\alpha + \beta}, \tag{6.45}$$

and

$$
\begin{aligned}
Var[Y] &= E[v_Y(X)] + Var[e_Y(X)] \\
&= E[nX(1 - X)] + Var[nX] \\
&= n\frac{\alpha\beta}{(\alpha + \beta)(\alpha + \beta + 1)} + n^2\frac{\alpha\beta}{(\alpha + \beta)^2(\alpha + \beta + 1)} \\
&= \frac{n\alpha\beta(\alpha + \beta + n)}{(\alpha + \beta)^2(\alpha + \beta + 1)}.
\end{aligned}
\tag{6.46}
$$

These expressions give some insight into the beta-binomial. Like the binomial, the beta-binomial counts the number of successes in n trials, and has expected value np for $p = \alpha/(\alpha + \beta)$. Consider the variances of a binomial and beta-binomial:

$$Var[\text{Binomial}] = np(1 - p) \quad \text{and} \quad Var[\text{Beta-Binomial}] = np(1 - p)\frac{\alpha + \beta + n}{\alpha + \beta + 1}. \tag{6.47}$$

Thus the beta-binomial has a larger variance, so it can be used to model situations in which the data are more disperse than the binomial, e.g., if the n trials are n offspring in the same litter, and success is survival. The larger $\alpha + \beta$, the closer the beta-binomial is to the binomial.

You might wish to check the mean and variance in the normal example 6.4.2.

The same procedure works for vectors:

$$E[\mathbf{Y}] = E[e_{\mathbf{Y}}(\mathbf{X})] \quad \text{where} \quad e_{\mathbf{Y}}(\mathbf{x}) = E[\mathbf{Y} \mid \mathbf{X} = \mathbf{x}], \tag{6.48}$$

and

$$Cov[\mathbf{Y}] = E[v_{\mathbf{Y}}(\mathbf{X})] + Cov[e_{\mathbf{Y}}(\mathbf{X})], \quad \text{where} \quad v_{\mathbf{Y}}(\mathbf{x}) = Cov[\mathbf{Y} \mid \mathbf{X} = \mathbf{x}]. \tag{6.49}$$

In particular, considering a single covariance,

$$Cov[Y_1, Y_2] = E[c_{Y_1, Y_2}(\mathbf{X})] + Cov[e_{Y_1}(\mathbf{X}), e_{Y_2}(\mathbf{X})], \tag{6.50}$$

where

$$c_{Y_1, Y_2}(\mathbf{x}) = Cov[Y_1, Y_2 \mid \mathbf{X} = \mathbf{x}]. \tag{6.51}$$

6.4.4 Fruit flies

Arnold (1981) presents an experiment concerning the genetics of *Drosophila pseudoobscura*, a type of fruit fly. We are looking at a particular *locus* (place) on a pair of chromosomes. The locus has two possible *alleles* (values): TL $\equiv$ *TreeLine* and CU $\equiv$ *Cuernavaca*. Each individual has two of these, one on each chromosome. The individual's *genotype* is the pair of alleles it has. Thus the genotype could be (TL,TL), (TL,CU), or (CU,CU). (There is no distinction made between (CU,TL) and (TL,CU).)

The objective is to estimate $\theta \in (0, 1)$, the proportion of CU in the population. In this experiment, the researchers randomly collected 10 adult males. Unfortunately, one cannot determine the genotype of the adult fly just by looking at him. One can determine the genotype of young flies, though. So the researchers bred each of these ten flies with a (different) female known to be (TL,TL), and analyzed two of the offspring from each mating. Each offspring receives one allele from each parent. Thus if the mother's alleles are (A_1, A_2) and the father's are (B_1, B_2), each offspring has four (maybe not distinct) possibilities:

$$
\begin{array}{c|cc}
 & \multicolumn{2}{c}{\text{Father}} \\
\text{Mother} & B_1 & B_2 \\
\hline
A_1 & (A_1, B_1) & (A_1, B_2) \\
A_2 & (A_2, B_1) & (A_2, B_2)
\end{array}
\qquad (6.52)
$$

In this case, there are three relevant, fairly simple, tables:

$$
\begin{array}{c|cc}
 & \multicolumn{2}{c}{\text{Father}} \\
\text{Mother} & TL & TL \\
\hline
TL & (TL, TL) & (TL, TL) \\
TL & (TL, TL) & (TL, TL)
\end{array}
\qquad
\begin{array}{c|cc}
 & \multicolumn{2}{c}{\text{Father}} \\
\text{Mother} & CU & TL \\
\hline
TL & (TL, CU) & (TL, TL) \\
TL & (TL, CU) & (TL, TL)
\end{array}
$$

$$
\begin{array}{c|cc}
 & \multicolumn{2}{c}{\text{Father}} \\
\text{Mother} & CU & CU \\
\hline
TL & (TL, CU) & (TL, CU) \\
TL & (TL, CU) & (TL, CU)
\end{array}
\qquad (6.53)
$$

The actual genotypes of the sampled offspring are next:

Father	Offsprings' genotypes
1	(TL, TL) & (TL, TL)
2	(TL, TL) & (TL, CU)
3	(TL, TL) & (TL, TL)
4	(TL, TL) & (TL, TL)
5	(TL, CU) & (TL, CU)
6	(TL, TL) & (TL, CU)
7	(TL, CU) & (TL, CU)
8	(TL, TL) & (TL, TL)
9	(TL, CU) & (TL, CU)
10	(TL, TL) & (TL, TL)

(6.54)

The probability distribution of these outcomes is governed by the population proportion θ of CUs under the following assumptions:

1. The ten chosen fathers are a simple random sample from the population.

2. The chance that a given father has 0, 1 or 2 CUs in his genotype follows the *Hardy-Weinberg* laws, which means that the number of CUs for each father is like flipping a coin twice independently, with probability of heads being θ.

3. For a given mating, the two offspring are each equally likely to get either of the fathers two alleles (as well as a TL from the mother), and what the two offspring get are independent.

Since each genotype is uniquely determined by the number of CUs it has (0, 1, or 2), we can represent the ith father by X_i, the number of CUs in his genotype. The mothers are all (TL, TL), so each offspring receives a TL from the mother, and randomly one of the alleles from the father. Let Y_{ij} be the indicator of whether the jth offspring of father i receives a CU from the father. That is,

$$Y_{ij} = \begin{cases} 0 & \text{if} \quad \text{Offspring } j \text{ from father } i \text{ is } (TL, TL) \\ 1 & \text{if} \quad \text{Offspring } j \text{ from father } i \text{ is } (TL, CU) \end{cases} \quad (6.55)$$

Then each "family" has three random variables, (X_i, Y_{i1}, Y_{i2}). We will assume that these triples are independent, in fact,

$$(X_1, Y_{11}, Y_{12}), \ldots, (X_n, Y_{n1}, Y_{n2}) \quad \text{are iid.} \quad (6.56)$$

Assumption #2, the Hardy-Weinberg law, implies that

$$X_i \sim \text{Binomial}(2, \theta), \quad (6.57)$$

because each father in effect randomly chooses two alleles from the population. Next, we specify the conditional distribution of the offspring given the father, i.e., $(Y_{i1}, Y_{i2}) \mid X_i = x_i$. If $x_i = 0$, then the father is (TL, TL), so the offspring will all receive a TL from the father, as in the first table in (6.53):

$$P[(Y_{i1}, Y_{i2}) = (0, 0) \mid X_i = 0] = 1. \quad (6.58)$$

Similarly, if $x_i = 2$, the father is (CU,CU), so the offspring will all receive a CU from the father, as in the third table in (6.53):

$$P[(Y_{i1}, Y_{i2}) = (1, 1) \mid X_i = 2] = 1. \quad (6.59)$$

Finally, if $x_i = 1$, the father is (TL,CU), which means each offspring has a 50-50 chance of receiving a CU from the father, as in the second table in (6.53):

$$P[(Y_{i1}, Y_{i2}) = (y_1, y_2) \mid X_i = 1] = \frac{1}{4} \text{ for } (y_1, y_2) \in \{(0,0), (0,1), (1,0), (1,1)\}. \quad (6.60)$$

This conditional distribution can be written more compactly by noting that $x_i/2$ is the chance that an offspring receives a CU, so that

$$(Y_{i1}, Y_{i2}) \mid X_i = x_i \sim \text{ iid } \text{Bernoulli}\left(\frac{x_i}{2}\right), \quad (6.61)$$

which using (4.56) yields the conditional pmf

$$f_{Y|X}(y_{i1}, y_{i2} \mid x_i) = \left(\frac{x_i}{2}\right)^{y_{i1} + y_{i2}} \left(1 - \frac{x_i}{2}\right)^{2 - y_{i1} - y_{i2}}. \quad (6.62)$$

The goal of the experiment is to estimate θ, but without knowing the X_i's. Thus the estimation has to be based on just the Y_{ij}'s. The marginal means are easy to find:

$$E[Y_{ij}] = E[e_{Y_{ij}}(X_i)], \quad \text{where } e_{Y_{ij}}(x_i) = E[Y_{ij} \mid X_i = x_i] = \frac{x_i}{2}, \qquad (6.63)$$

because conditionally Y_{ij} is Bernoulli. Then $E[X_i] = 2\theta$, hence

$$E[Y_{ij}] = \frac{2\theta}{2} = \theta. \qquad (6.64)$$

Nice! Then an obvious estimator of θ is the sample mean of all the Y_{ij}'s, of which there are $2n = 20$:

$$\widehat{\theta} = \frac{\sum_{i=1}^{n} \sum_{j=1}^{2} Y_{ij}}{2n}. \qquad (6.65)$$

This estimator is called the **Dobzhansky estimator**. To find the estimate, we just count the number of CUs in (6.54), which is 8, hence the estimate of θ is 0.4.

What is the variance of the estimate? Are Y_{i1} and Y_{i2} unconditionally independent? What is $Var[Y_{ij}]$? What is the marginal pmf of (Y_{i1}, Y_{i2})? See Exercises 6.8.15 and 6.8.16.

6.5 Conditional from the joint

Often one has a joint distribution, but is primarily interested in the conditional, e.g., many experiments involve collecting health data from a population, and interest centers on the conditional distribution of certain outcomes, such as longevity, conditional on other variables, such as sex, age, cholesterol level, activity level, etc. In Bayesian inference, one can find the joint distribution of the data and the parameter, and from that find the conditional distribution of the parameter given the data. Measure theory guarantees that for any joint distribution of $(\mathbf{X}, \mathbf{Y})$, there exists a conditional distribution of $\mathbf{Y} \mid \mathbf{X} = \mathbf{x}$ for each $\mathbf{x} \in \mathcal{X}$. It may not be unique, but any conditional distribution that combines with the marginal of $\mathbf{X}$ to yield the original joint distribution is a valid conditional distribution. If densities exists, then $f_{\mathbf{Y}|\mathbf{X}}(\mathbf{y} \mid \mathbf{x})$ is a valid conditional density if

$$f(\mathbf{x}, \mathbf{y}) = f_{\mathbf{Y}|\mathbf{X}}(\mathbf{y} \mid \mathbf{x}) f_{\mathbf{X}}(\mathbf{x}) \qquad (6.66)$$

for all $(\mathbf{x}, \mathbf{y}) \in \mathcal{W}$. Thus, given a joint density f, one can integrate out $\mathbf{y}$ to obtain the marginal $f_{\mathbf{X}}$, then define the conditional by

$$f_{\mathbf{Y}|\mathbf{X}}(\mathbf{y} \mid \mathbf{x}) = \frac{f(\mathbf{x}, \mathbf{y})}{f_{\mathbf{X}}(\mathbf{x})} = \frac{\text{Joint}}{\text{Marginal}} \quad \text{if } f_{\mathbf{X}}(\mathbf{x}) > 0. \qquad (6.67)$$

It does not matter how the conditional density is defined when $f_{\mathbf{X}}(\mathbf{x}) = 0$, as long as it is a density on $\mathcal{Y}_{\mathbf{x}}$, because in reconstructing the joint f in (6.66), it is multiplied by 0. Also, the conditional of X given $Y = y$ is the ratio of the joint to the marginal of Y. This formula works for pdfs, pmfs, and the mixed kind.

6.5.1 Coins

In Example 6.4.1 on the coins, the joint density of (X, Y) is

$$f(x, y) = \binom{n}{y} x^y (1 - x)^{n-y}, \qquad (6.68)$$

and the marginal distribution of Y, the number of heads, is, as in (6.29),

$$f_Y(y) = \frac{1}{n+1}, \ y = 0, \ldots, n. \tag{6.69}$$

Thus the conditional posterior distribution of X, the chance of heads, given Y, the number of heads, is

$$
\begin{aligned}
f_{X|Y}(x \mid y) = \frac{f(x,y)}{f_Y(y)} &= (n+1) \binom{n}{y} x^y (1-x)^{n-y} \\
&= \frac{(n+1)!}{y!(n-y)!} x^y (1-x)^{n-y} \\
&= \frac{\Gamma(n+2)}{\Gamma(y+1)\Gamma(n-y+1)} x^y (1-x)^{n-y} \\
&= \text{Beta}(y+1, n-y+1) \text{ pdf.}
\end{aligned} \tag{6.70}
$$

For example, if the experiment yields $Y = 3$ heads, then one's guess of what the probability of heads is for this particular coin is described by the Beta$(4,8)$ distribution. A reasonable guess could be the posterior mean,

$$E[X \mid Y = 3] = E[\text{Beta}(4,8)] = \frac{4}{4+8} = \frac{1}{3}. \tag{6.71}$$

Note that this is *not* the sample proportion of heads, 0.3, although it is close. The posterior mode (the x that maximizes the pdf) and posterior median are also reasonable point estimates. A more informative quantity might be a probability interval:

$$P[0.1093 < X < 0.6097 \mid Y = 3] = 95\%. \tag{6.72}$$

So there is a 95% chance that the chance of heads is somewhere between 0.11 and 0.61. (These numbers were found using the qbeta function in R.) It is not a very tight interval, but there is not much information in just ten flips.

6.5.2 Bivariate normal

If one can recognize the form of a particular density for Y within a joint density, then it becomes unnecessary to explicitly find the marginal of X and divide the joint by the marginal. For example, the $N(\mu, \sigma^2)$ density can be written

$$
\begin{aligned}
\phi(z \mid \mu, \sigma^2) &= \frac{1}{\sqrt{2\pi}\sigma} e^{-\frac{1}{2}\frac{(z-\mu)^2}{\sigma^2}} \\
&= c(\mu, \sigma^2) e^{z\frac{\mu}{\sigma^2} - z^2 \frac{1}{2\sigma^2}}.
\end{aligned} \tag{6.73}
$$

That is, we factor the pdf into a constant we do not care about at the moment, that depends on the fixed parameters, and the important component containing all the z-action.

Now consider the bivariate normal,

$$(X, Y) \sim N(\boldsymbol{\mu}, \boldsymbol{\Sigma}), \text{ where } \boldsymbol{\mu} = (\mu_X, \mu_Y) \text{ and } \boldsymbol{\Sigma} = \begin{pmatrix} \sigma_X^2 & \sigma_{XY} \\ \sigma_{XY} & \sigma_Y^2 \end{pmatrix}. \tag{6.74}$$

Assuming Σ is invertible, the joint pdf is (as in (5.46))

$$f(x,y) = \frac{1}{2\pi} \frac{1}{\sqrt{|\Sigma|}} e^{-\frac{1}{2}((x,y)-\mu)\Sigma^{-1}((x,y)-\mu)'}. \tag{6.75}$$

We try to factor the conditional pdf of $Y \mid X = x$ as

$$f_{Y|X}(y \mid x) = \frac{f(x,y)}{f_X(x)} = c(x,\mu,\Sigma)g(y,x,\mu,\Sigma), \tag{6.76}$$

where c has as many factors as possible that are free of y (including $f_X(x)$), and g has everything else. Exercise 6.8.5 shows that c can be chosen so that

$$g(y,x,\mu,\Sigma) = e^{y\gamma_1 + y^2\gamma_2}, \tag{6.77}$$

where

$$\gamma_1 = \frac{\sigma_{XY}(x-\mu_X) + \mu_Y\sigma_Y^2}{|\Sigma|} \quad \text{and} \quad \gamma_2 = -\frac{1}{2}\frac{\sigma_X^2}{|\Sigma|}. \tag{6.78}$$

Now compare the g in (6.77) to the exponential term in (6.73). Since the latter is a normal pdf, and the space of Z in (6.73) is the same as that of Y in (6.77), the pdf in (6.76) must be normal, where the parameters (μ,σ^2) are found by matching γ_1 to μ/σ^2 and γ_2 to $-1/(2\sigma^2)$. Doing so, we obtain

$$\sigma^2 = \sigma_Y^2 - \frac{\sigma_{XY}^2}{\sigma_X^2} \quad \text{and} \quad \mu = \mu_Y + \frac{\sigma_{XY}}{\sigma_X^2}(x-\mu_X). \tag{6.79}$$

That is,

$$Y \mid X = x \sim N\left(\mu_Y + \frac{\sigma_{XY}}{\sigma_X^2}(x-\mu_X), \sigma_Y^2 - \frac{\sigma_{XY}^2}{\sigma_X^2}\right). \tag{6.80}$$

Because we know the normal pdf, we could work backwards to find the c in (6.76), but there is no need to do that.

This is a normal linear model, as in (6.6), where

$$Y \mid X = x \sim N(\alpha + \beta x, \sigma_e^2) \tag{6.81}$$

with

$$\beta = \frac{\sigma_{XY}}{\sigma_X^2}, \quad \alpha = \mu_Y - \beta\mu_X, \quad \text{and} \quad \sigma_e^2 = \sigma_Y^2 - \frac{\sigma_{XY}^2}{\sigma_X^2}. \tag{6.82}$$

These equations should be familiar from linear regression. An alternative method for deriving conditional distribution in the multivariate normal is given in Section 7.7.

6.6 Bayes theorem: Reversing the conditionals

We have already essentially derived Bayes theorem, but it is important enough to deserve its own section. The theorem takes the conditional density of $\mathbf{Y}$ given $\mathbf{X} = \mathbf{x}$ and the marginal distribution of $\mathbf{X}$, and produces the conditional density of $\mathbf{X}$ given $\mathbf{Y} = \mathbf{y}$. It uses the formula *conditional = joint/marginal*, where the marginal is found by integrating out the $\mathbf{y}$ from the joint.

Theorem 6.3. Bayes. *Suppose* $\mathbf{Y} \mid \mathbf{X} = \mathbf{x}$ *has density* $f_{\mathbf{Y}\mid\mathbf{X}}(\mathbf{y} \mid \mathbf{x})$ *and* $\mathbf{X}$ *has marginal density* $f_{\mathbf{X}}(\mathbf{x})$. *Then,*

$$f_{\mathbf{X}\mid\mathbf{Y}}(\mathbf{x} \mid \mathbf{y}) = \frac{f_{\mathbf{Y}\mid\mathbf{X}}(\mathbf{y} \mid \mathbf{x}) f_{\mathbf{X}}(\mathbf{x})}{\int_{\mathcal{X}_y} f_{\mathbf{Y}\mid\mathbf{X}}(\mathbf{y} \mid \mathbf{z}) f_{\mathbf{X}}(\mathbf{z}) d\mathbf{z}}. \tag{6.83}$$

The integral will be a summation if $\mathbf{X}$ *is discrete.*

Proof. With $f(\mathbf{x}, \mathbf{y})$ being the joint density,

$$
\begin{aligned}
f_{\mathbf{X}\mid\mathbf{Y}}(\mathbf{x} \mid \mathbf{y}) &= \frac{f(\mathbf{x}, \mathbf{y})}{f_{\mathbf{Y}}(\mathbf{y})} \\
&= \frac{f_{\mathbf{Y}\mid\mathbf{X}}(\mathbf{y} \mid \mathbf{x}) f_{\mathbf{X}}(\mathbf{x})}{\int_{\mathcal{X}_y} f(\mathbf{z}, \mathbf{y}) d\mathbf{z}} \\
&= \frac{f_{\mathbf{Y}\mid\mathbf{X}}(\mathbf{y} \mid \mathbf{x}) f_{\mathbf{X}}(\mathbf{x})}{\int_{\mathcal{X}_y} f_{\mathbf{Y}\mid\mathbf{X}}(\mathbf{y} \mid \mathbf{z}) f_{\mathbf{X}}(\mathbf{z}) d\mathbf{z}}.
\end{aligned}
\tag{6.84}
$$

$\square$

Bayes theorem is often used with sets. Let $\mathcal{A} \subset \mathcal{X}$ and $\mathcal{B}_1, \ldots, \mathcal{B}_K$ be a partition of $\mathcal{X}$, i.e.,

$$\mathcal{B}_i \cap \mathcal{B}_j = \varnothing \text{ for } i \neq j, \text{ and } \cup_{k=1}^{K} \mathcal{B}_k = \mathcal{X}. \tag{6.85}$$

Then

$$P[\mathcal{B}_k \mid \mathcal{A}] = \frac{P[\mathcal{A} \mid \mathcal{B}_k] P[\mathcal{B}_k]}{\sum_{l=1}^{K} P[\mathcal{A} \mid \mathcal{B}_l] P[\mathcal{B}_l]}. \tag{6.86}$$

6.6.1   AIDS virus

A common illustration of Bayes theorem involves testing for some medical condition, e.g., a blood test for the AIDS virus. Suppose the test is 99% accurate. If a random person's test is positive, does that mean the person is 99% sure of having the virus? Let $A_+ =$ "test is positive", $A_- =$ "test is negative", $B_+ =$ "person has the virus" and $B_- =$ "person does not have the virus." Then we know the conditionals

$$P[A_+ \mid B_+] = 0.99 \text{ and } P[A_- \mid B_-] = 0.99, \tag{6.87}$$

but they are not of interest. We want to know the reverse conditional, $P[B_+ \mid A_+]$, the chance of having the virus given the test is positive. There is no way to figure this probability out without the marginal of B, that is, the marginal chance a random person has the virus. Let us say that $P[B_+] = 1/10,000$. Now we can use Bayes theorem (6.86)

$$
\begin{aligned}
P[B_+ \mid A_+] &= \frac{P[A_+ \mid B_+] P[B_+]}{P[A_+ \mid B_+] P[B_+] + P[A_+ \mid B_-] P[B_-]} \\
&= \frac{0.99 \times \frac{1}{10000}}{0.99 \times \frac{1}{10000} + 0.01 \times \frac{9999}{10000}} \\
&\approx 0.0098.
\end{aligned}
\tag{6.88}
$$

Thus the chance of having the virus, given the test is positive, is only about 1/100. That is lower than one might expect, but it is substantially higher than the overall chance of 1/10000. (This example is a bit simplistic in that random people do not take the test, but more likely people who think they may be at risk.)

6.6.2 Beta posterior for the binomial

The coin example in Section 6.5.1 can be generalized by using a beta in place of the uniform. In a Bayesian framework, we suppose the probability of success, θ, has a prior distribution Beta(α, β), and $Y \mid \Theta = \theta$ is Binomial(n, θ). (So now x has become θ.) The prior is supposed to represent knowledge or belief about what the θ is before seeing the data Y. To find the posterior, or what we are to think after seeing the data $Y = y$, we need the conditional distribution of Θ given $Y = y$.

The joint density is

$$f(\theta, y) = f_{Y \mid \Theta}(y \mid \theta) f_\Theta(\theta)$$
$$= \binom{n}{y} \theta^y (1 - \theta)^{n-y} \beta(\alpha, \beta) \theta^{\alpha-1} (1 - \theta)^{\beta-1}$$
$$= c(y, \alpha, \beta) \theta^{y+\alpha-1} (1 - \theta)^{n-y+\beta-1}. \tag{6.89}$$

Because we are interested in the pdf of Θ, we put everything not depending on θ in the constant. But the part that does depend on θ is the meat of a Beta$(\alpha + y, \beta + n - y)$ density, hence that is the posterior:

$$\Theta \mid Y = y \sim \text{Beta}(\alpha + y, \beta + n - y). \tag{6.90}$$

That is, we do not explicitly have to find the marginal pmf of Θ, which we did do in (6.70).

6.7 Conditionals and independence

If X and Y are independent, then it makes sense that the distribution of one does not depend on the value of the other, which is true.

Lemma 6.4. *The random vectors* $\mathbf{X}$ *and* $\mathbf{Y}$ *are independent if and only if (a version of) the conditional distribution of* $\mathbf{Y}$ *given* $\mathbf{X} = \mathbf{x}$ *does not depend on* $\mathbf{x}$.

The parenthetical "a version of" is there to make the statement precise, since it is possible (e.g., if $\mathbf{X}$ is continuous) for the conditional distribution to depend on $\mathbf{x}$ for a few $\mathbf{x}$ without changing the joint distribution.

When there are densities, this result follows directly:

$$\text{Independence} \implies f(\mathbf{x}, \mathbf{y}) = f_{\mathbf{X}}(\mathbf{x}) f_{\mathbf{Y}}(\mathbf{y}) \text{ and } \mathcal{W} = \mathcal{X} \times \mathcal{Y}$$
$$\implies f_{\mathbf{Y} \mid \mathbf{X}}(\mathbf{y} \mid \mathbf{x}) = \frac{f(\mathbf{x}, \mathbf{y})}{f_{\mathbf{X}}(\mathbf{x})} = f_{\mathbf{Y}}(\mathbf{y}) \text{ and } \mathcal{Y}_{\mathbf{x}} = \mathcal{Y}, \tag{6.91}$$

which does not dependent on $\mathbf{x}$. The other way, if the conditional distribution of $\mathbf{Y}$ given $\mathbf{X} = \mathbf{x}$ does not depend on $\mathbf{x}$, then $f_{\mathbf{Y} \mid \mathbf{X}}(\mathbf{y} \mid \mathbf{x}) = f_{\mathbf{Y}}(\mathbf{y})$ and $\mathcal{Y}_{\mathbf{x}} = \mathcal{Y}$, hence

$$f(\mathbf{x}, \mathbf{y}) = f_{\mathbf{Y} \mid \mathbf{X}}(\mathbf{y} \mid \mathbf{x}) f_{\mathbf{X}}(\mathbf{x}) = f_{\mathbf{Y}}(\mathbf{y}) f_{\mathbf{X}}(\mathbf{x}) \text{ and } \mathcal{W} = \mathcal{X} \times \mathcal{Y}. \tag{6.92}$$

6.7.1 Independence of residuals and X

Suppose (X, Y) is bivariate normal as in (6.74),

$$(X, Y) \sim N\left((\mu_X, \mu_Y), \begin{pmatrix} \sigma_X^2 & \sigma_{XY} \\ \sigma_{XY} & \sigma_Y^2 \end{pmatrix} \right). \tag{6.93}$$

We then have that

$$Y \mid X = x \sim N(\alpha + \beta x, \sigma_e^2), \tag{6.94}$$

where the parameters are given in (6.82). The **residual** is $Y - \alpha - \beta X$. What is its conditional distribution? First, for fixed x,

$$[Y - \alpha - \beta X \mid X = x] =^{\mathcal{D}} [Y - \alpha - \beta x \mid X = x]. \tag{6.95}$$

This equation means that when we are conditioning on $X = x$, the conditional distribution stays the same if we fix $X = x$, which follows from the original definition of conditional expected value in (6.15). We know that subtracting the mean from a normal leaves a normal with mean 0 and the same variance, hence

$$Y - \alpha - \beta x \mid X = x \sim N(0, \sigma_e^2). \tag{6.96}$$

But the right-hand side has no x, hence $Y - \alpha - \beta X$ is independent of X, and has marginal distribution $N(0, \sigma_e^2)$.

6.8   Exercises

Exercise 6.8.1. Suppose (X, Y) has pdf $f(x, y)$, and that the conditional pdf of $Y \mid X = x$ does not depend on x. That is, there is a function $g(y)$ such that $f(x, y)/f_X(x) = g(y)$ for all $x \in \mathcal{X}$. Show that $g(y)$ is the marginal pdf of Y. [Hint: Find the joint pdf in terms of the conditional and marginal of X, then integrate out x.]

Exercise 6.8.2. A study was conducted on people near Newcastle on Tyne in 1972-74 (Appleton, French, and Vanderpump, 1996), and followed up twenty years later. We will focus on 1314 women in the study. The three variables we will consider are Z: age group (three values); X: whether they smoked or not (in 1974); and Y: whether they were still alive in 1994. Here are the frequencies:

Age group	Young (18 − 34)		Middle (35 − 64)		Old (65+)	
Smoker?	Yes	No	Yes	No	Yes	No
Died	5	6	92	59	42	165
Lived	174	213	262	261	7	28

(6.97)

(a) Treating proportions in the table as probabilities, find

$$P[Y = \text{Lived} \mid X = \text{Smoker}] \quad \text{and} \quad P[Y = \text{Lived} \mid X = \text{Nonsmoker}]. \tag{6.98}$$

Who were more likely to live, smokers or nonsmokers? (b) Find $P[X = \text{Smoker} \mid Z = z]$ for $z = $ Young, Middle, and Old. What do you notice? (c) Find

$$P[Y = \text{Lived} \mid X = \text{Smoker} \ \& \ Z = z] \tag{6.99}$$

and

$$P[Y = \text{Lived} \mid X = \text{Nonsmoker} \ \& \ Z = z] \tag{6.100}$$

for $z = $ Young, Middle, and Old. Adjusting for age group, who were more likely to live, smokers or nonsmokers? (d) Conditionally on age, the relationship between smoking and living is negative for each age group. Is it true that marginally (not conditioning on age), the relationship between smoking and living is negative? What is the explanation? (Simpson's paradox.)

Exercise 6.8.3. Suppose in a large population, the proportion of people who are infected with the HIV virus is $\epsilon = 1/100,000$. (In the example in Section 6.6.1, this proportion was $1/10,000$.) People can take a blood test to see whether they have the virus. The test is 99% accurate: The chance the test is positive given the person has the virus is 99%, and the chance the test is negative given the person does not have the virus is also 99%. Suppose a randomly chosen person takes the test. (a) What is the chance that this person does have the virus given that the test is positive? Is this close to 99%? (b) What is the chance that this person does have the virus given that the test is negative? Is this close to 1%? (c) Do the probabilities in (a) and (b) sum to 1?

Exercise 6.8.4. (a) Find the mode for the $\text{Beta}(\alpha, \beta)$ distribution. (b) What is the value for the $\text{Beta}(4, 8)$? How does it compare to the posterior mean in (6.71)?

Exercise 6.8.5. Consider (X, Y) being bivariate normal, $N(\pmb{\mu}, \pmb{\Sigma})$, as in (6.74). (a) Show that the exponent in (6.75) can be written

$$-\frac{1}{2|\pmb{\Sigma}|}(\sigma_Y^2(x - \mu_X) - 2\sigma_{XY}(x - \mu_X)(y - \mu_Y) + \sigma_X^2(y - \mu_Y)^2). \tag{6.101}$$

(b) Consider (6.101) as a quadratic in y, so that x and the parameters are constants. Show that y has coefficient $\gamma_1 = (\sigma_{XY}(x - \mu_X) + \mu_Y\sigma_Y^2)/|\pmb{\Sigma}|$ and y^2 has coefficient $\gamma_2 = -\sigma_X^2/(2|\pmb{\Sigma}|)$, as in (6.78). (c) Argue that the conditional pdf of $Y \mid X = x$ can be written as in (6.76) and (6.77), i.e.,

$$f_{Y|X}(y \mid x) = c(x, \pmb{\mu}, \pmb{\Sigma})e^{y\gamma_1 + y^2\gamma_2}. \tag{6.102}$$

(You do not have to explicitly find the function c, though you are welcome to do so.)

Exercise 6.8.6. Set

$$\frac{\mu}{\sigma^2} = \frac{\sigma_{XY}(x - \mu_X) + \mu_Y\sigma_Y^2}{|\pmb{\Sigma}|} \quad \text{and} \quad -\frac{1}{2\sigma^2} = -\frac{1}{2}\frac{\sigma_X^2}{|\pmb{\Sigma}|}. \tag{6.103}$$

Solve for μ and σ^2. The answers should be as in (6.79).

Exercise 6.8.7. Suppose $X \sim \text{Gamma}(\alpha, \lambda)$. Then $E[X] = \alpha/\lambda$ and $Var[X] = \alpha/\lambda^2$. (a) Find $E[X^2]$. (b) Is $E[1/X] = 1/E[X]$? (c) Find $E[1/X]$. It is finite for which values of α? (d) Find $E[1/X^2]$. It is finite for which values of α? (e) Now suppose $Y \sim \text{Gamma}(\beta, \delta)$, and it is independent of the X above. Let $R = X/Y$. Find $E[R]$, $E[R^2]$, and $Var[R]$.

Exercise 6.8.8. Suppose $X \mid \Theta = \theta \sim \text{Poisson}(c\theta)$, where c is a fixed constant. Also, marginally, $\Theta \sim \text{Gamma}(\alpha, \lambda)$. (a) Find the joint density $f(x, \theta)$ of (X, Θ). It can be written as $d\theta^{\alpha^* - 1}e^{-\lambda^*\theta}$ for some α^* and λ^*, where the d may depend on x, c, λ, α, but not on θ. What are α^* and λ^*? (b) Find the conditional distribution of $\Theta \mid X = x$. (c) Find $E[\Theta \mid X = x]$.

Exercise 6.8.9. In 1954, a large experiment was conducted to test the effectiveness of the Salk vaccine for preventing polio. A number of children were randomly assigned to two groups, one group receiving the vaccine, and a control group receiving a placebo. The number of children contracting polio (denoted x) in each group was then recorded. For the vaccine group, $n_V = 200745$ and $x_V = 57$; for the control group, $n_C = 201229$ and $x_C = 142$. That is, 57 of the 200745 children getting the vaccine contracted polio, and 142 of the 201229 children getting the placebo contracted polio. Let Θ_V and Θ_C be the polio rates per 100,000 children for the population vaccine and control groups, and suppose that (X_V, Θ_V) is independent of (X_C, Θ_C). Furthermore, suppose

$$X_V \mid \Theta_V = \theta_V \sim \text{Poisson}(c_V\,\theta_V) \quad \text{and} \quad X_C \mid \Theta_C = \theta_C \sim \text{Poisson}(c_C\,\theta_C), \qquad (6.104)$$

where the c's are the n's divided by 100,000, and marginally, $\Theta_V \sim \text{Gamma}(\alpha, \lambda)$ and $\Theta_C \sim \text{Gamma}(\alpha, \lambda)$ (i.e., they have the same prior). It may be reasonable to take priors with mean about 25, and standard deviation also about 25. (a) What do α and λ need to be so that $E[\Theta_V] = E[\Theta_C] = 25$ and $Var[\Theta_V] = Var[\Theta_C] = 25^2$? (b) What are the posterior distributions for the Θ's based on the above numbers? (That is, find $\Theta_V \mid X_V = 57$ and $\Theta_C \mid X_C = 142$.) (c) Find the posterior means and standard deviations of the Θ's. (d) We are hoping that the vaccine has a lower rate than the control. What is $P[\Theta_V < \Theta_C \mid X_V = 57, X_C = 142]$? (Sketch the two posterior pdfs.) (e) Consider the ratio of rates, $R = \Theta_V/\Theta_C$. Find $E[R \mid X_V = 57, X_C = 142]$ and $\sqrt{Var[R \mid X_V = 57, X_C = 142]}$. (f) True or false: The vaccine probably cuts the rate of polio by at least half (i.e., $P[R < 0.5 \mid X_V = 57, X_C = 142] > 0.5$). (g) What do you conclude about the effectiveness of the vaccine?

Exercise 6.8.10. Suppose $(Y_1, Y_2, \ldots, Y_{K-1}) \sim \text{Dirichlet}(\alpha_1, \ldots, \alpha_K)$, where $K > 4$. (a) Argue that

$$\frac{Y_1 Y_4}{Y_2 Y_3} =_{\mathcal{D}} \frac{X_1 X_4}{X_2 X_3}, \qquad (6.105)$$

where the X_i's are independent gammas. What are the parameters of the X_i's? (b) Find the following expected values, if they exist:

$$E\left[\frac{Y_1 Y_4}{Y_2 Y_3}\right] \quad \text{and} \quad E\left[\left(\frac{Y_1 Y_4}{Y_2 Y_3}\right)^2\right]. \qquad (6.106)$$

For which values of the α_i's does the first expected value exist? For which does the second exist?

Exercise 6.8.11. Suppose

$$(Z_1, Z_2, Z_3, Z_4) \mid (P_1, P_2, P_3, P_4) = (p_1, p_2, p_3, p_4) \sim \text{Multinomial}(n, (p_1, p_2, p_3, p_4)), \qquad (6.107)$$

and

$$(P_1, P_2, P_3) \sim \text{Dirichlet}(\alpha_1, \alpha_2, \alpha_3, \alpha_4). \qquad (6.108)$$

Note that $P_4 = 1 - P_1 - P_2 - P_3$. (a) Find the conditional distribution of

$$(P_1, P_2, P_3) \mid (Z_1, Z_2, Z_3, Z_4) = (z_1, z_2, z_3, z_4). \qquad (6.109)$$

(b) Data from the General Social Survey of 1991 included a comparison of men's and women's belief in the afterlife. See Agresti (2013). Assume the data is multinomial with four categories, arranged as follows:

Gender	Belief in afterlife	
	Yes	No or Undecided
Females	Z_1	Z_2
Males	Z_3	Z_4

(6.110)

The odds that a female believes in the afterlife is then P_1/P_2, and the odds that a male believes in the afterlife is P_3/P_4. The ratio of these odds is called the **odds ratio**:

$$\text{Odds Ratio} = \frac{P_1 P_4}{P_2 P_3}. \tag{6.111}$$

Then using Exercise 6.8.10, it can be shown that

$$[\text{Odds Ratio} \mid (Z_1, Z_2, Z_3, Z_4) = (z_1, z_2, z_3, z_4)] =^D \frac{X_1 X_4}{X_2 X_3}, \tag{6.112}$$

where the X_i's are independent, $X_i \sim \text{Gamma}(\beta_i, 1)$. Give the β_i's (which should be functions of the α_i's and z_i's). (c) Show that gender and belief in afterlife are independent if and only if Odds Ratio $= 1$. (d) The actual data from the survey are

	Belief in afterlife	
Gender	Yes	No or Undecided
Females	435	147
Males	375	134

(6.113)

Take the prior with all $\alpha_i = 1/2$. Find the posterior expected value and posterior standard deviation of the odds ratio. What do you conclude about the difference between men and women here?

Exercise 6.8.12. Suppose X_1 and X_2 are independent, $X_1 \sim \text{Poisson}(\lambda_1)$ and $X_2 \sim \text{Poisson}(\lambda_2)$. Let $T = X_1 + X_2$, which is $\text{Poisson}(\lambda_1 + \lambda_2)$. (a) Find the joint space and joint pmf of (X_1, T). (b) Find the conditional space and conditional pmf of $X_1 \mid T = t$. (c) Letting $p = \lambda_1/(\lambda_1 + \lambda_2)$, write the conditional pmf from part (b) in terms of p (and t). What is the conditional distribution?

Exercise 6.8.13. Now suppose X_1, X_2, X_3 are iid $\text{Poisson}(\lambda)$, and $T = X_1 + X_2 + X_3$. What is the distribution of $(X_1, X_2, X_3) \mid T = t$? What are the conditional mean vector and conditional covariance matrix of $(X_1, X_2, X_3) \mid T = t$?

Exercise 6.8.14. Suppose Z_1, Z_2, Z_3, Z_4 are iid Bernoulli(p) variables, and let Y be their sum. (a) Find the conditional space and pmf of (Z_1, Z_2, Z_3, Z_4) given $Y = 2$. (b) Find $E[Z_1 \mid Y = 2]$ and $Var[Z_1 \mid Y = 2]$. (c) Find $Cov[Z_1, Z_2 \mid Y = 2]$. (d) Let $\bar{Z}$ be the sample mean of the Z_i's. Find $E[\bar{Z} \mid Y = 2]$ and $Var[\bar{Z} \mid Y = 2]$.

Exercise 6.8.15. This problem is based on the fruit fly example, Section 6.4.1. Suppose $(Y_1, Y_2) \mid X = x$ are (conditionally) independent Bernoulli($x/2$)'s, and $X \sim \text{Binomial}(2, \theta)$. (a) Show that the marginal distribution of (Y_1, Y_2) is given by

$$f_{Y_1, Y_2}(0, 0) = \frac{1}{2}(1 - \theta)(2 - \theta);$$

$$f_{Y_1, Y_2}(0, 1) = f_{Y_1, Y_2}(1, 0) = \frac{1}{2}\theta(1 - \theta);$$

$$f_{Y_1, Y_2}(1, 1) = \frac{1}{2}\theta(1 + \theta). \tag{6.114}$$

(b) Find $Cov[Y_1, Y_2]$, their marginal covariance. Are Y_1 and Y_2 marginally independent? (c) The offsprings' genotypes are observed, but not the fathers'. But given

the offspring, one can guess what the father is by obtaining the conditional distribution of X given (Y_1, Y_2). In the following, take $\theta = 0.4$. (i) If the offspring has
$(y_1, y_2) = (0, 0)$, what would be your best guess for the father's x? (Best in the sense
of having the highest conditional probability.) Are you sure of your guess? (ii) If the
offspring has $(y_1, y_2) = (0, 1)$, what would be your best guess for the father's x? Are
you sure of your guess? (iii) If the offspring has $(y_1, y_2) = (1, 1)$, what would be your
best guess for the father's x? Are you sure of your guess?

Exercise 6.8.16. Continue the fruit flies example in Exercise 6.8.15. Let $W = Y_1 + Y_2$,
so that $W \mid X = x \sim \text{Binomial}(2, x/2)$ and $X \sim \text{Binomial}(2, \theta)$. (a) Find $E[W \mid X = x]$
and $Var[W \mid X = x]$. (b) Now find $E[W]$, $E[X^2]$, and $Var[W]$. (c) Which is larger, the
marginal variance of W or the marginal variance of X, or are they the same? (d) Note
that the Dobzhansky estimator $\widehat{\theta}$ in (6.65) is $\overline{W}/2$ based on $W_1, \ldots, W_n$ iid versions of
W. What is $Var[\widehat{\theta}]$ as a function of θ?

Exercise 6.8.17. (Student's t).

Definition 6.5. *Let Z and U be independent, $Z \sim N(0, 1)$ and $U \sim \text{Gamma}(v/2, 1/2)$
$(= \chi^2_v$ if v is an integer), and $T = Z/\sqrt{U/v}$. Then T is called Student's t on v degrees of
freedom, written $T \sim t_v$.*

(a) Show that $E[T] = 0$ (if $v > 1$) and $Var[T] = v/(v - 2)$ (if $v > 2$). [Hint: Note that
since Z and U are independent, $E[T] = E[Z]E[\sqrt{v/U}]$ if both expectations are finite.
Show that $E[1/\sqrt{U}]$ is finite if $v > 1$. Similarly for $E[T^2]$, where you can use the fact
from Exercise 6.8.7 that $E[1/U] = 1/(v - 2)$ for $v > 2$.] (b) The joint distribution of
(T, U) can be represented by

$$T \mid U = u \sim N(0, v/u) \quad \text{and} \quad U \sim \text{Gamma}(v/2, 1/2). \qquad (6.115)$$

Write down the joint pdf. (c) Integrate out the u from the joint pdf to show that the
marginal pdf of T is

$$f_v(t) = \frac{\Gamma((v + 1)/2)}{\Gamma(v/2)\sqrt{v\pi}} \frac{1}{(1 + t^2/v)^{(v+1)/2}}. \qquad (6.116)$$

For what values of v is this density valid?

Exercise 6.8.18. Suppose that $X \sim N(\mu, 1)$, and $Y = X^2$, so that $Y \sim \chi^2_1(\mu^2)$, a
noncentral χ^2 on one degree of freedom. (The pdf of Y was derived in Exercise
1.7.14.) Show that the moment generating function of Y is

$$M_Y(y) = \frac{1}{\sqrt{1 - 2t}} e^{\mu^2 t/(1 - 2t)}. \qquad (6.117)$$

[Hint: Find the mgf of Y by finding $E[e^{tX^2}]$ directly. Write out the integral with the pdf
of X, then complete the square in the exponent, where you will see another normal
pdf with a different variance.]

Exercise 6.8.19. Suppose that conditionally,

$$W \mid K = k \sim \chi^2_{v+2k}, \qquad (6.118)$$

and marginally, $K \sim \text{Poisson}(\lambda)$. (a) Find the unconditional mean and variance of W. (You can take the means and variances of the χ^2 and Poisson from Tables 1.1 and 1.2.) (b) What is the conditional mgf of W given $K = k$? (You don't need to derive it, just write it down based on what you know about the mgf of the $\chi^2_m =$ Gamma$(m/2, 1/2)$.) For what values of t is it finite? (c) Show that the unconditional mgf of W is

$$M_W(t) = \frac{1}{(1-2t)^{\nu/2}} \, e^{2\lambda t/(1-2t)}. \tag{6.119}$$

(d) Suppose $Y \sim \chi^2_1(\mu^2)$ as in Exercise 6.8.18. The mgf for Y is of the form in (6.119). What are the corresponding ν and λ? What are the mean and variance of the Y? [Hint: Use part (a).]

Exercise 6.8.20. *In this problem you should use your intuition, rather than try to formally construct conditional densities.* Suppose U_1, U_2, U_3 are iid Uniform$(0, 1)$, and let $U_{(3)}$ be the maximum of the three. Consider the conditional distribution of $U_1 \mid U_{(3)} = .9$. (a) Is the conditional distribution continuous? (b) Is the conditional distribution discrete? (c) What is the conditional space? (d) What is $P[U_1 = .9 \mid U_{(3)} = .9]$? (e) What is $E[U_1 \mid U_{(3)} = .9]$?

Exercise 6.8.21. Imagine a box containing three cards. One is red on both sides, one is green on both sides, and one is red on one side and green on the other. You close your eyes and randomly pick a card and lay it on the table. You open your eyes and notice that the side facing up is red. What is the chance the side facing down is red, too? [Hint: Let Y_1 be the color of the side facing up, and Y_2 the color of the side facing down. Then find the joint pmf of (Y_1, Y_2), and from that $P[Y_2 = \text{red} \mid Y_1 = \text{red}]$.]

Chapter 7

The Multivariate Normal Distribution

7.1 Definition

Almost all data are multivariate, that is, entail more than one variable. There are two general-purpose multivariate models: the **multivariate normal** for continuous data, and the **multinomial** for categorical data. There are many specialized multivariate distributions, but these two are the only ones that are used in all areas of statistics. We have seen the bivariate normal in Section 5.4.1, and the multinomial is introduced in Section 2.5.3.

The multivariate normal is by far the most commonly used distribution for continuous multivariate data. Which is not to say that all data are distributed normally, nor that all techniques assume such. Rather, one usually either assumes normality, or makes few assumptions at all and relies on asymptotic results. Some of the nice properties of the multivariate normal:

- It is completely determined by the means, variances, and covariances.

- Elements are independent if and only if they are uncorrelated.

- Marginals of multivariate normals are multivariate normal.

- An affine transformation of a multivariate normal is multivariate normal.

- Conditionals of a multivariate normal are multivariate normal.

- The sample mean is independent of the sample covariance matrix in an iid normal sample.

The multivariate normal arises from iid **standard normals**, that is, iid $N(0,1)$'s. Suppose $\mathbf{Z} = (Z_1, \ldots, Z_M)$ is an $1 \times M$ vector of iid $N(0,1)$'s. Because they are independent, all the covariances are zero, so that

$$E[\mathbf{Z}] = \mathbf{0}_M \text{ and } Cov[\mathbf{Z}] = \mathbf{I}_M. \tag{7.1}$$

A general multivariate normal distribution can have any (legitimate) mean and covariance, achieved through the use of affine transformations. Here is the definition.

Definition 7.1. *Let the $1 \times p$ vector $\mathbf{Y}$ be defined by*

$$\mathbf{Y} = \boldsymbol{\mu} + \mathbf{Z}\mathbf{B}', \tag{7.2}$$

where $\mathbf{B}$ is $p \times M$, $\boldsymbol{\mu}$ is $1 \times p$, and $\mathbf{Z}$ is a $1 \times M$ vector of iid standard normals. Then $\mathbf{Y}$ is **multivariate normal with mean $\boldsymbol{\mu}$ and covariance matrix $\boldsymbol{\Sigma} \equiv \mathbf{B}\mathbf{B}'$, written**

$$\mathbf{Y} \sim N_p(\boldsymbol{\mu}, \boldsymbol{\Sigma}). \tag{7.3}$$

We often drop the subscript "p" from the notation. The definition is for a row vector $\mathbf{Y}$. A multivariate normal column vector is defined the same way, then transposed. That is, if $\mathbf{Z}$ is a $K \times 1$ vector of iid $N(0,1)$'s, then $\mathbf{Y} = \boldsymbol{\mu} + \mathbf{B}\mathbf{Z}$ is $N(\boldsymbol{\mu}, \boldsymbol{\Sigma})$ as well, where now $\boldsymbol{\mu}$ is $p \times 1$. In this book, I'll generally use a row vector if the elements are measurements of different variables on the same observation, and a column vector if they are measurements of the same variable on different observations, but there is no strict demarcation. The thinking is that a typical $n \times p$ data matrix has n observations, represented by the rows, and p variables, represented by the columns. In fact, a multivariate normal matrix is simply a long multivariate normal vector chopped up and arranged into a matrix.

From (7.1) and (7.2), the usual affine transformation results from Section 2.3 show that $E[\mathbf{Y}] = \boldsymbol{\mu}$ and $Cov[\mathbf{Y}] = \mathbf{B}\mathbf{B}' = \boldsymbol{\Sigma}$. The definition goes further, implying that the distribution depends on $\mathbf{B}$ only through the $\mathbf{B}\mathbf{B}'$. For example, consider the two matrices

$$\mathbf{B}_1 = \begin{pmatrix} 1 & 1 & 0 \\ 0 & 1 & 2 \end{pmatrix} \text{ and } \mathbf{B}_2 = \begin{pmatrix} \frac{3}{\sqrt{5}} & \frac{1}{\sqrt{5}} \\ 0 & \sqrt{5} \end{pmatrix}, \tag{7.4}$$

so that

$$\mathbf{B}_1\mathbf{B}_1' = \mathbf{B}_2\mathbf{B}_2' = \begin{pmatrix} 2 & 1 \\ 1 & 5 \end{pmatrix} \equiv \boldsymbol{\Sigma}. \tag{7.5}$$

Thus the definition says that both

$$\mathbf{Y}_1 = (Z_1 + Z_2, Z_2 + 2Z_3) \text{ and } \mathbf{Y}_2 = \left(\frac{3}{\sqrt{5}}Z_1 + \frac{1}{\sqrt{5}}Z_2, \sqrt{5}Z_2 \right) \tag{7.6}$$

are $N(\mathbf{0}_2, \boldsymbol{\Sigma})$. They clearly have the same mean and covariance matrix, but it is not obvious they have the exact same distribution, especially as they depend on different numbers of Z_i's. To see the distributions are the same, we have to look at the mgf. We have already found the mgf for the bivariate ($p = 2$) normal in (5.50), and the proof here is the same. The answer is

$$M_{\mathbf{Y}}(\mathbf{t}) = e^{\boldsymbol{\mu}\mathbf{t}' + \frac{1}{2}\mathbf{t}\boldsymbol{\Sigma}\mathbf{t}'}, \quad \mathbf{t} \in \mathbb{R}^p. \tag{7.7}$$

Thus the distribution of $\mathbf{Y}$ does depend on just $\boldsymbol{\mu}$ and $\boldsymbol{\Sigma}$, so that because $\mathbf{Y}_1$ and $\mathbf{Y}_2$ have the same $\mathbf{B}_i\mathbf{B}_i'$ (and the same mean), they have the same distribution.

Can $\boldsymbol{\mu}$ and $\boldsymbol{\Sigma}$ be anything, or are there restrictions? Any $\boldsymbol{\mu}$ is possible since there are no restrictions on it in the definition. The covariance matrix $\boldsymbol{\Sigma}$ can be $\mathbf{B}\mathbf{B}'$ for any $p \times M$ matrix $\mathbf{B}$. Note that M is arbitrary, too. Clearly, $\boldsymbol{\Sigma}$ must be symmetric, but we already knew that. It must also be **nonnegative definite**, which we define now.

Definition 7.2. *A symmetric $p \times p$ matrix Ω is **nonnegative definite** if*

$$c\Omega c' \geq 0 \ \text{for all } 1 \times p \text{ vectors } c. \tag{7.8}$$

*The Ω is **positive definite** if*

$$c\Omega c' > 0 \ \text{for all } 1 \times p \text{ vectors } c \neq 0. \tag{7.9}$$

Note that $c\mathbf{BB}'c' = \|c\mathbf{B}\|^2 \geq 0$, which means that Σ must be nonnegative definite. But from (2.54),

$$c\Sigma c' = Cov(\mathbf{Y}c') = Var(\mathbf{Y}c') \geq 0, \tag{7.10}$$

because all variances are nonnegative. That is, any covariance matrix has to be non-negative definite, not just multivariate normal ones.

So we know that Σ must be symmetric and nonnegative definite. Are there any other restrictions, or for any symmetric nonnegative definite matrix is there a corresponding $\mathbf{B}$? Yes. In fact, there are potentially many such *square roots* $\mathbf{B}$. See Exercise 7.8.5. A nice one is the symmetric square root which will be seen in (7.15).

We conclude that the multivariate normal distribution is defined for any (μ, Σ), where $\mu \in \mathbb{R}^p$ and Σ is symmetric and positive definite, i.e., it can be any valid covariance matrix.

7.1.1 Spectral decomposition

To derive a symmetric square root, as well as perform other useful tasks, we need the following decomposition. Recall from (5.51) that a $p \times p$ matrix Γ is orthogonal if

$$\Gamma'\Gamma = \Gamma\Gamma' = \mathbf{I}_p. \tag{7.11}$$

Theorem 7.3. *Spectral decomposition for symmetric matrices.* *If Ω is a symmetric $p \times p$ matrix, then there exists a $p \times p$ orthogonal matrix Γ and a unique $p \times p$ diagonal matrix Λ with diagonals $\lambda_1 \geq \lambda_2 \geq \cdots \geq \lambda_p$ such that*

$$\Omega = \Gamma\Lambda\Gamma'. \tag{7.12}$$

Exercise 7.8.1 shows that the columns of Γ are eigenvectors of Ω, with corresponding eigenvalues λ_i's. Here are some more handy facts about symmetric Ω and its spectral decomposition.

- Ω is positive definite if and only if all λ_i's are positive, and nonnegative definite if and only if all λ_i's are nonnegative. (Exercises 7.8.2 and 7.8.3.)

- The trace and determinant are, respectively,

$$\text{trace}(\Omega) = \sum \lambda_i \ \text{and} \ |\Omega| = \prod \lambda_i. \tag{7.13}$$

  The trace of a square matrix is the sum of its diagonals. (Exercise 7.8.4.)

- Ω is invertible if and only if its eigenvalues are nonzero, in which case its inverse is

$$\Omega^{-1} = \Gamma\Lambda^{-1}\Gamma'. \tag{7.14}$$

  Thus the inverse has the same eigenvectors, and eigenvalues $1/\lambda_i$. (Exercise 7.8.5.)

- If $\boldsymbol{\Omega}$ is nonnegative definite, then with $\boldsymbol{\Lambda}^{1/2}$ being the diagonal matrix with diagonal elements $\sqrt{\lambda_i}$,

$$\boldsymbol{\Omega}^{1/2} = \boldsymbol{\Gamma}\boldsymbol{\Lambda}^{1/2}\boldsymbol{\Gamma}' \tag{7.15}$$

is a symmetric square root of $\boldsymbol{\Omega}$, that is, it is symmetric and $\boldsymbol{\Omega}^{1/2}\boldsymbol{\Omega}^{1/2} = \boldsymbol{\Omega}$. (Follows from Exercise 7.8.5.)

The last item was used in the previous section to guarantee that any covariance matrix has a square root.

7.2 Some properties of the multivariate normal

We will prove the next few properties using the representation (7.2). They can also be easily shown using the mgf.

7.2.1 Affine transformations

Affine transformations of multivariate normals are also multivariate normal, because any affine transformation of a multivariate normal vector is an affine transformation of an affine transformation of a standard normal vector, and an affine transformation of an affine transformation is also an affine transformation. That is, suppose $\mathbf{Y} \sim N_p(\boldsymbol{\mu}, \boldsymbol{\Sigma})$, and $\mathbf{W} = \mathbf{c} + \mathbf{Y}\mathbf{D}'$ for $q \times p$ matrix $\mathbf{D}$ and $q \times 1$ vector $\mathbf{c}$. Then we know that for some $\mathbf{B}$ with $\mathbf{B}\mathbf{B}' = \boldsymbol{\Sigma}$, $\mathbf{Y} = \boldsymbol{\mu} + \mathbf{Z}\mathbf{B}'$, where $\mathbf{Z}$ is a vector of iid standard normals. Hence

$$\mathbf{W} = \mathbf{c} + \mathbf{Y}\mathbf{D}' = \mathbf{c} + (\boldsymbol{\mu} + \mathbf{Z}\mathbf{B}')\mathbf{D}' = \mathbf{c} + \boldsymbol{\mu}\mathbf{D}' + \mathbf{Z}(\mathbf{D}\mathbf{B})'. \tag{7.16}$$

Then by Definition 7.1,

$$\mathbf{W} \sim N(\mathbf{c} + \boldsymbol{\mu}\mathbf{D}', \mathbf{D}\mathbf{B}\mathbf{B}'\mathbf{D}') = N(\mathbf{c} + \boldsymbol{\mu}\mathbf{D}', \mathbf{D}\boldsymbol{\Sigma}\mathbf{D}'). \tag{7.17}$$

Of course, the mean and covariance result we already knew.

As a simple but important special case, suppose $X_1, \ldots, X_n$ are iid $N(\mu, \sigma^2)$, so that $\mathbf{X} = (X_1, \ldots, X_n)' \sim N(\mu\mathbf{1}_n, \mathbf{I}_n)$, where $\mathbf{1}_n$ is the $n \times 1$ vector of all 1's. Then $\overline{X} = \mathbf{c}\mathbf{X}$ where $\mathbf{c} = (1/n)\mathbf{1}_n'$. Thus

$$\overline{X} \sim N\left(\frac{1}{n}\mathbf{1}_n'\mu\mathbf{1}_n, \frac{1}{n}\mathbf{1}_n'\sigma^2\mathbf{I}_n\frac{1}{n}\mathbf{1}_n\right) = N\left(\mu, \frac{1}{n}\sigma^2\right), \tag{7.18}$$

since $\mathbf{1}_n'\mathbf{1}_n = n$. This result checks with what we found in (4.45) using mgfs.

7.2.2 Marginals

Because marginals are special cases of affine transformations, marginals of multivariate normals are also multivariate normal. One needs just to pick off the appropriate means and covariances. So If $\mathbf{Y} = (Y_1, \ldots, Y_5)$ is $N_5(\boldsymbol{\mu}, \boldsymbol{\Sigma})$, and $\mathbf{W} = (Y_2, Y_5)$, then

$$\mathbf{W} \sim N_2\left((\mu_2, \mu_5), \begin{pmatrix} \sigma_{22} & \sigma_{25} \\ \sigma_{52} & \sigma_{55} \end{pmatrix}\right). \tag{7.19}$$

Here, σ_{ij} is the ij^{th} element of $\boldsymbol{\Sigma}$, so that $\sigma_{ii} = \sigma_i^2$. See Exercise 7.8.6.

7.2.3 Independence

In Section 3.3, we showed that independence of two random variables means that their covariance is 0, but that a covariance of 0 does not imply independence. But, with multivariate normals, it does: if (X, Y) is bivariate normal, and $Cov(X, Y) = 0$, then X and Y are independent. The next theorem proves a generalization of this independence to sets of variables. We will use (2.110), where for vectors $\mathbf{X} = (X_1, \ldots, X_p)$ and $\mathbf{Y} = (Y_1, \ldots, Y_q)$,

$$Cov[\mathbf{X}, \mathbf{Y}] = \begin{pmatrix} Cov[X_1, Y_1] & Cov[X_1, Y_2] & \cdots & Cov[X_1, Y_q] \\ Cov[X_2, Y_1] & Cov[X_2, Y_2] & \cdots & Cov[X_2, Y_q] \\ \vdots & \vdots & \ddots & \vdots \\ Cov[X_p, Y_1] & Cov[X_p, Y_2] & \cdots & Cov[X_p, Y_q] \end{pmatrix}, \tag{7.20}$$

the matrix of all possible covariances of one element from $\mathbf{X}$ and one from $\mathbf{Y}$.

Theorem 7.4. *Suppose*

$$\mathbf{W} = (\mathbf{X}, \mathbf{Y}) \tag{7.21}$$

is multivariate normal, where $\mathbf{X}$ *is* $1 \times p$ *and* $\mathbf{Y}$ *is* $1 \times q$*. If* $Cov[\mathbf{X}, \mathbf{Y}] = \mathbf{0}$ *(i.e.,* $Cov[X_i, Y_j] = 0$ *for all* i, j*), then* $\mathbf{X}$ *and* $\mathbf{Y}$ *are independent.*

Proof. For simplicitly, we will assume the mean of $\mathbf{W}$ is 0. Because covariances between the X_i's and Y_j's are zero,

$$Cov(\mathbf{W}) = \begin{pmatrix} Cov(\mathbf{X}) & \mathbf{0} \\ \mathbf{0} & Cov(\mathbf{Y}) \end{pmatrix}. \tag{7.22}$$

(The $\mathbf{0}$'s denote matrices of the appropriate size with all elements zero.) Both of those individual covariance matrices have square roots, hence there are matrices $\mathbf{B}$, $p \times p$, and $\mathbf{C}$, $q \times q$, such that

$$Cov[\mathbf{X}] = \mathbf{BB}' \quad \text{and} \quad Cov[\mathbf{Y}] = \mathbf{CC}'. \tag{7.23}$$

Thus

$$Cov[\mathbf{W}] = \mathbf{AA}' \quad \text{where} \quad \mathbf{A} = \begin{pmatrix} \mathbf{B} & \mathbf{0} \\ \mathbf{0} & \mathbf{C} \end{pmatrix}. \tag{7.24}$$

Then by definition, we know that we can represent the distribution of $\mathbf{W}$ with $\mathbf{Z}$ being a $1 \times (p + q)$ vector of iid standard normals,

$$(\mathbf{X}, \mathbf{Y}) = \mathbf{W} = \mathbf{ZA}' = (\mathbf{Z}_1, \mathbf{Z}_2) \begin{pmatrix} \mathbf{B} & \mathbf{0} \\ \mathbf{0} & \mathbf{C} \end{pmatrix}' = (\mathbf{Z}_1 \mathbf{B}', \mathbf{Z}_2 \mathbf{C}'), \tag{7.25}$$

where $\mathbf{Z}_1$ and $\mathbf{Z}_2$ are $1 \times p$ and $1 \times q$, respectively. But that means that

$$\mathbf{X} = \mathbf{Z}_1 \mathbf{B}' \quad \text{and} \quad \mathbf{Y} = \mathbf{Z}_2 \mathbf{C}'. \tag{7.26}$$

Because $\mathbf{Z}_1$ is independent of $\mathbf{Z}_2$, we have that $\mathbf{X}$ is independent of $\mathbf{Y}$. □

Note that this result can be extended to partitions of $\mathbf{W}$ into more than two groups. That is, if $\mathbf{W} = (\mathbf{X}^{(1)}, \ldots, \mathbf{X}^{(K)})$, where each $\mathbf{X}^{(k)}$ is a vector, then

$$\mathbf{W} \text{ multivariate normal and } Cov[\mathbf{X}^{(k)}, \mathbf{X}^{(l)}] = \mathbf{0} \text{ for all } k \neq l$$

$$\implies \mathbf{X}^{(1)}, \ldots, \mathbf{X}^{(K)} \text{ are mutually independent.} \tag{7.27}$$

Especially, if the covariance matrix of a multivariate normal vector is diagonal, then all of the elements are mutually independent.

7.3   PDF

The multivariate normal has a pdf only if the covariance is invertible (i.e., positive definite). In that case, its pdf is easy to find using the same procedure used to find the pdf of the bivariate normal in Section 5.4.1. Suppose $\mathbf{Y} \sim N_p(\boldsymbol{\mu}, \boldsymbol{\Sigma})$ where $\boldsymbol{\Sigma}$ is a $p \times p$ positive definite matrix. Let $\mathbf{B}$ be the symmetric square root of $\boldsymbol{\Sigma}$ as in (7.15), which is also positive definite. (Why?) Then, if $\mathbf{Y}$ is a row vector, $\mathbf{Y} = \boldsymbol{\mu} + \mathbf{ZB}'$, where $\mathbf{Z} \sim N(\mathbf{0}_p, \mathbf{I}_p)$. Follow the steps as in (5.42) to (5.46). The only real difference is that because we have p Z_i's, the power of the 2π is $p/2$ instead of 1. Thus

$$f_{\mathbf{Y}}(\mathbf{y}) = \frac{1}{(2\pi)^{p/2}} \frac{1}{\sqrt{|\boldsymbol{\Sigma}|}} e^{-\frac{1}{2}(\mathbf{y}-\boldsymbol{\mu})\boldsymbol{\Sigma}^{-1}(\mathbf{y}-\boldsymbol{\mu})'}. \tag{7.28}$$

If $\mathbf{Y}$ is a column vector, then the $(\mathbf{y} - \boldsymbol{\mu})$ and $(\mathbf{y} - \boldsymbol{\mu})'$ are switched.

7.4   Sample mean and variance

Often one desires a confidence interval for the population mean. Specifically, suppose $X_1, \ldots, X_n$ are iid $N(\mu, \sigma^2)$. By (7.18), $\overline{X} \sim N(\mu, \sigma^2/n)$, so that

$$Z = \frac{\overline{X} - \mu}{\sigma/\sqrt{n}} \sim N(0,1). \tag{7.29}$$

This Z is called a **pivotal quantity**, meaning it has a known distribution even though its definition includes unknown parameters. Then

$$P\left[-1.96 < \frac{\overline{X} - \mu}{\sigma/\sqrt{n}} < 1.96\right] = 0.95, \tag{7.30}$$

or, untangling the equations to get μ in the middle,

$$P\left[\overline{X} - 1.96 \frac{\sigma}{\sqrt{n}} < \mu < \overline{X} + 1.96 \frac{\sigma}{\sqrt{n}}\right] = 0.95. \tag{7.31}$$

Thus,

$$\left(\overline{X} - 1.96 \frac{\sigma}{\sqrt{n}}, \overline{X} + 1.96 \frac{\sigma}{\sqrt{n}}\right) \quad \text{is a 95\% confidence interval for } \mu, \tag{7.32}$$

at least if σ is known.

But what if σ is not known? Then you estimate it, which will change the distribution, that is,

$$\frac{\overline{X} - \mu}{\hat{\sigma}/\sqrt{n}} \sim ?? \tag{7.33}$$

The sample variance for a sample $x_1, \ldots, x_n$ is

$$s^2 = \frac{\sum(x_i - \overline{x})^2}{n}, \quad \text{or is it } s_*^2 = \frac{\sum(x_i - \overline{x})^2}{n-1}? \tag{7.34}$$

Rather than worry about that question now, we will find the joint distribution of the mean with the numerator:

$$(\overline{X}, U), \quad \text{where } U = \sum(X_i - \overline{X})^2. \tag{7.35}$$

To start, note that the deviations $x_i - \overline{x}$ are linear functions of the x_i's, as of course is $\overline{x}$, and we know how to deal with linear combinations of normals. That is, letting $\mathbf{X} = (X_1, \ldots, X_n)'$ be the column vector of observations, because the elements are iid,

$$\mathbf{X} \sim N(\mu \mathbf{1}_n, \sigma^2 \mathbf{I}_n), \tag{7.36}$$

where $\mathbf{1}_n$ is the $n \times 1$ vector of all 1's. Then as in Exercise 2.7.5 and equation (7.18), the mean can be written $\overline{X} = (1/n)\mathbf{1}_n'\mathbf{X}$, and the deviations

$$\begin{pmatrix} X_1 - \overline{X} \\ X_2 - \overline{X} \\ \vdots \\ X_n - \overline{X} \end{pmatrix} = \mathbf{X} - \mathbf{1}_n\overline{X} = \mathbf{I}_n\mathbf{X} - \frac{1}{n}\mathbf{1}_n\mathbf{1}_n'\mathbf{X} = \mathbf{H}_n\mathbf{X}, \tag{7.37}$$

where

$$\mathbf{H}_n = \mathbf{I}_n - \frac{1}{n}\mathbf{1}_n\mathbf{1}_n' \tag{7.38}$$

is the $n \times n$ **centering** matrix. It is called the centering matrix because for any $n \times 1$ vector $\mathbf{a}$, $\mathbf{H}_n\mathbf{a}$ subtracts the mean of the elements from each element, centering the values at 0. Note that if all the elements are the same, centering will set everything to 0, i.e.,

$$\mathbf{H}_n\mathbf{1}_n = \mathbf{0}_n. \tag{7.39}$$

Also, if the mean of the elements already is 0, centering does nothing, which in particular means that $\mathbf{H}_n(\mathbf{H}_n\mathbf{a}) = \mathbf{H}_n\mathbf{a}$, or

$$\mathbf{H}_n\mathbf{H}_n = \mathbf{H}_n. \tag{7.40}$$

Such a matrix is called **idempotent**. It is not difficult to verify (7.40) directly using the definition (7.38) and multiplying things out. In fact, $\mathbf{I}_n$ and $(1/n)\mathbf{1}_n\mathbf{1}_n'$ are also idempotent.

Back to the task. To analyze the mean and deviations together, we stack them:

$$\begin{pmatrix} \overline{X} \\ X_1 - \overline{X} \\ \vdots \\ X_n - \overline{X} \end{pmatrix} - \begin{pmatrix} \frac{1}{n}\mathbf{1}_n' \\ \mathbf{H}_n \end{pmatrix} \mathbf{X}. \tag{7.41}$$

Equation (7.41) gives explicitly that the vector containing the mean and deviations is a linear transformation of a multivariate normal, hence the vector is multivariate normal. The mean and covariance are

$$E\left[\begin{pmatrix} \overline{X} \\ X_1 - \overline{X} \\ \vdots \\ X_n - \overline{X} \end{pmatrix} \right] = \begin{pmatrix} \frac{1}{n}\mathbf{1}_n' \\ \mathbf{H}_n \end{pmatrix} \mu \mathbf{1}_n = \begin{pmatrix} \frac{1}{n}\mathbf{1}_n'\mathbf{1}_n \\ \mathbf{H}_n\mathbf{1}_n \end{pmatrix} \mu = \begin{pmatrix} \mu \\ \mathbf{0}_n \end{pmatrix} \tag{7.42}$$

and

$$
Cov\left[\begin{pmatrix} \overline{X} \\ X_1 - \overline{X} \\ \vdots \\ X_n - \overline{X} \end{pmatrix}\right] = \begin{pmatrix} \frac{1}{n}\mathbf{1}'_n \\ \mathbf{H}_n \end{pmatrix} \sigma^2 \mathbf{I}_n \begin{pmatrix} \frac{1}{n}\mathbf{1}'_n \\ \mathbf{H}_n \end{pmatrix}'
$$

$$
= \sigma^2 \begin{pmatrix} \frac{1}{n^2}\mathbf{1}'_n\mathbf{1}_n & \frac{1}{n}\mathbf{1}'_n\mathbf{H}_n \\ \frac{1}{n}\mathbf{H}_n\mathbf{1}_n & \mathbf{H}_n\mathbf{H}_n \end{pmatrix}
$$

$$
= \sigma^2 \begin{pmatrix} \frac{1}{n} & \mathbf{0}'_n \\ \mathbf{0}_n & \mathbf{H}_n \end{pmatrix}. \tag{7.43}
$$

Look at the $\mathbf{0}_n$'s: The covariances between $\overline{X}$ and the deviations $\mathbf{H}_n\mathbf{X}$ are zero, hence with the multivariate normality means they are independent. Further, we can read off the distributions:

$$
\overline{X} \sim N\left(\mu, \frac{1}{n}\sigma^2\right) \quad \text{and} \quad \mathbf{H}_n\mathbf{X} \sim N(\mathbf{0}_n, \sigma^2\mathbf{H}_n). \tag{7.44}
$$

The first we already knew. But because $\overline{X}$ and $\mathbf{H}_n\mathbf{X}$ are independent, and $U = \|\mathbf{H}_n\mathbf{X}\|^2$ is a function of just $\mathbf{H}_n\mathbf{X}$,

$$
\overline{X} \text{ and } U = \|\mathbf{H}_n\mathbf{X}\|^2 = \sum(X_i - \overline{X})^2 \text{ are independent.} \tag{7.45}
$$

The next section goes through development of the χ^2 distribution, which eventually (Lemma 7.6) shows that U/σ^2 is χ^2_{n-1}.

7.5 Chi-square distribution

In Exercises 1.7.13, 1.7.14, and 2.7.11, we defined the central and noncentral chi-square distributions on one degree of freedom. Here we look at the more general chi-squares.

Definition 7.5. *Suppose the $\nu \times 1$ vector $\mathbf{Z} \sim N(\mathbf{0}, \mathbf{I}_\nu)$. Then*

$$
W = Z_1^2 + \cdots + Z_\nu^2 = \mathbf{Z}'\mathbf{Z} \tag{7.46}
$$

*has the **central chi-square distribution** on ν degrees of freedom, written*

$$
W \sim \chi^2_\nu. \tag{7.47}
$$

Often one drops the "central" when referring to this distribution, unless trying to distinguish it from the noncentral chi-square coming in Section 7.5.3. (It is also often called "chi-squared.")

The expected value and variance of a central chi-square are easy to find since we know that for $Z \sim N(0,1)$, $E[Z^2] = Var[Z] = 1$, and $E[Z^4] = 3$ since the kurtosis κ_4 is 0. Thus $Var[Z^2] = E[Z^4] - E[Z^2]^2 = 2$. For $W \sim \chi^2_\nu$, by (7.46),

$$
E[W] = \nu E[Z^2] = \nu \quad \text{and} \quad Var[W] = \nu Var[Z^2] = 2\nu. \tag{7.48}
$$

Also, if $W_1, \ldots, W_k$ are independent, with $W_k \sim \chi^2_{\nu_k}$, then

$$W_1 + \cdots + W_K \sim \chi^2_{\nu_1 + \cdots + \nu_K}, \tag{7.49}$$

because each W_k is a sum of ν_k independent standard normal squares, so the sum of the W_k's is the sum of all $\nu_1 + \cdots + \nu_K$ independent standard normal squares.

For the pdf, recall that in Exercise 1.7.13, we showed that a χ^2_1 is a Gamma(1/2, 1/2). Thus a χ^2_ν is a sum of ν independent such gammas. These gammas all have the same rate $\lambda = 1/2$, hence we just add up the α's, which are all $1/2$, hence

$$\chi^2_\nu = \text{Gamma}\left(\frac{\nu}{2}, \frac{1}{2}\right), \tag{7.50}$$

as can be ascertained from Table 1.1 on page 7. This representation is another way to verify (7.48) and (7.49).

Now suppose $\mathbf{Y} \sim N(\boldsymbol{\mu}, \boldsymbol{\Sigma})$, where $\mathbf{Y}$ is $p \times 1$. We can do a multivariate standardization analogous to the univariate one in (7.29) if $\boldsymbol{\Sigma}$ is invertible:

$$\mathbf{Z} = \boldsymbol{\Sigma}^{-\frac{1}{2}}(\mathbf{Y} - \boldsymbol{\mu}). \tag{7.51}$$

Here, we will take $\boldsymbol{\Sigma}^{-1/2}$ to be inverse of the symmetric square root of $\boldsymbol{\Sigma}$ as in (7.15), though any square root will do. Since $\mathbf{Z}$ is a linear transformation of $\mathbf{X}$, it is multivariate normal. It is easy to see that $E[\mathbf{Z}] = \mathbf{0}$. For the covariance:

$$Cov[\mathbf{Z}] = \boldsymbol{\Sigma}^{-\frac{1}{2}} Cov[\mathbf{Y}] \boldsymbol{\Sigma}^{-\frac{1}{2}} = \boldsymbol{\Sigma}^{-\frac{1}{2}} \boldsymbol{\Sigma} \boldsymbol{\Sigma}^{-\frac{1}{2}} = \mathbf{I}_p. \tag{7.52}$$

Then

$$\mathbf{Z}'\mathbf{Z} = (\mathbf{Y} - \boldsymbol{\mu})' \boldsymbol{\Sigma}^{-\frac{1}{2}} \boldsymbol{\Sigma}^{-\frac{1}{2}} (\mathbf{Y} - \boldsymbol{\mu}) = (\mathbf{Y} - \boldsymbol{\mu})' \boldsymbol{\Sigma}^{-1} (\mathbf{Y} - \boldsymbol{\mu}) \sim \chi^2_p \tag{7.53}$$

by (7.46) and (7.47). Random variables of the form $(\mathbf{y} - \mathbf{a})'\mathbf{C}(\mathbf{y} - \mathbf{a})$ are called **quadratic forms**. We can use this random variable as a pivotal quantity, so that if $\boldsymbol{\Sigma}$ is known, then

$$\{\boldsymbol{\mu} \mid (\mathbf{y} - \boldsymbol{\mu})' \boldsymbol{\Sigma}^{-1} (\mathbf{y} - \boldsymbol{\mu}) \le \chi^2_{p,\alpha}\} \tag{7.54}$$

is a $100 \times (1 - \alpha)\%$ confidence region for $\boldsymbol{\mu}$, where $\chi^2_{p,\alpha}$ is the $(1 - \alpha)^{th}$ quantile of the χ^2_q. This region is an ellipsoid.

7.5.1 Noninvertible covariance matrix

We would like to apply this result to the $\mathbf{H}_n\mathbf{X}$ vector in (7.44), but we cannot use (7.54) directly because the covariance matrix of $\mathbf{H}_n\mathbf{X}$, $\sigma^2\mathbf{H}_n$, is not invertible. In general, if $\boldsymbol{\Sigma}$ is not invertible, then instead of its regular inverse, we use the **Moore-Penrose inverse**, which is a pseudoinverse, meaning it is not a real inverse but in some situations acts like one. To define it for nonnegative definite symmetric matrices, first let $\boldsymbol{\Sigma} = \boldsymbol{\Gamma}\boldsymbol{\Lambda}\boldsymbol{\Gamma}'$ be the spectral decomposition (7.12) of $\boldsymbol{\Sigma}$. If $\boldsymbol{\Sigma}$ is not invertible, then some of the diagonals (eigenvalues) of $\boldsymbol{\Lambda}$ will be zero. Suppose there are ν positive eigenvalues. Since the λ_i's are in order from largest to smallest, we have that

$$\lambda_1 \ge \lambda_2 \ge \cdots \ge \lambda_\nu > 0 = \lambda_{\nu+1} = \cdots = \lambda_p. \tag{7.55}$$

The Moore-Penrose inverse uses a formula similar to that in (7.14), but in the inner matrix, takes reciprocal of just the positive λ_i's. That is, let $\boldsymbol{\Lambda}_1$ be the $\nu \times \nu$ diagonal matrix with diagonals $\lambda_1, \ldots, \lambda_\nu$. Then the Moore-Penrose inverse of Σ and its square root are defined to be

$$\Sigma^+ = \Gamma \begin{pmatrix} \Lambda_1^{-1} & 0 \\ 0 & 0 \end{pmatrix} \Gamma' \text{ and } (\Sigma^+)^{\frac{1}{2}} = \Gamma \begin{pmatrix} \Lambda_1^{-\frac{1}{2}} & 0 \\ 0 & 0 \end{pmatrix} \Gamma', \qquad (7.56)$$

respectively. See Exercise 12.7.10 for general matrices.

Now if $\mathbf{Y} \sim N(\boldsymbol{\mu}, \Sigma)$, we let

$$\mathbf{Z}^+ = (\Sigma^+)^{\frac{1}{2}} (\mathbf{Y} - \mu). \qquad (7.57)$$

Again, $E[\mathbf{Z}^+] = \mathbf{0}$, but the covariance is

$$\begin{aligned}
Cov[\mathbf{Z}^+] &= (\Sigma^+)^{\frac{1}{2}} \Sigma (\Sigma^+)^{\frac{1}{2}} \\
&= \Gamma \begin{pmatrix} \Lambda_1^{-\frac{1}{2}} & 0 \\ 0 & 0 \end{pmatrix} \Gamma' \Gamma \begin{pmatrix} \Lambda_1 & 0 \\ 0 & 0 \end{pmatrix} \Gamma' \Gamma \begin{pmatrix} \Lambda_1^{-\frac{1}{2}} & 0 \\ 0 & 0 \end{pmatrix} \Gamma' \\
&= \Gamma \begin{pmatrix} \Lambda_1^{-\frac{1}{2}} & 0 \\ 0 & 0 \end{pmatrix} \begin{pmatrix} \Lambda_1 & 0 \\ 0 & 0 \end{pmatrix} \begin{pmatrix} \Lambda_1^{-\frac{1}{2}} & 0 \\ 0 & 0 \end{pmatrix} \Gamma' \\
&= \Gamma \begin{pmatrix} \mathbf{I}_\nu & 0 \\ 0 & 0 \end{pmatrix} \Gamma'. \qquad (7.58)
\end{aligned}$$

To get rid of the final two orthogonal matrices, we set

$$\mathbf{Z} = \Gamma' \mathbf{Z}^+ \sim N\left(\mathbf{0}, \begin{pmatrix} \mathbf{I}_\nu & 0 \\ 0 & 0 \end{pmatrix}\right). \qquad (7.59)$$

Note that the elements of $\mathbf{Z}$ are independent, the first ν of them are $N(0,1)$, and the last $p - \nu$ are $N(0,0)$, which means they are identically 0. Hence we have

$$(\mathbf{Z}^+)' \mathbf{Z}^+ = \mathbf{Z}' \mathbf{Z} = Z_1^2 + \ldots + Z_\nu^2 \sim \chi_\nu^2. \qquad (7.60)$$

To summarize,

$$(\mathbf{Y} - \mu)' \Sigma^+ (\mathbf{Y} - \mu) \sim \chi_\nu^2, \quad \nu = \#\{\text{Positive eigenvalues of } \Sigma\}. \qquad (7.61)$$

Note that this formula is still valid if Σ is invertible, because then $\Sigma^+ = \Sigma^{-1}$.

7.5.2   Idempotent covariance matrix

Recall that a matrix $\mathbf{H}$ is idempotent if $\mathbf{HH} = \mathbf{H}$. It turns out that the Moore-Penrose inverse of a symmetric idempotent matrix $\mathbf{H}$ is $\mathbf{H}$ itself. To see this fact, let $\mathbf{H} = \Gamma \Lambda \Gamma'$ be the spectral decomposition, and write

$$\begin{aligned}
\mathbf{HH} = \mathbf{H} &\Longrightarrow \Gamma \Lambda \Gamma' \Gamma \Lambda \Gamma' = \Gamma \Lambda \Gamma' \\
&\Longrightarrow \Gamma \Lambda^2 \Gamma' = \Gamma \Lambda \Gamma' \\
&\Longrightarrow \Lambda^2 = \Lambda. \qquad (7.62)
\end{aligned}$$

Since Λ is diagonal, we must have that $\lambda_i^2 = \lambda_i$ for each i. But then each λ_i must be 0 or 1. Letting ν be the number that are 1, since the eigenvalues have the positive ones first,

$$\mathbf{H} = \boldsymbol{\Gamma} \begin{pmatrix} \mathbf{I}_\nu & \mathbf{0} \\ \mathbf{0} & \mathbf{0} \end{pmatrix} \boldsymbol{\Gamma}', \tag{7.63}$$

hence (7.56) with $\Lambda_1 = \mathbf{I}_\nu$ shows that the Moore-Penrose inverse of $\mathbf{H}$ is $\mathbf{H}$. Also, by (7.13), the trace of a matrix is the sum of its eigenvalues, hence in this case

$$\text{trace}(\mathbf{H}) = \nu. \tag{7.64}$$

Thus (7.61) shows that

$$\mathbf{Y} \sim N(\boldsymbol{\mu}, \mathbf{H}) \implies (\mathbf{Y} - \boldsymbol{\mu})'(\mathbf{Y} - \boldsymbol{\mu}) \sim \chi^2_{\text{trace}(\mathbf{H})}, \tag{7.65}$$

where we use the fact that $(\mathbf{Y} - \boldsymbol{\mu})$ and $\mathbf{H}(\mathbf{Y} - \boldsymbol{\mu})$ have the same distribution.

Finally turn to (7.44), where we started with $\mathbf{X} \sim N(\mu \mathbf{1}_n, \sigma^2 \mathbf{I}_n)$, and derived in (7.44) that $\mathbf{H}_n \mathbf{X} \sim N(\mathbf{0}, \sigma^2 \mathbf{H}_n)$. Then by (7.65) with $\mathbf{Y} = (1/\sigma^2)\mathbf{H}_n \mathbf{X}$, we have

$$\frac{1}{\sigma}\mathbf{H}_n \mathbf{X} \sim N(\mathbf{0}, \mathbf{H}_n) \implies \frac{1}{\sigma^2}\mathbf{X}'\mathbf{H}_n \mathbf{X} \sim \chi^2_{\text{trace}(\mathbf{H}_n)}. \tag{7.66}$$

Exercise 7.8.11 shows that

$$\mathbf{X}'\mathbf{H}_n \mathbf{X} = \sum_{i=1}^{n}(X_i - \overline{X})^2 \quad \text{and} \quad \text{trace}(\mathbf{H}_n) = n - 1. \tag{7.67}$$

Together with (7.44) and (7.45), we have the following.

Lemma 7.6. *If $X_1, \ldots, X_n$ are iid $N(\mu, \sigma^2)$, then $\overline{X}$ and $\sum(X_i - \overline{X})^2$ are independent, with*

$$\overline{X} \sim N\left(\mu, \frac{1}{n}\sigma^2\right) \quad \text{and} \quad \sum(X_i - \overline{X})^2 \sim \sigma^2 \chi^2_{n-1}. \tag{7.68}$$

($U \sim \sigma^2 \chi^2_\nu$ means that $U/\sigma^2 \sim \chi^2_\nu$.)

Since $E[\chi^2_\nu] = \nu$, $E[\sum(X_i - \overline{X})^2] = (n-1)\sigma^2$. Thus of the two sample variance formulas in (7.34), only the second is unbiased, meaning it has expected value σ^2:

$$E[S_*^2] = E\left[\frac{\sum(X_i - \overline{X})^2}{n-1}\right] = \frac{(n-1)\sigma^2}{n-1} = \sigma^2. \tag{7.69}$$

(Which doesn't mean it is better than S^2.)

7.5.3 Noncentral chi-square distribution

Definition 7.5 of the central chi-square assumed that the mean of the normal vector was zero. The **noncentral** chi-square allows arbitrary means:

Definition 7.7. *Suppose $\mathbf{Z} \sim N(\boldsymbol{\gamma}, \mathbf{I}_\nu)$. Then*

$$W = \mathbf{Z}'\mathbf{Z} = \|\mathbf{Z}\|^2 \tag{7.70}$$

*has the **noncentral chi-squared distribution** on ν degrees of freedom with noncentrality parameter $\Delta = \|\boldsymbol{\gamma}\|^2$, written*

$$W \sim \chi^2_\nu(\Delta). \tag{7.71}$$

Note that the central χ^2 is the noncentral chi-square with $\Delta = 0$. See Exercise 7.8.20 for the mgf, and Exercise 7.8.22 for the pdf, of the noncentral chi-square.

This definition implies that the distribution depends on the parameter γ through just the noncentrality parameter. That is, if Z and Z^* are multivariate normal with the same covariance matrix I_ν but different means γ and γ^*, repectively, that $\|Z\|^2$ and $\|Z^*\|^2$ have the same distribution as long as $\|\gamma\|^2 = \|\gamma^*\|^2$. Is that claim plausible? The key is that if Γ is an orthogonal matrix, then $\|\Gamma Z\|^2 = \|Z\|^2$. Thus Z and ΓZ would lead to the same chi-squared. Take $Z \sim N(\gamma, I_\nu)$, and let Γ be the orthogonal matrix such that

$$\Gamma\gamma = \begin{pmatrix} \|\gamma\| \\ 0 \\ \vdots \\ 0 \end{pmatrix} = \begin{pmatrix} \sqrt{\Delta} \\ 0 \\ \vdots \\ 0 \end{pmatrix}. \tag{7.72}$$

Any orthogonal matrix whose first row is $\gamma/\|\gamma\|$ will work. Then

$$\|Z\|^2 =^{\mathcal{D}} \|\Gamma Z\|^2, \tag{7.73}$$

and the latter clearly depends on just Δ, which shows that the definition is fine. (Of course, inspecting the mgf or pdf will also prove the claim.)

Analogous to (7.49), it can be shown that if W_1 and W_2 are independent, then

$$W_1 \sim \chi^2_{\nu_1}(\Delta_1) \text{ and } W_2 \sim \chi^2_{\nu_2}(\Delta_2) \implies W_1 + W_2 \sim \chi_{\nu_1 + \nu_2}(\Delta_1 + \Delta_2). \tag{7.74}$$

For the mean and variance, we start with $Z_i \sim N(\gamma_i, 1)$, so that $Z_i^2 \sim \chi^2_1(\gamma_i^2)$. Exercise 6.8.19 finds the mean and variance of such:

$$E[\chi^2_1(\gamma_i^2)] = E[Z_i^2] = 1 + \gamma_i^2 \text{ and } Var[\chi^2_1(\gamma_i^2)] = Var[Z_i^2] = 2 + 4\gamma_i^2. \tag{7.75}$$

Thus for $Z \sim N(\gamma, I_\nu)$, $W \sim \chi^2_\nu(\|\gamma\|^2)$, hence

$$E[W] = E[\|Z\|^2] = \sum_{i=1}^\nu E[Z_i^2] = \nu + \sum_{i=1}^\nu \gamma_i^2 = \nu + \|\gamma\|^2, \text{ and}$$

$$Var[W] = Var[\|Z\|^2] = \sum_{i=1}^\nu Var[Z_i^2] = 2\nu + 4\sum_{i=1}^\nu \gamma_i^2 = 2\nu + 4\|\gamma\|^2. \tag{7.76}$$

7.6 Student's t distribution

Here we answer the question of how to find a confidence interval for μ as in (7.32) when σ is unknown. The pivotal quantity we use is

$$T = \frac{\overline{X} - \mu}{S_* / \sqrt{n}}, \quad S_*^2 = \frac{\sum(X_i - \overline{X})^2}{n-1}. \tag{7.77}$$

Exercise 6.8.17 introduced the Student's t, finding its mean, variance, and pdf. For our purposes here, if $Z \sim N(0, 1)$ and $U \sim \chi^2_\nu$, where Z and U are independent, then

$$T = \frac{Z}{\sqrt{U/\nu}} \sim t_\nu, \tag{7.78}$$

Student's t on ν degrees of freedom.

From Lemma 7.6, we have the condition for (7.78) satisfied for $\nu = n - 1$ by setting

$$Z = \frac{\overline{X} - \mu}{\sigma / \sqrt{n}} \quad \text{and} \quad U = \frac{\Sigma(X_i - \overline{X})^2}{\sigma^2}. \tag{7.79}$$

Then

$$T = \frac{(\overline{X} - \mu)/(\sigma/\sqrt{n})}{\sqrt{\frac{\Sigma(X_i - \overline{X})^2/\sigma^2}{n-1}}} = \frac{\overline{X} - \mu}{S_*/\sqrt{n}} \sim t_{n-1}. \tag{7.80}$$

A 95% confidence interval for μ is

$$\overline{X} \pm t_{n-1,0.025} \frac{S_*}{\sqrt{n}}, \tag{7.81}$$

where $t_{\nu,\alpha/2}$ is the cutoff point that satisfies

$$P[-t_{\nu,\alpha/2} < t_\nu < t_{\nu,\alpha/2}] = 1 - \alpha. \tag{7.82}$$

7.7 Linear models and the conditional distribution

Expanding on the simple linear model in (6.33), we consider the conditional model

$$\mathbf{Y} \mid \mathbf{X} = \mathbf{x} \sim N(\alpha + \mathbf{x}\beta, \Sigma_e) \quad \text{and} \quad \mathbf{X} \sim N(\mu_X, \Sigma_{XX}), \tag{7.83}$$

where $\mathbf{Y}$ is $1 \times q$, $\mathbf{X}$ is $1 \times p$, and β is a $p \times q$ matrix. As in Section 6.7.1,

$$\mathbf{E} = \mathbf{Y} - \alpha - \mathbf{X}\beta \quad \text{and} \quad \mathbf{X} \tag{7.84}$$

are independent and multivariate normal, and their joint distribution is also multivariate normal:

$$(\mathbf{X}, \mathbf{E}) \sim N\left((\mu_X, \mathbf{0}_q), \begin{pmatrix} \Sigma_{XX} & 0 \\ 0 & \Sigma_e \end{pmatrix}\right). \tag{7.85}$$

(Here, $\mathbf{0}_q$ is a row vector.)

To find the joint distribution of $\mathbf{X}$ and $\mathbf{Y}$, we note that $(\mathbf{X}, \mathbf{Y})$ is a affine transformation of $(\mathbf{X}, \mathbf{E})$, hence is multivariate normal. Specifically,

$$(\mathbf{X}, \mathbf{Y}) = (\mathbf{0}_p, \alpha) + (\mathbf{X}, \mathbf{E}) \begin{pmatrix} \mathbf{I}_p & \beta \\ 0 & \mathbf{I}_q \end{pmatrix}$$

$$\sim N\left((\mu_X, \mu_Y), \begin{pmatrix} \Sigma_{XX} & \Sigma_{XY} \\ \Sigma_{YX} & \Sigma_{YY} \end{pmatrix}\right), \tag{7.86}$$

where

$$(\mu_X, \mu_Y) = (\mathbf{0}_p, \alpha) + (\mu_X, \mathbf{0}_q) \begin{pmatrix} \mathbf{I}_p & \beta \\ 0 & \mathbf{I}_q \end{pmatrix} = (\mu_X, \alpha + \mu_X\beta) \tag{7.87}$$

and

$$\begin{pmatrix} \Sigma_{XX} & \Sigma_{XY} \\ \Sigma_{YX} & \Sigma_{YY} \end{pmatrix} = \begin{pmatrix} \mathbf{I}_p & 0 \\ \beta' & \mathbf{I}_q \end{pmatrix} \begin{pmatrix} \Sigma_{XX} & 0 \\ 0 & \Sigma_e \end{pmatrix} \begin{pmatrix} \mathbf{I}_p & \beta \\ 0 & \mathbf{I}_q \end{pmatrix}$$

$$= \begin{pmatrix} \Sigma_{XX} & \Sigma_{XX}\beta \\ \beta'\Sigma_{XX} & \Sigma_e + \beta'\Sigma_{XX}\beta \end{pmatrix}. \tag{7.88}$$

We invert the above process to find the conditional distribution of $\mathbf{Y}$ given $\mathbf{X}$ from the joint. First, we solve for $\boldsymbol{\alpha}$, $\boldsymbol{\beta}$, and $\boldsymbol{\Sigma}_e$ in terms of the $\boldsymbol{\mu}$'s and $\boldsymbol{\Sigma}$'s in (7.88):

$$\boldsymbol{\beta} = \boldsymbol{\Sigma}_{XX}^{-1}\boldsymbol{\Sigma}_{XY}, \quad \boldsymbol{\alpha} = \boldsymbol{\mu}_Y - \boldsymbol{\mu}_X\boldsymbol{\beta}, \quad \text{and} \quad \boldsymbol{\Sigma}_e = \boldsymbol{\Sigma}_{YY} - \boldsymbol{\Sigma}_{YX}\boldsymbol{\Sigma}_{XX}^{-1}\boldsymbol{\Sigma}_{XY}. \tag{7.89}$$

Lemma 7.8. *Suppose*

$$(\mathbf{X}, \mathbf{Y}) \sim N\left((\boldsymbol{\mu}_X, \boldsymbol{\mu}_Y), \begin{pmatrix} \boldsymbol{\Sigma}_{XX} & \boldsymbol{\Sigma}_{XY} \\ \boldsymbol{\Sigma}_{YX} & \boldsymbol{\Sigma}_{YY} \end{pmatrix}\right), \tag{7.90}$$

where $\boldsymbol{\Sigma}_{XX}$ is invertible. Then

$$\mathbf{Y} \mid \mathbf{X} = \mathbf{x} \sim N(\boldsymbol{\alpha} + \mathbf{x}\boldsymbol{\beta}, \boldsymbol{\Sigma}_e), \tag{7.91}$$

where $\boldsymbol{\alpha}$, $\boldsymbol{\beta}$ and $\boldsymbol{\Sigma}_e$ are given in (7.89).

The lemma deals with row vectors $\mathbf{X}$ and $\mathbf{Y}$. For the record, here is the result for column vectors:

$$\begin{pmatrix} \mathbf{X} \\ \mathbf{Y} \end{pmatrix} \sim N\left(\begin{pmatrix} \boldsymbol{\mu}_X \\ \boldsymbol{\mu}_Y \end{pmatrix}, \begin{pmatrix} \boldsymbol{\Sigma}_{XX} & \boldsymbol{\Sigma}_{XY} \\ \boldsymbol{\Sigma}_{YX} & \boldsymbol{\Sigma}_{YY} \end{pmatrix}\right) \implies \mathbf{Y} \mid \mathbf{X} = \mathbf{x} \sim N(\boldsymbol{\alpha} + \boldsymbol{\beta}\mathbf{x}, \boldsymbol{\Sigma}_e), \tag{7.92}$$

where $\boldsymbol{\Sigma}_e$ is as in (7.89), but now

$$\boldsymbol{\beta} = \boldsymbol{\Sigma}_{YX}\boldsymbol{\Sigma}_{XX}^{-1} \quad \text{and} \quad \boldsymbol{\alpha} = \boldsymbol{\mu}_Y - \boldsymbol{\beta}\boldsymbol{\mu}_X. \tag{7.93}$$

Chapter 12 goes more deeply into linear regression.

7.8   Exercises

Exercises 7.8.1 to 7.8.5 are based on the $p \times p$ symmetric matrix $\boldsymbol{\Omega}$ with spectral decomposition $\boldsymbol{\Gamma}\boldsymbol{\Lambda}\boldsymbol{\Gamma}'$ as in Theorem 7.3 on page 105.

Exercise 7.8.1. Let $\boldsymbol{\gamma}_1, \ldots, \boldsymbol{\gamma}_p$ be the columns of $\boldsymbol{\Gamma}$. The $p \times 1$ vector $\mathbf{v}$ is a **eigenvector** of $\boldsymbol{\Omega}$ with corresponding **eigenvalue** a if $\boldsymbol{\Omega}\mathbf{v} = a\mathbf{v}$. Show that for each i, $\boldsymbol{\gamma}_i$ is an eigenvector of $\boldsymbol{\Omega}$. What is the eigenvalue corresponding to $\boldsymbol{\gamma}_i$? [Hint: Show that $\boldsymbol{\Gamma}'\boldsymbol{\gamma}_i$ has one element equal to one, and the rest zero.]

Exercise 7.8.2. (a) For $1 \times p$ vector $\mathbf{c}$, show that $\mathbf{c}\boldsymbol{\Omega}\mathbf{c}' = \sum_{i=1}^{p} b_i^2 \lambda_i$ for some vector $\mathbf{b} = (b_1, \ldots, b_p)$. [Hint: Let $\mathbf{b} = \mathbf{c}\boldsymbol{\Gamma}$.] (b) Suppose $\lambda_i > 0$ for all i. Argue that $\boldsymbol{\Omega}$ is positive definite. (c) Suppose $\lambda_i \leq 0$ for some i. Find a $\mathbf{c} \neq \mathbf{0}$ such that $\mathbf{c}\boldsymbol{\Omega}\mathbf{c}' \leq 0$. [Hint: You can use one of the columns of $\boldsymbol{\Omega}$, transposed.] (d) Do parts (b) and (c) show that $\boldsymbol{\Omega}$ is positive definite if and only if all $\lambda_i > 0$?

Exercise 7.8.3. (a) Suppose $\lambda_i \geq 0$ for all i. Argue that $\boldsymbol{\Omega}$ is nonnegative definite. (b) Suppose $\lambda_i < 0$ for some i. Find a $\mathbf{c}$ such that $\mathbf{c}\boldsymbol{\Omega}\mathbf{c}' < 0$. (c) Do parts (a) and (b) show that $\boldsymbol{\Omega}$ is nonnegative definite if and only if all $\lambda_i \geq 0$?

Exercise 7.8.4. (a) Show that $|\boldsymbol{\Omega}| = \prod_{i=1}^{p} \lambda_i$. [Use can use that fact that $|\mathbf{AB}| = |\mathbf{A}||\mathbf{B}|$ for square matrices. Also, recall from (5.55) that the determinant of an orthogonal matrix is ± 1.] (b) Show that $\text{trace}(\boldsymbol{\Omega}) = \sum_{i=1}^{p} \lambda_i$. [Hint: Use the fact that $\text{trace}(\mathbf{AB}') = \text{trace}(\mathbf{B}'\mathbf{A})$, if the matrices are the same dimensions.]

Exercise 7.8.5. (a) Show that if all the λ_i's are nonzero, then $\Omega^{-1} = \Gamma\Lambda^{-1}\Gamma'$. (b) Suppose Ω is nonnegative definite, and Ψ is any $p \times p$ orthogonal matrix. Show that $\mathbf{B} = \Gamma\Lambda^{1/2}\Psi'$ is a square root of Ω, that is, $\Omega = \mathbf{BB}'$.

Exercise 7.8.6. Suppose $\mathbf{Y}$ is 1×5, $\mathbf{Y} \sim N(\mu, \Sigma)$. Find the matrix $\mathbf{B}$ so that $\mathbf{YB}' = (Y_2, Y_5) \equiv \mathbf{W}$. Show that the distribution of $\mathbf{W}$ is given in (7.19).

Exercise 7.8.7. Let

$$\Sigma = \begin{pmatrix} 5 & 2 \\ 2 & 4 \end{pmatrix}. \tag{7.94}$$

(a) Find the upper triangular matrix $\mathbf{A}$,

$$\mathbf{A} = \begin{pmatrix} a & b \\ 0 & c \end{pmatrix}, \tag{7.95}$$

such that $\Sigma = \mathbf{AA}'$, where both a and c are positive. (b) Find $\mathbf{A}^{-1}$ (which is also upper triangular). (c) Now suppose $\mathbf{X} = (X_1, X_2)'$ (a column vector) is multivariate normal with mean $(0, 0)'$ and covariance matrix Σ from above. Let $\mathbf{Y} = \mathbf{A}^{-1}\mathbf{X}$. What is $Cov[\mathbf{Y}]$? (d) Are Y_1 and Y_2 independent?

Exercise 7.8.8. Suppose

$$\mathbf{X} = \begin{pmatrix} X_1 \\ X_2 \\ X_3 \end{pmatrix} \sim N_3 \left(\begin{pmatrix} \mu \\ \mu \\ \mu \end{pmatrix}, \sigma^2 \begin{pmatrix} 1 & \rho & \rho \\ \rho & 1 & \rho \\ \rho & \rho & 1 \end{pmatrix} \right). \tag{7.96}$$

(So all the means are equal, all the variances are equal, and all the covariance are equal.) Let $\mathbf{Y} = \mathbf{AX}$, where

$$\mathbf{A} = \begin{pmatrix} 1 & 1 & 1 \\ 1 & -1 & 0 \\ 1 & 1 & -2 \end{pmatrix}. \tag{7.97}$$

(a) Find $E[\mathbf{Y}]$ and $Cov[\mathbf{Y}]$. (b) True or false: (i) $\mathbf{Y}$ is multivariate normal; (ii) Y_1, Y_2 and Y_3 are identically distributed; (iii) The Y_i's are pairwise independent; (iv) The Y_i's are mutually independent.

Exercise 7.8.9. True or false: (a) If $X \sim N(0, 1)$ and $Y \sim N(0, 1)$, and $Cov[X, Y] = 0$, then X and Y are independent. (b) If $Y \mid X = x \sim N(0, 4)$ and $X \sim \text{Uniform}(0, 1)$, then X and Y are independent. (c) Suppose $X \sim N(0, 1)$ and $Z \sim N(0, 1)$ are independent, and $Y = \text{Sign}(Z)|X|$ (where $\text{Sign}(x)$ is $+1$ if $x > 0$, and -1 if $x < 0$, and $\text{Sign}(0) = 0$). True or false: (i) $Y \sim N(0, 1)$; (ii) (X, Y) is bivariate normal; (iii) $Cov[X, Y] = 0$; (iv) (X, Z) is bivariate normal. (d) If (X, Y) is bivariate normal, and $Cov[X, Y] = 0.5$, then X and Y are independent. (e) If $X \sim N(0, 1)$ and $Y \sim N(0, 1)$, and $Cov[X, Y] = 0.5$, then (X, Y) is bivariate normal. (f) Suppose (X, Y, Z) is multivariate normal, and $Cov[X, Y] = Cov[X, Z] = Cov[Y, Z] = 0$. True or false: (i) X, Y and Z are pairwise independent; (ii) X, Y and Z are mutually independent.

Exercise 7.8.10. Suppose

$$\begin{pmatrix} X \\ Y \end{pmatrix} \sim N \left(\begin{pmatrix} 0 \\ 0 \end{pmatrix}, \begin{pmatrix} 1 & \rho \\ \rho & 1 \end{pmatrix} \right), \tag{7.98}$$

and let

$$W = \begin{pmatrix} XY \\ X^2 \\ Y^2 \end{pmatrix}.$$ (7.99)

The goal of the exercise is to find $Cov[W]$. (a) What are $E[XY]$, $E[X^2]$, and $E[Y^2]$? (b) Both X^2 and Y^2 are distributed χ_1^2. What are $Var[X^2]$ and $Var[Y^2]$? (c) What is the conditional distribution $Y \mid X = x$? (d) To find $Var[XY]$, first condition on $X = x$. Find $E[XY \mid X = x]$ and $Var[XY \mid X = x]$. Then find $Var[XY]$. [Hint: Use (6.43).] (e) Now $Cov[X^2, XY]$ can be written $Cov[E[X^2|X], E[XY|X]] + E[Cov[X^2, XY|X]]$. What is $E[X^2|X = x]$? Find $Cov[X^2, XY]$, which is then same as $Cov[Y^2, XY]$. (f) Finally, for $Cov[X^2, Y^2]$, first find $Cov[X^2, Y^2 \mid X = x]$ and $E[Y^2 \mid X = x]$. Thus what is $Cov[X^2, Y^2]$?

Exercise 7.8.11. Suppose X is $n \times 1$ and H_n is the $n \times n$ centering matrix, $H_n = I_n - (1/n)1_n 1_n'$. (a) Show that $X'H_n X = \sum_{i=1}^{n}(X_i - \overline{X})^2$. (b) Show that $trace(H_n) = n - 1$.

Exercise 7.8.12. Suppose $X_1, \ldots, X_n$ are iid $N(\mu, \sigma^2)$, and $Y_1, \ldots, Y_m$ are iid $N(\gamma, \sigma^2)$, and the X_i's and Y_i's are independent. Let $U = \sum(X_i - \overline{X})^2$, $V = \sum(Y_i - \overline{Y})^2$. (a) Are $\overline{X}, \overline{Y}, U$ and V mutually independent? (b) Let $D = \overline{X} - \overline{Y}$. What is the distribution of D? (c) The distribution of $U + V$ is σ^2 times what distribution? What are the degrees of freedom? (d) What is an unbiased estimate of σ^2? (It should depend on both U and V.) (e) Let W be that unbiased estimator of σ^2. Find the function of D, W, n, m, μ, and γ that is distributed as a Student's t. (f) Now take $n = m = 5$. A 95% confidence interval for $\mu - \gamma$ is then $D \pm c \times se(\widehat{\mu - \gamma})$. What are c and $se(\widehat{\mu - \gamma})$?

Exercise 7.8.13. Suppose $Y_1, \ldots, Y_n \mid B = \beta$ are independent, where

$$Y_i \mid B = \beta \sim N(\beta x_i, \sigma^2).$$ (7.100)

The x_i's are assumed to be known fixed quantities. Also, marginally, $B \sim N(0, \sigma_0^2)$. The σ^2 and σ_0^2 are assumed known. (a) The conditional pdf of $(Y_1, \ldots, Y_n)$ can be written

$$f_{Y \mid B}(y_1, \ldots, y_n \mid B = \beta) = a e^{-\frac{1}{2}\beta^2 C} e^{\beta D}$$ (7.101)

for some C and D, where the a does not depend on β at all. What are C and D? (They should be functions of the x_i's, y_i's, and σ^2.) (b) Similarly, the marginal pdf of B can be written

$$f_B(\beta) = a^* e^{-\frac{1}{2}\beta^2 L} e^{\beta M},$$ (7.102)

where a^* does not depend on β. What are L and M? (c) The joint pdf of $(Y_1, \ldots, Y_n, B)$ is thus

$$f(y_1, \ldots, y_n, \beta) = a a^* e^{-\frac{1}{2}\beta^2 R} e^{\beta S},$$ (7.103)

What are R and S? (d) What is the posterior distribution of B, $B \mid (Y_1, \ldots, Y_n) = (y_1, \ldots, y_n)$? (It should be normal.) What are the posterior mean and variance of B? (e) Let $\widehat{\beta} = \sum x_i Y_i / \sum x_i^2$. Show that

$$\widehat{\beta} \mid B = \beta \sim N\left(\beta, \frac{\sigma^2}{\sum x_i^2}\right).$$ (7.104)

(The mean and variance were found in Exercise 2.7.6.) Is the posterior mean of B equal to $\hat{\beta}$? Is the posterior variance of B equal to the conditional variance of $\hat{\beta}$? (f) Find the limits of the posterior mean and variance of B as the prior variance, σ_0^2, goes to ∞. Is the limit of the posterior mean of B equal to $\hat{\beta}$? Is the limit of the posterior variance of B equal to the conditional variance of $\hat{\beta}$?

Exercise 7.8.14. Consider the Bayesian model with

$$Y \mid M = \mu \sim N(\mu, \sigma^2) \quad \text{and} \quad M \sim N(\mu_0, \sigma_0^2), \tag{7.105}$$

where $\mu_0, \sigma^2 > 0$, and $\sigma_0^2 >$ are known. (a) Making the appropriate identifications in (7.83), show that

$$(M, Y) \sim N\left((\mu_0, \mu_0), \begin{pmatrix} \sigma_0^2 & \sigma_0^2 \\ \sigma_0^2 & \sigma^2 + \sigma_0^2 \end{pmatrix}\right). \tag{7.106}$$

(b) Next, show that

$$M \mid Y = y \sim N\left(\frac{\sigma^2 \mu_0 + \sigma_0^2 y}{\sigma^2 + \sigma_0^2}, \frac{\sigma^2 \sigma_0^2}{\sigma^2 + \sigma_0^2}\right). \tag{7.107}$$

(c) The **precision** of a random variable is the inverse of the variance. Let $w^2 = 1/\sigma^2$ and $w_0^2 = 1/\sigma_0^2$ be the precisions for the distributions in (7.105). Show that for the conditional distribution in (7.107), $E[M \mid Y = y] = (w_0^2 \mu_0 + w^2 y)/(w_0^2 + w^2)$, a weighted average of the prior mean and observation, weighted by their respective precisions. Also show that the conditional precision of $M \mid Y = y$ is the sum of the two precisions, $w_0^2 + w^2$. (d) Now suppose $Y_1, \ldots, Y_n$ given $M = \mu$ are iid $N(\mu, \sigma^2)$, and M is distributed as above. What is the conditional distribution of $\overline{Y} \mid M = \mu$? Show that

$$M \mid \overline{Y} = \bar{y} \sim N\left(\frac{\sigma^2 \mu_0 + n\sigma_0^2 \bar{y}}{\sigma^2 + n\sigma_0^2}, \frac{\sigma^2 \sigma_0^2}{\sigma^2 + n\sigma_0^2}\right). \tag{7.108}$$

(e) Find $l_{\bar{y}}$ and $u_{\bar{y}}$ such that

$$P[l_{\bar{y}} < M < u_{\bar{y}} \mid \overline{Y} = \bar{y}] = 0.95. \tag{7.109}$$

That interval is a 95% probability interval for μ. [See (6.72).]

Exercise 7.8.15. Consider a multivariate analog of the posterior mean in Exercise 7.8.14. Here,

$$\mathbf{Y} \mid \mathbf{M} = \boldsymbol{\mu} \sim N_p(\boldsymbol{\mu}, \boldsymbol{\Sigma}) \quad \text{and} \quad \mathbf{M} \sim N_p(\boldsymbol{\mu}_0, \boldsymbol{\Sigma}_0), \tag{7.110}$$

where $\boldsymbol{\Sigma}$, $\boldsymbol{\Sigma}_0$, and $\boldsymbol{\mu}_0$ are known, and the two covariance matrices are invertible. (a) Show that the joint distribution of $(\mathbf{Y}, \mathbf{M})$ is multivariate normal. What are the parameters? (They should be multivariate analogs of those in (7.106).) (b) Show that the conditional distribution of $\mathbf{M} \mid \mathbf{Y} = \mathbf{y}$ is multivariate normal with

$$E[\mathbf{M} \mid \mathbf{Y} = \mathbf{y}] = \boldsymbol{\mu}_0 + (\mathbf{y} - \boldsymbol{\mu}_0)(\boldsymbol{\Sigma}_0 + \boldsymbol{\Sigma})^{-1}\boldsymbol{\Sigma}_0 \tag{7.111}$$

and

$$Cov[\mathbf{M} \mid \mathbf{Y} = \mathbf{y}] = \boldsymbol{\Sigma}_0 - \boldsymbol{\Sigma}_0(\boldsymbol{\Sigma}_0 + \boldsymbol{\Sigma})^{-1}\boldsymbol{\Sigma}_0. \tag{7.112}$$

(c) Let the precision matrices be defined by $\Omega = \Sigma^{-1}$ and $\Omega_0 = \Sigma_0^{-1}$. Show that

$$\Sigma_0 - \Sigma_0(\Sigma_0 + \Sigma)^{-1}\Sigma_0 = (\Omega_0 + \Omega)^{-1}. \tag{7.113}$$

[Hint: Start by setting $(\Sigma_0 + \Sigma)^{-1} = \Omega(\Omega_0 + \Omega)^{-1}\Omega_0$, then try to simplify. At some point you may want to note that $I_p - \Omega(\Omega_0 + \Omega)^{-1} = \Omega_0(\Omega_0 + \Omega)^{-1}$.] (d) Use similar calculations on $E[\mathbf{M} \mid \mathbf{Y} = \mathbf{y}]$ to finally obtain that

$$\mathbf{M} \mid \mathbf{Y} = \mathbf{y} \sim N(\widehat{\mu}^*, (\Omega_0 + \Omega)^{-1}), \quad \text{where } \widehat{\mu}^* = (\mu_0 \Omega_0 + \mathbf{y}\Omega)(\Omega_0 + \Omega)^{-1}. \tag{7.114}$$

Note the similarity to the univariate case in (7.108). (e) Show that marginally, $\mathbf{Y} \sim N(\mu_0, \Sigma_0 + \Sigma)$.

Exercise 7.8.16. The distributional result in Exercise 7.8.15 leads to a simple answer to a particular complete-the-square problem. The joint distribution of $(\mathbf{Y}, \mathbf{M})$ in that exercise can be expressed two ways, depending on which variable is conditioned upon first. That is, using simplified notation,

$$f(\mathbf{y}, \mu) = f(\mathbf{y} \mid \mu)f(\mu) = f(\mu \mid \mathbf{y})f(\mathbf{y}). \tag{7.115}$$

Focussing on just the terms in the exponents in the last two expression in (7.115), show that

$$(\mathbf{y} - \mu)\Omega(\mathbf{y} - \mu)' + (\mu - \mu_0)\Omega_0(\mu - \mu_0)$$
$$= (\mu - \widehat{\mu}^*)(\Omega_0 + \Omega)(\mu - \widehat{\mu}^*) + (\mathbf{y} - \mu_0)(\Omega_0^{-1} + \Omega^{-1})^{-1}(\mathbf{y} - \mu_0)' \tag{7.116}$$

for $\widehat{\mu}^*$ in (7.114). Thus the right-hand-side completes the square in terms of μ.

Exercise 7.8.17. Exercise 6.8.17 introduced Student's t distribution. This exercise treats a multivariate version. Suppose

$$\mathbf{Z} \sim N(0_p, I_p) \quad \text{and} \quad U \sim \text{Gamma}(\nu/2, 1/2), \quad \text{and set } \mathbf{T} = \frac{1}{(U/\nu)}\mathbf{Z}. \tag{7.117}$$

Then $\mathbf{T}$ has the standard multivariate Student's t distribution on ν degrees of freedom, written $\mathbf{T} \sim t_{p,\nu}$. Note that $\mathbf{T}$ can be either a row vector or column vector. (a) Show that the joint distribution of $(\mathbf{T}, U)$ can be represented by

$$\mathbf{T} \mid U = u \sim N(0, (\nu/u)I_p) \quad \text{and} \quad U \sim \text{Gamma}(\nu/2, 1/2). \tag{7.118}$$

Write down the joint pdf. (b) Show that $E[T] = 0$ (if $\nu > 1$) and $Cov[\mathbf{T}] = (\nu/(\nu - 2))I_p$ (if $\nu > 2$). Are the elements of $\mathbf{T}$ uncorrelated? Are they independent? (c) Show that the marginal pdf of T is

$$f_{\nu,p}(t) = \frac{\Gamma((\nu + p)/2)}{\Gamma(\nu/2)(\sqrt{\nu\pi})^p} \frac{1}{(1 + \|\mathbf{t}\|^2/\nu)^{(\nu+p)/2}}. \tag{7.119}$$

Exercise 7.8.18. If U and V are independent, with $U \sim \chi_\nu^2$ and $V \sim \chi_\mu^2$, then

$$W = \frac{U/\nu}{V/\mu} \tag{7.120}$$

has the $F_{\nu,\mu}$ distribution. (a) Let $X = \nu W/(\mu + \nu W)$. Argue that X is Beta(α,β), and give the parameters in terms of μ and ν. (b) From Exercise 5.6.2, we know that $Y = X/(1 - X)$ has pdf $f_Y(y) = cy^{\alpha-1}/(1+y)^{\alpha+\beta}$, where c is the constant for the beta. Give W as a function of Y. (c) Show that the pdf of W is

$$h(w \mid \nu, \mu) = \frac{\Gamma((\nu + \mu)/2)}{\Gamma(\nu/2)\Gamma(\mu/2)}(\nu/\mu)^{\nu/2}\frac{w^{\nu/2-1}}{(1 - (\nu/\mu)w)^{(\nu+\mu)/2}}. \tag{7.121}$$

[Hint: Use the Jacobian technique to find the pdf of W from that of Y in part (b).] (d) Suppose $T \sim t_k$. Argue that T^2 is F, and give the degrees of freedom for the F. [What is the definition of Student's t?]

Exercise 7.8.19. Suppose $X_1, \ldots, X_n$ are independent $N(\mu_X, \sigma_X^2)$'s, and $Y_1, \ldots, Y_m$ are independent $N(\mu_Y, \sigma_Y^2)$'s, where the X_i's are independent of the Y_i's. Also, let

$$S_X^2 = \frac{\sum_{i=1}^n (X_i - \overline{X})^2}{n - 1} \quad \text{and} \quad S_Y^2 = \frac{\sum_{i=1}^m (Y_i - \overline{Y})^2}{m - 1}. \tag{7.122}$$

(a) For what constant τ, depending on σ_X^2 and σ_Y^2, is

$$F = \tau \frac{S_X^2}{S_Y^2} \tag{7.123}$$

distributed as an $F_{\nu,\mu}$? Give ν, μ. This F is a pivotal quantity. (b) Find l and u, as functions of S_X^2, S_Y^2, such that

$$P[l < \frac{\sigma_X^2}{\sigma_Y^2} < u] = 95\%. \tag{7.124}$$

Exercise 7.8.20. Suppose $Z_1, \ldots, Z_\nu$ are independent, with $Z_i \sim N(\mu_i, 1)$, so that

$$W = Z_1^2 + \cdots + Z_\nu^2 \sim \chi_\nu^2(\Delta), \quad \Delta = \|\mu\|^2, \tag{7.125}$$

as in Definition 7.5. From Exercise 6.8.19, we know that the mgf of the Z_i^2 is

$$(1 - 2t)^{-1/2} e^{\mu_i^2 t/(1-2t)}. \tag{7.126}$$

(a) What is the mgf of W? Does it depend on the μ_i's through just Δ? (b) Consider the distribution on (U, X) given by

$$U \mid X = k \sim \chi_{\nu+2k}^2 \quad \text{and} \quad X \sim \text{Poisson}(\lambda), \tag{7.127}$$

so that U is a Poisson(λ) mixture of $\chi_{\nu+2k}^2$'s. By Exercise 6.8.19, we have that the marginal mgf of U is

$$M_U(t) = (1 - 2t)^{-\nu/2} e^{\lambda 2t/(1-2t)}. \tag{7.128}$$

By matching the mgf in (7.128) with that of W in part (a), show that $W \sim \chi_\nu^2(\Delta)$ is a Poisson($\Delta/2$) mixture of $\chi_{\nu+2k}^2$'s.

Exercise 7.8.21. This and the next two exercises use **generalized hypergeometric functions**. For nonnegative integers p and q, the function $_pF_q$ is a function of real (or complex) y, with parameters $\boldsymbol{\alpha} = (\alpha_1, \ldots, \alpha_p)$ and $\boldsymbol{\beta} = (\beta_1, \ldots, \beta_q)$, given by

$$_pF_q(\boldsymbol{\alpha}\,;\,\boldsymbol{\beta}\,;\,y) = \sum_{k=0}^{\infty} \left[\left(\prod_{i=1}^{p} \frac{\Gamma(\alpha_i + k)}{\Gamma(\alpha_i)} \right) \left(\prod_{j=1}^{q} \frac{\Gamma(\beta_j)}{\Gamma(\beta_j + k)} \right) \frac{y^k}{k!} \right]. \tag{7.129}$$

If either p or q is zero, then the corresponding product of gammas is just 1. Depending on the values of the y and the parameters, the function may or may not converge. (a) Show that $_0F_0(-\,;\,-\,;\,y) = e^y$, where the "$-$" is a placeholder for the nonexistent $\boldsymbol{\alpha}$ or $\boldsymbol{\beta}$. (b) Show that for $|y| < 1$ and $\alpha > 0$, $_1F_0(\alpha\,;\,-\,;\,y) = (1-y)^{-\alpha}$. [Hint: Expand $(1-z)^{-\alpha}$ in a Taylor series about $z = 0$ (so a Maclaurin series), and use the fact that $\Gamma(a+1) = a\Gamma(a)$ as in Exercise 1.7.10(b).] (c) Show that the mgf of $Z \sim \text{Beta}(\alpha, \beta)$ is $M_Z(t) = {}_1F_1(\alpha\,;\,\alpha+\beta\,;\,y)$. The $_1F_1$ is called the **confluent hypergeometric function**. [Hint: In the integral for $E[e^{Zt}]$, expand the exponential in its Mclaurin series.]

Exercise 7.8.22. From Exercise 7.8.20, we have that $W \sim \chi_\nu^2(\Delta)$ has the same distribution as U in (7.127), where $\lambda = \Delta/2$. (a) Show that the marginal pdf of W is

$$f(w \mid \nu, \Delta) = g(w \mid \nu)e^{-\Delta/2} \sum_{k=0}^{\infty} \frac{\Gamma(\nu/2)}{\Gamma(\nu/2+k)} \frac{1}{k!} \left(\frac{\Delta w}{4} \right)^k, \tag{7.130}$$

where $g(w \mid \nu)$ is the pdf of the central χ_ν^2. (b) Show that the pdf in part (a) can be written

$$f(w \mid \nu, \Delta) = g(w \mid \nu)e^{-\Delta/2}{}_0F_1(-\,;\,\nu/2\,;\,\Delta w/4). \tag{7.131}$$

Exercise 7.8.23. If U and V are independent, with $U \sim \chi_\nu^2(\Delta)$ and $V \sim \chi_\mu^2$, then

$$Y = \frac{U/\nu}{V/\mu} \sim F_{\nu,\mu}(\Delta), \tag{7.132}$$

the **noncentral F** with degrees of freedom (ν, μ) and noncentrality parameter Δ. (So that if $\Delta = 0$, Y is central F as in (7.120).) The goal of this exercise is to derive the pdf of the noncentral F. (a) From Exercise 7.8.20, we know that the distribution of U can be represented as in (7.127) with $\lambda = \Delta/2$. Let $Z = U/(U+V)$. The conditional distribution of $Z \mid X = k$ is then a beta. What are its parameters? (b) Write down the marginal pdf of Z. Show that it can be written as

$$f_Z(z) = c(z\,;\,a,b)\,e^{-\Delta/2}{}_1F_1((\nu+\mu)/2\,;\,\nu/2\,;\,w), \tag{7.133}$$

where $c(z\,;\,a,b)$ is the Beta(a,b) pdf, and $_1F_1$ is defined in (7.129). Give the a, b, and w in terms of ν, μ, Δ, and z. (c) Since $Z = \nu Y/(\mu+\nu Y)$, we can find the pdf of Y from (7.133) using the same Jacobian as in Exercise 7.8.18(b). Show that the pdf of Y can be written as

$$f_Y(y) = h(y \mid \nu, \mu)e^{-\Delta/2}{}_1F_1((\nu+\mu)/2\,;\,\nu/2\,;\,w), \tag{7.134}$$

where h is the pdf of $F_{\nu,\mu}$ in (7.121) and w is the same as in part (b) but written in terms of y.

Exercise 7.8.24. The material in Section 7.7 can be used to find a useful matrix identity. Suppose $\boldsymbol{\Sigma}$ is a $(p+q) \times (p+q)$ symmetric matrix whose inverse $\mathbf{C} \equiv \boldsymbol{\Sigma}^{-1}$ exists. Partition these matrices into blocks as in (7.88), so that

$$\begin{pmatrix} \boldsymbol{\Sigma}_{XX} & \boldsymbol{\Sigma}_{XY} \\ \boldsymbol{\Sigma}_{YX} & \boldsymbol{\Sigma}_{YY} \end{pmatrix} \quad \& \quad \mathbf{C} = \begin{pmatrix} \mathbf{C}_{XX} & \mathbf{C}_{XY} \\ \mathbf{C}_{YX} & \mathbf{C}_{YY} \end{pmatrix}, \tag{7.135}$$

where $\boldsymbol{\Sigma}_{XX}$ and $\mathbf{C}_{XX}$ are $p \times p$, $\boldsymbol{\Sigma}_{YY}$ and $\mathbf{C}_{YY}$ are $q \times q$, etc. With $\boldsymbol{\beta} = \boldsymbol{\Sigma}_{XX}^{-1}\boldsymbol{\Sigma}_{XY}$, let

$$\mathbf{A} = \begin{pmatrix} \mathbf{I}_p & 0 \\ -\boldsymbol{\beta}' & \mathbf{I}_q \end{pmatrix}. \tag{7.136}$$

(a) Find $\mathbf{A}^{-1}$. [Hint: Just change the sign on the $\boldsymbol{\beta}$.] (b) Show that

$$\mathbf{A}\boldsymbol{\Sigma}\mathbf{A}' = \begin{pmatrix} \boldsymbol{\Sigma}_{XX} & 0 \\ 0 & \boldsymbol{\Sigma}_e \end{pmatrix} \quad \text{where} \quad \boldsymbol{\Sigma}_e = \boldsymbol{\Sigma}_{YY} - \boldsymbol{\Sigma}_{YX}\boldsymbol{\Sigma}_{XX}^{-1}\boldsymbol{\Sigma}_{XY} \tag{7.137}$$

from (7.89). (c) Take inverses on the two sides of the first equation in (7.137) to show that

$$\mathbf{C} = \mathbf{A}' \begin{pmatrix} \boldsymbol{\Sigma}_{XX}^{-1} & 0 \\ 0 & \boldsymbol{\Sigma}_e^{-1} \end{pmatrix} \mathbf{A} = \begin{pmatrix} \boldsymbol{\Sigma}_{XX}^{-1} + \boldsymbol{\Sigma}_{XX}^{-1}\boldsymbol{\Sigma}_{XY}\boldsymbol{\Sigma}_e^{-1}\boldsymbol{\Sigma}_{YX}\boldsymbol{\Sigma}_{XX}^{-1} & -\boldsymbol{\Sigma}_{XX}^{-1}\boldsymbol{\Sigma}_{XY}\boldsymbol{\Sigma}_e^{-1} \\ -\boldsymbol{\Sigma}_e^{-1}\boldsymbol{\Sigma}_{YX}\boldsymbol{\Sigma}_{XX}^{-1} & \boldsymbol{\Sigma}_e^{-1} \end{pmatrix}. \tag{7.138}$$

In particular, $\mathbf{C}_{YY} = \boldsymbol{\Sigma}_e^{-1}$.

Chapter 8

Asymptotics: Convergence in Probability and Distribution

So far we have been concerned with finding the exact distribution of random variables and functions of random variables. Especially in estimation or hypothesis testing, functions of data can become quite complicated, so it is necessary to find approximations to their distributions. One way to address such difficulties is to look at what happens when the sample size is large, or actually, as it approaches infinity. In many cases, nice asymptotic results are available, and they give surprisingly good approximations even when the sample size is nowhere near infinity.

8.1 Set-up

We assume that we have a sequence of random variables, or random vectors. That is, for each n, we have a random $p \times 1$ vector $\mathbf{W}_n$ with space $\mathcal{W}_n (\subset \mathbb{R}^p)$ and probability distribution P_n. There need not be any particular relationship between the $\mathbf{W}_n$'s for different n's, but in the most common situation we will deal with, $\mathbf{W}_n$ is some function of iid $\mathbf{X}_1, \dots, \mathbf{X}_n$, so as $n \to \infty$, the function is based on more and more observations.

The two types of convergence we will consider are convergence in probability to a constant (Section 8.2) and convergence in distribution to a random vector (Section 8.4).

8.2 Convergence in probability to a constant

A sequence of constants a_n approaching the constant c means that as $n \to \infty$, a_n gets arbitrarily close to c; technically, for any $\epsilon > 0$, eventually $|a_n - c| < \epsilon$. That definition does not immediately transfer to random variables. For example, suppose $\overline{X}_n$ is the mean of n iid $N(\mu, \sigma^2)$'s. We will see that the law of large numbers says that as $n \to \infty$, $\overline{X}_n \to \mu$. But that cannot be always true, since no matter how large n is, the space of $\overline{X}_n$ is $\mathbb{R}$. On the other hand, the probability is high that $\overline{X}_n$ is close to μ. That is, for any $\epsilon > 0$,

$$P_n[|\overline{X}_n - \mu| < \epsilon] = P[|N(0,1)| < \sqrt{n}\epsilon/\sigma] = \Phi(\sqrt{n}\epsilon/\sigma) - \Phi(-\sqrt{n}\epsilon/\sigma), \qquad (8.1)$$

where Φ is the distribution function of $Z \sim N(0,1)$. Now let $n \to \infty$. Because Φ is a distribution function, the first Φ on the right in (8.1) goes to 1, and the second goes to 0, so that

$$P_n[|\overline{X}_n - \mu| < \epsilon] \longrightarrow 1. \tag{8.2}$$

Thus $\overline{X}_n$ isn't for sure close to μ, but is with probability 0.9999999999 (assuming n is large enough). Now for the definition.

Definition 8.1. *The sequence of random variables W_n* **converges in probability** *to the constant c, written*

$$W_n \longrightarrow^{\mathcal{P}} c, \tag{8.3}$$

if for every $\epsilon > 0$,

$$P_n[|W_n - c| < \epsilon] \longrightarrow 1. \tag{8.4}$$

If $\mathbf{W}_n$ is a sequence of random $p \times 1$ vectors, and $\mathbf{c}$ is a $p \times 1$ constant vector, then $\mathbf{W}_n \to^{\mathcal{P}} \mathbf{c}$ if for every $\epsilon > 0$,

$$P_n[\|\mathbf{W}_n - \mathbf{c}\| < \epsilon] \longrightarrow 1. \tag{8.5}$$

It turns out that $\mathbf{W}_n \to^{\mathcal{P}} \mathbf{c}$ if and only if each component $W_{ni} \to^{\mathcal{P}} c_i$, where $\mathbf{W}_n = (W_{n1}, \ldots, W_{np})'$.

As an example, suppose $X_1, \ldots, X_n$ are iid Beta(2,1), which has space (0,1), pdf $f_X(x) = 2x$, and distribution function $F(x) = x^2$ for $0 < x < 1$. Denote the minimum of the X_i's by $X_{(1)}$. You would expect that as the number of observations between 0 and 1 increase, the minimum would get pushed down to 0. So the question is whether $X_{(1)} \to^{\mathcal{P}} 0$. To prove it, take any $1 > \epsilon > 0$ (Why is it ok for us to ignore $\epsilon \geq 1$?), and look at

$$P_n[|X_{(1)} - 0| < \epsilon] = P_n[X_{(1)} < \epsilon], \tag{8.6}$$

because $X_{(1)}$ is positive. That final probability is $F_{(1)}(\epsilon)$, where $F_{(1)}$ is the distribution function of $X_{(1)}$. Now the minimum is larger than ϵ if and only if all the observations are larger than ϵ, and since the observations are independent, we can write

$$
\begin{aligned}
F_{(1)}(\epsilon) &= 1 - P_n[X_{(1)} \geq \epsilon] \\
&= 1 - P[X_1 \geq \epsilon]^n \\
&= 1 - (1 - F_X(\epsilon))^n \\
&= 1 - (1 - \epsilon^2)^n \longrightarrow 1 \text{ as } n \to \infty.
\end{aligned}
\tag{8.7}
$$

(Alternatively, we could use the formula for the pdf of order statistics in (5.95).) Thus

$$\min\{X_1, \ldots, X_n\} \longrightarrow^{\mathcal{P}} 0. \tag{8.8}$$

The examples in (8.1) and (8.7) are unusual in that we can calculate the probabilities exactly. It is more common that some inequalities are used, such as Chebyshev's in the next section.

8.3 Chebyshev's inequality and the law of large numbers

The most basic result for convergence in probability is the following.

Lemma 8.2. *Weak law of large numbers (WLLN).* *If $X_1, \ldots, X_n$ are iid with (finite) mean μ, then*

$$\overline{X}_n \longrightarrow^P \mu. \tag{8.9}$$

See Theorem 2.2.9 in Durrett (2010) for a proof. We will prove a slightly weaker version, where we assume that the variance of the X_i's is finite as well. First we need an inequality.

Lemma 8.3. *Chebyshev's inequality.* *For random variable W and $\epsilon > 0$,*

$$P[|W| \geq \epsilon] \leq \frac{E[W^2]}{\epsilon^2}. \tag{8.10}$$

Proof.

$$
\begin{aligned}
E[W^2] &= E[W^2 I[|W| < \epsilon]] + E[W^2 I[|W| \geq \epsilon]] \\
&\geq E[W^2 I[|W| \geq \epsilon]] \\
&\geq \epsilon^2 \, E[I[|W| \geq \epsilon]] \\
&= \epsilon^2 \, P[|W| \geq \epsilon].
\end{aligned}
\tag{8.11}
$$

Then (8.10) follows. □

A similar proof can be applied to any nondecreasing function $\phi(w) : [0, \infty) \to \mathbb{R}$ to show that

$$P[|W| \geq \epsilon] \leq \frac{E[\phi(|W|)]}{\phi(\epsilon)}. \tag{8.12}$$

Chebyshev's inequality uses $\phi(w) = w^2$. The general form is called **Markov's inequality**.

Suppose $X_1, \ldots, X_n$ are iid with mean μ and variance $\sigma^2 < \infty$. Then using Chebyshev's inequality with $W = \overline{X}_n - \mu$, we have that for any $\epsilon > 0$,

$$P[|\overline{X}_n - \mu| \geq \epsilon] \leq \frac{Var[\overline{X}_n]}{\epsilon^2} = \frac{\sigma^2}{n\epsilon^2} \longrightarrow 0. \tag{8.13}$$

Thus $P[|\overline{X}_n - \mu| \leq \epsilon] \to 1$, and $\overline{X}_n \to^P \mu$.

The weak law of large numbers can be applied to means of functions of the X_i's. For example, if $E[X_i^2] < \infty$, then

$$\frac{1}{n} \sum_{i=1}^{n} X_i^2 \longrightarrow^P E[X_i^2] = \mu^2 + \sigma^2, \tag{8.14}$$

because the $X_1^2, \ldots, X_n^2$ are iid with mean $\mu^2 + \sigma^2$.

Not only means of functions, but functions of the mean are of interest. For example, if the X_i's are iid Exponential(λ), then the mean is $1/\lambda$, so that

$$\overline{X}_n \longrightarrow^P \frac{1}{\lambda}. \tag{8.15}$$

But we really want to estimate λ. Does

$$\frac{1}{\overline{X}_n} \longrightarrow^P \lambda \ ? \tag{8.16}$$

We could find the mean and variance of $1/\overline{X}_n$, but more simply we note that if $\overline{X}_n$ is close to $1/\lambda$, $1/\overline{X}_n$ must be close to λ, because the function $1/w$ is continuous. Formally, we have the following mapping result.

Lemma 8.4. *If* $W_n \rightarrow^P c$, *and* $g(w)$ *is a function continuous at* $w = c$, *then*

$$g(W_n) \longrightarrow^P g(c). \tag{8.17}$$

Proof. By definition of continuity, for every $\epsilon > 0$, there exists a $\delta > 0$ such that

$$|w - c| < \delta \Longrightarrow |g(w) - g(c)| < \epsilon. \tag{8.18}$$

Thus the event on the right happens at least as often as that on the left, i.e.,

$$P_n[|W_n - c| < \delta] \le P_n[|g(W_n) - g(c)| < \epsilon]. \tag{8.19}$$

The definition of $\rightarrow^P$ means that $P_n[|W_n - c| < \delta] \rightarrow 1$ for any $\delta > 0$, but $P_n[|g(W_n) - g(c)| < \epsilon]$ is larger, hence

$$P_n[|g(W_n) - g(c)| < \epsilon] \longrightarrow 1, \tag{8.20}$$

proving (8.17). □

Thus the answer to (8.16) is "Yes." Such an estimator is said to be **consistent**. This lemma also works for vector $\mathbf{W}_n$, that is, if $g(\mathbf{w})$ is continuous at $\mathbf{c}$, then

$$\mathbf{W}_n \longrightarrow^P \mathbf{c} \Longrightarrow g(\mathbf{W}_n) \longrightarrow^P g(\mathbf{c}). \tag{8.21}$$

For example, suppose $X_1, \ldots, X_n$ are iid with mean μ and variance $\sigma^2 < \infty$. Then by (8.9) and (8.14),

$$\mathbf{W}_n = \begin{pmatrix} \overline{X}_n \\ \frac{1}{n} \sum X_i^2 \end{pmatrix} \longrightarrow^P \begin{pmatrix} \mu \\ \sigma^2 + \mu^2 \end{pmatrix}. \tag{8.22}$$

Letting $g(w_1, w_2) = w_2 - w_1^2$, we have $g(\overline{x}_n, \sum x_i^2/n) = s_n^2$, the sample variance (with denominator n), hence

$$s_n^2 \longrightarrow^P (\sigma^2 + \mu^2) - \mu^2 = \sigma^2. \tag{8.23}$$

Also, $S_n \rightarrow^P \sigma$.

8.3.1 Regression through the origin

Consider regression through the origin, that is, $(X_1, Y_2), \ldots, (X_n, Y_n)$ are iid,

$$E[Y_i \mid X_i = x_i] = \beta x_i, \quad Var[Y_i \mid X_i = x_i] = \sigma_e^2, \quad E[X_i] = \mu_X, \quad Var[X_i] = \sigma_X^2 > 0. \tag{8.24}$$

We will see later (Exercise 12.7.18) that the least squares estimate of β is

$$\widehat{\beta}_n = \frac{\sum_{i=1}^n x_i y_i}{\sum_{i=1}^n x_i^2}. \tag{8.25}$$

Is this a consistent estimator? We know by (8.14) that

$$\frac{1}{n}\sum_{i=1}^{n} X_i^2 \longrightarrow^{\mathcal{P}} \mu_X^2 + \sigma_X^2. \tag{8.26}$$

Also, $X_1Y_1, \ldots, X_nY_n$ are iid, and

$$E[X_iY_i \mid X_i = x_i] = x_i\, E[Y_i \mid X_i = x_i] = \beta x_i^2, \tag{8.27}$$

hence

$$E[X_iY_i] = E[\beta X_i^2] = \beta(\mu_X^2 + \sigma_X^2). \tag{8.28}$$

Thus the WLLN shows that

$$\frac{1}{n}\sum_{i=1}^{n} X_iY_i \longrightarrow^{\mathcal{P}} \beta(\mu_X^2 + \sigma_X^2). \tag{8.29}$$

Now consider

$$\mathbf{W}_n = \left(\frac{1}{n}\sum_{i=1}^{n} X_iY_i, \frac{1}{n}\sum_{i=1}^{n} X_i^2\right) \quad \text{and} \quad \mathbf{c} = (\beta(\mu_X^2 + \sigma_X^2), \mu_X^2 + \sigma_X^2), \tag{8.30}$$

so that $\mathbf{W}_n \to^{\mathcal{P}} \mathbf{c}$. The function $g(w_1, w_2) = w_1/w_2$ is continuous at $\mathbf{w} = \mathbf{c}$, hence (8.21) shows that

$$g(\mathbf{W}_n) = \frac{\frac{1}{n}\sum_{i=1}^{n} X_iY_i}{\frac{1}{n}\sum_{i=1}^{n} X_i^2} \longrightarrow^{\mathcal{P}} g(\mathbf{c}) = \frac{\beta(\mu_X^2 + \sigma_X^2)}{\mu_X^2 + \sigma_X^2}, \tag{8.31}$$

that is,

$$\widehat{\beta}_n = \frac{\sum_{i=1}^{n} X_iY_i}{\sum_{i=1}^{n} X_i^2} \longrightarrow^{\mathcal{P}} \beta. \tag{8.32}$$

So, yes, the least squares estimator is consistent.

8.4   Convergence in distribution

Convergence to a constant is helpful, but generally more information is needed, as for confidence intervals. E.g, if we can say that

$$\frac{\widehat{\theta} - \theta}{SE(\widehat{\theta})} \approx N(0,1), \tag{8.33}$$

then an approximate 95% confidence interval for θ would be

$$\widehat{\theta} \pm 2 \times SE(\widehat{\theta}), \tag{8.34}$$

where the "2" is approximately 1.96. Thus we need to find the approximate distribution of a random variable. In the asymptotic setup, we need the notion of W_n converging to a random variable. It is formalized by looking at the respective distribution function for each possible value, almost.

Definition 8.5. *Suppose W_n is a sequence of random variables, and W is a random variable. Let F_n be the distribution function of W_n, and F be the distribution function of W. Then W_n converges in distribution to W if*

$$F_n(w) \longrightarrow F(w) \tag{8.35}$$

for every $w \in \mathbb{R}$ at which F is continuous. This convergence is written

$$W_n \longrightarrow^{\mathcal{D}} W. \tag{8.36}$$

For example, go back to $X_1, \ldots, X_n$ iid Beta(2,1), but now let

$$W_n = n\, X_{(1)}, \tag{8.37}$$

where again $X_{(1)}$ is the minimum of the n observations. The minimum itself goes to 0, but by multiplying by n it may not. The distribution function of W_n is $F_{W_n}(w) = 0$ if $w \leq 0$ and if $w > 0$, and using calculations as in (8.7) with $\epsilon = w/n$,

$$F_{W_n}(w) = P[W_n \leq w] = P[X_{(1)} \leq w/n] = 1 - (1 - (w/n)^2)^n. \tag{8.38}$$

Now let $n \to \infty$. We will use the fact that for a sequence c_n,

$$\lim_{n \to \infty} \left(1 - \frac{c_n}{n}\right)^n = e^{-\lim_{n \to \infty} c_n} \tag{8.39}$$

if the limit exists. Applying this equation to (8.38), we have $c_n = w^2/n$, which goes to 0. Thus $F_{W_n}(w) \to 1 - 1 = 0$. This limit is not distribution function, hence $nX_{(1)}$ does not have a limit in distribution. What happens is that $nX_{(1)}$ is going to ∞. So multiplying by n is too strong. What about $V_n = \sqrt{n}X_{(1)}$? Then we can show that for $v > 0$,

$$F_{V_n}(v) = 1 - (1 - v^2/n)^n \longrightarrow 1 - e^{-v^2}, \tag{8.40}$$

since $c_n = v^2$ in (8.39). For $v \leq 0$, $F_{V_n}(v) = 0$, since $X_{(1)} > 0$. Thus the limit is 0, too. Hence

$$F_{V_n}(v) \longrightarrow \begin{cases} 1 - e^{-v^2} & \text{if} \quad v > 0 \\ 0 & \text{if} \quad v \leq 0 \end{cases}. \tag{8.41}$$

Is the right-hand side a distribution function of some random variable? Yes, indeed. Thus $\sqrt{n}X_{(1)}$ does have a limit in distribution, the distribution function being given in (8.41).

For another example, suppose $X_n \sim$ Binomial$(n, \lambda/n)$ for some fixed $\lambda > 0$. The distribution function of X_n is

$$F_n(x) = \begin{cases} 0 & \text{if} \quad x < 0 \\ \sum_{i=0}^{\text{floor}(x)} f(i \mid n, \lambda/n) & \text{if} \quad 0 \leq x \leq n \\ 1 & \text{if} \quad n < x \end{cases}, \tag{8.42}$$

where floor(x) is the largest integer less than or equal to x, and $f(i \mid n, \lambda/n)$ is the Binomial$(n, \lambda/n)$ pmf. Taking the limit as $n \to \infty$ of F_n requires taking the limit of the f's, so we will do that first. For a positive integer $i \leq n$,

$$f\left(i \mid n, \frac{\lambda}{n}\right) = \frac{n!}{i!(n-i)!}\left(\frac{\lambda}{n}\right)^i \left(1 - \frac{\lambda}{n}\right)^{n-i}$$

$$= \frac{\lambda^i}{i!} \frac{n(n-1)\cdots(n-i+1)}{n^i}\left(1 - \frac{\lambda}{n}\right)^n \left(1 - \frac{\lambda}{n}\right)^{-i}. \tag{8.43}$$

Now i is fixed and $n \to \infty$. Consider the various factors on the right. The first has no n's. The second has i terms on the top, and n^i on the bottom, so can be written

$$\frac{n(n-1)\cdots(n-i+1)}{n^i} = \frac{n}{n}\frac{n-1}{n}\cdots\frac{n-i+1}{n}$$

$$= 1\left(1-\frac{1}{n}\right)\cdots\left(1-\frac{i-1}{n}\right) \to 1. \qquad (8.44)$$

The third term goes to $e^{-\lambda}$, and the fourth goes to 1. Thus for any positive integer i,

$$f\left(i\,\middle|\,n,\frac{\lambda}{n}\right) \longrightarrow e^{-\lambda}\frac{\lambda^i}{i!}, \qquad (8.45)$$

which is the pmf of the Poisson(λ). Going back to the F_n in (8.42), note that no matter how large x is, as $n \to \infty$, eventually $x < n$, so that the third line never comes into play. Thus

$$F_n(x) \longrightarrow \begin{cases} 0 & \text{if } x < 0 \\ \sum_{i=0}^{\text{floor}(x)} e^{-\lambda}\frac{\lambda^i}{i!} & \text{if } 0 \le x \end{cases}. \qquad (8.46)$$

But that is the distribution function of the Poisson(λ), i.e.,

$$\text{Binomial}\left(n,\frac{\lambda}{n}\right) \longrightarrow^{\mathcal{D}} \text{Poisson}(\lambda). \qquad (8.47)$$

8.4.1 Points of discontinuity of F

In the definition, the convergence (8.35) does not need to hold at w's for which F is not continuous. This relaxation exists because sometimes the limit of the F_n's will have points at which the function is continuous from the left but not the right, whereas F's need to be continuous from the right. For example, take $W_n = Z + 1/n$, where $Z \sim \text{Bernoulli}(1/2)$. It seems reasonable that $W_n \longrightarrow^{\mathcal{D}} \text{Bernoulli}(1/2)$. Let F_n be the distribution function for W_n:

$$F_n(w) = \begin{cases} 0 & \text{if } w < 1/n \\ 1/2 & \text{if } 1/n \le w < 1+1/n \\ 1 & \text{if } 1+1/n \le w. \end{cases} \qquad (8.48)$$

Now let $n \to \infty$, so that

$$F_n(w) \longrightarrow \begin{cases} 0 & \text{if } w \le 0 \\ 1/2 & \text{if } 0 < w \le 1 \\ 1 & \text{if } 1 < w \end{cases}. \qquad (8.49)$$

That limit is **not** a distribution function, though it would be if the $\le$'s and $<$'s were switched, in which case it would be the F for a Bernoulli$(1/2)$. Luckily, the definition allows the limit to be wrong at points of discontinuity, which are 0 and 1 in this example, so we **can** say that $W_n \longrightarrow^{\mathcal{D}} \text{Bernoulli}(1/2)$.

8.4.2　Converging to a constant random variable

It could be that $W_n \longrightarrow^{\mathcal{D}} W$, where $P[W = c] = 1$, that is, W is a constant random variable. (Sounds like an oxymoron.) But that looks like W_n is converging to a constant. It is:

$$W_n \longrightarrow^{\mathcal{D}} W \text{ where } P[W = c] = 1 \text{ if and only if } W_n \longrightarrow^{\mathcal{P}} c. \qquad (8.50)$$

Let $F_n(w)$ be the distribution function of W_n, and F the distribution function of W, so that

$$F(w) = \begin{cases} 0 & \text{if } w < c \\ 1 & \text{if } c \leq w \end{cases}. \qquad (8.51)$$

For any $\epsilon > 0$,

$$P[|W_n - c| \leq \epsilon] = P[W_n \leq c + \epsilon] - P[W_n < c - \epsilon]. \qquad (8.52)$$

Also,

$$P[W_n \leq c - 3\epsilon/2] \leq P[W_n < c - \epsilon] \leq P[W_n \leq c - \epsilon/2], \qquad (8.53)$$

hence, because $F_n(w) = P[W_n \leq w]$,

$$F_n(c + \epsilon) - F_n(c - \epsilon/2) \leq P[|W_n - c| \leq \epsilon] \leq F_n(c + \epsilon) - F_n(c - 3\epsilon/2). \qquad (8.54)$$

We use this equation to show (8.50).

1. First suppose that $W_n \to^{\mathcal{D}} W$, so that $F_n(w) \to F(w)$ if $w \neq c$. Then applying the convergence to $w = c + \epsilon$ and $c - \epsilon/2$,

$$F_n(c + \epsilon) \to F(c + \epsilon) = 1 \text{ and}$$
$$F_n(c - \epsilon/2) \to F(c - \epsilon/2) = 0. \qquad (8.55)$$

   Then (8.54) shows that $P[|W_n - c| \leq \epsilon] \to 1$, proving that $W_n \to^{\mathcal{P}} c$.

2. Next, suppose $W_n \longrightarrow^{\mathcal{P}} c$. Then $P[|W_n - c| \leq \epsilon] \to 1$, hence from (8.54),

$$F_n(c + \epsilon) - F_n(c - 3\epsilon/2) \longrightarrow 1. \qquad (8.56)$$

   The only way that can happen is for

$$F_n(c + \epsilon) \to 1 \text{ and } F_n(c - 3\epsilon/2) \to 0. \qquad (8.57)$$

   But since that holds for any $\epsilon > 0$, for any $w < c$, $F_n(w) \to 0$, and for any $w > c$, $F_n(w) \to 1$. That is, $F_n(w) \to F(w)$ for $w \neq c$, proving that $W_n \to^{\mathcal{D}} W$.

8.5　Moment generating functions

It may be difficult to find distribution functions and their limits. Often moment generating functions are easier to work with. Recall that if two random variables have the same moment generating function that is finite in a neighborhood of 0, then they have the same distribution. It is true also of limits, that is, if the mgfs of a sequence of random variables converge to a mgf, then the random variable converge. See Section 30 of Billingsley (1995) for a proof.

Lemma 8.6. *Suppose $W_1, W_2, \ldots$ is a sequence of random variables, where $M_n(t)$ is the mgf of W_n, and suppose W is a random variable with mgf $M(t)$. If for some $\epsilon > 0$, $M_n(t) < \infty$ and $M(t) < \infty$ for all n and all $|t| < \epsilon$,*

$$W_n \longrightarrow^{\mathcal{D}} W \quad \text{if and only if} \quad M_n(t) \longrightarrow M(t) \quad \text{for all } |t| < \epsilon. \tag{8.58}$$

Looking again at the binomial example in (8.42), if $X_n \sim \text{Binomial}(n, \lambda/n)$, then using (2.83) we see that its mgf is

$$M_n(t) = \left(\left(1 - \frac{\lambda}{n}\right) + \frac{\lambda}{n} e^t \right)^n = \left(1 + \frac{-\lambda + \lambda e^t}{n}\right)^n. \tag{8.59}$$

Letting $n \to \infty$,

$$M_n(t) \longrightarrow e^{-\lambda + \lambda e^t}, \tag{8.60}$$

which is the mgf of a Poisson(λ), showing again that (8.47) holds.

8.6 Central limit theorem

We know that sample means tend to the population mean, if the latter exists. But we can obtain more information with a distributional limit. In the normal case, we know that the sample mean is normal, and with appropriate normalization, it is standard normal, i.e., Normal(0,1). A **central limit theorem** is one that says a properly normalized sample mean approaches normality even if the original variables are not normal.

Start with $X_1, \ldots, X_n$ iid with mean 0 and variance 1, and mgf $M_X(t)$, which is finite for $|t| < \epsilon$ for some $\epsilon > 0$. The variance of $\overline{X}_n$ is $1/n$, so to normalize it we multiply by $\sqrt{n}$:

$$W_n = \sqrt{n}\,\overline{X}_n. \tag{8.61}$$

To find the asymptotic distribution of W_n, we first find its mgf, $M_n(t)$:

$$
\begin{aligned}
M_n(t) = E[e^{tW_n}] &= E[e^{t\sqrt{n}\,\overline{X}_n}] \\
&= E[e^{(t/\sqrt{n}) \sum X_i}] \\
&= E[e^{(t/\sqrt{n}) X_i}]^n \\
&= M_X(t/\sqrt{n})^n.
\end{aligned} \tag{8.62}
$$

Now $M_n(t)$ is finite if $M_X(t/\sqrt{n})$ is, and $M_X(t/\sqrt{n}) < \infty$ if $|t/\sqrt{n}| < \epsilon$, which is certainly true if $|t| < \epsilon$. That is, $M_n(t) < \infty$ if $|t| < \epsilon$.

To find the limit of M_n, first take logs:

$$\log(M_n(t)) = n \log(M_X(t/\sqrt{n})) = n c_X(t/\sqrt{n}), \tag{8.63}$$

where $c_X(t) = \log(M_X(t))$ is the cumulant generating function for a single X_i. Expand c_X in a Taylor series about $t = 0$:

$$c_X(t) = c_X(0) + t\,c_X'(0) + \frac{t^2}{2} c_X''(t^*), \quad t^* \text{ between 0 and } t. \tag{8.64}$$

But $c_X(0) = 0$, and $c'_X(0) = E[X] = 0$, by assumption. Thus substituting $t/\sqrt{n}$ for t in (8.64) yields

$$c_X(t/\sqrt{n}) = \frac{t^2}{2n} c''_X(t^*_n), \quad t^*_n \text{ between 0 and } t/\sqrt{n}, \tag{8.65}$$

hence by (8.63),

$$\log(M_n(t)) = \frac{t^2}{2} c''_X(t^*_n). \tag{8.66}$$

The mgf $M_X(t)$ has all its derivatives as long as $|t| < \epsilon$, which means so does c_X. In particular, $c''_X(t)$ is continuous at $t = 0$. As $n \to \infty$, t^*_n gets squeezed between 0 and $t/\sqrt{n}$, hence $t^*_n \to 0$, and

$$\log(M_n(t)) = \frac{t^2}{2} c''_X(t^*_n) \to \frac{t^2}{2} c''_X(0) = \frac{t^2}{2} Var[X_i] = \frac{t^2}{2}, \tag{8.67}$$

because we have assumed that $Var[X_i] = 1$. Finally,

$$M_n(t) \longrightarrow e^{t^2/2}, \tag{8.68}$$

which is the mgf of a N(0,1), i.e.,

$$\sqrt{n}\, \overline{X}_n \longrightarrow^{\mathcal{D}} N(0,1). \tag{8.69}$$

There are many central limit theorems, depending on various assumptions, but the most basic is the following.

Theorem 8.7. Central limit theorem. *Suppose $X_1, X_2, \ldots$ are iid with mean 0 and variance 1. Then (8.69) holds.*

What we proved using (8.68) required the mgf be finite in a neighborhood of 0. This theorem does not need mgfs, only that the variance is finite. A slight generalization of the theorem has $X_1, X_2, \ldots$ iid with mean μ and variance $\sigma^2, 0 < \sigma^2 < \infty$, and concludes that

$$\sqrt{n}\, (\overline{X}_n - \mu) \longrightarrow^{\mathcal{D}} N(0,\sigma^2). \tag{8.70}$$

E.g., see Theorem 27.1 of Billingsley (1995).

8.6.1 Supersizing

Convergence in distribution immediately translates to multivariate random variables. That is, suppose $\mathbf{W}_n$ is a $p \times 1$ random vector with distribution function F_n. Then

$$\mathbf{W}_n \longrightarrow^{\mathcal{D}} \mathbf{W} \tag{8.71}$$

for some $p \times 1$ random vector $\mathbf{W}$ with distribution function F if

$$F_n(\mathbf{w}) \longrightarrow F(\mathbf{w}) \tag{8.72}$$

for all points $\mathbf{w} \in \mathbb{R}$ at which $F(\mathbf{w})$ is continuous.

If $M_n(\mathbf{t})$ is the mgf of $\mathbf{W}_n$, and $M(\mathbf{t})$ is the mgf of $\mathbf{W}$, and these mgfs are all finite for $\|\mathbf{t}\| < \epsilon$ for some $\epsilon > 0$, then

$$\mathbf{W}_n \longrightarrow^{\mathcal{D}} \mathbf{W} \text{ iff } M_n(\mathbf{t}) \longrightarrow M(\mathbf{t}) \text{ for all } \|\mathbf{t}\| < \epsilon. \tag{8.73}$$

Now for the central limit theorem. Suppose $X_1, X_2, \ldots$ are iid random vectors with mean μ, finite covariance matrix Σ, and mgf $M_X(t) < \infty$ for $\|t\| < \epsilon$. Set

$$\mathbf{W}_n = \sqrt{n}(\overline{\mathbf{X}}_n - \mu). \tag{8.74}$$

Let $\mathbf{a}$ be any $p \times 1$ vector, and write

$$\mathbf{a}'\mathbf{W}_n = \sqrt{n}(\mathbf{a}'\overline{\mathbf{X}}_n - \mathbf{a}'\mu) = \sqrt{n}\left(\frac{1}{n}\sum_{i=1}^{n}\mathbf{a}'\mathbf{X}_i - \mathbf{a}'\mu\right). \tag{8.75}$$

Thus $\mathbf{a}'\mathbf{W}_n$ is the normalized sample mean of the $\mathbf{a}'\mathbf{X}_i$'s, and the regular central limit theorem (actually, equation 8.70) can be applied, where $\sigma^2 = Var[\mathbf{a}'\mathbf{X}_i] = \mathbf{a}'\Sigma\mathbf{a}$:

$$\mathbf{a}'\mathbf{W}_n \longrightarrow^{\mathcal{D}} N(0, \mathbf{a}'\Sigma\mathbf{a}). \tag{8.76}$$

But then that means the mgfs converge in (8.76): Letting $\mathbf{t} = t\mathbf{a}$,

$$E[e^{t(\mathbf{a}'\mathbf{W}_n)}] \longrightarrow e^{\frac{t^2}{2}\mathbf{a}'\Sigma\mathbf{a}}, \tag{8.77}$$

for $t\|\mathbf{a}\| < \epsilon$. Now switch notation so that $\mathbf{a} = \mathbf{t}$ and $t = 1$, and we have that

$$M_n(\mathbf{t}) = E[e^{\mathbf{t}'\mathbf{W}_n}] \longrightarrow e^{\frac{1}{2}\mathbf{t}'\Sigma\mathbf{t}}, \tag{8.78}$$

which holds for any $\|\mathbf{t}\| < \epsilon$. The right hand side is the mgf of a $N(\mathbf{0}, \Sigma)$, so

$$\mathbf{W}_n = \sqrt{n}(\overline{\mathbf{X}}_n - \mu) \longrightarrow^{\mathcal{D}} N(\mathbf{0}, \Sigma). \tag{8.79}$$

Example. Suppose $X_1, X_2, \ldots$ are iid with mean μ and variance σ^2. One might be interested in the joint distribution of the sample mean and variance, after some normalization. When the data are normal, we know the answer exactly, of course, but what about otherwise? We won't answer that question quite yet, but take a step by looking at the joint distribution of the sample means of the X_i's and the X_i^2's. We will assume that $Var[X_i^2] < \infty$. We start with

$$\mathbf{W}_n = \sqrt{n}\left(\frac{1}{n}\sum_{i=1}^{n}\begin{pmatrix} X_i \\ X_i^2 \end{pmatrix} - \begin{pmatrix} \mu \\ \mu^2 + v^2 \end{pmatrix}\right). \tag{8.80}$$

Then the central limit theorem says that

$$\mathbf{W}_n \longrightarrow^{\mathcal{D}} N\left(\mathbf{0}_2, Cov\begin{pmatrix} X_i \\ X_i^2 \end{pmatrix}\right). \tag{8.81}$$

Look at that covariance. We know $Var[X_i] = \sigma^2$. Also,

$$Var[X_i^2] = E[X_i^4] - E[X_i^2]^2 = \mu_4' - (\mu^2 + \sigma^2)^2, \tag{8.82}$$

and

$$Cov[X_i, X_i^2] = E[X_i^3] - E[X_i]E[X_i^2] = \mu_3' - \mu(\mu^2 + \sigma^2), \tag{8.83}$$

where $\mu'_k = E[X^k_i]$, the raw k^{th} moment from (2.55).

It's not pretty, but the final answer is

$$\sqrt{n}\left(\frac{1}{n}\sum_{i=1}^{n}\left(\begin{array}{c}X_i\\X_i^2\end{array}\right)-\left(\begin{array}{c}\mu\\\mu^2+\sigma^2\end{array}\right)\right)$$

$$\longrightarrow^{\mathcal{D}} N\left(0_2,\left(\begin{array}{cc}\sigma^2 & \mu'_3-\mu(\mu^2+\sigma^2)\\\mu'_3-\mu(\mu^2+\sigma^2) & \mu'_4-(\mu^2+\sigma^2)^2\end{array}\right)\right). \quad (8.84)$$

8.7 Exercises

Exercise 8.7.1. Suppose $X_1, X_2, \ldots$ all have $E[X_i] = \mu$ and $Var[X_i] = \sigma^2 < \infty$, and they are uncorrelated. Show that $\overline{X}_n \to \mu$ in probability.

Exercise 8.7.2. Consider the random variable W_n with

$$P[W_n = 0] = 1 - \frac{1}{n} \quad \text{and} \quad P[W_n = a_n] = \frac{1}{n} \quad (8.85)$$

for some constants a_n, $n = 1, 2, \ldots$. (a) For each given sequence a_n, find the limits as $n \to \infty$, when existing, for W_n, $E[W_n]$, and $Var[W_n]$. (i) $a_n = 1/n$. (ii) $a_n = 1$. (iii) $a_n = \sqrt{n}$. (iv) $a_n = n$. (v) $a_n = n^2$. (b) For which sequences a_n in part (a) can one use Chebyshev's inequality to find the limit of W_n in probability? (c) Does $W_n \to^{\mathcal{P}} c$ imply that $E[W_n] \to c$?

Exercise 8.7.3. Suppose $X_n \sim$ Binomial$(n, 1/2)$, and let $W_n = X_n/n$. (a) What is the limit of W_n in probability? (b) Suppose n is even. Find $P[W_n = 1/2]$. Does this probability approach 1 as $n \to \infty$?

Exercise 8.7.4. Let f be a function on $(0, 1)$ with $\int_0^1 f(u)du < \infty$. Let $U_1, U_2, \ldots$ be iid Uniform$(0,1)$'s, and let $X_n = (f(U_1) + \cdots + f(U_n))/n$. Show that $X_n \to \int_0^1 f(u)du$ in probability.

Exercise 8.7.5. Suppose $X_1, \ldots, X_n$ are iid $N(\mu, 1)$. Find the exact probability $P[|\overline{X}_n - \mu| > \epsilon]$, and the bound given by Chebyshev's inequality, for $\epsilon = 0.1$ for various values of n. Is the bound very close to the exact probability?

Exercise 8.7.6. Suppose

$$\left(\begin{array}{c}X_1\\Y_1\end{array}\right), \ldots, \left(\begin{array}{c}X_n\\Y_n\end{array}\right) \text{ are iid } N\left(\left(\begin{array}{c}\mu_X\\\mu_Y\end{array}\right),\left(\begin{array}{cc}\sigma_X^2 & \rho\sigma_X\sigma_Y\\\rho\sigma_X\sigma_Y & \sigma_Y^2\end{array}\right)\right). \quad (8.86)$$

Assume $\sigma_X^2 > 0$ and $\sigma_Y^2 > 0$, and let $S_X^2 = \sum_{i=1}^{n}(X_i - \overline{X}_n)^2/n$ and $S_Y^2 = \sum_{i=1}^{n}(Y_i - \overline{Y}_n)^2/n$. Find the limits in probability of the following. (Actually, the answers do not depend on the normality assumption.) (a) $\sum_{i=1}^{n} X_i Y_i/n$. (b) $S_{XY} = \sum_{i=1}^{n}(X_i - \overline{X})(Y_i - \overline{Y})/n$. (c) $R_n = S_{XY}/(S_X S_Y)$. (d) What if instead of dividing by n for S_X^2, S_Y^2, and S_{XY}, we divide by $n - 1$?

Exercise 8.7.7. The distribution of a random variable Y is called a **mixture** of two distributions if its distribution function is $F_Y(y) = (1 - \epsilon)F_1(y) + \epsilon F_2(y)$, $y \in \mathbb{R}$, where F_1 and F_2 are distribution functions, and $0 < \epsilon < 1$. The idea is that with probability $1 - \epsilon$, Y has distribution F_1, and with probability ϵ, it has distribution F_2. Now let Y_n have the following mixture distribution: $N(\mu, 1)$ with probability $1 - \epsilon_n$, and $N(n, 1)$ with probability ϵ_n, where $\epsilon_n \in (0,1)$. (a) Write down the distribution function of Y_n in terms of Φ, the distribution function of a $N(0,1)$. (b) Let $\epsilon_n \to 0$ as $n \to \infty$. What is the limit of the distribution function in (a)? What does that say about the distribution of Y, the limit of the Y_n's? (c) What is $E[Y]$? Find $E[Y_n]$ and its limit when (i) $\epsilon_n = 1/\sqrt{n}$. (ii) $\epsilon_n = 1/n$. (iii) $\epsilon_n = 1/n^2$. (d) Does $Y_n \to^{\mathcal{D}} Y$ imply that $E[Y_n] \to E[Y]$?

Exercise 8.7.8. Suppose X_n is Geometric($1/n$). What is the limit in distribution of X_n/n? [Hint: First find the distribution function of $Y_n = X_n/n$.]

Exercise 8.7.9. Suppose $U_1, U_2, \ldots, U_n$ are iid Uniform(0,1), and $U_{(1)}$ is their minimum. Then $U_{(1)}$ has distribution function $F_n(u) = 1 - (1 - u)^n$, $u \in (0,1)$, as in (8.7). (What is $F_n(u)$ for $u \le 0$ or $u \ge 1$?) (a) What is the limit of $F_n(u)$ as $n \to \infty$ for $u \in (0,1)$? (b) What is the limit for $u \le 0$? (c) What is the limit for $u \ge 1$? (d) Thus the limit of $F_n(u)$ is the distribution function (at least for $u \ne 0$) of what random variable? Choose among (i) a constant random variable, with value 0; (ii) a constant random variable, with value 1; (iii) a Uniform(0,1); (iv) an Exponential(1); (v) none of the above.

Exercise 8.7.10. Continue with the setup in Exercise 8.7.9. Let $V_n = nU_{(1)}$, and let $G_n(v)$ be its distribution function. (a) For $v \in (0,n)$, $G_n(v) = P[V_n \le v] = P[U_{(1)} \le c] = F_n(c)$ for some c. What is c (as a function of v, n)? (b) Find $G_n(v)$. (c) What is the limit of $G_n(v)$ as $n \to \infty$ for $v > 0$? (d) That limit is the distribution function of what distribution? Choose among (i) a constant random variable, with value 0; (ii) a constant random variable, with value 1; (iii) a Uniform(0,1); (iv) an Exponential(1); (v) none of the above.

Exercise 8.7.11. Continue with the setup in Exercises 8.7.9 and 8.7.10. (a) Find the distribution function of $\sqrt{n}\, U_{(1)}$. What is its limit for $y > 0$? What is the limit in distribution of $\sqrt{n}\, U_{(1)}$, if it exists? (b) Same question, but for $n^2 U_{(1)}$.

Exercise 8.7.12. Suppose $X_1, X_2, \ldots, X_n$ are iid Exponential(1). Let $W_n = b_n X_{(1)}$, where $X_{(1)}$ is the minimum of the X_i's and $b_n > 0$. (a) Find the distribution function $F_{(1)}$ of $X_{(1)}$, and show that $F_{W_n}(w) = F_{(1)}(w/b_n)$ is the distribution function of W_n. (b) For each of the following sequences, decide whether W_n goes in probability to a constant, goes in distribution to a non-constant random variable, or does not have a limit in distribution: (i) $b_n = 1$; (ii) $b_n = \log(n)$; (iii) $b_n = n$; (iv) $b_n = n^2$. If W_n goes to a constant, give the constant, and if it goes to a random variable, specify the distribution function. [Hint: Find the limit of $F_{W_n}(w)$ for each fixed w. Note that since W_n is always positive, we automatically have that $F_{W_n}(w) = 0$ for $w < 0$.]

Exercise 8.7.13. Again, $X_1, X_2, \ldots, X_n$ are iid Exponential(1). Let $U_n = X_{(n)} - a_n$. (a) Find the distribution function of U_n, $F_n(u)$. (b) For each of the following sequences, decide whether U_n goes in probability to a constant, goes in distribution to a non-constant random variable, or does not have a limit in distribution: (i) $a_n = 1$; (ii)

$a_n = \log(n)$; (iii) $a_n = n$; (iv) $a_n = n^2$. If U_n goes to a constant, give the constant, and if it goes to a random variable, specify the distribution function. [Hint: The distribution function of U_n in each case will be of the form $(1 - c_n/n)^n$, for which you can use (8.39). Also, recall the Gumbel distribution, from (5.103). Note that you must deal with any u unless $a_n = 1$, since for any $x > 0$, eventually $x - a_n < u$.]

Exercise 8.7.14. This question considers the asymptotic distribution of the sample maximum, $X_{(n)}$, based on $X_1, \ldots, X_n$ iid. Specifically, what do constants a_n and b_n need to be so that $W_n = b_n X_{(n)} - a_n \to^D W$, where W is a non-constant random variable? The constants, and W, will depend on the distribution of the X_i's. In the parts below, find the appropriate a_n, b_n, and W for the X_i's having the given distribution. [Find the distribution function of the W_n, and see what a_n and b_n have to be for that to go to a non-trivial distribution function.] (a) Exponential(1). (b) Uniform(0,1). (c) Beta$(\alpha, 1)$. (This uses the same a_n and b_n as in part (b).) (d) Gumbel(0), from (5.103). (e) Logistic.

Exercise 8.7.15. Here, $X_1, \ldots, X_n$, are iid Laplace(0, 1), so they have mgf $M(s) = 1/(1 - s^2)$. Let $Z_n = \sqrt{n}\, \overline{X} = \sum X_i / \sqrt{n}$. (a) What is the mgf $M_{\sum X_i}(s)$ of $\sum X_i$? (b) What is the mgf $M_{Z_n}(t)$ of Z_n? (c) What is the limit of $M_{Z_n}(t)$ as $n \to \infty$? (d) What random variable is that the mgf of?

Chapter 9

Asymptotics: Mapping and the Δ-Method

The law of large numbers and central limit theorem are useful on their own, but they can be combined in order to find convergence for many more interesting situations. In this chapter we look at mapping and the Δ-method. Lemma 8.4 on page 128 and equation (8.21) deal with mapping, where $\mathbf{W}_n \to^{\mathcal{P}} \mathbf{c}$ implies that $g(\mathbf{W}_n) \to^{\mathcal{P}} g(\mathbf{c})$ if g is continuous at $\mathbf{c}$. General mapping results allow mixing convergences in probability and in distribution.

The Δ-method extends the central limit theorem to normalized functions of the sample mean.

9.1 Mapping

The next lemma lists a number of mapping results, all of which relate to one another.

Lemma 9.1. Mapping.

1. *If $W_n \to^{\mathcal{D}} W$, and $g : \mathbb{R} \to \mathbb{R}$ is continuous at all points in $\mathcal{W}$, the space of W, then*

$$g(W_n) \longrightarrow^{\mathcal{D}} g(W). \tag{9.1}$$

2. *Similarly, for multivariate $\mathbf{W}_n$ ($p \times 1$) and g ($q \times 1$): If $\mathbf{W}_n \to^{\mathcal{D}} \mathbf{W}$, and $g : \mathbb{R}^p \to \mathbb{R}^q$ is continuous at all points in $\mathcal{W}$, then*

$$g(\mathbf{W}_n) \longrightarrow^{\mathcal{D}} g(\mathbf{W}). \tag{9.2}$$

3. *The next results constitute what is usually called **Slutsky's theorem**, or sometimes **Cramér's theorem**. Suppose that*

$$Z_n \longrightarrow^{\mathcal{P}} c \ \text{ and } \ W_n \longrightarrow^{\mathcal{D}} W. \tag{9.3}$$

Then

$$Z_n + W_n \longrightarrow^{\mathcal{D}} c + W, \ \ Z_n W_n \longrightarrow^{\mathcal{D}} cW, \tag{9.4}$$

and

$$\text{if } c \neq 0, \ \frac{W_n}{Z_n} \longrightarrow^{\mathcal{D}} \frac{W}{c}. \tag{9.5}$$

4. *Generalizing Slutsky, if* $Z_n \longrightarrow^P c$ *and* $W_n \longrightarrow^D W$, *and* $g : \mathbb{R} \times \mathbb{R} \to \mathbb{R}$ *is continuous at* $\{c\} \times W$, *then*

$$g(Z_n, W_n) \longrightarrow^D g(c, W). \tag{9.6}$$

5. *Finally, the multivariate version of #4. If* $\mathbf{Z}_n \longrightarrow^P \mathbf{c}$ $(p_1 \times 1)$ *and* $\mathbf{W}_n \longrightarrow^D \mathbf{W}$ $(p_2 \times 1)$, *and* $g : \mathbb{R}^{p_1} \times \mathbb{R}^{p_2} \to \mathbb{R}^q$ *is continuous at* $\{\mathbf{c}\} \times W$, *then*

$$g(\mathbf{Z}_n, \mathbf{W}_n) \longrightarrow^D g(\mathbf{c}, \mathbf{W}). \tag{9.7}$$

All the other four follow from #2, because convergence in probability is the same as convergence in distribution to a constant random variable. Also, #2 is just the multivariate version of #1, so basically they are all the same. They appear various places in various forms, though. The idea is that as long as the function is continuous, the limit of the function is the function of the limit. See Theorem 29.2 of Billingsley (1995) for an even more general result.

Together with the law of large numbers and central limit theorem, these mapping results can prove a huge number of useful approximations. The t statistic is one example. Let $X_1, \ldots, X_n$ be iid with mean μ and variance $\sigma^2 \in (0, \infty)$. Student's t statistic is defined as

$$T_n = \sqrt{n}\, \frac{\overline{X}_n - \mu}{S_{*n}}, \quad \text{where } S_{*n}^2 = \frac{\sum_{i=1}^n (X_i - \overline{X}_n)^2}{n-1}. \tag{9.8}$$

We know that if the data are normal, this $T_n \sim t_{n-1}$ exactly, but if the data are not normal, who knows what the distribution is. We can find the limit, though, using Slutsky. Take

$$Z_n = S_{*n} \quad \text{and} \quad W_n = \sqrt{n}\, (\overline{X}_n - \mu). \tag{9.9}$$

Then

$$Z_n \longrightarrow^P \sigma \quad \text{and} \quad W_n \longrightarrow^D N(0, \sigma^2) \tag{9.10}$$

from (8.22) and the central limit theorem, respectively. Because $\sigma^2 > 0$, the final component of (9.5) shows that

$$T_n = \frac{W_n}{Z_n} \longrightarrow^D \frac{N(0, \sigma^2)}{\sigma} = N(0, 1). \tag{9.11}$$

Thus for large n, T_n is approximately $N(0, 1)$ even if the data are not normal, and

$$\overline{X}_n \pm 2\, \frac{S_{*n}}{\sqrt{n}} \tag{9.12}$$

is an approximate 95% confidence interval for μ. Notice that this result doesn't say anything about small n, especially it doesn't say that the t is better than the z when the data are not normal. Other studies have shown that the t is fairly robust, so it can be used at least when the data are approximately normal. Actually, heavy tails for the X_i's means light tails for the T_n, so the z might be better than t in that case.

9.1.1   Regression through the origin

Recall the example in Section 8.3.1. We know $\widehat{\beta}_n$ is a consistent estimator of β, but what about its asymptotic distribution? That is,

$$\sqrt{n}\,(\widehat{\beta}_n - \beta) \longrightarrow^{\mathcal{D}} ?? \tag{9.13}$$

We need to do some manipulation to get it into a form where we can use the central limit theorem, etc. To that end,

$$\begin{aligned}
\sqrt{n}\,(\widehat{\beta}_n - \beta) &= \sqrt{n}\left(\frac{\sum_{i=1}^{n} X_i Y_i}{\sum_{i=1}^{n} X_i^2} - \beta\right) \\
&= \sqrt{n}\,\frac{\sum_{i=1}^{n} X_i Y_i - \beta \sum_{i=1}^{n} X_i^2}{\sum_{i=1}^{n} X_i^2} \\
&= \sqrt{n}\,\frac{\sum_{i=1}^{n}(X_i Y_i - \beta X_i^2)}{\sum_{i=1}^{n} X_i^2} \\
&= \frac{\sqrt{n}\,\sum_{i=1}^{n}(X_i Y_i - \beta X_i^2)/n}{\sum_{i=1}^{n} X_i^2/n}.
\end{aligned} \tag{9.14}$$

The numerator in the last expression contains the sample mean of the $(X_i Y_i - \beta X_i^2)$'s. Conditionally, from (8.24),

$$E[X_i Y_i - \beta X_i^2 \mid X_i = x_i] = x_i \beta x_i - \beta x_i^2 = 0, \;\; Var[X_i Y_i - \beta X_i^2 \mid X_i = x_i] = x_i^2\,\sigma_e^2, \tag{9.15}$$

so that unconditionally,

$$E[X_i Y_i - \beta X_i^2] = 0, \;\; Var[X_i Y_i - \beta X_i^2] = E[X_i^2\,\sigma^2] + Var[0] = \sigma_e^2(\sigma_X^2 + \mu_X^2). \tag{9.16}$$

Thus the central limit theorem shows that

$$\sqrt{n}\,\sum_{i=1}^{n}(X_i Y_i - \beta X_i^2)/n \longrightarrow^{\mathcal{D}} N(0, \sigma_e^2(\sigma_X^2 + \mu_X^2)). \tag{9.17}$$

We already know from (8.29) that $\sum X_i^2/n \to^{P} \sigma_X^2 + \mu_X^2$, hence by Slutsky (9.5),

$$\sqrt{n}\,(\widehat{\beta}_n - \beta) \longrightarrow^{\mathcal{D}} \frac{N(0, \sigma_e^2(\sigma_X^2 + \mu_X^2))}{\sigma_X^2 + \mu_X^2} = N\left(0, \frac{\sigma_e^2}{\sigma_X^2 + \mu_X^2}\right). \tag{9.18}$$

9.2   Δ-method

The central limit theorem deals with sample means, but often we are interested in some function of the mean, such as the $g(\overline{X}_n) = 1/\overline{X}_n$ in (8.16). One way to linearize a function is to use a one-step Taylor series. Thus if $\overline{X}_n$ is close to its mean μ, then $g(\overline{X}_n) \approx g(\mu) + (\overline{X}_n - \mu)g'(\mu)$, and the central limit theorem can be applied to the right-hand side. This method is called the Δ-**method**, which we formally define next in more generality (i.e., we do not need to base it on the sample mean).

Lemma 9.2. Δ-method. *Suppose*

$$\sqrt{n}\,(Y_n - \mu) \longrightarrow^{\mathcal{D}} W, \tag{9.19}$$

and the function $g : \mathbb{R} \to \mathbb{R}$ *has a continuous derivative at* μ. *Then*

$$\sqrt{n}\,(g(Y_n) - g(\mu)) \longrightarrow^{\mathcal{D}} g'(\mu)\,W. \tag{9.20}$$

Proof. Taylor series yields

$$g(X_n) = g(\mu) + (Y_n - \mu)g'(\mu_n^*), \quad \mu_n^* \text{ is between } Y_n \text{ and } \mu, \tag{9.21}$$

hence

$$\sqrt{n}\,(g(Y_n) - g(\mu)) = \sqrt{n}\,(Y_n - \mu)\,g'(\mu_n^*). \tag{9.22}$$

We wish to show that $g'(\mu_n^*) \to^{\mathcal{P}} g'(\mu)$, but first need to show that $Y_n \to^{\mathcal{P}} \mu$. Now

$$Y_n - \mu = [\sqrt{n}\,(Y_n - \mu)] \times \frac{1}{\sqrt{n}}$$

$$\longrightarrow^{\mathcal{D}} W \times 0 \quad (\text{because } \frac{1}{\sqrt{n}} \longrightarrow^{\mathcal{P}} 0)$$

$$= 0. \tag{9.23}$$

That is, $Y_n - \mu \to^{\mathcal{P}} 0$, hence $Y_n \to^{\mathcal{P}} \mu$. Because μ_n^* is trapped between Y_n and μ, $\mu_n^* \to^{\mathcal{P}} \mu$, which by continuity of g' means that

$$g'(\mu_n^*) \longrightarrow^{\mathcal{P}} g'(\mu). \tag{9.24}$$

Applying Slutsky (9.5) to (9.22), by (9.19) and (9.24),

$$\sqrt{n}\,(Y_n - \mu)\,g'(\mu_n^*) \longrightarrow^{\mathcal{D}} g'(\mu)\,W, \tag{9.25}$$

which via (9.22) proves (9.20). □

Usually, the limiting W is normal, so that under the conditions on g, we have that

$$\sqrt{n}\,(Y_n - \mu) \longrightarrow^{\mathcal{D}} N(0, \sigma^2) \implies \sqrt{n}\,(g(Y_n) - g(\mu)) \longrightarrow^{\mathcal{D}} N(0, g'(\mu)^2 \sigma^2). \tag{9.26}$$

9.2.1   Median

Here we apply Lemma 9.2 to the sample median. We have $X_1, \ldots, X_n$ iid with continuous distribution function F and pdf f. Let η be the median, so that $F(\eta) = 1/2$, and assume that the pdf f is positive and continuous at η. For simplicity we take n odd and set $k_n = (n+1)/2$, so that $X_{(k_n)}$, the k_n^{th} order statistic, is the median. Exercise 9.5.4 shows that for $U_1, \ldots, U_n$ iid Uniform(0,1),

$$\sqrt{n}\,(U_{(k_n)} - \tfrac{1}{2}) \longrightarrow N(0, \tfrac{1}{4}). \tag{9.27}$$

Thus for a function g with continuous derivative at $1/2$, the Δ-method (9.26) shows that

$$\sqrt{n}\,(g(U_{(k_n)}) - g(\tfrac{1}{2})) \longrightarrow N(0, \tfrac{1}{4} g'(\tfrac{1}{2})^2). \tag{9.28}$$

Let $g(u) = F^{-1}(u)$. Then

$$g(U_{(k_n)}) = F^{-1}(U_{(k_n)}) =^{\mathcal{D}} X_{(k_n)} \quad \text{and} \quad g(\tfrac{1}{2}) = F^{-1}(\tfrac{1}{2}) = \eta. \tag{9.29}$$

We also use the fact that

$$g'(u) = \frac{1}{F'(F^{-1}(u))} \Longrightarrow g'(\tfrac{1}{2}) = \frac{1}{f(\eta)}. \tag{9.30}$$

Thus making the substitutions in (9.28), we obtain

$$\sqrt{n}\,(X_{(k_n)} - \eta) \longrightarrow N\left(0, \frac{1}{4f(\eta)^2}\right). \tag{9.31}$$

If n is even and one takes the average of the two middle values as the median, then one can show that the asymptotic results are the same.

Recall location-scale families of distributions in Section 4.2.4. Consider just location families, so that for given pdf $f(x)$, the family of pdfs in the model consist of the $f_\mu(x) = f(x - \mu)$ for $\mu \in \mathbb{R}$. We restrict to f's that are symmetric about 0, hence the median of $f_\mu(x)$ is μ, as is the mean if the mean exists. In these cases, both the sample mean and sample median are reasonable estimates of μ. Which is better? The exact answer may be difficult (though if $n = 1$ they are both the same), but we can use asymptotics to approximately compare the variances when n is large. That is, we know $\sqrt{n}(\overline{X}_n - \mu)$ is asymptotically $N(0, \sigma^2)$ if the $\sigma^2 = Var[X_i] < \infty$, and (9.31) provides that the asymptotic distribution of $\sqrt{n}(\text{Median}_n - \mu)$ is $N(0, \tau^2)$, where $\tau^2 = 1/(4f(0)^2)$ since $f_\mu(\mu) = f(0)$. If $\sigma^2 > \tau^2$, then asymptotically the median is better, and vice versa. The ratio σ^2/τ^2 is called the **asymptotic relative efficiency of the median to the mean**. Table (9.32) gives these values for various choices of f.

Base distribution	σ^2	τ^2	σ^2/τ^2	
Normal$(0,1)$	1	$\pi/2$	$2/\pi \approx 0.6366$	
Cauchy	∞	$\pi^2/4$	∞	(9.32)
Laplace	2	1	2	
Uniform$(-1,1)$	$1/3$	1	$1/3$	
Logistic	$\pi^2/3$	4	$\pi^2/12 \approx 0.8225$	

The mean is better for the normal, uniform, and logistic, but the median is better for the Laplace and, especially, for the Cauchy. Generally, the thinner the tails of the distribution, the relatively better the mean performs.

9.3 Variance stabilizing transformations

Often, the variance of an estimator depends on the value of the parameter being estimated. For example, if $X_n \sim \text{Binomial}(n, p)$, then with $\hat{p}_n = X_n/n$,

$$Var[\hat{p}_n] = \frac{p(1-p)}{n}. \tag{9.33}$$

In regression situations, one usually desires the dependent Y_i's to have the same variance for each i, but if these Y_i's are binomial, or Poisson, the variance will not be constant. Also, confidence intervals are easier if the standard error does not need to be estimated.

By taking an appropriate function of the estimator, we may be able to achieve approximately constant variance. Such a function is called a *variance stabilizing transformation*. Formally, if $\widehat{\theta}_n$ is an estimator of θ, then we wish to find a g such that

$$\sqrt{n}\,(g(\widehat{\theta}_n) - g(\theta)) \longrightarrow^{\mathcal{D}} N(0, 1). \tag{9.34}$$

The "1" for the variance is arbitrary. The important thing is that it does not depend on θ.

In the binomial example, the variance stabilizing g would satisfy

$$\sqrt{n}\,(g(\widehat{p}_n) - g(p)) \longrightarrow^{\mathcal{D}} N(0, 1). \tag{9.35}$$

We know that

$$\sqrt{n}\,(\widehat{p}_n - p) \longrightarrow^{\mathcal{D}} N(0, p(1 - p)), \tag{9.36}$$

and by the Δ-method,

$$\sqrt{n}\,(g(\widehat{p}_n) - g(p)) \longrightarrow^{\mathcal{D}} N(0, g'(p)^2 p(1 - p)). \tag{9.37}$$

What should g be so that that variance is 1? We need to solve

$$g'(p) = \frac{1}{\sqrt{p(1 - p)}}, \tag{9.38}$$

so that

$$g(p) = \int_0^p \frac{1}{\sqrt{y(1 - y)}}\,dy. \tag{9.39}$$

First, let $u = \sqrt{y}$, so that $y = u^2$ and $dy = 2u\,du$, and

$$g(p) = \int_0^{\sqrt{p}} \frac{1}{u\sqrt{1 - u^2}}\,2u\,du = 2\int_0^{\sqrt{p}} \frac{1}{\sqrt{1 - u^2}}\,du. \tag{9.40}$$

The integral is $\arcsin(u)$, which means the variance stabilizing transformation is

$$g(p) = 2\,\arcsin(\sqrt{p}). \tag{9.41}$$

Note that adding a constant to g won't change the derivative. The approximation suggested by (9.35) is then

$$2\,\arcsin\left(\sqrt{\widehat{p}_n}\right) \approx N\left(2\,\arcsin(\sqrt{p}), \frac{1}{n}\right). \tag{9.42}$$

An approximate 95% confidence interval for $2\,\arcsin(\sqrt{p})$ is

$$2\,\arcsin\left(\sqrt{\widehat{p}_n}\right) \pm \frac{2}{\sqrt{n}}. \tag{9.43}$$

That interval can be inverted to obtain the interval for p, that is, apply $g^{-1}(u) = \sin(u/2)^2$ to both ends:

$$p \in \left(\sin\left(\arcsin\left(\sqrt{\widehat{p}_n}\right) - \frac{1}{\sqrt{n}}\right)^2, \; \sin\left(\arcsin\left(\sqrt{\widehat{p}_n}\right) + \frac{1}{\sqrt{n}}\right)^2\right). \tag{9.44}$$

Brown, Cai, and DasGupta (2001) show that this interval, but with $\hat{p}_n$ replaced by $(x + 3/8)/(n + 3/4)$ as proposed by Anscombe (1948), is quite a bit better than the usual approximate interval,

$$\hat{p}_n \pm 2\sqrt{\frac{\hat{p}_n(1 - \hat{p}_n)}{n}}. \tag{9.45}$$

9.4 Multivariate Δ-method

It might be that we have a function of several random variables to deal with, or more generally we have several functions of several variables. For example, we might be interested in the mean and variance simultaneously. So we start with a sequence of $p \times 1$ random vectors, whose asymptotic distribution is multivariate normal:

$$\sqrt{n}\,(\mathbf{Y}_n - \boldsymbol{\mu}) \longrightarrow^{\mathcal{D}} N(\mathbf{0}_n, \boldsymbol{\Sigma}), \tag{9.46}$$

and a function $g : \mathbb{R}^p \to \mathbb{R}^q$. We cannot just take the derivative of g, since there are pq of them. What we need is the entire matrix of derivatives, just as for finding the Jacobian. Letting

$$g(\mathbf{y}) = \begin{pmatrix} g_1(\mathbf{y}) \\ g_2(\mathbf{y}) \\ \vdots \\ g_q(\mathbf{y}) \end{pmatrix} \quad \text{and} \quad \mathbf{y} = \begin{pmatrix} y_1 \\ y_2 \\ \vdots \\ y_p \end{pmatrix}, \tag{9.47}$$

define the $q \times p$ matrix

$$\mathbf{D}(\mathbf{y}) = \begin{pmatrix} \frac{\partial}{\partial w_1} g_1(\mathbf{y}) & \frac{\partial}{\partial w_2} g_1(\mathbf{y}) & \cdots & \frac{\partial}{\partial w_p} g_1(\mathbf{y}) \\ \frac{\partial}{\partial w_1} g_2(\mathbf{y}) & \frac{\partial}{\partial w_2} g_2(\mathbf{y}) & \cdots & \frac{\partial}{\partial w_p} g_2(\mathbf{y}) \\ \vdots & \vdots & \ddots & \vdots \\ \frac{\partial}{\partial w_1} g_q(\mathbf{y}) & \frac{\partial}{\partial w_2} g_q(\mathbf{y}) & \cdots & \frac{\partial}{\partial w_p} g_q(\mathbf{y}) \end{pmatrix}. \tag{9.48}$$

Lemma 9.3. *Multivariate Δ-method. Suppose (9.46) holds, and $\mathbf{D}(\mathbf{y})$ in (9.48) is continuous at $\mathbf{y} = \boldsymbol{\mu}$. Then*

$$\sqrt{n}\,(g(\mathbf{Y}_n) - g(\boldsymbol{\mu})) \longrightarrow^{\mathcal{D}} N(\mathbf{0}_n, \mathbf{D}(\boldsymbol{\mu})\boldsymbol{\Sigma}\mathbf{D}(\boldsymbol{\mu})'). \tag{9.49}$$

The $\boldsymbol{\Sigma}$ is $p \times p$ and $\mathbf{D}$ is $q \times p$, so that the covariance in (9.49) is $q \times q$, as it should be. Some examples follow.

9.4.1 Mean, variance, and coefficient of variation

Go back to the example that ended with (8.84): $X_1, \ldots, X_n$ are iid with mean μ, variance σ^2, $E[X_i^3] = \mu_3'$ and $E[X_i^4] = \mu_4'$. Ultimately, we wish to find the asymptotic distribution of $\overline{X}_n$ and S_n^2, so we start with that of $(\sum X_i, \sum X_i^2)'$:

$$\mathbf{Y}_n = \frac{1}{n} \sum_{i=1}^{n} \begin{pmatrix} X_i \\ X_i^2 \end{pmatrix}, \quad \boldsymbol{\mu} = \begin{pmatrix} \mu \\ \mu^2 + \sigma^2 \end{pmatrix}, \tag{9.50}$$

and

$$\Sigma = \begin{pmatrix} \sigma^2 & \mu_3' - \mu(\mu^2 + \sigma^2) \\ \mu_3' - \mu(\mu^2 + \sigma^2) & \mu_4' - (\mu^2 + \sigma^2)^2 \end{pmatrix}. \tag{9.51}$$

Since $S_n^2 = \sum X_i^2/n - \overline{X}_n^2$,

$$\begin{pmatrix} \overline{X}_n \\ S_n^2 \end{pmatrix} = g(\overline{X}_n, \sum X_i^2/n) = \begin{pmatrix} g_1(\overline{X}_n, \sum X_i^2/n) \\ g_2(\overline{X}_n, \sum X_i^2/n) \end{pmatrix}, \tag{9.52}$$

where

$$g_1(y_1, y_2) = y_1 \quad \text{and} \quad g_2(y_1, y_2) = y_2 - y_1^2. \tag{9.53}$$

Then

$$\mathbf{D}(y) = \begin{pmatrix} \frac{\partial y_1}{\partial y_1} & \frac{\partial y_1}{\partial y_2} \\ \frac{\partial(y_2 - y_1^2)}{\partial y_1} & \frac{\partial(y_2 - y_1^2)}{\partial y_2} \end{pmatrix} = \begin{pmatrix} 1 & 0 \\ -2y_1 & 1 \end{pmatrix}. \tag{9.54}$$

Also,

$$g(\mu) = \begin{pmatrix} \mu \\ \sigma^2 + \mu^2 - \mu^2 \end{pmatrix} = \begin{pmatrix} \mu \\ \sigma^2 \end{pmatrix}, \quad \mathbf{D}(\mu) = \begin{pmatrix} 1 & 0 \\ -2\mu & 1 \end{pmatrix}, \tag{9.55}$$

and

$$\mathbf{D}(\mu)\Sigma\mathbf{D}(\mu)' = \begin{pmatrix} 1 & 0 \\ -2\mu & 1 \end{pmatrix} \begin{pmatrix} \sigma^2 & \mu_3' - \mu(\mu^2 + \sigma^2) \\ \mu_3' - \mu(\mu^2 + \sigma^2) & \mu_4' - (\mu^2 + \sigma^2)^2 \end{pmatrix} \begin{pmatrix} 1 & -2\mu \\ 0 & 1 \end{pmatrix}$$

$$= \begin{pmatrix} \sigma^2 & \mu_3' - \mu^3 - 3\mu\sigma^2 \\ \mu_3' - \mu^3 - 3\mu\sigma^2 & \mu_4' - \sigma^4 - 4\mu\mu_3' + 3\mu^4 + 6\mu^2\sigma^2 \end{pmatrix}.$$

Yikes! Notice in particular that the sample mean and variance are not asymptotically independent necessarily.

Before we go on, let's assume the data are normal. In that case,

$$\mu_3' = \mu^3 + 3\mu\sigma^2 \quad \text{and} \quad \mu_4' = 3\sigma^4 + 6\sigma^2\mu^2 + \mu^4, \tag{9.56}$$

(left to the reader), and, magically,

$$\mathbf{D}(\mu)\Sigma\mathbf{D}(\mu)' = \begin{pmatrix} \sigma^2 & 0 \\ 0 & 2\sigma^4 \end{pmatrix}, \tag{9.57}$$

hence

$$\sqrt{n}\left(\begin{pmatrix} \overline{X}_n \\ S_n^2 \end{pmatrix} - \begin{pmatrix} \mu \\ \sigma^2 \end{pmatrix} \right) \longrightarrow^{\mathcal{D}} N\left(0_2, \begin{pmatrix} \sigma^2 & 0 \\ 0 & 2\sigma^4 \end{pmatrix} \right). \tag{9.58}$$

Actually, that is not surprising, since we know the variance of $\overline{X}_n$ is σ^2/n, and that of S_n^2 is the variance of a χ^2_{n-1} times $(n-1)\sigma^2/n$, which is $2(n-1)^3\sigma^4/n^2$. Multiplying those by n and letting $n \to \infty$ yields the diagonals σ^2 and $2\sigma^4$. Also the mean and variance are independent, so their covariance is 0.

From these, we can find the *coefficient of variance*, or the noise-to-signal ratio,

$$cv = \frac{\sigma}{\mu} \quad \text{and sample version} \quad \widehat{cv} = \frac{S_n}{\overline{X}_n}. \tag{9.59}$$

Then $\hat{cv} = h(\overline{X}_n, S_n^2)$ where $h(w_1, w_2) = \sqrt{w_2}/w_1$. The derivatives here are

$$\mathbf{D}_h(w_1, w_2) = \left(-\frac{\sqrt{w_2}}{w_1^2}, \frac{1}{2\sqrt{w_2}\,w_1}\right) \implies \mathbf{D}_h(\mu, \sigma^2) = \left(-\frac{\sigma}{\mu^2}, \frac{1}{2\sigma\mu}\right). \tag{9.60}$$

Then, assuming that $\mu \neq 0$,

$$\mathbf{D}_h \mathbf{\Sigma} \mathbf{D}_h' = \left(-\frac{\sigma}{\mu^2}, \frac{1}{2\sigma\mu}\right) \begin{pmatrix} \sigma^2 & 0 \\ 0 & 2\sigma^4 \end{pmatrix} \begin{pmatrix} -\frac{\sigma}{\mu^2} \\ \frac{1}{2\sigma\mu} \end{pmatrix}$$

$$= \frac{\sigma^4}{\mu^4} + \frac{1}{2}\frac{\sigma^2}{\mu^2}. \tag{9.61}$$

The asymptotic distribution is then

$$\sqrt{n}\,(\hat{cv} - cv) \longrightarrow^{\mathcal{D}} N\left(0, \frac{\sigma^4}{\mu^4} + \frac{1}{2}\frac{\sigma^2}{\mu^2}\right) = N(0, cv^2(cv^2 + \tfrac{1}{2})). \tag{9.62}$$

For example, data on $n = 102$ female students' heights had a mean of 65.56 and a standard deviation of 2.75, so the $\hat{cv} = 2.75/65.56 = 0.0419$. We can find a confidence interval by estimating the variance in (9.62) in the obvious way:

$$\left(\hat{cv} \pm 2\,\frac{|\hat{cv}|\sqrt{0.5 + \hat{cv}^2}}{\sqrt{n}}\right) = (0.0419 \pm 2 \times 0.0029) = (0.0361, 0.0477). \tag{9.63}$$

For the men, the mean is 71.25 and the sd is 2.94, so their cv is 0.0413. That's practically the same as for the women. The men's standard error of $\hat{cv}$ is 0.0037 (the $n = 64$), so a confidence interval for the difference between the women and men is

$$(0.0419 - 0.0413 \pm 2\sqrt{0.0029^2 + 0.0037^2}) = (0.0006 \pm 0.0094). \tag{9.64}$$

Clearly 0 is in that interval, so there does not appear to be any difference between the coefficients of variation.

9.4.2 Correlation coefficient

Consider the bivariate normal sample,

$$\begin{pmatrix} X_1 \\ Y_1 \end{pmatrix}, \ldots, \begin{pmatrix} X_n \\ Y_n \end{pmatrix} \text{ are iid } \sim N\left(0_2, \begin{pmatrix} 1 & \rho \\ \rho & 1 \end{pmatrix}\right). \tag{9.65}$$

The sample correlation coefficient in this case (we don't need to subtract the means) is

$$R_n = \frac{\sum_{i=1}^n X_i Y_i}{\sqrt{\sum_{i=1}^n X_i^2 \sum_{i=1}^n Y_i^2}}. \tag{9.66}$$

From Exercise 8.7.6(c), we know that $R_n \to^{\mathcal{P}} \rho$. What about the asymptotic distribution? Notice that R_n can be written as a function of three sample means,

$$R_n = g\left(\frac{1}{n}\sum_{i=1}^n X_i Y_i, \frac{1}{n}\sum_{i=1}^n X_i^2, \frac{1}{n}\sum_{i=1}^n Y_i^2\right) \text{ where } g(w_1, w_2, w_3) = \frac{w_1}{\sqrt{w_1 w_2}}. \tag{9.67}$$

First, apply the central limit theorem to the three means:

$$\sqrt{n} \left(\begin{pmatrix} \frac{1}{n}\sum_{i=1}^{n} X_i Y_i \\ \frac{1}{n}\sum_{i=1}^{n} X_i^2 \\ \frac{1}{n}\sum_{i=1}^{n} Y_i^2 \end{pmatrix} - \mu \right) \longrightarrow^{\mathcal{D}} N(0_3, \Sigma). \tag{9.68}$$

Now $E[X_i Y_i] = \rho$ and $E[X_i^2] = E[Y_i^2] = 1$, hence

$$\mu = \begin{pmatrix} \rho \\ 1 \\ 1 \end{pmatrix}. \tag{9.69}$$

The covariance is a little more involved. Exercise 7.8.10 shows that

$$\Sigma = Cov \begin{pmatrix} X_i Y_i \\ X_i^2 \\ Y_i^2 \end{pmatrix} = \begin{pmatrix} 1+\rho^2 & 2\rho & 2\rho \\ 2\rho & 2 & 2\rho^2 \\ 2\rho & 2\rho^2 & 2 \end{pmatrix}. \tag{9.70}$$

Exercise 9.5.10 applies the Δ-method to (9.68) to obtain

$$\sqrt{n} \, (R_n - \rho) \longrightarrow^{\mathcal{D}} N(0, (1-\rho^2)^2). \tag{9.71}$$

9.4.3   Affine transformations

If A is a $q \times p$ matrix, then it is easy to get the asymptotic distribution of $AX_n + b$:

$$\sqrt{n} \, (X_n - \mu) \longrightarrow^{\mathcal{D}} N(0_n, \Sigma) \implies \sqrt{n} \, (AX_n + b - (A\mu + b)) \longrightarrow^{\mathcal{D}} N(0_n, A\Sigma A'), \tag{9.72}$$

because for the function $g(w) = Aw$, $D(w) = A$.

9.5   Exercises

Exercise 9.5.1. Suppose $X_1, X_2, \ldots$ are iid with $E[X_i] = 0$ and $Var[X_i] = \sigma^2 < \infty$. Show that

$$\frac{\sum_{i=1}^{n} X_i}{\sqrt{\sum_{i=1}^{n} X_i^2}} \longrightarrow^{\mathcal{D}} N(0,1). \tag{9.73}$$

Exercise 9.5.2. Suppose

$$\begin{pmatrix} X_1 \\ Y_1 \end{pmatrix}, \cdots, \begin{pmatrix} X_n \\ Y_n \end{pmatrix} \text{ are iid } N\left(\begin{pmatrix} 0 \\ 0 \end{pmatrix}, \begin{pmatrix} \sigma_X^2 & \sigma_{XY} \\ \sigma_{XY} & \sigma_Y^2 \end{pmatrix} \right). \tag{9.74}$$

Then $Y_i \mid X_i = x_i \sim N(\alpha + \beta x_i, \sigma_e^2)$. (What is α?) As in Section 8.3.1 let

$$\widehat{\beta}_n = \frac{\sum_{i=1}^{n} X_i Y_i}{\sum_{i=1}^{n} X_i^2}, \tag{9.75}$$

and set $W_n = \sqrt{n}(\widehat{\beta}_n - \beta)$. (a) Give the V_i for which

$$W_n = \sqrt{n} \left(\frac{\sum_{i=1}^{n} V_i}{\sum_{i=1}^{n} X_i^2} \right), \tag{9.76}$$

where V_i depends on X_i, Y_i, and β. (b) Find $E[V_i \mid X_i = x_i]$ and $Var[V_i \mid X_i = x_i]$. (c) Find $E[V_i]$ and $\sigma_V^2 = Var[V_i]$. (d) Are the V_i's iid? (e) For what a does

$$\frac{\sum_{i=1}^{n} V_i}{n^a} \longrightarrow^{\mathcal{D}} N(0, \sigma_V^2)? \tag{9.77}$$

(f) Find the b and $c > 0$ for which

$$\frac{\sum_{i=1}^{n} X_i^2}{n^b} \longrightarrow^{P} c. \tag{9.78}$$

(g) Using parts (e) and (f), $W_n \longrightarrow^{\mathcal{D}} N(0, \sigma_B^2)$ for what σ_B^2? What theorem is needed to prove the result?

Exercise 9.5.3. Here, $Y_k \sim \text{Beta}(k, k)$. We are interested in the limit of $\sqrt{k}(Y_k - 1/2)$ as $k \to \infty$. Start by representing Y_k with gammas, or more particularly, Exponential(1)'s. So let $X_1, \ldots, X_{2k}$ be iid Exponential(1). Then

$$Y_k = \frac{X_1 + \cdots + X_k}{X_1 + \cdots + X_k + X_{k+1} + \cdots + X_{2k}}. \tag{9.79}$$

(a) Is this representation correct? (b) Now write

$$Y_k - \tfrac{1}{2} = c \, \frac{U_1 + \cdots + U_k}{V_1 + \cdots + V_k} = c \, \frac{\overline{U}_k}{\overline{V}_k}, \tag{9.80}$$

where $U_i = X_i - X_{k+i}$ and $V_i = X_i + X_{k+i}$. What is c? (c) Find $E[U_i]$, $Var[U_i]$, and $E[V_i]$. (d) So

$$\sqrt{k} \, (Y_k - \tfrac{1}{2}) = c \, \frac{\sqrt{k} \, \overline{U}_k}{\overline{V}_k}. \tag{9.81}$$

What is the asymptotic distribution of $\sqrt{k} \, \overline{U}_k$ as $k \to \infty$? What theorem do you use? (e) What is the limit of $\overline{V}_k$ in probablity as $k \to \infty$? What theorem do you use? (f) Finally, $\sqrt{k} \, (Y_k - 1/2) \longrightarrow^{\mathcal{D}} N(0, v)$. What is v? (It is a number.) What theorem do you use?

Exercise 9.5.4. Suppose $U_1, \ldots, U_n$ are iid Uniform(0,1), and n is odd. Then the sample median is $U_{(k_n)}$ for $k_n = (n+1)/2$, hence $U_{(k_n)} \sim \text{Beta}(k_n, k_n)$. Letting $k = k_n$ and $Y_k = U_{(k_n)}$ in Exercise 9.5.3, we have that

$$\sqrt{k_n}(U_{(k_n)} - \tfrac{1}{2}) \longrightarrow N(0, v). \tag{9.82}$$

(a) What is the limit of n/k_n? (b) Show that

$$\sqrt{n}(U_{(k_n)} - \tfrac{1}{2}) \longrightarrow N(0, \tfrac{1}{4}). \tag{9.83}$$

What theorem did you use?

Exercise 9.5.5. Suppose $T \sim t_v$, Student's t. Exercise 6.8.17 shows that $E[T] = 0$ if $v > 1$ and $Var[T] = v/(v-2)$ if $v > 2$, as well as gives the pdf. This exercise is based on $X_1, X_2, \ldots, X_n$ iid with pdf $f_v(x - \mu)$, where f_v is the t_v pdf. (a) Find the asymptotic efficiency of the median relative to the mean for $v = 1, 2, \ldots, 7$. [You might want to use the function dt in R to find the density of the t_v.] (b) When v is small, which is better, the mean or the median? (c) When v is large, which is better, the mean or the median? (d) For which v is the asymptotic relative efficiency closest to 1?

Exercise 9.5.6. Suppose $U_1, \ldots, U_n$ are iid from the location family with f_α being the Beta(α, α) pdf, so that the pdf of each U_i is $f_\alpha(u - \eta)$ for $\eta \in \mathbb{R}$. (a) What must be added to the sample mean or sample median in order to obtain an unbiased estimator of η? (b) Find the asymptotic variances of the two estimators in part (a) for $\alpha = 1/10, 1/2, 1, 2, 10$. Also, find the asymptotic relative efficiency of the median to the mean for each α. What do you see?

Exercise 9.5.7. Suppose $X_1, \ldots, X_n$ are iid Poisson(λ). (a) What is the asymptotic distribution of $\sqrt{n}(\overline{X}_n - \lambda)$ as $n \to \infty$? (b) Consider the function g, such that $\sqrt{n}(g(\overline{X}_n) - g(\lambda)) \to^{\mathcal{D}} N(0,1)$. What is its derivative, $g'(w)$? (c) What is $g(w)$?

Exercise 9.5.8. (a) Suppose $X_1, \ldots, X_n$ are iid Exponential(θ). Since $E[X_i] = 1/\theta$, we might consider using $1/\overline{X}_n$ an estimator of θ. (a) What is the asymptotic distribution of $\sqrt{n}(1/\overline{X}_n - \theta)$? (b) Find $g(w)$ so that $\sqrt{n}(g(1/\overline{X}_n) - g(\theta)) \to^{\mathcal{D}} N(0,1)$.

Exercise 9.5.9. Suppose $X_1, \ldots, X_n$ are iid Gamma$(p, 1)$ for $p > 0$. (a) Find the asymptotic distribution of $\sqrt{n}(\overline{X}_n - p)$. (b) Find a statistic B_n (depending on the data alone) such that $\sqrt{n}(\overline{X}_n - p)/B_n \to^{\mathcal{D}} N(0,1)$. (c) Based on the asymptotic distribution in part (b), find an approximate 95% confidence interval for p. (d) Next, find a function g such that $\sqrt{n}(g(\overline{X}_n) - g(p)) \to^{\mathcal{D}} N(0,1)$. (e) Based on the asymptotic distribution in part (d), find an approximate 95% confidence interval for p. (f) Let $\overline{x}_n = 100$ with $n = 25$. Compute the confidence intervals using parts (c) and (e). Are they reasonably similar?

Exercise 9.5.10. Suppose

$$\begin{pmatrix} X_1 \\ Y_1 \end{pmatrix}, \cdots, \begin{pmatrix} X_n \\ Y_n \end{pmatrix} \quad \text{are iid } N\left(\begin{pmatrix} 0 \\ 0 \end{pmatrix}, \begin{pmatrix} 1 & \rho \\ \rho & 1 \end{pmatrix} \right). \tag{9.84}$$

Also, let

$$\mathbf{W}_i = \begin{pmatrix} X_i Y_i \\ X_i^2 \\ Y_i^2 \end{pmatrix}. \tag{9.85}$$

Equations (9.68) to (9.70) exhibit the asymptotic distribution of

$$\sqrt{n}\left(\overline{\mathbf{W}}_n - \begin{pmatrix} \rho \\ 1 \\ 1 \end{pmatrix} \right) \tag{9.86}$$

as $n \to \infty$, where $\overline{\mathbf{W}}_n$ is the sample mean of the $\mathbf{W}_i$'s. Take $g(w_1, w_2, w_3) = w_1/\sqrt{w_2 w_3}$ as in (9.67) so that

$$R_n = g(\overline{\mathbf{W}}_n) = \frac{\sum X_i Y_i}{\sqrt{\sum X_i^2} \sqrt{\sum Y_i^2}}. \tag{9.87}$$

(a) Find the vector of derivatives of g, $\mathbf{D}_g$, evaluated at $\mathbf{w} = (\rho, 1, 1)'$. (b) Use the Δ-method to show that

$$\sqrt{n}(R_n - \rho) \longrightarrow N(0, (1 - \rho^2)^2). \tag{9.88}$$

as $n \to \infty$.

Exercise 9.5.11. Suppose R_n is the sample correlation coefficient from an iid sample of bivariate normals as in Exercise 9.5.10. Consider the function h, such that $\sqrt{n}(h(R_n) - h(\rho)) \longrightarrow N(0,1)$. (a) Find $h'(w)$. (b) Show that

$$h(w) = \frac{1}{2} \log \left(\frac{1+w}{1-w} \right). \tag{9.89}$$

[Hint: You might want to use partial fractions, that is, write $h'(w)$ as $A/(1-w) + B/(1+w)$. What are A and B?] The statistic $h(R_n)$ is called **Fisher's z**. (c) In a sample of $n = 712$ students, the sample correlation coefficient between $x =$ shoe size and $y = \log(\#$ of pairs of shoes owned) is $r = -0.500$. Find an approximate 95% confidence interval for $h(\rho)$ (using "±2"). (Assuming these data are a simple random sample from a large normal population.) (d) What is the corresponding approximate 95% confidence interval for ρ? Is 0 in the interval? What do you conclude? (e) For just men, the sample size is 227 and the correlation between x and y is 0.0238. For just women, the sample size is 485 and the correlation between x and y is -0.0669. Find the confidence intervals for the population correlations for the men and women using the method in part (d). Is 0 in either or both of those intervals? What do you conclude?

Exercise 9.5.12. Suppose $\mathbf{X}_n \sim$ Multinomial$(n, \mathbf{p})$ where $\mathbf{p} = (p_1, p_2, p_3, p_4)'$ (so $K = 4$). Then $E[\mathbf{X}_n] = n\mathbf{p}$ and, from Exercise 2.7.7,

$$Cov[\mathbf{X}_n] = n\Sigma \text{ where } \Sigma = \begin{pmatrix} p_1 & 0 & 0 & 0 \\ 0 & p_2 & 0 & 0 \\ 0 & 0 & p_3 & 0 \\ 0 & 0 & 0 & p_4 \end{pmatrix} - \mathbf{pp}'. \tag{9.90}$$

(a) Suppose $\mathbf{Z}_1, \ldots, \mathbf{Z}_n$ are iid Multinomial $(1, \mathbf{p})$. Show that $\mathbf{X}_n = \mathbf{Z}_1 + \cdots + \mathbf{Z}_n$ is Multinomial$(n, \mathbf{p})$. [Hint: Use mgfs from Section 2.5.3.] (b) Argue that by part (a), the central limit theorem can be applied to show that

$$\sqrt{n}(\widehat{\mathbf{p}}_n - \mathbf{p}) \longrightarrow^{\mathcal{D}} N(\mathbf{0}, \Sigma). \tag{9.91}$$

(c) Arrange the p_i's in a 2×2 table:

$$\begin{array}{c|c} p_1 & p_2 \\ \hline p_3 & p_4 \end{array} \tag{9.92}$$

In Exercise 6.8.11 we saw the odds ratio. Here we look at the **log odds ratio**, given by $\log((p_1/p_2)/(p_3/p_4))$. The higher it is, the more positively associated being is row 1 is with being in column 1. For $\mathbf{w} = (w_1, \ldots, w_4)'$ with each $w_i \in (0,1)$, let $g(\mathbf{w}) = \log(w_1 w_4/(w_2 w_3))$, so that $g(\mathbf{p})$ is the log odds ratio for the 2×2 table. Show that in the independence case as in Exercise 6.8.11(c), $g(\mathbf{p}) = 0$. (d) Find $\mathbf{D}_g(\mathbf{w})$, the vector of derivatives of g, and and show that

$$\sqrt{n}\,(g(\widehat{\mathbf{p}}_n) - g(\mathbf{p})) \longrightarrow^{\mathcal{D}} N(0, \sigma_g^2), \tag{9.93}$$

where $\sigma_g^2 = 1/p_1 + 1/p_2 + 1/p_3 + 1/p_4$.

Exercise 9.5.13. In a statistics class, people were classified on how well they did on the combined homework, labs and inclass assignments (hi or lo), and how well they did on the exams (hi or lo). Thus each person was classified into one of four groups. Letting **p** be the population probabilities, arrange the vector in a two-by-two table as above. The table of observed counts is

Exams →	Lo	Hi
Homework ↓		
Lo	36	18
Hi	18	35

(9.94)

(a) Find the observed log odds ratio for these data and its estimated standard error. Find an approximate 95% confidence interval for $g(\mathbf{p})$. (b) What is the corresponding confidence interval for the odds ratio? What do you conclude?

The next two tables split the data by gender:

Women				Men		
Exams →	Lo	Hi		Exams →	Lo	Hi
Homework ↓				Homework ↓		
Lo	26	6		Lo	10	12
Hi	10	28		Hi	8	7

(9.95)

Assume the men and women are independent, each with their own multinomial distribution. (c) Find the difference between the women's and men's log odds ratios, and the standard error of that difference. What do you conclude about the difference between the women and men? (d) Looking at the women's and men's odds ratios separately, what do you conclude?

Part II

Statistical Inference

Chapter *10*

Statistical Models and Inference

Most, although not all, of the material so far has been straight probability calculations, that is, we are given a probability distribution, and try to figure out the implications (what **X** is likely to be, marginals, conditionals, moments, what happens asymptotically, etc.). Statistics generally concerns itself with the reverse problem, that is, observing the data **X** = **x**, and then having to guess aspects of the probability distribution that generated **x**. This "guessing" goes under the general rubric of **inference**. Four major aspects of inference are

- **Estimation**: What is the best guess of a particular parameter (vector), or function of the parameter? The estimate may be a point estimate, or a point estimate and measure of accuracy, or an interval or region, e.g., "The mean is in the interval (10.44,19.77)."

- **Hypothesis testing**: The question is whether a specific hypothesis, the **null hypothesis**, about the distribution is true, so that the inference is basically either "yes" or "no", along with an idea of how reliable the conclusion is.

- **Prediction**: One is interested in predicting a new observation, possibly depending on a covariate. For example, the data may consist of a number of (X_i, Y_i) pairs, and a new observation comes along, where we know the x but not the y, and wish to guess that y. We may be predicting a numerical variable, e.g., return on an investment, or a categorical variable, e.g., the species of a plant.

- **Model selection**: There may be several models under consideration, e.g., in multiple regression each subset of potential regressors defines a model. The goal would be to choose the best one, or a set of good ones.

The boundaries between these notions are not firm. One can consider prediction to be estimation, or model selection as an extension of hypothesis testing with more than two hypotheses. Whatever the goal, the first task is to specify the **statistical model**.

10.1 Statistical models

A probability model consists of a random **X** in space $\mathcal{X}$ and a probability distribution P. A **statistical model** also has **X** and space $\mathcal{X}$, but an entire **family** $\mathcal{P}$ of probability distributions on $\mathcal{X}$. By family we mean a set of distributions; the only restriction being that they are all distributions for the same $\mathcal{X}$. Such families can be quite general, e.g.,

$$\mathcal{X} = \mathbb{R}^n, \mathcal{P} = \{P \mid X_1, \dots, X_n \text{ are iid with finite mean and variance}\}. \tag{10.1}$$

This family includes all kinds of distributions (iid normal, gamma, beta, binomial), but not ones with the X_i's correlated, or distributed Cauchy (which has no mean or variance). Another possibility is the family with the X_i's iid with a continuous distribution.

Often, the families are parametrized by a finite-dimensional parameter $\boldsymbol{\theta}$, i.e.,

$$\mathcal{P} = \{P_{\boldsymbol{\theta}} \mid \boldsymbol{\theta} \in \mathcal{T}\}, \quad \text{where} \quad \mathcal{T} \subset \mathbb{R}^K. \tag{10.2}$$

The $\mathcal{T}$ is called the **parameter space**. We are quite familiar with parameters, but for statistical models we must be careful to specify the parameter space as well. For example, suppose X and Y are independent, $X \sim N(\mu_X, \sigma_X^2)$ and $Y \sim N(\mu_Y, \sigma_Y^2)$. Then the following parameter spaces lead to distinctly different models:

$$
\begin{aligned}
\mathcal{T}_1 &= \{(\mu_X, \sigma_X^2, \mu_Y, \sigma_Y^2) \in \mathbb{R} \times (0, \infty) \times \mathbb{R} \times (0, \infty)\}; \\
\mathcal{T}_2 &= \{(\mu_X, \sigma_X^2, \mu_Y, \sigma_Y^2) \mid \mu_X \in \mathbb{R}, \mu_Y \in \mathbb{R}, \sigma_X^2 = \sigma_Y^2 \in (0, \infty)\}; \\
\mathcal{T}_3 &= \{(\mu_X, \sigma_X^2, \mu_Y, \sigma_Y^2) \mid \mu_X \in \mathbb{R}, \mu_Y \in \mathbb{R}, \sigma_X^2 = \sigma_Y^2 = 1\}; \\
\mathcal{T}_4 &= \{(\mu_X, \sigma_X^2, \mu_Y, \sigma_Y^2) \mid \mu_X \in \mathbb{R}, \mu_Y \in \mathbb{R}, \mu_X > \mu_Y, \sigma_X^2 = \sigma_Y^2 \in (0, \infty)\}. \tag{10.3}
\end{aligned}
$$

The first model places no restrictions on the parameters, other than the variances are positive. The second one demands the two variances be equal. The third sets the variances to 1, which is equivalent to saying that the variances are known to be 1. The last one equates the variances, as well as specifying that the mean of X is larger than that of Y.

A **Bayesian model** includes a (prior) distribution on $\mathcal{P}$, which in the case of a parametrized model means a distribution on $\mathcal{T}$. In fact, the model could include a family of prior distributions, although we will not deal with that case explicitly.

Before we introduce inference, we take a brief look at how probability is interpreted.

10.2   Interpreting probability

In Section 1.2, we defined probability distributions mathematically, starting with some axioms. Everything else flowed from those axioms. But as in all mathematical objects, they do not in themselves have physical reality. In order to make practical use of the results, we must somehow connect the mathematical objects to the physical world. That is, how is one to interpret $P[A]$? In games of chance, people generally feel confident that they know what "the chance of heads" or "the chance of a full house" mean. But other probabilities may be less obvious, e.g., "the chance that it rains next Tuesday" or "the chance ____ and ____ get married" (fill in the blanks with any two people). Two popular interpretations are **frequency** and **subjective**. Both have many versions, and there are also many other interpretations, but much of this material is beyond the scope of the author. Here are sketches of the two.

Frequency. An experiment is presumed to be repeatable, so that one could conceivably repeat the experiment under the exact same conditions over and over again (i.e., infinitely often). Then the probability of a particular event A, $P[A]$, is the long-run proportion of times it occurs, as the experiment is repeated forever. That is, it is the long-run frequency A occurs. This interpretation implies that probability is objective in the sense that it is inherent in the experiment, not a product of one's beliefs. This interpretation works well for games of chance. One can imagine rolling a die or spinning a roulette wheel an "infinite" number of times. Population sampling also fits in well, as one could imagine repeatedly taking a random sample of 100 subjects from a given population. The frequentist interpretation can not be applied to situations that are not in principle repeatable, such as whether two people will get married, or whether a particular candidate will win an election. One would have to imagine redoing the world over and over *Groundhog Day*-like.

Subjective. The subjective approach allows each person to have a different probability, so that for a given person, $P[A]$ is that person's opinion of the probability of A. The only assumption is that each person's probabilities cohere, that is, satisfy the probability axioms. Subjective probability can be applied to any situation. For a repeatable experiment, people's subjective probabilities would tend to agree, whereas in other cases, such as the probability a certain team will win a particular game, their probabilities could differ widely.

Some subjectivists make the assumption that any given person's subjective probabilities can be elicited using a betting paradigm. For example, suppose the event in question is "Pat and Leslie will get married," the choices being "Yes" and "No," and we wish to elicit your probability of the event "Yes." We give you $10, and ask you for a number w, which will be used in two possible bets:

$$\text{Bet 1} \to \text{Win } \$w \text{ if "Yes", Lose } \$10 \text{ if "No";}$$
$$\text{Bet 2} \to \text{Lose } \$10 \text{ if "Yes", Win } \$(100/w) \text{ if "No".} \tag{10.4}$$

Some dastardly being will decide which of the bets you will take, so the w should be an amount for which you are willing to take either of those two bets. For example, if you choose $\$w = \$5$, then you are willing to accept a bet that pays only $5 if they do get married, and loses $10 if they don't; **and** you are willing to take a bet that wins $20 if they do not get married, and loses $10 if they do. These numbers suggest you expect they will get married. Suppose p is your subjective probability of "Yes." Then your willingness to take Bet 1 means you expect to not lose money:

$$\text{Bet 1} \to E[Winnings] = p(\$w) - (1-p)(\$10) \geq 0. \tag{10.5}$$

Same with Bet 2:

$$\text{Bet 2} \to E[Winnings] = -p(\$10) + (1-p)\$(100/w) \geq 0. \tag{10.6}$$

A little algebra translates those two inequaties into

$$p \geq \frac{10}{10+w} \text{ and } p \leq \frac{100/w}{10+100/w} = \frac{10}{10+w}, \tag{10.7}$$

which of course means that

$$p = \frac{10}{10+w}. \tag{10.8}$$

With $\$w = \$5$, your $p = 2/3$ that they will get married.

The betting approach is then an alternative to the frequency approach. Whether it is practical to elicit an entire probability distribution (i.e., $P[A]$ for all $A \subset \mathcal{X}$), and whether the result will satisfy the axioms, is questionable, but the main point is that there is in principle a grounding to a subjective probability.

10.3   Approaches to inference

Paralleling the interpretations of probability are the two main approaches to statistical inference: frequentist and Bayesian. Both aim to make inferences about θ based on observing the data $\mathbf{X} = \mathbf{x}$, but take different tacks.

Frequentist. The frequentist approach assumes that the parameter θ is fixed but unknown (that is, we know only that $\theta \in \mathcal{T}$). An inference is an **action**, which is a function

$$\delta : \mathcal{X} \longrightarrow \mathcal{A}, \tag{10.9}$$

for some **action space** $\mathcal{A}$. The action space depends on the type of inference desired. For example, if one wishes to estimate θ, then $\delta(\mathbf{x})$ would be the estimate, and $\mathcal{A} = \mathcal{T}$. Or δ may be a vector containing the estimate as well as an estimate of its variance, or it may be a two-dimensional vector representing a confidence interval, as in (7.32). In hypothesis testing, we often take $\mathcal{A} = \{0, 1\}$, where 0 means accept the null hypothesis, and 1 means reject it. The properties of a procedure δ, which would describe how good it is, are based on the behavior if the experiment were repeated over and over, with θ fixed. Thus an estimator δ of θ is **unbiased** if

$$E_\theta[\delta(\mathbf{X})] = \theta \text{ for all } \theta \in \mathcal{T}. \tag{10.10}$$

Or a confidence interval procedure $\delta(\mathbf{x}) = (l(\mathbf{x}), u(\mathbf{x}))$ has 95% coverage if

$$P_\theta[l(\mathbf{X}) < \theta < u(\mathbf{X})] \geq 0.95 \text{ for all } \theta \in \mathcal{T}. \tag{10.11}$$

Understand that the 95% does **not** refer to your particular interval, but rather to the infinite number of intervals that you imagine arising from repeating the experiment over and over. That is, without a prior, for fixed $\mathbf{x}$,

$$P[l(\mathbf{x}) < \theta < u(\mathbf{x})] \neq 0.95, \tag{10.12}$$

because there is nothing random in the probability statement. The actual probably is either 0 or 1, depending on whether θ is indeed between $l(\mathbf{x})$ and $u(\mathbf{x})$, but we typically do not know which value it is.

Bayesian. The frequentist approach does not tell you what to think of θ. It just produces a number or numbers, then reassures you by telling you what would happen if you repeated the experiment an infinite number of times. The Bayesian approach, by contrast, tells you what to think. More precisely, given your prior distribution on $\mathcal{T}$, which may be your subjective distribution, the Bayes approach tells you how to update your opinion upon observing $\mathbf{X} = \mathbf{x}$. The update is of course the posterior, which we know how to find using Bayes theorem (Theorem 6.3 on page 94). The posterior $f_{\Theta|\mathbf{X}}(\theta \mid \mathbf{x})$ is the inference, or at least all inferences are derived from it. For example, an estimate could be the posterior mean, median, or mode. A 95% probability interval is any interval $(l_\mathbf{x}, u_\mathbf{x})$ such that

$$P[l_\mathbf{x} < \Theta < u_\mathbf{x} \mid \mathbf{X} = \mathbf{x}] = 0.95. \tag{10.13}$$

A hypothesis test would calculate

$$P[\text{Null hypothesis is true} \mid \mathbf{X} = \mathbf{x}], \qquad (10.14)$$

or, if an accept/reject decision is desired, reject the null hypothesis if the posterior probability of the null is less than some cutoff, say 0.50 or 0.01.

A drawback to the frequentist approach is that we cannot say what we wish to say, such as the probability a null hypothesis is true, or the probability μ is between two numbers. Bayesians can make such statements, but as the cost of having to come up with a (usually subjective) prior. The subjectivity means that different people can come to different conclusions from the same data. (Imagine a tobacco company and a consumer advocate analyzing the same smoking data.) Fortunately, there are more or less well-accepted "objective" priors, and especially when the data is strong, different reasonable priors will lead to practically the same posteriors. From an implementation point of view, sometimes frequentist procedures are computationally easier, and sometimes Bayesian procedures are. It may not be philosophically pleasing, but it is not a bad idea to take an opportunistic view and use whichever approach best moves your understanding along.

There are other approaches to inference, such as the likelihood approach, the structural approach, the fiducial approach, and the fuzzy approach. These are all interesting and valuable, but seem a bit iffy to me.

The rest of the course goes more deeply into inference.

10.4 Exercises

Exercise 10.4.1. Suppose $X_1, \ldots, X_n$ are independent $N(\mu, \sigma_0^2)$, where $\mu \in \mathbb{R}$, $n = 25$, and $\sigma_0^2 = 9$. Also, suppose that $U \sim \text{Uniform}(0,1)$, and U is independent of the X_i's. The μ does not have a prior in this problem. Consider the following two confidence interval procedures for μ:

$$\text{Procedure 1:}\quad CI_1(\mathbf{x}, u) = \left(\bar{x} - 1.96 \, \frac{\sigma_0}{\sqrt{n}}, \bar{x} + 1.96 \, \frac{\sigma_0}{\sqrt{n}} \right);$$

$$\text{Procedure 2:}\quad CI_2(\mathbf{x}, u) = \begin{cases} \mathbb{R} & \text{if} \quad u \le .95 \\ \varnothing & \text{if} \quad u > .95 \end{cases}. \qquad (10.15)$$

The $\varnothing$ is the empty set. (a) Find $P[\mu \in CI_1(\mathbf{X}, U)]$. (b) Find $P[\mu \in CI_2(\mathbf{X}, U)]$. (c) Suppose $\bar{x} = 70$ and $u = 0.5$. Using Procedure 1, is $(68.824, 71.176)$ a 95% confidence interval for μ? (d) Given the data in part (c), using Procedure 2, is $(-\infty, \infty)$ a 95% confidence interval for μ? (e) Given the data in part (c), using Procedure 1, does $P[68.824 < \mu < 71.176] = 0.95$? (f) Given the data in part (c), using Procedure 2, does $P[\mu \in CI_2(\mathbf{x}, u)] = 0.95$? If not, what is the probability? (g) Suppose $\bar{x} = 70$ and $u = 0.978$. Using Procedure 2, does $P[\mu \in CI_2(\mathbf{x}, u)] = 0.95$? If not, what is the probability?

Exercise 10.4.2. Continue with the situation in Exercise 10.4.1, but now suppose there is a prior on μ, so that

$$\bar{X} \mid M = \mu \sim N(\mu, (0.6)^2),$$
$$M \sim N(66, 10^2), \qquad (10.16)$$

and U is independent of $(\mathbf{X}, M)$. (a) Find the posterior distribution $M \mid \overline{X} = 70, U = 0.978$. Does it depend on U? (b) Find $P[M \in CI_1(\mathbf{X}, U) \mid \overline{X} = 70, U = 0.978]$. (c) Find $P[M \in CI_2(\mathbf{X}, U) \mid \overline{X} = 70, U = 0.978]$. (d) Which 95% confidence interval procedure seems to give closest to 95% confidence that M is in the interval?

Exercise 10.4.3. Imagine you have to weigh a substance whose true weight is μ, in milligrams. There are two scales, an old mechanical one and a new electronic one. Both are unbiased, but the mechanical one has a measurement error of 3 milligrams, while the electronic one has a measurement error of only 1 milligram. Letting Y be the measurement, and S be the scale, we have that

$$Y \mid S = \text{mech} \sim N(\mu, 3^2),$$
$$Y \mid S = \text{elec} \sim N(\mu, 1). \tag{10.17}$$

There is a fifty-fifty chance you get to use the good scale, so marginally, $P[S = \text{mech}] = P[S = \text{elec}] = 1/2$. Consider the confidence interval procedure, $CI(y) = (y - 3, y + 3)$. (a) Find $P[\mu \in CI(Y) \mid S = \text{mech}]$. (b) Find $P[\mu \in CI(Y) \mid S = \text{elec}]$. (c) Find the unconditional probability, $P[\mu \in CI(Y)]$. (d) The interval $(y - 3, y + 3)$ is a $Q\%$ confidence interval for μ. What is Q? (e) Suppose the data are $(y, s) = (14, \text{mech})$. Using the CI above, what is the $Q\%$ confidence interval for μ? (f) Suppose the data are $(y, s) = (14, \text{elec})$. Using the CI above, what is the $Q\%$ confidence interval for μ? (g) What is the difference between the two intervals from (e) and (f)? (h) Are you equally confident in them?

Exercise 10.4.4. Continue with the situation in Exercise 10.4.3, but now suppose there is a prior on μ, so that

$$Y \mid M = \mu, S = \text{mech} \sim N(\mu, 3^2),$$
$$Y \mid M = \mu, S = \text{elec} \sim N(\mu, 1),$$
$$M \sim N(16, 15^2), \tag{10.18}$$

and $P[S = \text{mech}] = P[S = \text{elec}] = 1/2$, where M and S are independent. (a) Find the posterior $M \mid Y = 14, S = \text{mech}$. (b) Find the posterior $M \mid Y = 14, S = \text{elec}$. (c) Find $P[Y - 3 < M < Y + 3 \mid Y = 14, S = \text{mech}]$. (d) Find $P[Y - 3 < M < Y + 3 \mid Y = 14, S = \text{elec}]$. (e) Is $Q\%$ (Q is from Exercise 10.4.3 (d)) a good measure of confidence for the interval $(y - 3, y + 3)$ for the data $(y, s) = (14, \text{mech})$? For the data $(y, s) = (14, \text{elec})$?

Chapter 11

Estimation

11.1 Definition of estimator

We assume a model with parameter space $\mathcal{T}$, and suppose we wish to estimate some function g of θ,

$$g : \mathcal{T} \longrightarrow \mathbb{R}^q. \tag{11.1}$$

This function could be θ itself, or just part of θ. For example, if $X_1, \ldots, X_n$ are iid $N(\mu, \sigma^2)$, where $\theta = (\mu, \sigma^2) \in \mathbb{R} \times (0, \infty)$, some possible one-dimensional g's are

$$g(\mu, \sigma^2) = \mu;$$
$$g(\mu, \sigma^2) = \sigma;$$
$$g(\mu, \sigma^2) = \sigma/\mu = \text{coefficient of variation};$$
$$g(\mu, \sigma^2) = P[X_i \leq 10] = \Phi((10 - \mu)/\sigma), \tag{11.2}$$

where Φ is the distribution function for $N(0, 1)$.

Formally, an **estimator** is a function $\delta(\mathbf{x})$,

$$\delta : \mathcal{X} \longrightarrow \mathcal{A}, \tag{11.3}$$

where $\mathcal{A}$ is some space, presumably the space of $g(\theta)$, but not always. The estimator can be any function of $\mathbf{x}$, but *cannot depend on an unknown parameter.* Thus with $g(\mu, \sigma^2) = \sigma/\mu$ in the above example,

$$\delta(x_1, \ldots, x_n) = \frac{s}{\bar{x}}, \ [s^2 = \sum(x_i - \bar{x})^2/n] \text{ is an estimator,}$$

$$\delta(x_1, \ldots, x_n) = \frac{\sigma}{\bar{x}} \text{ is not an estimator.} \tag{11.4}$$

We often use the "hat" notation, so that if δ is an estimator of $g(\theta)$, we would write

$$\delta(\mathbf{x}) = \widehat{g(\theta)}. \tag{11.5}$$

Any function can be an estimator, but that does not mean it will be a particularly good estimator. There are basically two questions we must address: How does one

161

find reasonable estimators? How do we decide which estimators are good? This chapter looks at plug-in methods and Bayesian estimation. Chapter 12 considers least squares and similar procedures as applied to linear regression. Chapter 13 presents maximum likelihood estimation, a widely applicable approach. Later chapters (19 and 20) deal with optimality of estimators.

11.2   Bias, standard errors, and confidence intervals

Making inferences about a parameter typically involves more than just a point estimate. One would also like to know how accurate the estimate is likely to be, or have
a reasonable range of values. One basic measure is **bias**, which is how far off the
estimator δ of $g(\theta)$ is on average:

$$\text{Bias}_\theta[\delta] = E_\theta[\delta(\mathbf{X})] - g(\theta). \tag{11.6}$$

For example, if $X_1, \ldots, X_n$ are iid with variance $\sigma^2 < \infty$, then $S^2 = \sum (X_i - \overline{X})^2/n$ has

$$E[S^2] = \frac{n-1}{n} \sigma^2 \implies \text{Bias}_{\sigma^2}[S^2] = -\frac{1}{n}\sigma^2. \tag{11.7}$$

Thus $\delta(\mathbf{X}) = S^2$ is a biased estimator of σ^2. Instead, if we divide by $n-1$, we have
the unbiased estimator S_*^2. (We saw this result for the normal in (7.69).)
    A little bias is not a big deal, but one would not like a huge amount of bias.
Another basic measure of accuracy is the **standard error** of an estimator:

$$se_\theta[\delta] = \sqrt{Var_\theta[\delta]} \quad or \quad \sqrt{\widehat{Var_\theta[\delta]}}, \tag{11.8}$$

that is, it is the theoretical standard deviation of the estimator, or an estimator thereof.
In Exercise 11.7.1 we will see that the mean square error, $E_\theta[(\delta(\mathbf{X}) - g(\theta))^2]$, combines
the bias and standard error, being the bias squared plus the variance. Section 19.3
delves more formally into optimality considerations.
    Confidence intervals (as in (10.11)) or probability intervals (as in (10.13)) will of-
ten be more informative than simple point estimates, even with the standard errors.
A common approach to deriving confidence intervals uses what are called pivotal
quantities, as introduced in (7.29). A pivotal quantity is a function of the data and the
parameter, whose distribution does not depend on the parameter. That is, suppose
$T(\mathbf{X}; \theta)$ has a distribution that is known or can be approximated. Then for given α,
we can find constants A and B, free of θ, such that

$$P[A < T(\mathbf{X}; \theta) < B] = (\text{or} \approx) 1 - \alpha. \tag{11.9}$$

If the pivotal quantity works, then we can invert the event so that our estimand $g(\theta)$
is in the middle, and statistics (free of θ) define the interval:

$$A < T(\mathbf{x}; \theta) < B \Leftrightarrow l(\mathbf{x}) < g(\theta) < u(\mathbf{x}). \tag{11.10}$$

Then $(l(\mathbf{x}), u(\mathbf{x}))$ is a (maybe approximate) $100(1-\alpha)\%$ confidence interval for $g(\theta)$.
The quintessential pivotal quantities are the z statistic in (7.29) and t statistic in (7.77).
    In many situations, the exact distribution of an estimator is difficult to find, so that
asymptotic considerations become useful. For a sequence of estimators $\delta_1, \delta_2, \ldots$, an
analog of unbiasedness is **consistency**, where δ_n is a consistent estimator of $g(\theta)$ if

$$\delta_n \longrightarrow^P g(\theta) \quad \text{for all } \theta \in \mathcal{T}. \tag{11.11}$$

Note that consistency and unbiasedness are distinct notions. Consistent estimators do not have to be unbiased: S_n^2 is a consistent estimator of σ^2 in the iid case, but is not unbiased. Also, an unbiased estimator need not be consistent (can you think of an example?), though an estimator that is unbiased and has variance going to zero is consistent, by Chebyshev's inequality (Lemma 8.3).

Likewise, whereas the exact standard error may not be available, we may have a proxy, $se_n(\delta_n)$, for which

$$\frac{\delta_n - g(\theta)}{se_n(\delta_n)} \longrightarrow^{\mathcal{D}} N(0,1). \tag{11.12}$$

Then an approximate confidence interval for $g(\theta)$ is

$$\delta_n \pm 2\, se_n(\delta_n). \tag{11.13}$$

Here the Δ-method (Section 9.2) often comes in useful.

11.3 Plug-in methods: Parametric

Often the parameter of interest has an obvious sample analog, or is a function of some parameters that have obvious analogs. For example, if $X_1, \ldots, X_n$ are iid, then it may be reasonable to estimate $\mu = E[X_i]$ by $\overline{X}$, $\sigma^2 = Var[X_i]$ by S^2, and the coefficient of variation by $S/\overline{X}$ (see (11.2) and (11.4)).

An obvious estimator of $P[X_i \leq 10]$ is

$$\delta(\mathbf{x}) = \frac{\#\{x_i \leq 10\}}{n}. \tag{11.14}$$

A parametric model may suggest other options. For example, if the data are iid $N(\mu, \sigma^2)$, then $P[X_i \leq 10] = \Phi((10 - \mu)/\sigma)$, so that we can plug in the mean and standard deviation estimates to obtain the alternative estimator

$$\delta^*(\mathbf{x}) = \Phi\left(\frac{10 - \overline{x}}{s}\right). \tag{11.15}$$

Or suppose the X_i's are iid Beta(α, β), with $(\alpha, \beta) \in (0, \infty) \times (0, \infty)$. Then from Table 1.1 on page 7, the population mean and variance are

$$\mu = \frac{\alpha}{\alpha + \beta} \quad \text{and} \quad \sigma^2 = \frac{\alpha\beta}{(\alpha \mid \beta)^2(\alpha + \beta + 1)}. \tag{11.16}$$

The sample quantities $\overline{x}$ and s^2 are estimates of those functions of α and β, hence the estimates $\widehat{\alpha}$ and $\widehat{\beta}$ of α and β would be the solutions to

$$\overline{x} = \frac{\widehat{\alpha}}{\widehat{\alpha} + \widehat{\beta}} \quad \text{and} \quad s^2 = \frac{\widehat{\alpha}\widehat{\beta}}{(\widehat{\alpha} + \widehat{\beta})^2(\widehat{\alpha} + \widehat{\beta} + 1)}, \tag{11.17}$$

or after some algebra,

$$\widehat{\alpha} = \overline{x}\left(\frac{\overline{x}(1 - \overline{x})}{s^2} - 1\right) \quad \text{and} \quad \widehat{\beta} = (1 - \overline{x})\left(\frac{\overline{x}(1 - \overline{x})}{s^2} - 1\right). \tag{11.18}$$

The estimators in (11.18) are special plug-in estimators, called **method of moments** estimators, because the estimates of the parameters are chosen to match the population moments with their sample versions. Method of moment estimators are not necessarily strictly defined. For example, in the Poisson(λ), both the mean and variance are λ, so that $\widehat{\lambda}$ could be $\bar{x}$ or s^2. Also, one has to choose moments that work. For example, if the data are iid $N(0, \sigma^2)$, and we wish to estimate σ, the mean is useless because one cannot do anything to match $0 = \bar{x}$.

Finding standard errors for plug-in estimators often involves another plug-in. For example, we know that when estimating the mean with iid observations, $Var[\overline{X}] = \sigma^2/n$ if $Var[X_i] = \sigma^2 < \infty$, hence we can use $s/\sqrt{n}$ as the standard error. Or in the Poisson case, $se_n(\overline{X}) = \sqrt{\overline{X}/n}$, since the mean and variance are the same.

11.3.1   Coefficient of variation

Suppose we have a sample of iid $N(\mu, \sigma^2)$'s, and we wish to find a confidence interval for the coefficient of variation in (11.2), $cv = g(\mu, \sigma^2) = \sigma/\mu$, assuming $\mu \neq 0$. Let $\delta_n = \delta(x_1, \ldots, x_n) = s/\bar{x}$ (or we could use $s_*/\bar{x}$). In (9.62) we used the Δ-method to show that

$$\sqrt{n}(\delta_n - cv) \longrightarrow^{\mathcal{D}} N(0, cv^2(cv^2 + \tfrac{1}{2})). \tag{11.19}$$

Thus we can estimate the standard error of δ_n by plugging δ_n into the standard deviation:

$$se_n(\delta_n) = \frac{1}{\sqrt{n}} |\delta_n| \sqrt{\delta_n^2 + \tfrac{1}{2}}. \tag{11.20}$$

We can then use Lemma 9.1 on page 139 to show that

$$\sqrt{n} \frac{\delta_n - cv}{|\delta_n| \sqrt{\delta_n^2 + \tfrac{1}{2}}} \longrightarrow^{\mathcal{D}} N(0, 1), \tag{11.21}$$

hence $\delta_n \pm 2\, se_n(\delta_n)$ is an approximate 95% confidence interval for cv, as in (11.13).

11.4   Plug-in methods: Nonparametric

A nonparametric model is one that cannot be defined using a finite number of parameters. Here we will focus on the model where $\mathbf{X}_1, \ldots, \mathbf{X}_n$ are iid, each with space $\mathcal{X} \subset \mathbb{R}^p$ and distribution function F, where we assume that $F \in \mathcal{F}$, for $\mathcal{F}$ being a large class of distribution functions. For example, it could contain all continuous distribution functions, or all with finite mean, or all with unimodal pdfs.

This F is the ultimate parameter, and we are interested in estimating *functionals* on $\mathcal{F}$,

$$\theta : \mathcal{F} \longrightarrow \mathbb{R}^q. \tag{11.22}$$

That is, the argument for the function θ is F. The mean, median, variance, $P[X_i \leq 10] = F(10)$, etc., are all such functionals for appropriate $\mathcal{F}$'s. Notice that the α and β of a beta distribution are not such functionals, because they are not defined for non-beta distributions. The $\theta(F)$ is then the population parameter θ.

The plug-in estimate is obtained by plugging in an estimate of the parameter F. Most generally for the iid case, this estimate $\widehat{F}_n$ is the **empirical distribution function,** defined by

$$\widehat{F}_n(x) = \frac{\#\{x_i \mid x_i \le x\}}{n} \qquad (11.23)$$

in the univariate case. If the data are p-variate, with $\mathbf{x}_i = (x_{i1}, \dots, x_{ip})$, then

$$\widehat{F}_n(\mathbf{x}) = \frac{\#\{\mathbf{x}_i \mid x_{ij} \le x_j \text{ for all } j = 1, \dots, p\}}{n}, \qquad (11.24)$$

where $\mathbf{x} = (x_1, \dots, x_n)$. This $\widehat{F}_n$ is the distribution function for the random vector $\mathbf{X}^*$, which has space

$$\mathcal{X}^* = \{\text{the distinct values among } \mathbf{x}_1, \dots, \mathbf{x}_n\} \qquad (11.25)$$

and probabilities

$$P^*[\mathbf{X}^* = \mathbf{x}^*] = \frac{\#\{\mathbf{x}_i \mid \mathbf{x}_i = \mathbf{x}^*\}}{n}, \quad \mathbf{x}^* \in \mathcal{X}^*. \qquad (11.26)$$

Thus $\mathbf{X}^*$ is generated by randomly choosing one of the observations $\mathbf{x}_i$. For example, if the sample is $3, 5, 2, 6, 2, 2$, then $\mathcal{X}^* = \{2, 3, 5, 6\}$ and

$$P[X^* = 2] = \frac{1}{2}, \quad P[X^* = 3] = P[X^* = 5] = P[X^* = 6] = \frac{1}{6}. \qquad (11.27)$$

Back to estimating the $\theta(F)$ in (11.22), the plug-in estimator is then

$$\widehat{\theta} = \widehat{\theta}(\mathbf{X}_1, \dots, \mathbf{X}_n) = \theta(\widehat{F}_n). \qquad (11.28)$$

If θ is the mean, $\theta(\widehat{F}_n) = \overline{x}$, and if θ is the variance, $\theta(\widehat{F}_n) = s^2 = \sum(x_i - \overline{x})^2/n$. As in (11.2), the coefficient of variation can be estimated by the sample version, $s/\overline{x}$. The parameter $P[X_i \le 10] = F(10)$ is estimated by $\widehat{F}_n(10)$, which is the same as (11.14).

In the next section we look at estimating standard errors and confidence intervals of such estimators.

11.5 Plug-in methods: Bootstrap

Continue with the setup in the previous section. We will introduce the **bootstrap** procedure to estimate the bias and standard error of $\theta(\widehat{F}_n)$, and confidence intervals for $\theta(F)$. In general, the bootstrap can be used to estimate the distribution of a function T of the data and the distribution function:

$$T(\mathbf{X}_1, \dots, \mathbf{X}_n; F), \quad \text{where } \mathbf{X}_1, \dots, \mathbf{X}_n \text{ are iid } \sim F. \qquad (11.29)$$

We will focus on taking

$$T(\mathbf{X}_1, \dots, \mathbf{X}_n; F) = \theta(\widehat{F}_n) - \theta(F) = \widehat{\theta} - \theta(F). \qquad (11.30)$$

Note that the bias of our estimator is the expected value of T, and the standard error is the standard deviation of T. In addition, this statistic can be used like a pivotal quantity as in (11.10), so that if we can find A and B such that

$$P[A < T(\mathbf{X}_1, \dots, \mathbf{X}_n; F) < B] = P[A < \widehat{\theta} - \theta(F) < B] = 1 - \alpha, \qquad (11.31)$$

then a $100(1 - \alpha)\%$ confidence interval for $\theta(F)$ is

$$(\widehat{\theta} - B, \widehat{\theta} - A). \tag{11.32}$$

The bootstrap estimate of the distribution of T is also a plug-in estimator, where we plug the empirical distribution function in for the F, and the data in T are replaced by iid observations drawn from $\widehat{F}_n$. That is, the bootstrap estimate of the distribution of T in (11.30) is the distribution of

$$T(\mathbf{X}_1^*, \ldots, \mathbf{X}_n^*; \widehat{F}_n), \quad \text{where } \mathbf{X}_1^*, \ldots, \mathbf{X}_n^* \text{ are iid} \sim \widehat{F}_n. \tag{11.33}$$

Note that the $\mathbf{X}_i^*$'s distribution is equivalent to drawing n observations with replacement from the data $\{x_1, \ldots, x_n\}$.

In principle it is easy to find the bootstrap distribution since there is only a finite number of possible such draws. In practice, though, the number of possibilities is combinatorially large (see Exercise 11.7.9), so we usually estimate the estimate of the distribution by taking a number, say K, of random samples of the $\mathbf{X}_i^*$'s. That is, the kth bootstrap sample $\{x_{k1}^*, \ldots, x_{kn}^*\}$ is obtained by randomly drawing n observations with replacement from the original data $\{x_1, \ldots, x_n\}$. The kth realization of the T in (11.33) is then

$$t_k^* = T(x_{k1}^*, \ldots, x_{kn}^*; \widehat{F}_n) = \widehat{\theta}(x_{k1}^*, \ldots, x_{kn}^*) - \widehat{\theta}(x_1, \ldots, x_n). \tag{11.34}$$

We then estimate A and B in (11.31) and (11.32) with a and b, respectively, where a is the 0.025^{th} quantile, and b is the 0.975^{th} quantile, of the bootstrapped T's, $\{t_1^*, \ldots, t_K^*\}$.

11.5.1 Sample mean and median

Take the univariate data $X_1, \ldots, X_n$ to be iid F, and $\theta(F) = E[X_i]$, so that $\theta(\widehat{F}_n) = \overline{X}_n$, the usual sample mean. Then

$$T(X_1, \ldots, X_n; F) = \overline{X}_n - \theta(F). \tag{11.35}$$

The bootstrap estimate of the distribution of T is the distribution of

$$T^* = T(X_1^*, \ldots, X_n^*; \widehat{F}_n) = \overline{X}_n^* - \overline{x}_n, \tag{11.36}$$

where $\overline{X}_n^*$ is the sample mean of the bootstrapped sample.

It is easy to find the mean and variance of T^*. Since $X_1^*, \ldots, X_n^*$ are iid observations from a distribution with mean $\overline{x}_n$ and variance $s_n^2 = \sum(x_i - \overline{x}_n)^2/n$, we have

$$E[T^*] = 0 \quad \text{and} \quad Var[T^*] = \frac{s_n^2}{n}. \tag{11.37}$$

Thus the bootstrap estimate of the bias is 0 (which we know is the actual bias), and the bootstrap estimate of the standard error of the sample mean is $s_n/\sqrt{n}$, which is also reasonable. For a 95% confidence interval, we first need to find a and b so that

$$P[a < \overline{X}_n^* - \overline{x}_n < b] \approx 0.95, \tag{11.38}$$

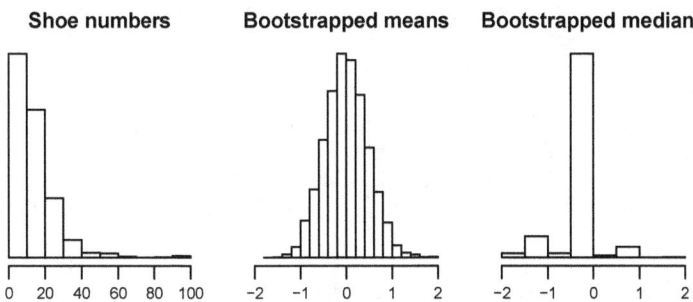

Figure 11.1: Bootstrapping the shoes.

the approximation being needed because the distribution is discrete. To find the exact a and b, we would need to find the sample means for all possible bootstrap samples, then line them up in order to find the 2.5 and 97.5 percentiles. Unless n is very small, that approach is not feasible, so we resample.

To illustrate, each of a sample of 712 students were asked, "How many pairs of shoes do you own?" The first histogram in Figure 11.1 shows the original data, where "shoe number" means number of pairs owned. We look at both the mean and median, hoping to estimate a confidence interval for the population quantities. The sample mean is $\bar{x}_n = 15.2079$ and the $\text{Median}\{x_1, \ldots, x_n\} = 12$. The two bootstrap quantities are

$$T^*_{\text{Mean}} = \overline{X}^*_n - \bar{x}_n \quad \text{and} \quad T^*_{\text{Median}} = \text{Median}\{X^*_1, \ldots, X^*_n\} - \text{Median}\{x_1, \ldots, x_n\}.$$
(11.39)

The second and third histograms in Figure 11.1 are the histograms of $K = 5000$ bootstrapped values of the T^*_{Mean}'s and T^*_{Median}'s, respectively. That is, for the mean, we took 5000 bootstrap samples of 712 observations without replacement from the original data. For the kth bootstrap sample, we found the mean, then subtracted the original sample mean from the bootstrap mean to obtain t^*_k. Then $t^*_1, \ldots, t^*_K$ are used to estimate the mean, standard deviation, and quantiles a and b of T^*_{Mean}. A similar procedure was used for the median. The following table contains the results:

	Mean	Standard deviation	(a, b)	
T^*_{Mean}	0.0013	0.4710	$(-0.8919, 0.9425)$	(11.40)
T^*_{Median}	-0.0769	0.4511	$(-1.25, 1.25)$	

As before, a and b are the 0.025^{th} and 0.975^{th} sample quantiles of the bootstrapped T^*'s, respectively. The intervals (a, b) contain about 95% of the bootstrapped samples (exactly 95% for the mean, but 98.1% for the median).

The means of the bootstrapped quantities estimate the biases of the estimators. In this case, we see both biases are quite small, as expected for the mean. The standard deviations are our estimates of the standard errors of the estimators. Both the sample mean and sample median have standard errors a little below 0.5 pairs of shoes (which is one shoe). For the sample mean, we could also use the usual standard error estimate of $s_n / \sqrt{n}$, which in this case is 0.4692, about the same as the bootstrap estimate. For the median, we do not really have a ready alternative for estimating the standard error. In Section 9.2.1 we used the Δ-method to find an estimate, but that procedure required knowing the pdf. Here, we do not even have a pdf, since the data are discrete.

Turn to confidence intervals. The histogram of the bootstrapped means looks very close to a normal curve, suggesting that $\bar{x}_n \pm 2\,\mathrm{se}(\bar{x}_n)$ would be a reasonable confidence interval. The bootstrapped estimate of the confidence interval using the quantiles a and b would be $(\bar{x}_n - b, \bar{x}_n - a)$ as in (11.32). The two intervals are very similar:

$$\bar{x}_n \pm 2\,\mathrm{se}(\bar{x}_n) = (14.27, 16.15) \quad \text{and} \quad (\bar{x}_n - b, \bar{x}_n - a) = (14.27, 16.10). \qquad (11.41)$$

For the median, the histogram does not look particularly normal, though it does appear reasonably symmetric. In fact, it is a discrete distribution, with values only at integers and half-integers (since the data is all integers). Thus I took the a and b at quarter-integers, i.e., ± 1.25. The bootstrap estimate of the confidence interval for the median is then

$$(\mathrm{Median}\{x_1, \ldots, x_n\} - b, \mathrm{Median}\{x_1, \ldots, x_n\} - a) = (10.75, 13.25). \qquad (11.42)$$

This analysis suggests that the population mean number of pairs is between about 14 and 16, while the median is between about 11 to 13. Not surprisingly, given the positive skewness of the data, the mean is larger than the median.

11.5.2   Using R

Bootstrapping directly is fairly easy in R, though there are also packages with enhanced capabilities. Here, we will illustrate bootstrapping the median. The observations are in the vector x. To find 5000 bootstrapped values of the median, we use

```
medstar <- NULL # Vector to collect the bootstrapped medians
for(k in 1:5000) {
        xstar <- sample(x,replace=T) # Obtains one bootstrap sample
        medstar <- c(medstar,median(xstar))
}
```

The t_k^*'s are found by subtracting the sample median from the bootstrapped medians, from which we can estimate the bias, standard error, and confidence interval:

```
tstar <- medstar - median(x)
mean(tstar) # Estimates the bias
sd(tstar) # Estimates the standard error
ab <- quantile(tstar,c(0.025,0.975)) # Obtains a and b
median(x) - ab[2:1] # Finds the confidence interval.
```

Then if x contains the sample of 712 shoe numbers, the output of the above code is -0.0769 for the bias, 0.4511 for the standard error, and (11,13) for the confidence interval, since $a = -1$ and $b = 1$. These numbers will be slightly different each time we run the code. Note that in (11.40), I made the continuity correction expanding the (a, b) interval by 0.25 on each side because of the discreteness of the distribution of the t_k^*'s.

11.6 Posterior distribution

In the Bayesian framework, the inference is the posterior, because the posterior has all the information about the parameter given the data. Estimation of θ then becomes a matter of summarizing the distribution. In Section 6.6.2, we have $X \mid \Theta = \theta \sim$ Binomial(n, θ) with prior $\Theta \sim$ Beta(α, β), so that

$$\Theta \mid X = x \sim \text{Beta}(\alpha + x, \beta + n - x). \tag{11.43}$$

One estimate of the parameter is the posterior mean,

$$E[\Theta \mid X = x] = \frac{\alpha + x}{\alpha + \beta + n}. \tag{11.44}$$

Another is the posterior mode, the value with the highest density, which turns out to be

$$Mode[\Theta \mid X = x] = \frac{\alpha + x - 1}{\alpha + \beta + n - 2}. \tag{11.45}$$

One can think of α as the prior number of successes and $\alpha + \beta$ as the prior sample size, the larger $\alpha + \beta$, the more weight the prior has in the posterior.

A 95% **probability interval** for the parameter given the data is (l_x, u_x), where the endpoints are chosen to satisfy

$$P[l_x < \Theta < u_x \mid X = x] = 0.95. \tag{11.46}$$

There are many choices. the simplest may be to take $P[\Theta < l_x \mid X = x] = P[\Theta > u_x \mid X = x] = 0.025$. Commonly they are chosen so that the interval contains the points with the highest posterior density, but still with 95% probability.

11.6.1 Normal mean

Exercise 7.8.14 treats the normal mean case when the variance is known, and the prior on the mean is normal. That is, if

$$Y_1, \ldots, Y_n \mid M = \mu \text{ are iid } N(\mu, \sigma^2), \quad M \sim N(\mu_0, \sigma_0^2), \tag{11.47}$$

where μ_0, σ_0^2, and σ^2 are known, then

$$M \mid \overline{Y} = \overline{y} \sim N\left(\frac{\omega_0 \mu_0 + n\omega \overline{y}}{\omega_0 + n\omega}, \frac{1}{\omega_0 + n\omega}\right), \tag{11.48}$$

where $\omega_0 = 1/\sigma_0^2$ and $\omega = 1/\sigma^2$ are the precisions. Note that the posterior mean is a weighted average of the prior mean and the sample mean, with weights proportional to their precisions. In particular, if we take a very flat prior, meaning take σ_0^2 to be very large or ω_0 to be near 0, then the posterior distribution for M is very close to $N(\overline{y}, \sigma^2/n)$. Thus the usual frequentist confidence interval can be a good approximation of a probability interval. That is, consider the probability

$$P\left[\overline{y} - 1.96 \frac{\sigma}{\sqrt{n}} < M < \overline{y} + 1.96 \frac{\sigma}{\sqrt{n}} \,\middle|\, \overline{Y} = \overline{y}\right] \tag{11.49}$$

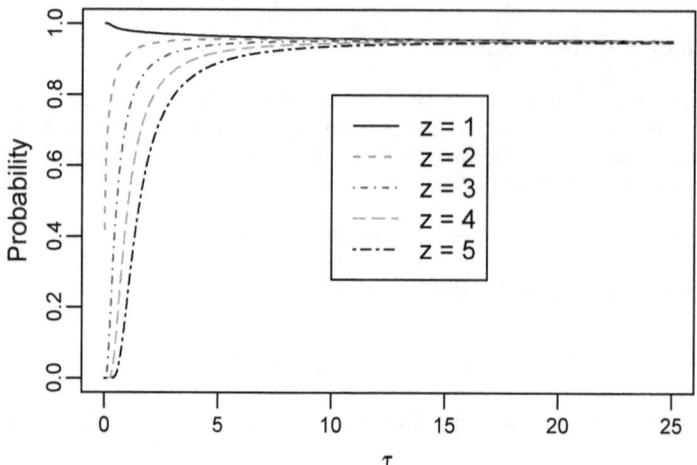

Figure 11.2: The posterior probability that the mean is in the 95% confidence interval as a function of τ for values of $z = 1, 2, 3, 4, 5$. See (11.50).

as a function of the prior parameters μ_0 and σ_0^2. Exercise 11.7.16 shows that this probability can be written

$$\Phi\left(\frac{\frac{z}{1+\tau} + 1.96}{\sqrt{\frac{\tau}{1+\tau}}}\right) - \Phi\left(\frac{\frac{z}{1+\tau} - 1.96}{\sqrt{\frac{\tau}{1+\tau}}}\right), \text{ where } z = \sqrt{n}\,\frac{\bar{y} - \mu_0}{\sigma} \text{ and } \tau = \frac{n\sigma_0^2}{\sigma^2}. \quad (11.50)$$

The z is a measure of how far the sample mean is from the prior mean, and τ is the ratio of the prior variance to the conditional variance of $\overline{Y}$ given $M = \mu$. Figure 11.2 graphs some of these probabilities. What we see is that even if the prior guess of the mean is off by five standard errors, the confidence interval settles to a probability close to 95% by the time the prior variance is 15 times that of the sample variance. In any case, as $\sigma_0^2 \to \infty$, the probability goes to 95%. Thus we are justified in taking as flat a prior as we wish.

In each of the binomial example in (11.43) and the normal one in (11.48), the prior and posterior distributions are of the same form (beta in the former case, normal in the latter), but the posterior has updated parameters depending on the observed data. Such priors are called **conjugate priors**. Exercise 11.7.14 discovers the conjugate prior for the exponential. Fink (1997) is a compendium of conjugate priors, and Diaconis and Ylvisaker (1979) take a more theoretical look at them.

Consider the normal case when σ^2 is not known. We will use the precisions in the notation, so that

$$Y_1, \ldots, Y_n \mid M = \mu, \Omega = \omega \text{ are iid } N(\mu, 1/\omega). \quad (11.51)$$

We need to place a joint prior on (μ, ω). Start by inspecting the conditional pdf:

$$f(\mathbf{y} \mid \mu, \omega) = c\,\omega^{n/2}e^{-\frac{1}{2}\omega(n(\bar{y}-\mu)^2+\Sigma(y_i-\bar{y})^2)}, \quad (11.52)$$

where we use

$$\sum(y_i - \mu)^2 = n(\bar{y} - \mu)^2 + \sum(y_i - \bar{y})^2. \tag{11.53}$$

A conjugate prior would have the same form as the likelihood in terms of the parameters. If you look at the μ, you see what looks like a normal-type pdf. If you ignore the $n(\bar{y} - \mu)^2$ term, the likelihood looks like a gamma pdf in ω. The problem is that the ω also appears in the μ term. It turns out that the form can be replicated using a two-stage prior:

$$M \mid \Omega = \omega \sim N(\mu_0, 1/(k_0\omega)) \quad \text{and} \quad \Omega \sim \text{Gamma}(\nu_0/2, \lambda_0/2), \tag{11.54}$$

where $\mu_0, k_0, \nu_0,$ and λ_0 are known prior parameters, the last three being positive. (Note that if ν_0 is an integer, $\Omega \sim (1/\lambda_0)\chi^2_{\nu_0}$.) The joint prior pdf can be written

$$f(\mu, \omega) = c\,\omega^{(\nu_0+1)/2-1}e^{-\frac{1}{2}\omega(k_0(\mu-\mu_0)^2+\lambda_0)}. \tag{11.55}$$

The form of the posterior is obtained by multiplying the likelihood and prior. It is easy to see what happens to the power of ω at the front. In the exponent, we again have $-\omega/2$ times a quadratic in μ, but it takes little work to complete the square. The answer (Exercise 11.7.20) is

$$f(\mu, \omega \mid \mathbf{y}) = c^*\omega^{(\nu_0+n+1)/2-1}e^{-\frac{1}{2}\omega((n+k_0)(\mu-\mu^*)^2+\widehat{\lambda}^*)}, \tag{11.56}$$

where

$$\widehat{\mu}^* = \frac{k_0\mu_0 + n\bar{y}}{k_0 + n} \quad \text{and} \quad \widehat{\lambda}^* = \lambda_0 + \sum(y_i - \bar{y})^2 + \frac{nk_0}{n + k_0}(\bar{y} - \mu_0)^2. \tag{11.57}$$

Note that μ^* is the analog of the posterior mean in (11.48). Matching the prior parameters in (11.55) to their analogs in (11.56), we have

$$\nu_0 \to \nu_0 + n, \quad k_0 \to k_0 + n, \quad \mu_0 \to \mu^*, \quad \text{and} \quad \lambda_0 \to \widehat{\lambda}^*. \tag{11.58}$$

Formally we have the posterior distribution of (M, Ω). The following lemma helps to understand this prior. See Exercise 11.7.19 for the proof.

Lemma 11.1. *Suppose (M, Ω) has distribution as given in (11.54). Then*

$$E[M] = \mu_0, \quad Var[M] = \frac{\lambda_0}{k_0(\nu_0 - 2)}, \quad \text{and} \quad E\left[\frac{1}{\Omega}\right] = \frac{\lambda_0}{\nu_0 - 2}, \tag{11.59}$$

the last two if $\nu_0 > 2$. Also, marginally

$$\sqrt{\nu_0}\,\frac{M - \mu_0}{\sqrt{\lambda_0/k_0}} \sim t_{\nu_0}, \tag{11.60}$$

Student's t on ν_0 degrees of freedom. See (6.115). (Note: The density is well-defined even for non-integer values of $\nu_0 > 0$.)

Plugging in the posterior parameters, we can see that $E[M \mid \mathbf{Y} = \mathbf{y}] = \mu^*$ of (11.57), which is close to $\bar{y}$ if k_0 is near zero. For the variance,

$$E\left[\frac{1}{\Omega} \,\middle|\, \mathbf{Y} = \mathbf{y}\right] = \frac{\lambda_0 + \sum(y_i - \bar{y})^2 + \frac{nk_0}{n+k_0}(\bar{y} - \mu_0)^2}{\nu_0 + n - 2} \approx s^2. \tag{11.61}$$

If λ_0, k_0, and ν_0 are near zero, then the posterior mean of σ^2 is very close to the sample variance (dividing by $n-2$). Likewise, the posterior variance of the mean is

$$Var[M \mid \mathbf{Y} = \mathbf{y}] = \frac{\lambda_0 + \Sigma(y_i - \bar{y})^2 + \frac{nk_0}{n+k_0}(\bar{y} - \mu_0)^2}{(k_0 + n)(\nu_0 + n - 2)} \approx \frac{s^2}{n}. \tag{11.62}$$

The marginal posterior distribution of M is given by

$$\sqrt{\nu_0 + n} \; \frac{M - \hat{\mu}^*}{\sqrt{(\lambda_0 + \Sigma(y_i - \bar{y})^2 + \frac{nk_0}{n+k_0}(\bar{y} - \mu_0)^2)/(n + k_0)}} \sim t_{\nu_0 + n}, \tag{11.63}$$

where that variable is close to the usual t statistic $\sqrt{n}(M - \bar{y})/s$.

11.6.2   Improper priors

As always with Bayesian inference, the choice of a prior is important. If there are substantive reasons to choose a particular prior or a range of priors, then there is justification for using those. Often, researchers wish to have priors that are chosen automatically. Such priors should have minimal influence on the inference (i.e., the posteriors), and be generally accepted by other researchers.

In the normal and binomial situations in the previous parts of this section, we saw that by taking prior parameter values close to the boundary of their spaces, the posterior quantities became close to the sample quantities we would use in frequentist evaluations. This happens because the posteriors are becoming less informative. E.g., if the prior on the mean is $N(\mu_0, \sigma_0^2)$ and σ_0^2 is very large, then the prior is expressing very little knowledge of the value of the parameter. It could be anything! If the prior parameter actually reaches the limit, e.g., taking $\sigma_0^2 = \infty$, then the prior is no longer a probability distribution. In this case, it would be uniform on the real line, which has total mass $+\infty$. Such priors are called **improper**. Jeffreys (1961) made an early effort to specify reasonable automatic priors, often called **noninformative** or **reference** priors, in a number of specific situations. If the parameter space is $\mathbb{R}$, he suggests using the uniform prior, and if the parameter σ has space$(0, \infty)$, he suggests the density $d\sigma/\sigma$, which is also improper. For the binomial p, Jeffreys likes Beta(1/2,1/2).

Interestingly, improper priors often lead to proper posteriors, where the posterior pdf is defined formally as if the prior were proper. For example, if we use the uniform prior on $\mu \in \mathbb{R}$, the posterior density is

$$f_{M \mid \mathbf{Y}}(\mu \mid \mathbf{y}) = \frac{f(\mathbf{y} \mid \mu)}{\int_{-\infty}^{\infty} f(\mathbf{y} \mid \mu^*) d\mu^*}. \tag{11.64}$$

If this denominator does not converge, then this procedure is not useful. Even if it does, there may in general be problems with improper priors. See Kass and Wasserman (1996) for an overview of improper priors, including examples where they lead to difficulties. It is safest to use proper priors, but using improper priors in situations where they have been shown to work well is fine.

11.7   Exercises

Exercise 11.7.1. For any estimator, let $\text{MSE}_\theta[\delta] = E_\theta[(\delta(\mathbf{X}) - g(\theta))^2]$, the mean square error. Show that $\text{MSE}_\theta[\delta] = \text{Bias}_\theta[\delta]^2 + Var_\theta[\delta]$ if the variance is finite. [Hint: Write $\delta(\mathbf{X}) - g(\theta) = \delta(\mathbf{X}) - E_\theta[\delta] + E_\theta[\delta] - g(\theta)$ and expand the square.]

Exercise 11.7.2. True or false. (The convergences are all as $n \to \infty$.)   (a) If an estimator is consistent, then it is unbiased. (b) If an estimator is unbiased, then it is consistent. (c) If an estimator is consistent, then its bias goes to zero. (d) If an estimator's bias goes to zero, then it is consistent. (e) If an estimator's variance goes to zero, then it is consistent. (f) If an estimator is unbiased and its variance goes to zero, then it is consistent. (g) If an estimator's variance and bias both go to zero, then it is consistent. (h) If an estimator is consistent, then its variance goes to zero. (i) If W_n is asymptotically $N(0, \sigma^2)$, then $E[W_n]$ goes to 0. (j) If W_n is asymptotically $N(0, \sigma^2)$, then $Var[W_n]$ goes to σ^2.

Exercise 11.7.3. Let $X_1, \ldots, X_n$ be iid $N(\mu, 1)$. We are interested in estimating μ^2. Consider the two estimators

$$\delta_n^{(1)} = \overline{x}_n^2 - \frac{1}{n} \text{ and } \delta_n^{(2)} = \frac{1}{n} \sum_{i=1}^{n} x_i^2 - 1. \tag{11.65}$$

(a) Show that the two estimators are unbiased. (b) Show that the two estimators are consistent. (c) Find the τ_1^2 for which $\sqrt{n}(\delta_n^{(1)} - \mu^2) \to^D N(0, \tau_1^2)$. (d) Find the τ_2^2 for which $\sqrt{n}(\delta_n^{(2)} - \mu^2) \to^D N(0, \tau_2^2)$. (e) The asymptotic relative efficiency of $\delta_n^{(1)}$ to $\delta_n^{(2)}$ is τ_2^2/τ_1^2. Find this ratio as a function of μ. What is it at $\mu = 0$? (f) Is one of the estimators always better than the other?

Exercise 11.7.4. Suppose $X_1, \ldots, X_n$ are iid Gamma(α, λ), where $(\alpha, \lambda) \in (0, \infty) \times (0, \infty)$. Find method-of-moment estimators for α and λ.

Exercise 11.7.5. Let $X_1, \ldots, X_n$ be iid Beta$(\alpha, 1)$, $\alpha > 0$. (a) Consider the method of moments estimator, $\widehat{\alpha}_n$, of α based on $\overline{X}_n$. What is it? Is it consistent? (b) What is $Var[X_i]$? (c) Let

$$W_n = \sqrt{n} \, (\widehat{\alpha}_n - \alpha) \longrightarrow^D N(0, \sigma^2(\alpha)). \tag{11.66}$$

What is $\sigma^2(\alpha)$? (d) Does $\sigma^2(\widehat{\alpha}_n) \to^P \sigma^2(\alpha)$? (e) Does

$$\sqrt{n} \, \frac{\widehat{\alpha}_n - \alpha}{\sigma(\widehat{\alpha}_n)} \longrightarrow^D N(0, 1)? \tag{11.67}$$

(f) Using part (e), find an approximate 95% confidence interval for α when $n = 100$ and $\overline{x}_n = 3/4$. (g) Imagine instead the X_i's are Beta(α, α). What is the method of moments estimator of α based on $\overline{X}_n$ for this model?

Exercise 11.7.6. Suppose that the number of telephone calls coming in to a call center follows a Poisson process with a rate of λ calls per minute, where $\lambda \in (0, \infty)$. That is, if X is the number of calls coming in over the course of t minutes, then $X \sim$ Poisson$(t\lambda)$. (a) Assuming t is known, what is a reasonable estimate of λ based on X? (b) Assuming t is known, what is a reasonable estimate of the parameter $\theta = P_\lambda[\text{no calls in the next two minutes}]$ based on X? (First find θ in terms of λ.)

Exercise 11.7.7. Suppose $(X_1, M_1), \ldots, (X_n, M_n)$ are independent pairs, where for each i, $X_i \mid M_i = \mu_i \sim N(\mu_i, 1)$, and $M_i \sim N(0, \sigma^2)$. The μ_i's are unobserved and to be estimated. The parameter $\sigma^2 \in (0, \infty)$ is unknown. (a) What is the marginal distribution of $\mathbf{X} = (X_1, \ldots, X_n)$? [The X_i's are marginally iid.] (b) Find an estimate of σ^2 based on $X_1, \ldots, X_n$. (c) Find $E[M_i \mid X_i = x_i]$ as x_i times a function of σ^2. (d) Now use the estimate of σ^2 from part (b) to find an estimate of $E[M_i \mid X_i = x_i]$ as in part (c). The result is called an **empirical Bayes estimator** of μ_i. It is also a **shrinkage estimator** of μ_i, since it takes x_i and shrinks it a bit (or a lot, depending on $\sum x_i^2$).

Exercise 11.7.8. Suppose $X_1, \ldots, X_n$ are iid with distribution function F_X, $Y_1, \ldots, Y_m$ are iid with distribution function F_Y, and the X_i's are independent of the Y_i's. Consider estimating the parameter $\theta = P[X > Y]$, where X and Y are independent, $X \sim F_X$ and $Y \sim F_Y$. The data has $n = 4$ and $m = 5$, where the X_i's are the fastest speeds (in MPH) a sample of men have ever driven, and the Y_i's are the fastest speeds a sample of women have ever driven. Here are the values: Men: 135, 110, 110, 80; Women: 75, 100, 95, 90, 90. Give an estimate of θ. [Hint: The parameter is the probability that a randomly chosen man has driven faster than a randomly chosen women. The estimate is the analog for the given samples.]

Exercise 11.7.9. Consider the sample $x_1, \ldots, x_n$, where all the x_i's are distinct. How many different sets of bootstrap samples, $\{x_1^*, \ldots, x_n^*\}$ are there? (If $n = 3$, then the sets $\{x_1, x_3, x_1\}$ and $\{x_3, x_1, x_1\}$ are the same.) Feller (1968) introduced the "stars and bars" technique for certain combinatorial calculations. Here, we can represent a bootstrap sample by specifying how many times each x_i is in the sample. Thus if there are k_i of the x_i's, then $k_1 + \cdots + k_n = n$. Now write out a sequence of stars and bars as follows: Write down k_1 stars, then one bar, then k_2 stars, then one bar, $\ldots$, then k_{n-1} stars, then one bar, and finally k_n stars. You should end up with n stars and $n - 1$ bars. For example, if $n = 7$ and the bootstrap sample is $\{x_1, x_1, x_3, x_5, x_5, x_5, x_6\}$, the picture will be

$$* \; * \; | \; | \; * \; | \; | \; * \; * \; * \; | \; * \; | \qquad\qquad (11.68)$$

Note that the possible arrangements of n stars and $n - 1$ bars are in one-to-one correspondence with possible bootstrap samples. (a) Argue that there are $\binom{2n-1}{n}$ such arrangements. (b) How many bootstrap samples are there for $n = 10$, $n = 100$, and $n = 1000$? [You may wish to use the $\log(\Gamma(x))$ function in R, lgamma.] (c) For which n is the number of bootstrap samples approximately one googol (10^{100})?

Note: The data used in Exercises 11.7.10 through 11.7.12 can be found as R matrices in the file `http://istics.net/r/data_chapter_11.txt`.

Exercise 11.7.10. Henson, Rogers, and Reynolds (1996) performed an experiment to see if caffeine has a negative effect on short-term visual memory. High school students were randomly chosen: 9 from eighth grade, 10 from tenth grade, and 9 from twelfth grade. Each person was tested once after having caffeinated Coke, and once after having decaffeinated Coke. After each drink, the person was given ten seconds to try to memorize twenty small, common objects, then allowed a minute to write down as many as could be remembered. The main question of interest is whether people remembered more objects after the Coke without caffeine than after the Coke with caffeine. The data (in the R matrix caffeine) are

Grade 8		Grade 10		Grade 12	
Without	With	Without	With	Without	With
5	6	6	3	7	7
9	8	9	11	8	6
6	5	4	4	9	6
8	9	7	6	11	7
7	6	6	8	5	5
6	6	7	6	9	4
8	6	6	8	9	7
6	8	9	8	11	8
6	7	10	7	10	9
		10	6		

$$(11.69)$$

For each grade, consider the differences between the number of objects remembered without caffeine and the number remembered with caffeine, $X_i = $ Without $-$ With. The sample means of these three differences are 0, 0.70, and 2.22 for grade 8, 10, and 12, respectively. For each grade, find an approximate 95% confidence interval for the population mean using 5000 bootstrap samples. For which grades, if any, does caffeine seem to effect the number of objects remembered?

Exercise 11.7.11. Consider the data on the number of pairs of shoes people owned in in Section 11.5.1, but now separate men and women. See the R matrix shoes. The means and medians are given in the table:

	n	Mean	Median
Women	485	19.408	15
Men	227	6.233	5

$$(11.70)$$

(a) Consider the parameter θ being the ratio of the population mean for the women to that for the men. Find the plug-in estimate of θ based on the data in (11.70). For the two-sample situation here, a bootstrap sample consists of one bootstrap sample from the women, and one from the men, then the bootstrap quantity of interest is the ratio of sample means of the bootstrap samples. In the following, use 5000 bootstrap samples. (b) Find the bootstrap estimate of the bias of the estimate in part (a). Is there much bias? (c) Find the bootstrap estimate of the 95% confidence interval for θ. (d) Repeat the process for the ratio of medians.

Exercise 11.7.12. Continue with the data from Exercises 11.7.11 and 9.5.11 (c). As in the latter, let $x = $ shoe size and $y = \log(\#$ of pairs of shoes owned), so that their sample correlation is $r = -0.500$. The approximate 95% confidence interval for ρ based on the Δ-method was found to be $(-.5540, -.4417)$. Find the approximate 95% confidence interval using the bootstrap. Does it differ much from what we found earlier? (Here, we want to bootstrap (x_i, y_i) pairs. In R, we can bootstrap the indices with i<−sample(712,replace=T), which yields a vector of 712 indices between 1 and 712. Then the bootstrapped correlation is cor(x[i],y[i]).)

Exercise 11.7.13. Consider the bootstrap in the binomial case. We observe $X \sim$ Binomial(n, p), thinking of it as the sum of n iid Bernoulli(p)'s, $Z_1, \ldots, Z_n$. Then a bootstrap sample is $Z_1^*, \ldots, Z_n^*$. Let $X^* = Z_1^* + \cdots + Z_n^*$. (a) Given the data $X = x$, what is the exact distribution of X^*? (b) What is the bootstrap estimate of the bias of the estimate $\hat{p} = X/n$ of p? (c) What is the bootstrap estimate of the standard error of $\hat{p}$?

Exercise 11.7.14. Suppose $X_1, \ldots, X_n$ are iid Exponential(λ) for $\lambda \in (0, \infty)$. (a) Show that the pdf can be written as $f(x_1, \ldots, x_n \mid \lambda) = \lambda^n e^{-\lambda T}$ for some T, a function of the X_i's. What is T? (b) Consider the pdf for $\lambda \in (0, \infty)$ with parameters ν and τ such that for some c,

$$\rho(\lambda \mid \nu, \tau) = c(\nu, \tau) \lambda^\nu e^{-\lambda \tau} \tag{11.71}$$

is a legitimate density for $\nu > -1$ and $\tau > 0$. What is the distribution? (c) Now consider the Bayesian model where

$$X_1, \ldots, X_n \mid \Lambda = \lambda \text{ are iid Exponential}(\lambda), \text{ and } \Lambda \sim \rho(\lambda \mid \nu_0, \tau_0) \tag{11.72}$$

for some $\nu_0 > -1$ and $\tau_0 > 0$. Show the posterior density of Λ given the data is $\rho(\lambda \mid \tau^*, \nu^*)$ for τ^* a function of τ_0 and T, and and ν^* a function of ν_0 and n. What is the posterior distribution? Thus the conjugate prior for the exponential has density ρ. (d) What is the posterior mean of Λ given the data? What does this posterior mean approach as $\nu_0 \to -1$ and $\tau_0 \to 0$? Is the $\rho(-1, 0)$ density a proper one?

Exercise 11.7.15. Here the data are $X_1, \ldots, X_n$ iid Poisson(λ). (a) $\sqrt{n}(\overline{X}_n - \lambda) \to^{\mathcal{D}} N(0, v)$ for what v? (b) Find the approximate 95% confidence intervals for λ when $\overline{X} = 1$ and $n = 10, 100$, and 1000 based on the result in part (a). (c) Using the relatively noninformative prior $\lambda \sim \text{Gamma}(1/2, 1/2)$, find the 95% probability intervals for λ when $\overline{X} = 1$ and $n = 10, 100$, and 1000. How large does n have to be for the frequency-based interval from part (b) to approximate the Bayesian interval well? (d) Using the relatively strong prior $\lambda \sim \text{Gamma}(50, 50)$, find the 95% probability intervals for λ when $\overline{X} = 1$ and $n = 10, 100$, and 1000. How large does n have to be for the frequency-based interval from part (b) to approximate this Bayesian interval well?

Exercise 11.7.16. Here we assume that $Y_1, \ldots, Y_n \mid M = \mu$ are iid $N(\mu, \sigma^2)$ and $M \sim N(\mu_0, \sigma_0^2)$, where σ^2 is known. Then from (11.48), we have $M \mid \overline{Y} = \overline{y} \sim N(\mu^*, \sigma^{*2})$, where with $\omega_0 = 1/\sigma_0^2$ and $\omega = 1/\sigma^2$,

$$\mu^* = \frac{\sigma^2 \mu_0 + n \sigma_0^2 \overline{y}}{\sigma^2 + n \sigma_0^2} \quad \text{and} \quad \sigma^{*2} = \frac{\sigma^2 \sigma_0^2}{\sigma^2 + n \sigma_0^2}. \tag{11.73}$$

(See also (7.108).) (a) Show that

$$P\left[\overline{y} - 1.96 \frac{\sigma}{\sqrt{n}} < M < \overline{y} + 1.96 \frac{\sigma}{\sqrt{n}} \,\middle|\, \overline{Y} = \overline{y}\right]$$

$$= \Phi\left(\frac{\frac{z}{1+\tau} + 1.96}{\sqrt{\frac{\tau}{1+\tau}}}\right) - \Phi\left(\frac{\frac{z}{1+\tau} - 1.96}{\sqrt{\frac{\tau}{1+\tau}}}\right), \tag{11.74}$$

where $z = \sqrt{n}(\overline{y} - \mu_0)/\sigma$, $\tau = n\sigma_0^2/\sigma^2$, and Φ is the distribution function of a $N(0, 1)$. (b) Show that for any fixed z, the probability in (11.74) goes to 95% as $\tau \to \infty$. (c) Show that if we use the improper prior that is uniform on $\mathbb{R}$ (as in (11.64)), the posterior distribution $\mu \mid \overline{Y} = \overline{y}$ is exactly $N(\overline{y}, \sigma^2/n)$, hence the confidence interval has a posterior probability of exactly 95%.

Exercise 11.7.17. Suppose $X \mid \Theta = \theta \sim \text{Binomial}(n, \theta)$. Consider the prior with density $1/(\theta(1-\theta))$ for $\theta \in (0,1)$. (It looks like a Beta(0,0), if there were such a thing.) (a) Show that this prior is improper. (b) Find the posterior distribution as in (11.64), $\Theta \mid X = x$, for this prior. For some values of x the posterior is valid. For others, it is not, since the denominator is infinite. For which values is the posterior valid? What is the posterior in these cases? (c) If the posterior is valid, what is the posterior mean of Θ? How does it compare to the usual estimator of θ?

Exercise 11.7.18. Consider the normal situation with known mean but unknown variance or precision. Suppose $Y_1, \ldots, Y_n \mid \Omega = \omega$ are iid $N(\mu, 1/\omega)$ with μ known. Take the prior on Ω to be $\text{Gamma}(\nu_0/2, \lambda_0/2)$ as in (11.54). (a) Show that

$$\Omega \mid \mathbf{Y} = \mathbf{y} \sim \text{Gamma}((\nu_0 + n)/2, (\lambda_0 + \sum(y_i - \mu)^2)/2). \tag{11.75}$$

(b) Find $E[1/\Omega \mid \mathbf{Y} = \mathbf{y}]$. What is the value as λ_0 and ν_0 approach zero? (c) Ignoring the constant in the density, what is the density of the gamma with both parameters equalling zero? Is this an improper prior? Is the posterior using this prior valid? If so, what is it?

Exercise 11.7.19. This exercise is to prove Lemma 11.1. So let

$$M \mid \Omega = \omega \sim N(\mu_0, 1/(k_0\omega)) \quad \text{and} \quad \Omega \sim \text{Gamma}(\nu_0/2, \lambda_0/2), \tag{11.76}$$

as in (11.54). (a) Let $Z = \sqrt{k_0\Omega}(M - \mu_0)$ and $U = \lambda_0\Omega$. Show that $Z \sim N(0,1)$, $U \sim \text{Gamma}(\nu_0/2, 1/2)$, and Z and U are independent. [What is the conditional distribution $Z \mid U = u$? Also, refer to Exercise 5.6.1.] (b) Argue that $T = Z/\sqrt{U/\nu_0} \sim t_{\nu_0}$, which verifies (11.60). [See Definition 6.5 in Exercise 6.8.17.] (c) Derive the mean and variance of M based on the known mean and variance of Student's t. [See Exercise 6.8.17(a).] (d) Show that $E[1/\Omega] = \lambda_0/(\nu_0 - 2)$ if $\nu_0 > 2$.

Exercise 11.7.20. Show that

$$n(\bar{y} - \mu)^2 + k_0(\mu - \mu_0)^2 = (n + k_0)(\mu - \hat{\mu}^*)^2 + \frac{nk_0}{n + k_0}(\bar{y} - \mu_0)^2, \tag{11.77}$$

which is necessary to show that (11.56) holds, where $\hat{\mu}^* = (k_0\mu_0 + n\bar{y})/(k_0 + n)$ as in (11.57).

Exercise 11.7.21. Take the Bayesian setup with a one-dimensional parameter, so that we are given the conditional distribution $X \mid \Theta = \theta$ and the (proper) prior distribution of Θ with space $\mathcal{T} \subset \mathbb{R}$. Let $\delta(x) = E[\Theta \mid X = x]$ be the Bayes estimate of θ. Suppose that δ is an unbiased estimator of θ, so that $E[\delta(X) \mid \Theta = \theta] = \theta$. Assume that the marginal and conditional variances of $\delta(X)$ and Θ are finite. (a) Using the formula for covariance based on conditioning on X (as in (6.50)), show that the unconditional covariance $Cov[\Theta, \delta(X)]$ equals the unconditional $Var[\delta(X)]$. (b) Using the same formula, but conditioning on Θ, show that $Cov[\Theta, \delta(X)] = Var[\Theta]$. (c) Show that (a) and (b) imply that the correlation between $\delta(X)$ and Θ is 1. Use the result in (2.32) and (2.33) to help show that in fact Θ and $\delta(X)$ are the same (i.e., $P[\Theta = \delta(X)] = 1$). (d) The conclusion in (c) means that the only time the Bayes estimator is unbiased is when it is exactly equal to the parameter. Can you think of any situations where this phenomenon would occur?

Chapter 12

Linear Regression

12.1 Regression

How is height related to weight? How are sex and age related to heart disease? What factors influence crime rate? Questions such as these have one **dependent** variable of interest, and one or more **explanatory** or **predictor** variables. The goal is to assess the relationship of the explanatory variables to the dependent variable. Examples:

Dependent Variable (Y)	Explanatory Variables (X's)
Weight	Height, gender
Cholesterol level	Fat intake, obesity, exercise
Heart function	Age, sex
Crime rate	Density, income, education
Bacterial count	Drug

We will generically denote an observation $(Y, \mathbf{X})$, where the dependent variable is Y, and the vector of p explanatory variables is $\mathbf{X}$. The overall goal is to find a function $g(\mathbf{X})$ that is a good predictor of Y. The (mean) regression function uses the average Y for a particular vector of values of $\mathbf{X} = \mathbf{x}$ as the predictor. Median regression models the median Y for given $\mathbf{X} = \mathbf{x}$. That is,

$$g(\mathbf{x}) = \begin{cases} E[Y \mid \mathbf{X} = \mathbf{x}] & \text{in mean regression} \\ \text{Median}[Y \mid \mathbf{X} = \mathbf{x}] & \text{in median regression} \end{cases} . \tag{12.1}$$

The median is less sensitive to large values, so may be a more robust measure. More generally, **quantile regression** (Koenker and Bassett, 1978) seeks to determine a particular quantile of Y given $\mathbf{X} = \mathbf{x}$. For example, Y may be a measure of water depth in a river, and one wishes to know the 90th percentile level given $\mathbf{X} = \mathbf{x}$ to help warn of flooding. (Typically, "regression" refers to mean regression, so that median or quantile regression needs the adjective.)

The function g may or may not be a simple function of the $\mathbf{x}$'s, and in fact we might not even know the exact form. Linear regression tries to approximate the conditional expected value by a linear function of $\mathbf{x}$:

$$g(\mathbf{x}) = \beta_0 + \beta_1 x_1 + \cdots + \beta_K x_K \approx E[Y \mid \mathbf{X} = \mathbf{x}]. \tag{12.2}$$

As we saw in Lemma 7.8 on page 116, if $(Y, \mathbf{X})$ is (jointly) multivariate normal, then $E[Y \mid \mathbf{X} = \mathbf{x}]$ is itself linear, in which case there is no need for an approximation in (12.2).

The rest of this chapter deals with estimation in linear regression. Rather than trying to model Y and $\mathbf{X}$ jointly, everything will be performed conditioning on $\mathbf{X} = \mathbf{x}$, so we won't even mention the distribution of $\mathbf{X}$. The next section develops the matrix notation needed in order to formally present the model. Section 12.3 discusses least squares estimation, which is associated with mean regression. In Section 12.5, we present some regularization, which modifies the objective functions to reign in the sizes of the estimated coefficients, possibly improving prediction. Section 12.6 looks more carefully at median regression, which uses least absolute deviations as its objective function.

12.2   Matrix notation

Here we write the linear model in a universal matrix notation. **Simple linear regression** has one explanatory x variable, such as trying to predict cholesterol level (Y) from fat intake (x). If there are n observations, then the linear model would be written

$$Y_i = \beta_0 + \beta_1 x_i + E_i, \quad i = 1, \ldots, n. \tag{12.3}$$

Imagine stacking these equations on top of each other. That is, we construct vectors

$$\mathbf{Y} = \begin{pmatrix} Y_1 \\ Y_2 \\ \vdots \\ Y_n \end{pmatrix}, \quad \mathbf{E} = \begin{pmatrix} E_1 \\ E_2 \\ \vdots \\ E_n \end{pmatrix}, \quad \text{and } \boldsymbol{\beta} = \begin{pmatrix} \beta_0 \\ \beta_1 \end{pmatrix}. \tag{12.4}$$

For $\mathbf{X}$, we need a vector for the x_i's, but also a vector of 1's, which are surreptitiously multiplying the β_0:

$$\mathbf{x} = \begin{pmatrix} 1 & x_1 \\ 1 & x_2 \\ \vdots & \vdots \\ 1 & x_n \end{pmatrix}. \tag{12.5}$$

Then the model in (12.3) can be written compactly as

$$\mathbf{Y} = \mathbf{x}\boldsymbol{\beta} + \mathbf{E}. \tag{12.6}$$

When there is more than one explanatory variable, we need an extra subscript for x, so that x_{i1} is the value for fat intake and x_{i2} is the exercise level, say, for person i:

$$Y_i = \beta_0 + \beta_1 x_{i1} + \beta_2 x_{i2} + E_i, \quad i = 1, \ldots, n. \tag{12.7}$$

With K variables, the model would be

$$Y_i = \beta_0 + \beta_1 x_{i1} + \cdots + \beta_K x_{iK} + E_i, \quad i = 1, \ldots, n. \tag{12.8}$$

The general model (12.8) has the form (12.6) with a longer β and wider x:

$$\mathbf{Y} = \mathbf{x}\beta + \mathbf{E} = \begin{pmatrix} 1 & x_{11} & x_{12} & \cdots & x_{1K} \\ 1 & x_{21} & x_{22} & \cdots & x_{2K} \\ \vdots & \vdots & \vdots & \cdots & \vdots \\ 1 & x_{n1} & x_{n2} & \cdots & x_{nK} \end{pmatrix} \begin{pmatrix} \beta_0 \\ \beta_1 \\ \beta_2 \\ \vdots \\ \beta_K \end{pmatrix} + \mathbf{E}. \tag{12.9}$$

We will generally assume that the x_{ij}'s are fixed constants, hence the E_i's and Y_i's are the random quantities. It may be that the x-values are fixed by the experimenter (e.g., denoting dosages or treatment groups assigned to subjects), or $(\mathbf{Y}, \mathbf{X})$ has a joint distribution, but the analysis proceeds conditional on $\mathbf{Y}$ given $\mathbf{X} = \mathbf{x}$, and (12.9) describes this conditional distribution. Assumptions on the E_i's in mean regression, moving from least to most specific, include the following:

1. $E[E_i] = 0$, so that $E[\mathbf{E}] = \mathbf{0}$ and $E[\mathbf{Y}] = \mathbf{x}\beta$.

2. The E_i's are uncorrelated, hence the Y_i's are uncorrelated.

3. The E_i's are homoscedastic, i.e., they all have the same variance, hence so do the Y_i's.

4. The E_i's are iid.

5. The E_i's are multivariate normal. If the previous assumptions also hold, we have
$$\mathbf{E} \sim N(\mathbf{0}, \sigma^2 \mathbf{I}_n) \text{ which implies that } \mathbf{Y} \sim N(\mathbf{x}\beta, \sigma^2 \mathbf{I}_n). \tag{12.10}$$

Median regression replaces #1 with $\text{Median}(E_1) = 0$, and generally dispenses with #2 and #3 (since moments are unnecessary). General quantile regression would set the desired quantile of E_i to 0.

12.3 Least squares

In regression, or any prediction situation, one approach to estimation chooses the estimate of the parameters so that the predictions are close to the Y_i's. Least squares is a venerable and popular criterion. In the regression case, the least squares estimates of the β_i's are the values b_i that minimize the objective function

$$obj(\mathbf{b}\,;\mathbf{y}) = \sum_{i=1}^{n}(y_i - (b_0 + b_1 x_{i1} + \cdots + b_K x_{iK}))^2 = \|\mathbf{y} - \mathbf{x}\mathbf{b}\|^2. \tag{12.11}$$

We will take $\mathbf{x}$ to be $n \times p$, where if as in (12.11) there are K predictors plus the intercept, $p = K + 1$. In general, $\mathbf{x}$ need not have an intercept term (i.e., column of 1's). Least squares is tailored to mean regression because for a sample $z_1, \ldots, z_n$, the sample mean is the value of m that minimize $\sum(z_i - m)^2$ over m. (See Exercise 12.7.1. Also, Exercise 2.7.25 has the result for random variables.)

Ideally, we'd solve $\mathbf{y} = \mathbf{x}\mathbf{b}$ for $\mathbf{b}$, so that if $\mathbf{x}$ is square and invertible, then $\widehat{\beta} = \mathbf{x}^{-1}\mathbf{y}$. It is more likely that $\mathbf{x}'\mathbf{x}$ is invertible, at least when $p < n$, in which case we multiply both sides by $\mathbf{x}'$:

$$\mathbf{x}'\mathbf{y} = \mathbf{x}'\mathbf{x}\widehat{\beta} \implies \widehat{\beta} = (\mathbf{x}'\mathbf{x})^{-1}\mathbf{x}'\mathbf{y} \text{ if } \mathbf{x}'\mathbf{x} \text{ is invertible.} \tag{12.12}$$

If $p > n$, i.e., there are more parameters to estimate than observations, $x'x$ will not be invertible. Noninvertibility will occur for $p \leq n$ when there are linear redundancies in the variables. For example, predictors of a student's score on the final exam may include scores on each of three midterms, plus the average of the three midterms. Or redundancy may be random, such as when there are several categorical predictors, and by chance all the people in the sample that are from Asia are female. Such redundancies can be dealt with by eliminating one or more of the variables. Alternatively, we can use the Moore-Penrose inverse from (7.56), though if $x'x$ is not invertible, the least squares estimate is not unique. See also Exercise 12.7.11, which uses the Moore-Penrose inverse of x itself.

Assume that $x'x$ is invertible. We show that $\widehat{\beta}$ as in (12.12) does minimize the least squares criterion. Write

$$\begin{aligned}
\|y - xb\|^2 &= \|(y - x\widehat{\beta}) + (x\widehat{\beta} - xb)\|^2 \\
&= \|y - x\widehat{\beta}\|^2 + \|x\widehat{\beta} - xb\|^2 + 2(y - x\widehat{\beta})'(x\widehat{\beta} - xb).
\end{aligned} \tag{12.13}$$

The estimated fit is $\widehat{y} = x\widehat{\beta}$, and the estimated error or residual vector is $\widehat{e} = y - \widehat{y} = y - x\widehat{\beta}$. By definition of $\widehat{\beta}$, we have that

$$\widehat{y} = P_x y, \quad P_x = x(x'x)^{-1}x', \quad \text{and} \quad \widehat{e} = Q_x y, \quad Q_x = I_n - P_x. \tag{12.14}$$

(For those who know about projections: This P_x is the projection matrix onto the space spanned by the column of x, and Q_x is the projection matrix on the orthogonal complement to the space spanned by the columns of x.)

Exercise 12.7.4 shows that

$$P_x \text{ and } Q_x \text{ are symmetric and idempotent, } P_x x = x, \text{ and } Q_x x = 0. \tag{12.15}$$

(Recall from (7.40) that idempotent means $P_x P_x = P_x$.) Thus the cross-product term in (12.13) can be eliminated:

$$(y - x\widehat{\beta})'(x\widehat{\beta} - xb) = (Q_x y)'x(\widehat{\beta} - b) = y'(Q_x x)(\widehat{\beta} - b) = 0. \tag{12.16}$$

Hence

$$obj(b \, ; \, y) = \|y - xb\|^2 = y'Q_x y + (\widehat{\beta} - b)'x'x(\widehat{\beta} - b). \tag{12.17}$$

Since $x'x$ is nonnegative definite and invertible, it must be positive definite. Thus the second summand on the right-hand side of (12.17) must be positive unless $b = \widehat{\beta}$, proving that the least squares estimate of β is indeed $\widehat{\beta}$ in (12.12). The minimum of the least squares objective function is the sum of squared errors:

$$SS_e \equiv \|\widehat{e}\|^2 = \|y - x\widehat{\beta}\|^2 = y'Q_x y. \tag{12.18}$$

Is $\widehat{\beta}$ a good estimator? It depends partially on which of the assumptions in (12.10) hold. If $E[E] = 0$, then $\widehat{\beta}$ is unbiased. If Σ is the covariance matrix of E, then

$$Cov[\widehat{\beta}] = (x'x)^{-1}x'\Sigma x(x'x)^{-1}. \tag{12.19}$$

If the E_i's are uncorrelated and homoscedastic, with common $Var[E_i] = \sigma^2$, then $\Sigma = \sigma^2 I_n$, so that $Cov[\widehat{\beta}] = \sigma^2 (x'x)^{-1}$. In this case, the least squares estimator is the

best linear unbiased estimator (BLUE) in that it has the lowest variance among the linear unbiased estimators, where a linear estimator is one of the form **LY** for constant matrix **L**. See Exercise 12.7.8. The strictest assumption, which adds multivariate normality, bestows the estimator with multivariate normality:

$$\mathbf{E} \sim N(\mathbf{0}, \sigma^2 \mathbf{I}_n) \implies \widehat{\beta} \sim N(\beta, \sigma^2 (\mathbf{x}'\mathbf{x})^{-1}). \tag{12.20}$$

If this normality assumption does hold, then the estimator is the best unbiased estimator of β, linear or not. See Exercise 19.8.16.

On the other hand, the least squares estimator can be notoriously non-robust. Just one or a few wild values among the y_i's can ruin the estimate. See Figure 12.2.

12.3.1 Standard errors and confidence intervals

For this section we will make the normal assumption that $\mathbf{E} \sim N(\mathbf{0}, \sigma^2 \mathbf{I}_n)$, though much of what we say works without normality. From (12.20), we see that

$$Var[\widehat{\beta}_i] = \sigma^2 [(\mathbf{x}'\mathbf{x})^{-1}]_{ii}, \tag{12.21}$$

where the last term is the ith diagonal of $(\mathbf{x}'\mathbf{x})^{-1}$. To estimate σ^2, note that

$$\mathbf{Q_x Y} \sim N(\mathbf{Q_x x}\beta, \sigma^2 \mathbf{Q_x Q_x'}) = N(\mathbf{0}, \sigma^2 \mathbf{Q_x}) \tag{12.22}$$

by (12.14). Exercise 12.7.4 shows that $\mathrm{trace}(\mathbf{Q_x}) = n - p$, hence we can use (7.65) to show that

$$SS_e = \mathbf{Y}' \mathbf{Q_x Y} \sim \sigma^2 \chi^2_{n-p}, \text{ which implies that } \widehat{\sigma}^2 = \frac{SS_e}{n-p} \tag{12.23}$$

is an unbiased estimate of σ^2, leading to

$$se(\widehat{\beta}_i) = \widehat{\sigma} \sqrt{[(\mathbf{x}'\mathbf{x})^{-1}]_{ii}} . \tag{12.24}$$

We have the ingredients to use Student's t for confidence intervals, but first we need the independence of $\widehat{\beta}$ and $\widehat{\sigma}^2$. Exercise 12.7.6 uses calculations similar to those in (7.43) to show that $\widehat{\beta}$ and $\mathbf{Q_x Y}$ are in fact independent. To summarize,

Theorem 12.1. *If* $\mathbf{Y} = \mathbf{x}\beta + \mathbf{E}$, $\mathbf{E} \sim N(\mathbf{0}, \sigma^2 \mathbf{I}_n)$, *and* $\mathbf{x}'\mathbf{x}$ *is invertible, then* $\widehat{\beta}$ *and* SS_e *are independent, with*

$$\widehat{\beta} \sim N(\beta, \sigma^2 (\mathbf{x}'\mathbf{x})^{-1}) \text{ and } SS_e \sim \sigma^2 \chi^2_{n-p}. \tag{12.25}$$

From this theorem we can derive (Exercise 12.7.7) that

$$\frac{\widehat{\beta}_i - \beta_i}{se(\widehat{\beta}_i)} \sim t_{n-p} \implies \widehat{\beta}_i \pm t_{n-p,\alpha/2} \, se(\widehat{\beta}_i) \tag{12.26}$$

is a $100(1 - \alpha)\%$ confidence interval for β_i.

12.4   Bayesian estimation

We start by assuming that σ^2 is known in the normal model (12.10). The conjugate prior for β is also normal:

$$\mathbf{Y} \,|\, \beta = \mathbf{b} \sim N(\mathbf{xb}, \sigma^2 \mathbf{I}_n) \quad \text{and} \quad \beta \sim N(\beta_0, \Sigma_0), \tag{12.27}$$

where β_0 and Σ_0 are known. We use Bayes theorem to find the posterior distribution of $\beta \,|\, \mathbf{Y} = \mathbf{y}$. We have that

$$\widehat{\beta} \,|\, \beta = \mathbf{b} \sim N(\mathbf{b}, \sigma^2 (\mathbf{x}'\mathbf{x})^{-1}) \tag{12.28}$$

for $\widehat{\beta}$ in (12.12) and (12.25). Note that this setup is the same as that for the multivariate normal mean vector in Exercise 7.8.15, where β is the $\mathbf{M}$ and $\widehat{\beta}$ is the $\mathbf{Y}$. The only difference is that here we are using column vectors, but the basic results remain the same. In this case, the prior precision is $\Omega_0 \equiv \Sigma_0^{-1}$, and the conditional precision is $\mathbf{x}'\mathbf{x}/\sigma^2$. Thus we immediately have

$$\beta \,|\, \widehat{\beta} = \widehat{\mathbf{b}} \sim N(\widehat{\beta}^*, (\Omega_0 + \mathbf{x}'\mathbf{x}/\sigma^2)^{-1}), \tag{12.29}$$

where

$$\widehat{\beta}^* = (\Omega_0 + \mathbf{x}'\mathbf{x}/\sigma^2)^{-1}(\Omega_0 \beta_0 + (\mathbf{x}'\mathbf{x}/\sigma^2)\widehat{\mathbf{b}}). \tag{12.30}$$

If the prior variance is very large, so that the precision $\Omega_0 \approx \mathbf{0}$, the posterior mean and covariance are approximately the least squares estimate and its covariance:

$$\beta \,|\, \widehat{\beta} = \widehat{\mathbf{b}} \approx N(\widehat{\mathbf{b}}, \sigma^2 (\mathbf{x}'\mathbf{x})^{-1}). \tag{12.31}$$

For less vague priors, one may specialize to $\beta_0 = \mathbf{0}$, with the precision proportional to $\mathbf{I}_n$. For convenience take $\Omega_0 = (\kappa/\sigma^2)\mathbf{I}_n$ for some κ, so that κ indicates the relative precision of the prior to that of one observation (E_i). The posterior then resolves to

$$\beta \,|\, \widehat{\beta} = \widehat{\mathbf{b}} \sim N\left((\kappa \mathbf{I}_p + \mathbf{x}'\mathbf{x})^{-1}\mathbf{x}'\mathbf{x}\widehat{\mathbf{b}}, \sigma^2 (\kappa \mathbf{I}_p + \mathbf{x}'\mathbf{x})^{-1}\right). \tag{12.32}$$

This posterior mean is the **ridge regression** estimator of β,

$$\widehat{\beta}_\kappa = (\mathbf{x}'\mathbf{x} + \kappa \mathbf{I}_p)^{-1}\mathbf{x}'\mathbf{y}, \tag{12.33}$$

which we will see in the next section.

If σ^2 is not known, then we can use the prior used for the normal mean in Section 11.6.1. Using the precision $\omega = 1/\sigma^2$,

$$\widehat{\beta} \,|\, \beta = \mathbf{b}, \Omega = \omega \sim N(\mathbf{b}, (1/\omega)(\mathbf{x}'\mathbf{x})^{-1}), \tag{12.34}$$

where the prior is given by

$$\beta \,|\, \Omega = \omega \sim N(\beta_0, (1/\omega)\mathbf{K}_0^{-1}) \quad \text{and} \quad \Omega \sim \Gamma(\nu_0/2, \lambda_0/2). \tag{12.35}$$

Here, $\mathbf{K}_0$ is an invertible symmetric $p \times p$ matrix, and ν_0 and λ_0 are positive. It is not too hard to see that

$$E[\beta] = \beta_0, \quad Cov[\beta] = \frac{\lambda_0}{\nu_0 - 2}\mathbf{K}_0^{-1}, \quad \text{and} \quad E\left[\frac{1}{\Omega}\right] = \frac{\lambda_0}{\nu_0 - 2}, \tag{12.36}$$

similar to (11.59). The last two equations need $\nu_0 > 2$.

Analogous to (11.57) and (11.58), the posterior has the same form, but updating the parameters as

$$\boldsymbol{\beta}_0 \to \widehat{\boldsymbol{\beta}}^* \equiv (\mathbf{K}_0 + \mathbf{x}'\mathbf{x})^{-1}(\mathbf{K}_0\boldsymbol{\beta}_0 + (\mathbf{x}'\mathbf{x})\widehat{\boldsymbol{\beta}}), \quad \mathbf{K}_0 \to \mathbf{K}_0 + \mathbf{x}'\mathbf{x}, \quad \nu_0 \to \nu_0 + n, \quad (12.37)$$

$$\text{and} \quad \lambda_0 \to \widehat{\lambda}^* \equiv \lambda_0 + SS_e + (\widehat{\boldsymbol{\beta}} - \boldsymbol{\beta}_0)'(\mathbf{K}_0^{-1} + (\mathbf{x}'\mathbf{x})^{-1})^{-1}(\widehat{\boldsymbol{\beta}} - \boldsymbol{\beta}_0). \quad (12.38)$$

See Exercise 12.7.12. If the prior parameters $\boldsymbol{\beta}_0$, λ_0, ν_0 and $\mathbf{K}_0$ are all close to zero, then the posterior mean and covariance matrix of $\boldsymbol{\beta}$ have approximations

$$E[\boldsymbol{\beta} \mid \mathbf{Y} = \mathbf{y}] = \widehat{\boldsymbol{\beta}}^* \approx \widehat{\boldsymbol{\beta}} \quad (12.39)$$

and

$$Cov[\boldsymbol{\beta} \mid \mathbf{Y} = \mathbf{y}] = \frac{\widehat{\lambda}^*}{\nu_0 + n - 2}(\mathbf{K}_0^{-1} + (\mathbf{x}'\mathbf{x})^{-1}) \approx \frac{SS_e}{n - 2}(\mathbf{x}'\mathbf{x})^{-1}, \quad (12.40)$$

close to the frequentist estimates.

The marginal distribution of $\boldsymbol{\beta}$ under the prior or posterior is a multivariate Student's t, which was introduced in Exercise 7.8.17. If $\mathbf{Z} \sim N_p(\mathbf{0}, \mathbf{I}_p)$ and $U \sim \text{Gamma}(\nu/2, 1/2)$, then

$$\mathbf{T} \equiv \frac{1}{\sqrt{U/\nu}}\mathbf{Z} \sim t_{p,\nu}, \quad (12.41)$$

is a standard p-variate Student's t on ν degrees of freedom. With the parameters in (12.37) and (12.38), it can be shown that a posteriori

$$\mathbf{T} = \frac{1}{\sqrt{\widehat{\lambda}^*/(\nu_0 + n)}}(\mathbf{K}_0 + \mathbf{x}'\mathbf{x})^{1/2}(\boldsymbol{\beta} - \widehat{\boldsymbol{\beta}}^*) \sim t_{p,\nu_0 + n}. \quad (12.42)$$

12.5 Regularization

Often regression estimates are used for prediction. Instead of being primarily interested in estimating the values of $\boldsymbol{\beta}$, one is interested in how the estimates can be used to predict new y_i's from new x-vectors. For example, we may have data on the progress of diabetes in a number of patients, along with a variety of their health and demographic variables (age, sex, BMI, etc.). Based on these observations, we would then like to predict the progress of diabetes for a number of new patients for whom we know the predictors.

Suppose $\widehat{\boldsymbol{\beta}}$ is an estimator based on observing $\mathbf{Y} = \mathbf{x}\boldsymbol{\beta} + \mathbf{E}$, and a new set of observations are contemplated that follow the same model, i.e., $\mathbf{Y}^{New} = \mathbf{z}\boldsymbol{\beta} + \mathbf{E}^{New}$, where $\mathbf{z}$ contains the predictor variables for the new observations. We know the $\mathbf{z}$. We do not observe the $\mathbf{Y}^{New}$, but would like to estimate it. The natural estimator would then be $\widehat{\mathbf{Y}}^{New} = \mathbf{z}\widehat{\boldsymbol{\beta}}$.

12.5.1 Ridge regression

When there are many possible predictors, it may be that leaving some of them out of the equation can improve the prediction, since the variance in their estimation overwhelms whatever predictive power they have. Or it may be that the prediction can be improved by shrinking the estimates somewhat. A systematic approach to such shrinking is to add a **regularization** term to the objective function. The **ridge regression** term is a penalty based on the squared length of the parameter vector:

$$obj_\kappa(\mathbf{b} ; \mathbf{y}) = \|\mathbf{y} - \mathbf{xb}\|^2 + \kappa \|\mathbf{b}\|^2. \tag{12.43}$$

The $\kappa \geq 0$ is a tuning parameter, indicating how much weight to give to the penalty. As long as $\kappa > 0$, the minimizing $\mathbf{b}$ in (12.43) would be tend to be closer to zero than the least squares estimate. The larger κ, the more the estimate would be shrunk.

There are two questions: How to find the optimal $\mathbf{b}$ given the κ, and how to choose the κ. For given κ, we can use a trick to find the estimator. For $\mathbf{b}$ being $p \times 1$, write

$$obj_\kappa(\mathbf{b} ; \mathbf{y}) = \|\mathbf{y} - \mathbf{xb}\|^2 + \|\mathbf{0}_p - (\sqrt{\kappa}\mathbf{I}_p)\mathbf{b}\|^2$$

$$= \left\| \begin{pmatrix} \mathbf{y} \\ \mathbf{0}_p \end{pmatrix} - \begin{pmatrix} \mathbf{x} \\ \sqrt{\kappa}\mathbf{I}_p \end{pmatrix} \mathbf{b} \right\|^2. \tag{12.44}$$

This objective function looks like the least squares criterion, where we have added p observations, all with y-value of zero, and the ith one has x-vector with all zeros except $\sqrt{\kappa}$ for the ith predictor. Thus the minimizer is the **ridge estimator**, which can be shown (Exercise 12.7.13) to be

$$\widehat{\beta}_\kappa = (\mathbf{x}'\mathbf{x} + \kappa\mathbf{I}_p)^{-1}\mathbf{x}'\mathbf{Y}. \tag{12.45}$$

Notice that this estimator appeared as a posterior mean in (12.33).

This estimator was originally proposed by Hoerl and Kennard (1970) as a method to ameliorate the effects of multicollinearity in the x's. Recall the covariance matrix of the least squares estimator is $\sigma^2(\mathbf{x}'\mathbf{x})^{-1}$. If the x's are highly correlated among themselves, then some of the diagonals of $(\mathbf{x}'\mathbf{x})^{-1}$ are likely to be very large, hence adding a small positive number to the diagonals of $\mathbf{x}'\mathbf{x}$ can drastically reduce the variances, without increasing bias too much. See Exercise 12.7.17.

One way to choose the κ is to try to estimate the effectiveness of the predictor for various values of κ. Imagine n new observations that have the same predictor values as the data, but whose $\mathbf{Y}^{New}$ is unobserved and independent of the data $\mathbf{Y}$. That is, we assume

$$\mathbf{Y} = \mathbf{x}\boldsymbol{\beta} + \mathbf{E} \quad \text{and} \quad \mathbf{Y}^{New} = \mathbf{x}\boldsymbol{\beta} + \mathbf{E}^{New}, \tag{12.46}$$

where $\mathbf{E}$ and $\mathbf{E}^{New}$ are independent,

$$E[\mathbf{E}] = E[\mathbf{E}^{New}] = \mathbf{0}_p, \quad \text{and} \quad Cov[\mathbf{E}] = Cov[\mathbf{E}^{New}] = \sigma^2\mathbf{I}_p. \tag{12.47}$$

It is perfectly reasonable to take the predictors of the new variables to be different than for the observed data, but the formulas are a bit simpler with the same $\mathbf{x}$.

Our goal is to estimate how well the prediction based on $\mathbf{Y}$ predicts the $\mathbf{Y}^{New}$. We would like to look at the prediction error, but since we do not observe the new data, we will assess the expected value of the sum of squares of prediction errors:

$$ESS_{pred,\kappa} = E[\|\mathbf{Y}^{New} - \mathbf{x}\widehat{\beta}_\kappa\|^2]. \tag{12.48}$$

We do observe the data, so a first guess at estimating the prediction error is the observed error,

$$SS_{e,\kappa} = \|\mathbf{Y} - \mathbf{x}\widehat{\beta}_\kappa\|^2. \tag{12.49}$$

The observed error should be an underestimate of the prediction error, since we chose the estimate of β specifically to fit the observed $\mathbf{Y}$. How much of an underestimate? The following lemma helps to find $ESS_{pred,\kappa}$ and $ESS_{e,\kappa} = E[SS_{e,\kappa}]$. Its proof is in Exercise 12.7.14.

Lemma 12.2. *If* $\mathbf{W}$ *has finite mean and variance,*

$$E[\|\mathbf{W}\|^2] = \|E[\mathbf{W}]\|^2 + \text{trace}(Cov[\mathbf{W}]). \tag{12.50}$$

We apply the lemma with $\mathbf{W} = \mathbf{Y}^{New} - \mathbf{x}\widehat{\beta}_\kappa$, and with $\mathbf{W} = \mathbf{Y} - \mathbf{x}\widehat{\beta}_\kappa$. Since $E[\mathbf{Y}] = E[\mathbf{Y}^{New}]$, the expected value parts of $ESS_{pred,\kappa}$ and $ESS_{err,\kappa}$ are equal, so we do not have to do anything further on them. The covariances are different. Write

$$\mathbf{x}\widehat{\beta}_\kappa = \mathbf{P}_\kappa \mathbf{Y}, \quad \text{where} \quad \mathbf{P}_\kappa = \mathbf{x}(\mathbf{x}'\mathbf{x} + \kappa \mathbf{I}_p)^{-1}\mathbf{x}'. \tag{12.51}$$

Then for the prediction error, since $\mathbf{Y}^{New}$ and $\mathbf{Y}$ are independent,

$$Cov[\mathbf{Y}^{New} - \mathbf{x}\widehat{\beta}_\kappa] = Cov[\mathbf{Y}^{New} - \mathbf{P}_\kappa \mathbf{Y}] = Cov[\mathbf{Y}^{New}] + \mathbf{P}_\kappa Cov[\mathbf{Y}]\mathbf{P}_\kappa$$
$$= \sigma^2(\mathbf{I}_p + \mathbf{P}_\kappa^2). \tag{12.52}$$

For the observed error,

$$Cov[\mathbf{Y} - \mathbf{P}_\kappa \mathbf{Y}] = Cov[(\mathbf{I}_p - \mathbf{P}_\kappa)\mathbf{Y}] = (\mathbf{I}_p - \mathbf{P}_\kappa)Cov[\mathbf{Y}](\mathbf{I}_p - \mathbf{P}_\kappa)$$
$$= \sigma^2(\mathbf{I}_p - \mathbf{P}_\kappa)^2$$
$$= \sigma^2(\mathbf{I}_p + \mathbf{P}_\kappa^2 - 2\mathbf{P}_\kappa). \tag{12.53}$$

Thus for the covariance parts, the observed error has that extra $-2\mathbf{P}_\kappa$ term, so that

$$ESS_{pred,\kappa} - ESS_{e,\kappa} = 2\sigma^2 \, \text{trace}(\mathbf{P}_\kappa). \tag{12.54}$$

For given κ, $\text{trace}(\mathbf{P}_\kappa)$ can be calculated. We can use the usual unbiased estimator for σ^2 in (12.23) to obtain an unbiased estimator of the prediction error:

$$\widehat{ESS}_{pred,\kappa} = SS_{e,\kappa} + 2\widehat{\sigma}^2 \, \text{trace}(\mathbf{P}_\kappa). \tag{12.55}$$

Exercise 12.7.16 presents an efficient formula for this estimate. It is then reasonable to use the estimates based on the κ that minimizes the estimated prediction error in (12.55).

12.5.2 Hurricanes

Jung, Shavitt, Viswanathan, and Hilbe (2014) collected data on the most dangerous hurricanes in the US since 1950. The data here are primarily taken from that article, but the maximum wind speed was added, and the cost of damage was updated to 2014 equivalencies (in millions of dollars). Also, we added two outliers, Katrina and Audrey, which had been left out. We are interested in predicting the number of deaths caused by the hurricane based on five variables: minimum air pressure, category, damage, wind speed, and gender of the hurricane's name. We took logs of the dependent variable (actually, log(deaths+1)) and the damage variable.

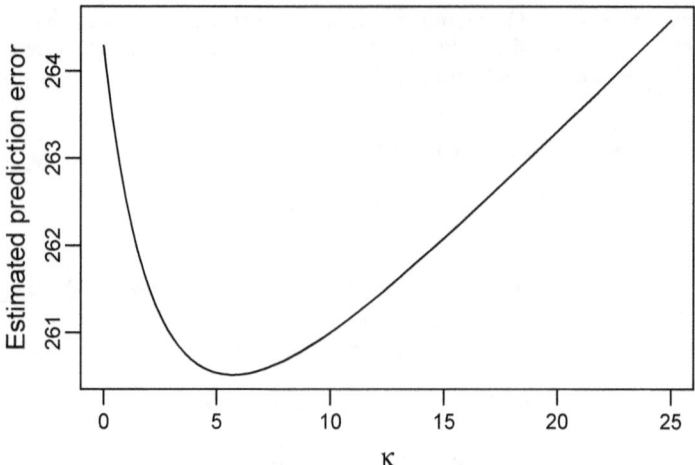

Figure 12.1: Estimated prediction error as a function of κ for ridge regression in the hurricane data.

In ridge regression, the κ is added to the diagonals of the $\mathbf{x}'\mathbf{x}$ matrix, which means that the effect of the ridge is stronger on predictors that have smaller sums of squares. In particular, the units in which the variables are measured has an effect on the results. To deal with this issue, we normalize all the predictors so that they have mean zero and variance 1. We also subtract the mean from the $\mathbf{Y}$ variable, so that we do not have to worry about an intercept.

Figure 12.1 graphs the estimated prediction error versus κ. Such graphs typically have a fairly sharp negative slope for small values of κ, then level off and begin to increase in κ. We searched over κ's at intervals of 0.1. The best we found was $\kappa = 5.9$. The estimated prediction error for that κ is 112.19. The least squares estimate ($\kappa = 0$) has an estimate of 113.83, so the best ridge estimate is a bit better than the least squares.

| | Slope | | SE | | t | |
	LS	Ridge	LS	Ridge	LS	Ridge
Pressure	−0.662	−0.523	0.255	0.176	−2.597	−2.970
Category	−0.498	−0.199	0.466	0.174	−1.070	−1.143
Damage	0.868	0.806	0.166	0.140	5.214	5.771
Wind	0.084	−0.046	0.420	0.172	0.199	−0.268
Gender	0.035	0.034	0.113	0.106	0.313	0.324

$$(12.56)$$

Table (12.56) contains the estimated β using least squares and ridge with the optimal κ. The first four predictors are related to the severity of the storm, so are highly intercorrelated. Gender is basically orthogonal to the others. Ridge regression tends to affect intercorrelated variable most, which we see. The category and wind estimates are cut in half. Pressure and damage are reduced, but not as much. Gender is hardly shrunk at all. The standard errors tend to be similarly reduced, leading to t statistics that have increased a bit. (See Exercise 12.7.15 for the standard errors.)

12.5.3   Subset selection: Mallows' C_p

In ridge regression, it is certainly possible to use different κ's for different variables, so that the regularization term in the objective function in (12.43) would be $\sum \kappa_i \beta_i^2$. An even more drastic proposal would be to have all such κ_i either 0 or ∞, that is, each parameter would either be left alone, or shrunk to 0. Which is a convoluted way to say we wish to use a subset of the predictors in the model. The main challenge is that if there are p predictors, then there are 2^p possible subsets. Fortunately, there are efficient algorithms to search through subsets, such as the leaps algorithm in R. See Lumley (2009).

Denote the matrix of a given subset of p^* of the predictors by $\mathbf{x}^*$, so that the model for this subset is

$$\mathbf{Y} = \mathbf{x}^* \boldsymbol{\beta}^* + \mathbf{E}, \quad E[\mathbf{E}] = \mathbf{0}, \quad Cov[\mathbf{E}] = \sigma^2 \mathbf{I}_n \tag{12.57}$$

where $\boldsymbol{\beta}^*$ is then $p^* \times 1$. We can find the usual least squares estimate of $\boldsymbol{\beta}^*$ as in (12.12), but with $\mathbf{x}^*$ in place of $\mathbf{x}$. To decide which subset to choose, or at least which are reasonable subsets to consider, we can again estimate the prediction sum of squares as in (12.55) for ridge regression. Calculations similar to those in (12.52) to (12.55) show that

$$\widehat{ESS}_{pred}^* = SS_e^* + 2\hat{\sigma}^2 \operatorname{trace}(\mathbf{P}_{\mathbf{x}^*}), \tag{12.58}$$

where

$$SS_e^* = \|\mathbf{Y} - \mathbf{x}^* \widehat{\boldsymbol{\beta}}^*\|^2 \quad \text{and} \quad \mathbf{P}_{\mathbf{x}^*} = \mathbf{x}^* (\mathbf{x}^{*\prime} \mathbf{x}^*)^{-1} \mathbf{x}^{*\prime}. \tag{12.59}$$

The $\hat{\sigma}^2$ is the estimate in (12.23) based on all the predictors. Exercise 12.7.4 shows that $\operatorname{trace}(\mathbf{P}_{\mathbf{x}^*}) = p^*$, the number of predictors. The resulting estimate of the prediction error is

$$\widehat{ESS}_{pred}^* = SS_e^* + 2p^* \hat{\sigma}^2, \tag{12.60}$$

which is equivalent to **Mallows' C_p** (Mallows, 1973), given by

$$C_p(\mathbf{x}^*) = \frac{\widehat{ESS}_{pred}^*}{\hat{\sigma}^2} - n = \frac{SS_e^*}{\hat{\sigma}^2} - n + 2p^*. \tag{12.61}$$

Back to the hurricane example, (12.62) has the estimated prediction errors for the ten best subsets. Each row denotes a subset, where the column under the variable's name indicates whether the variable is in that subset, 1=yes, 0=no.

Pressure	Category	Damage	Wind	Gender	SS_e^*	p^*	$\widehat{ESS}_{pred}^*$	
1	1	1	0	0	102.37	3	109.34	
1	0	1	1	0	103.63	3	110.59	
1	0	1	0	0	106.24	2	110.88	
1	1	1	0	1	102.26	4	111.55	
1	1	1	1	0	102.33	4	111.62	(12.62)
0	0	1	0	0	110.17	1	112.49	
1	0	1	1	1	103.54	4	112.83	
1	0	1	0	1	106.18	3	113.15	
1	1	1	1	1	102.21	5	113.83	
0	0	1	1	0	110.10	2	114.74	

We can see that the damage variable is in all the top 10 models, and pressure is in most of them. The other variables are each in 4 or 5 of them. The best model has pressure, category, and damage. The estimated prediction error for that model is 109.34, which is somewhat better than the best for ridge regression, 112.19. (It may not be a totally fair comparison, since the best ridge regression is found by a one-dimensional search over κ, while the subset regression is a discrete search.) See (12.69) for the estimated slopes in this model.

12.5.4   Lasso

Lasso is a technique similar to ridge regression, but features both shrinkage and subset selection all at once. It uses the sum of absolute values of the slopes as the regularization term. The objective function is

$$obj_\lambda(\mathbf{b}\,;\mathbf{y}) = \|\mathbf{y} - \mathbf{x}\mathbf{b}\|^2 + \lambda \sum |b_i| \tag{12.63}$$

for some $\lambda \geq 0$. There is no closed-form solution to the minimizer of that objective function (unless $\lambda = 0$), but convex programming techniques can be used. Efron, Hastie, Johnstone, and Tibshirani (2004) presents an efficient method to find the minimizers for all values of λ, implemented in the R package lars (Hastie and Efron, 2013). Hastie, Tibshirani, and Friedman (2009) contains an excellent treatment of lasso and other regularization procedures.

The solution in the simple $p = 1$ case gives some insight into what lasso is doing. See Exercise 12.7.18. The model is $Y_i = \beta x_i + E_i$. As in (12.17), but with just one predictor, we can write the objective function as

$$obj_\lambda(\mathbf{b}\,;\mathbf{y}) = SS_e + (b - \widehat{\beta})^2 \sum x_i^2 + \lambda|b|, \tag{12.64}$$

where $\widehat{\beta}$ is the least squares estimate. Thus the minimizing b also minimizes

$$h(b) = (b - \widehat{\beta})^2 + \lambda^*|b|, \quad \text{where } \lambda^* = \lambda / \sum x_i^2. \tag{12.65}$$

The function $h(b)$ is strictly convex, and goes to infinity if $|b|$ does, hence there is a unique minimum. If there is a b for which $h'(b) = 0$, then by convexity that b must be the minimizer. On the other hand, if there is no solution to $h'(b) = 0$, then since the minimizer cannot be at a point with a nonzero derivative, it must be where the derivative doesn't exist, which is at $b = 0$.

Now $h'(b) = 0$ implies that

$$b = \widehat{\beta} - \frac{\lambda^*}{2}\,\text{Sign}(b). \tag{12.66}$$

($\text{Sign}(b) = -1, 0$, or 1 as $b < 0, b = 0$, or $b > 0$.) Exercise 12.7.18 shows that there is such a solution if and only if $|\widehat{\beta}| \geq \lambda^*/2$, in which case the solution has the same sign as $\widehat{\beta}$. Hence

$$b = \begin{cases} 0 & \text{if } |\widehat{\beta}| < \lambda^*/2 \\ \widehat{\beta} - \frac{\lambda^*}{2}\,\text{Sign}(\widehat{\beta}) & \text{if } |\widehat{\beta}| \geq \lambda^*/2 \end{cases}. \tag{12.67}$$

Thus the lasso estimator starts with $\widehat{\beta}$, then shrinks it towards 0 by the amount $\lambda^*/2$, stopping at 0 if necessary. For $p > 1$, lasso generally shrinks all the least squares slopes, some of them (possibly) all the way to 0, but not in an obvious way.

As for ridge, we would like to use the best λ. Although there is not a simple analytic form for estimating the prediction error, Efron et al. (2004) suggests that the estimate (12.60) used in subset regression is a reasonable approximation:

$$\widehat{ESS}_{\lambda,pred} = SS_{e,\lambda} + 2p_\lambda \widehat{\sigma}^2, \qquad (12.68)$$

where p_λ is the number of non-zero slopes in the solution. For the hurricane data, the best $p_\lambda = 3$, with a $SS_{e\lambda} = 103.88$, which leads to $\widehat{ESS}_{\lambda,pred} = 110.85$. (The corresponding $\lambda = 0.3105$.)

Table (12.69) exhibits the estimated coefficients. Notice that lasso leaves out the two variables that the best subset regression does, and shrinks the remaining three. The damage coefficient is not shrunk very much, the category coefficient is cut by 2/3, similar to ridge, and pressure is shrunk by 1/3, versus about 1/5 for ridge. So indeed, lasso here combines ridge and subset regression. If asked, I'd pick either the lasso or subset regression as the best of these.

	Least squares	Ridge	Subset	Lasso	
Pressure	−0.6619	−0.5229	−0.6575	−0.4269	
Category	−0.4983	−0.1989	−0.4143	−0.1651	
Damage	0.8680	0.8060	0.8731	0.8481	
Wind	0.0838	−0.0460	0	0	(12.69)
Gender	0.0353	0.0342	0	0	
SS_e	102.21	103.25	102.37	103.88	
$\widehat{ESS}_{pred}$	113.83	112.19	109.34	110.85	

We note that there does not appear to be a standard approach to finding standard errors in lasso regression. Bootstrapping is a possibility, and Kyung, Gill, Ghosh, and Casella (2010) has a solution for Bayesian lasso, which closely approximates frequentist lasso.

12.6   Least absolute deviations

The least squares objective function minimizes the sum of squares of the residuals. As mentioned before, it is sensitive to values far from the center. M-estimators (Huber and Ronchetti (2011)) were developed as more robust alternatives, but ones that still provide reasonably efficient estimators. An M-estimator chooses m to minimize $\sum \rho(x_i, m)$ for some function ρ measuring the distance between x_i's and m. Special cases of M-estimators include those using $\rho(x_i, m) = \log(f(x_i - m))$ for some pdf f, which leads to the maximum likelihood estimates (see Section 13.6) for the location family with density f, and those based on L_q objective functions. The latter choose **b** to minimize $\sum |y_i - x_i \mathbf{b}|^q$, where x_i is the ith row of **x**. The least squares criterion is L_2, and the least absolute deviations criterion is L_1:

$$obj_1(\mathbf{b}\,;\mathbf{y}) = \sum |y_i - x_i \mathbf{b}|. \qquad (12.70)$$

We saw above that for a sample $y_1, \ldots, y_n$, the sample mean is the m that minimizes the sum of squares $\sum (y_i - m)^2$. Similarly, the sample median is the m that minimizes $\sum |y_i - m|$. (See Exercise 2.7.25 for the population version of this result.) Thus minimizing (12.70) is called **median regression**. There is no closed-form solution to finding the optimal **b**, but standard linear programming algorithms work

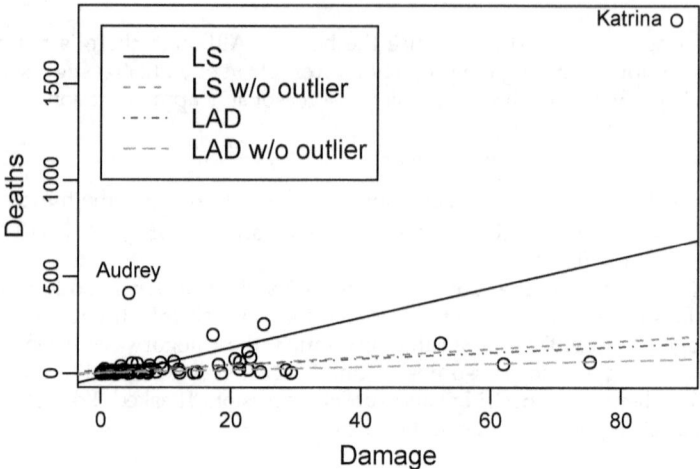

Figure 12.2: Estimated regression lines for deaths versus damage in the hurricane data. The lines were calculated using least squares and least absolute deviations, with and without the outlier, Katrina.

efficiently. We will use the R package quantreg by Koenker, Portnoy, Ng, Zeileis, Grosjean, and Ripley (2015).

To illustrate, we turn to the hurricane data, but take **Y** to be the number of deaths (not the log thereof), and the single **x** to be the damage in billions of dollars. Figure 12.2 plots the data. Notice there is one large outlier in the upper right, Katrina. The regular least squares line is the steepest, which is affected by the outlier. Redoing the least squares fit without the outlier changes the slope substantially, going from about 7.7 to 1.6. The least absolute deviations fit with all the data is very close to the least squares fit without the outlier. Removing the outlier changes this slope as well, but not by as much, 2.1 to 0.8. Thus least absolute deviations is much less sensitive to outliers than least squares.

The standard errors of the estimators of β are not obvious. Bassett and Koenker (1978) finds the asymptotic distribution under reasonable conditions. We won't prove, or use, the result, but present it because it has an interesting connection to the asymptotic distribution of the sample median found in (9.31). We assume we have a sequence of independent vectors $(Y_i, \mathbf{x}_i)$, where the $\mathbf{x}_i$'s are fixed, that follows the model (12.8), $Y_i = \mathbf{x}_i \boldsymbol{\beta} + E_i$. The E_i are assumed to be iid with continuous distribution F that has median 0 and a continuous and positive pdf $f(y)$ at $y = 0$. From (9.31), we have that

$$\sqrt{n}\,\text{Median}(E_1, \ldots, E_n) \longrightarrow^{\mathcal{D}} N\left(0, \frac{1}{4f(0)^2}\right). \tag{12.71}$$

We also need the sequence of $\mathbf{x}_i$'s to behave. Specifically, let $\mathbf{x}^{(n)}$ be the matrix with rows $\mathbf{x}_1, \ldots, \mathbf{x}_n$, and assume that

$$\frac{1}{n}\mathbf{x}^{(n)\prime}\mathbf{x}^{(n)} \longrightarrow \mathbf{D}, \tag{12.72}$$

where $\mathbf{D}$ is an invertible $p \times p$ matrix. Then if $\widehat{\boldsymbol{\beta}}_n$ is the unique minimizer of the objective function in (12.70) based on the first n observations,

$$\sqrt{n}(\widehat{\boldsymbol{\beta}}_n - \boldsymbol{\beta}) \longrightarrow^{\mathcal{D}} N\left(\mathbf{0}, \frac{1}{4f(0)^2}\mathbf{D}^{-1}\right). \tag{12.73}$$

Thus we can estimate the standard errors of the $\widehat{\beta}_i$'s as we did for least squares in (12.24), but using $1/(2f(0))$ in place of $\widehat{\sigma}$. But if we are not willing to assume we know the density f, then estimating this value can be difficult. There are other approaches, some conveniently available in the quantreg package. We'll use the bootstrap.

There are two popular methods for bootstrapping in regression. One considers the data to be iid $(p+1)$-vectors $(Y_i, \mathbf{X}_i)$, $i = 1, \ldots, n$, which implies that the Y_i's and $\mathbf{X}_i$'s have a joint distribution. A bootstrap sample involves choosing n of the vectors $(y_i, \mathbf{x}_i)$ with replacement, and finding the estimated coefficients for the bootstrap sample. This process is repeated a number of times, and the standard deviations of the resulting sets of coefficients become the bootstrap estimates of their standard errors. In this case, the estimated standard errors are estimating the *unconditional* standard errors, rather than the standard errors conditioning on $\mathbf{X} = \mathbf{x}$.

The other method is to fit the model to the data, then break each observation into the fit $(\widehat{y}_i = \mathbf{x}_i\widehat{\boldsymbol{\beta}})$ and residual $(\widehat{e}_i = y_i - \widehat{y}_i)$. A bootstrap sample starts by first choosing n from the estimated residuals without replacement. Call these values $e_1^*, \ldots, e_n^*$. Then the bootstrapped values of the dependent variable are $y_i^* = \widehat{y}_i + e_i^*$, $i = 1, \ldots, n$. That is, each bootstrapped observation has its own fit, but adds a randomly chosen residual. Then many such bootstrap sample are taken, and the standard errors are estimated as before. This process more closely mimics the conditional model we started with, but the estimated residuals that are bootstrapped are not quite iid as usually assumed in bootstrapping.

The table (12.74) uses the first bootstrap option to estimate the standard errors in the least absolute deviations regressions. Notice that as for the coefficients' estimates, the standard errors and t-statistics are much more affected by the outlier in least squares than in least absolute deviations.

	Estimate	Std. Error	t value	
Least squares	7.743	1.054	7.347	
Least squares w/o outlier	1.592	0.438	3.637	(12.74)
Least absolute deviations	2.093	1.185	1.767	
Least absolute deviations w/o outlier	0.803	0.852	0.943	

The other outlier, Audrey, does not have much of an effect on the estimates. Also, a better analysis uses logs of the variables, as above in (12.69). In that case, the outliers do not show up. Finally, we note that, not surprisingly, regularization is useful in least absolute deviation regression, though the theory is not as well developed as for least squares. Lasso is an option in quantreg.

There are many other methods for robustly estimating regression coefficients. Venables and Ripley (2002), pages 156–163, gives a practical introduction to some of them.

12.7   Exercises

Exercise 12.7.1. Show that for a sample $z_1, \ldots, z_n$, the quantity $\sum_{i=1}^{n}(z_i - m)^2$ is minimized over m by $m = \bar{z}$.

Exercise 12.7.2. Consider the simple linear regression model as in (12.3) through (12.5), where $\mathbf{Y} = \mathbf{x}\boldsymbol{\beta} + \mathbf{E}$ and $\mathbf{E} \sim N(\mathbf{0}, \sigma^2 \mathbf{I}_n)$ as in (12.10). Assume $n \geq 3$. (a) Show that

$$\mathbf{x}'\mathbf{x} = \begin{pmatrix} n & \sum x_i \\ \sum x_i & \sum x_i^2 \end{pmatrix}. \tag{12.75}$$

(b) Show that $|\mathbf{x}'\mathbf{x}| = n\sum(x_i - \bar{x})^2$. (c) Show that $\mathbf{x}'\mathbf{x}$ is invertible if and only if the x_i's are not all equal, and if it is invertible, that

$$(\mathbf{x}'\mathbf{x})^{-1} = \begin{pmatrix} \frac{1}{n} + \frac{\bar{x}^2}{\sum(x_i - \bar{x})^2} & -\frac{\bar{x}}{\sum(x_i - \bar{x})^2} \\ -\frac{\bar{x}}{\sum(x_i - \bar{x})^2} & \frac{1}{\sum(x_i - \bar{x})^2} \end{pmatrix}. \tag{12.76}$$

(d) Consider the mean of Y for given fixed value x_0 of the dependent variable. That is, let $\theta = \beta_0 + \beta_1 x_0$. The estimate is $\hat{\theta} = \hat{\beta}_0 + \hat{\beta}_1 x_0$, where $\hat{\beta}_0$ and $\hat{\beta}_1$ are the least squares estimates. Find the 2×1 vector $\mathbf{c}$ such that

$$\hat{\theta} = \mathbf{c}' \begin{pmatrix} \hat{\beta}_0 \\ \hat{\beta}_1 \end{pmatrix}. \tag{12.77}$$

(e) Show that

$$Var[\hat{\theta}] = \sigma^2 \left(\frac{1}{n} + \frac{(x_0 - \bar{x})^2}{\sum(x_i - \bar{x})^2} \right). \tag{12.78}$$

(f) A 95% confidence interval is $\hat{\theta} \pm t\, se(\hat{\theta})$, where the standard error uses the unbiased estimate of σ^2. What is the constant t?

Exercise 12.7.3. Suppose $(X_1, Y_1), \ldots, (X_n, Y_n)$ are iid pairs, with

$$\begin{pmatrix} X_i \\ Y_i \end{pmatrix} \sim N \left(\begin{pmatrix} \mu_X \\ \mu_Y \end{pmatrix}, \begin{pmatrix} \sigma_X^2 & \sigma_{XY} \\ \sigma_{XY} & \sigma_Y^2 \end{pmatrix} \right), \tag{12.79}$$

where $\sigma_X^2 > 0$ and $\sigma_Y^2 > 0$. Then the Y_i's conditional on the X_i's have a simple linear regression model:

$$\mathbf{Y} \mid \mathbf{X} = \mathbf{x} \sim N(\beta_0 \mathbf{1}_n + \beta_1 \mathbf{x}, \sigma^2 \mathbf{I}_n), \tag{12.80}$$

where $\mathbf{X} = (X_1, \ldots, X_n)'$ and $\mathbf{Y} = (Y_1, \ldots, Y_n)'$. Let $\rho = \sigma_{XY}/(\sigma_X \sigma_Y)$ be the population correlation coefficient. The Pearson sample correlation coefficient is defined by

$$r = \frac{\sum_{i=1}^n (x_i - \bar{x})(y_i - \bar{y})}{\sqrt{\sum_{i=1}^n (x_i - \bar{x})^2 \sum_{i=1}^n (y_i - \bar{y})^2}} = \frac{s_{XY}}{s_X s_Y}, \tag{12.81}$$

where s_{XY} is the sample covariance, $\sum(x_i - \bar{x})(y_i - \bar{y})/n$, and s_X^2 and s_Y^2 are the sample variances of the x_i's and y_i's, respectively. (a) Show that

$$\beta_1 = \rho \frac{\sigma_Y}{\sigma_X}, \quad \hat{\beta}_1 = r \frac{s_Y}{s_X}, \quad \text{and} \quad SS_e = n s_Y^2 (1 - r^2), \tag{12.82}$$

for SS_e as in (12.23). [Hint: For the SS_e result, show that $SS_e = \sum((y_i - \bar{y}) - \hat{\beta}_1(x_i - \bar{x}))^2$, then expand and simplify.] (b) Consider the Student's t-statistic for $\hat{\beta}_1$ in (12.26). Show that when $\rho = 0$, conditional on $\mathbf{X} = \mathbf{x}$, we have

$$T = \frac{\hat{\beta}_1 - \beta_1}{se(\hat{\beta}_i)} = \sqrt{n - 2} \frac{r}{\sqrt{1 - r^2}} \sim t_{n-2}. \tag{12.83}$$

(c) Argue that the distribution of T in part (b) is unconditionally t_{n-2} when $\rho = 0$, so that we can easily perform a test that the correlation is 0 based directly on r.

Exercise 12.7.4. Assume that $x'x$ is invertible, where x is $n \times p$. Take $P_x = x(x'x)^{-1}x'$, and $Q_x = I_n - P_x$ as in (12.14). (a) Show that P_x is symmetric and idempotent. (b) Show that Q_x is also symmetric and idempotent. (c) Show that $P_x x = x$ and $Q_x x = 0$. (d) Show that $\text{trace}(P_x) = p$ and $\text{trace}(Q_x) = n - p$.

Exercise 12.7.5. Verify that (12.13) through (12.16) lead to (12.17).

Exercise 12.7.6. Suppose $x'x$ is invertible and $E \sim N(0, \sigma^2 I_n)$. (a) Show that the fit $\widehat{Y} = P_x Y \sim N(x\beta, \sigma^2 P_x)$. (See (12.14).) (b) Show that $\widehat{Y}$ and the residuals $\widehat{E} = Q_x Y$ are independent. [Hint: What is $Q_x P_x$?] (c) Show that $\widehat{\beta}$ and $\widehat{E}$ are independent. [Hint: Show that $\widehat{\beta}$ is a function of just $\widehat{Y}$.] (d) We are assuming the E_i's are independent. Are the $\widehat{E}_i$'s independent?

Exercise 12.7.7. Assume that $E \sim N(0, \sigma^2 I_n)$ and that $x'x$ is invertible. Show that (12.23) through (12.25) imply that $(\widehat{\beta}_i - \beta_i)/se(\widehat{\beta}_i)$ is distributed t_{n-p}. [Hint: See (7.78) through (7.80).]

Exercise 12.7.8. A linear estimator of β is one of the form $\widehat{\beta}^* = LY$, where L is a $p \times n$ known matrix. Assume that $x'x$ is invertible. Then the least squares estimator $\widehat{\beta}$ is linear, with $L_0 = (x'x)^{-1}x'$. (a) Show that $\widehat{\beta}^*$ is unbiased if and only if $Lx = I_p$. (Does $L_0 x = I_p$?)

Next we wish to prove the **Gauss-Markov theorem**, which states that if $\widehat{\beta}^* = LY$ is unbiased, then

$$Cov[\widehat{\beta}^*] - Cov[\widehat{\beta}] \equiv M \text{ is nonnegative definite.} \tag{12.84}$$

For the rest of this exercise, assume that $\widehat{\beta}^*$ is unbiased. (b) Write $L = L_0 + (L - L_0)$, and show that

$$Cov[\widehat{\beta}^*] = Cov[(L_0 + (L - L_0))Y]$$
$$= Cov[L_0 Y] + Cov[(L - L_0)Y] + \sigma^2 L_0 (L - L_0)' + \sigma^2 (L - L_0) L_0'. \tag{12.85}$$

(c) Use part (a) to show that $L_0 (L - L_0)' = (L - L_0) L_0' = 0$. (d) Conclude that (12.84) holds with $M = Cov[(L - L_0)Y]$. Why is M nonnegative definite? (e) The importance of this conclusion is that the least squares estimator is **BLUE**: Best linear unbiased estimator. Show that (12.84) implies that for any $p \times 1$ vector c, $Var[c'\widehat{\beta}^*] \geq Var[c'\widehat{\beta}]$, and in particular, $Var[\widehat{\beta}_i^*] \geq Var[\widehat{\beta}_i]$ for any i.

Exercise 12.7.9. (This exercise is used in subsequent ones.) Given the $n \times p$ matrix x with $p \leq n$, let the spectral decomposition of $x'x$ be $\Psi \Lambda \Psi'$, so that Ψ is a $p \times p$ orthogonal matrix, and Λ is diagonal with diagonal elements $\lambda_1 \geq \lambda_2 \geq \cdots \geq \lambda_p \geq 0$. (See Theorem 7.3 on page 105.) Let r be the number of positive λ_i's, so that $\lambda_i > 0$ if $i \leq r$ and $\lambda_i = 0$ if $i > r$, and let Δ be the $r \times r$ diagonal matrix with diagonal elements $\sqrt{\lambda_i}$ for $i = 1, \ldots, r$. (a) Set $z = x\Psi$, and partition $z = (z_1, z_2)$, where z_1 is $n \times r$ and z_2 is $n \times (p - r)$. Show that

$$z'z = \begin{pmatrix} z_1'z_1 & z_1'z_2 \\ z_2'z_1 & z_2'z_2 \end{pmatrix} = \begin{pmatrix} \Delta^2 & 0 \\ 0 & 0 \end{pmatrix}, \tag{12.86}$$

hence $\mathbf{z}_2 = \mathbf{0}$. (b) Let $\mathbf{\Gamma}_1 = \mathbf{z}_1 \mathbf{\Delta}^{-1}$. Show that $\mathbf{\Gamma}_1' \mathbf{\Gamma}_1 = \mathbf{I}_r$, hence the columns of $\mathbf{\Gamma}_1$ are orthogonal. (c) Now with $\mathbf{z} = \mathbf{\Gamma}_1 (\mathbf{\Delta}, \mathbf{0})$, show that

$$\mathbf{x} = \mathbf{\Gamma}_1 \begin{pmatrix} \mathbf{\Delta} & \mathbf{0} \end{pmatrix} \mathbf{\Psi}'. \tag{12.87}$$

(d) Since the columns of the $n \times r$ matrix $\mathbf{\Gamma}_1$ are orthogonal, we can find an $n \times (n-r)$ matrix $\mathbf{\Gamma}_2$ such that $\mathbf{\Gamma} = (\mathbf{\Gamma}_1, \mathbf{\Gamma}_2)$ is an $n \times n$ orthogonal matrix. (You don't have to prove that, but you are welcome to.) Show that

$$\mathbf{x} = \mathbf{\Gamma} \begin{pmatrix} \mathbf{\Delta} & \mathbf{0} \\ \mathbf{0} & \mathbf{0} \end{pmatrix} \mathbf{\Psi}', \tag{12.88}$$

where the middle matrix is $n \times p$. (This formula is the **singular value decomposition** of $\mathbf{x}$. It says that for any $n \times p$ matrix $\mathbf{x}$, we can write (12.88), where $\mathbf{\Gamma}$ $(n \times n)$ and $\mathbf{\Psi}$ $(p \times p)$ are orthogonal, and $\mathbf{\Delta}$ $(r \times r)$ is diagonal with diagonal elements $\delta_1 \geq \delta_2 \geq \cdots \geq \delta_r > 0$. This exercise assumed $n \geq p$, but the $n < p$ case follows by transposing the formula and switching $\mathbf{\Gamma}$ and $\mathbf{\Psi}$.)

Exercise 12.7.10. Here we assume the matrix $\mathbf{x}$ has singular value decomposition (12.88). (a) Suppose $\mathbf{x}$ is $n \times n$ and invertible, so that the $\mathbf{\Delta}$ is $n \times n$. Show that

$$\mathbf{x}^{-1} = \mathbf{\Psi} \mathbf{\Delta}^{-1} \mathbf{\Gamma}'. \tag{12.89}$$

(b) Now let $\mathbf{x}$ be $n \times p$. When $\mathbf{x}$ is not invertible, we can use the Moore-Penrose inverse, which we saw in (7.56) for symmetric matrices. Here, it is defined to be the $p \times n$ matrix

$$\mathbf{x}^+ = \mathbf{\Psi} \begin{pmatrix} \mathbf{\Delta}^{-1} & \mathbf{0} \\ \mathbf{0} & \mathbf{0} \end{pmatrix} \mathbf{\Gamma}'. \tag{12.90}$$

Show that $\mathbf{x}\mathbf{x}^+\mathbf{x} = \mathbf{x}$. (c) If $\mathbf{x}'\mathbf{x}$ is invertible, then $\mathbf{\Delta}$ is $p \times p$. Show that in this case, $\mathbf{x}^+ = (\mathbf{x}'\mathbf{x})^{-1}\mathbf{x}'$. (d) Let $\mathbf{P}_\mathbf{x} = \mathbf{x}\mathbf{x}^+$ and $\mathbf{Q}_\mathbf{x} = \mathbf{I}_n - \mathbf{P}_\mathbf{x}$. Show that as in (12.15), $\mathbf{P}_\mathbf{x}\mathbf{x} = \mathbf{x}$ and $\mathbf{Q}_\mathbf{x}\mathbf{x} = \mathbf{0}$.

Exercise 12.7.11. Consider the model $\mathbf{Y} = \mathbf{x}\boldsymbol{\beta} + \mathbf{E}$ where $\mathbf{x}'\mathbf{x}$ may not be invertible. Let $\widehat{\boldsymbol{\beta}}^+ = \mathbf{x}^+\mathbf{y}$, where $\mathbf{x}^+$ is given in Exercise 12.7.10. Follow steps similar to (12.13) through (12.17) to show that $\widehat{\boldsymbol{\beta}}^+$ is a least squares estimate of $\boldsymbol{\beta}$.

Exercise 12.7.12. Let $\mathbf{Y} \,|\, \boldsymbol{\beta} = \mathbf{b}, \Omega = \omega \sim N(\mathbf{x}\mathbf{b}, (1/\omega)\mathbf{I}_n)$, and consider the prior $\boldsymbol{\beta} \,|\, \Omega = \omega \sim N(\boldsymbol{\beta}_0, \mathbf{K}_0^{-1}/\omega)$ and $\Omega \sim \text{Gamma}(\nu_0/2, \lambda_0/2)$. (a) Show that the conditional pdf can be written

$$f(\mathbf{y} \,|\, \mathbf{b}, \omega) = c\, \omega^{n/2} e^{-(\omega/2)((\widehat{\boldsymbol{\beta}} - \mathbf{b})'\mathbf{x}'\mathbf{x}(\widehat{\boldsymbol{\beta}} - \mathbf{b}) + SS_e)}. \tag{12.91}$$

[See (12.13) through (12.18).] (b) Show that the prior density is

$$\pi(\mathbf{b}, \omega) = d\, \omega^{(\nu_0 + p)/2 - 1} e^{-(\omega/2)((\mathbf{b} - \boldsymbol{\beta}_0)'\mathbf{K}_0(\mathbf{b} - \boldsymbol{\beta}_0) + \lambda_0)}. \tag{12.92}$$

(c) Multiply the conditional and prior densities. Show that the power of ω is now $\nu_0 + n + p$. (d) Show that

$$\begin{aligned}
(\widehat{\boldsymbol{\beta}} &- \mathbf{b})'\mathbf{x}'\mathbf{x}(\widehat{\boldsymbol{\beta}} - \mathbf{b}) + (\mathbf{b} - \boldsymbol{\beta}_0)'\mathbf{K}_0(\mathbf{b} - \boldsymbol{\beta}_0) \\
&= (\mathbf{b} - \widehat{\boldsymbol{\beta}}^*)'(\mathbf{K}_0 + \mathbf{x}'\mathbf{x})(\mathbf{b} - \widehat{\boldsymbol{\beta}}^*) + (\widehat{\boldsymbol{\beta}} - \boldsymbol{\beta}_0)'(\mathbf{K}_0^{-1} + (\mathbf{x}'\mathbf{x})^{-1})^{-1}(\widehat{\boldsymbol{\beta}} - \boldsymbol{\beta}_0),
\end{aligned} \tag{12.93}$$

for $\widehat{\beta}^*$ in (12.37). [Use (7.116).] (e) Show that the posterior density then has the same form as the prior density with respect to the **b** and ω, where the prior parameters are updated as in (12.37) and (12.38).

Exercise 12.7.13. Apply the formula for the least square estimate to the task of finding the **b** to minimize

$$\left\| \begin{pmatrix} \mathbf{y} \\ \mathbf{0}_p \end{pmatrix} - \begin{pmatrix} \mathbf{x} \\ \sqrt{\kappa}\mathbf{I}_p \end{pmatrix} \mathbf{b} \right\|^2. \tag{12.94}$$

Show that the minimizer is indeed the ridge estimator in (12.45), $(\mathbf{x}'\mathbf{x} + \kappa\mathbf{I}_p)^{-1}\mathbf{x}'\mathbf{y}$.

Exercise 12.7.14. Suppose **W** is $p \times 1$, and each W_i has finite mean and variance. (a) Show that $E[W_i^2] = E[W_i]^2 + Var[W_i]$. (b) Show that summing the individual terms yields $E[\|\mathbf{W}\|^2] = \|E[\mathbf{W}]\|^2 + \text{trace}(Cov[\mathbf{W}])$, which is Lemma 12.2.

Exercise 12.7.15. For $\widehat{\beta}_\kappa = (\mathbf{x}'\mathbf{x} + \kappa\mathbf{I}_p)^{-1}\mathbf{x}'\mathbf{Y}$ as in (12.45), show that

$$Cov[\widehat{\beta}_\kappa] = \sigma^2(\mathbf{x}'\mathbf{x} + \kappa\mathbf{I}_p)^{-1}\mathbf{x}'\mathbf{x}(\mathbf{x}'\mathbf{x} + \kappa\mathbf{I}_p)^{-1}. \tag{12.95}$$

(We can estimate σ^2 using the $\widehat{\sigma}^2$ as in (12.23) that we obtained from the least squares estimate.)

Exercise 12.7.16. This exercise provides a simple formula for the prediction error estimate in ridge regression for various values of κ. We will assume that $\mathbf{x}'\mathbf{x}$ is invertible. (a) Show that the invertibility implies that the singular value decomposition (12.88) can be written

$$\mathbf{x} = \Gamma \begin{pmatrix} \Delta \\ \mathbf{0} \end{pmatrix} \Psi', \tag{12.96}$$

i.e., the column of **0**'s in the middle matrix is gone. (b) Let $\widehat{\mathbf{e}}_\kappa$ be the estimated errors for ridge regression with given κ. Show that we can write

$$\widehat{\mathbf{e}}_\kappa = (\mathbf{I}_n - \mathbf{x}(\mathbf{x}'\mathbf{x} + \kappa\mathbf{I}_p)^{-1}\mathbf{x}')\mathbf{y}$$

$$= \Gamma \left(\mathbf{I}_n - \begin{pmatrix} \Delta \\ \mathbf{0} \end{pmatrix} (\Delta^2 + \kappa\mathbf{I}_p)^{-1} \begin{pmatrix} \Delta & \mathbf{0} \end{pmatrix} \right) \Gamma'\mathbf{y}. \tag{12.97}$$

(c) Let $\mathbf{g} = \Gamma'\mathbf{y}$. Show that

$$SS_{e,\kappa} = \|\widehat{\mathbf{e}}_\kappa\|^2 = \sum_{i=1}^{p} g_i^2 \frac{\kappa}{\delta_i^2 + \kappa} + \sum_{i=p+1}^{n} g_i^2. \tag{12.98}$$

Also, note that since the least squares estimate takes $\kappa = 0$, its sum of squared errors is $SS_{e,0} = \sum_{i=p+1}^{n} g_i^2$. (d) Take $\mathbf{P}_\kappa = \mathbf{x}(\mathbf{x}'\mathbf{x} + \kappa\mathbf{I}_p)^{-1}\mathbf{x}'$ as in (12.51). Show that

$$\text{trace}(\mathbf{P}_\kappa) = \sum \frac{\delta_i^2}{\delta_i^2 + \kappa}. \tag{12.99}$$

(e) Put parts (c) and (d) together to show that in (12.55), we have that

$$\widehat{ESS}_{pred,\kappa} = SS_{e,\kappa} + 2\widehat{\sigma}^2 \text{ trace}(\mathbf{P}_\kappa) = SS_{e,0} + \sum_{i=1}^{p} g_i^2 \frac{\kappa}{\delta_i^2 + \kappa} + 2\widehat{\sigma}^2 \sum \frac{\delta_i^2}{\delta_i^2 + \kappa}. \tag{12.100}$$

Hence once we find the g_i's and δ_i's, it is easy to calculate (12.100) for many values of κ.

Exercise 12.7.17. This exercise looks at the bias and variance of the ridge regression estimator. The ridge estimator for tuning parameter κ is $\widehat{\beta}_\kappa = (\mathbf{x}'\mathbf{x} + \kappa\mathbf{I}_p)^{-1}\mathbf{x}'\mathbf{Y}$ as in (12.45). Assume that $\mathbf{x}'\mathbf{x}$ is invertible, $E[\mathbf{E}] = 0$, and $Cov[\mathbf{E}] = \sigma^2\mathbf{I}_n$. The singular value decomposition of $\mathbf{x}$ in this case is given in (12.96). (a) Show that

$$E[\widehat{\beta}_\kappa] = (\mathbf{x}'\mathbf{x} + \kappa\mathbf{I}_p)^{-1}\mathbf{x}'\mathbf{x}\beta = \mathbf{\Psi}(\mathbf{\Delta}^2 + \kappa\mathbf{I}_p)^{-1}\mathbf{\Delta}^2\mathbf{\Psi}'\beta. \tag{12.101}$$

(b) Show that the bias of the ridge estimator can be written

$$\text{Bias}_\kappa = \mathbf{\Psi}\,\kappa(\mathbf{\Delta}^2 + \kappa\mathbf{I}_p)^{-1}\gamma, \quad \text{where } \gamma = \mathbf{\Psi}'\beta. \tag{12.102}$$

Also, show that the squared norm of the bias is

$$\|\text{Bias}_\kappa\|^2 = \sum\gamma_i^2\frac{\kappa}{(\delta_i^2 + \kappa)^2}. \tag{12.103}$$

(c) Use (12.95) to show that the covariance matrix of $\widehat{\beta}_\kappa$ can be written

$$Cov[\widehat{\beta}_\kappa] = \sigma^2\mathbf{\Psi}\mathbf{\Delta}^2(\mathbf{\Delta}^2 + \kappa\mathbf{I}_p)^{-2}\mathbf{\Psi}'. \tag{12.104}$$

Also, show that

$$\text{trace}(Cov[\widehat{\beta}_\kappa]) = \sigma^2\sum\frac{\delta_i^2}{(\delta_i^2 + \kappa)^2}. \tag{12.105}$$

(d) The total expected mean square error of the estimator $\widehat{\beta}_\kappa$ is defined to be $\text{MSE}_\kappa = E[\|\widehat{\beta}_\kappa - \beta\|^2]$. Use Lemma 12.2 with $\mathbf{W} = \widehat{\beta}_\kappa - \beta$ to show that

$$\text{MSE}_\kappa = \|\text{Bias}_\kappa\|^2 + \text{trace}(Cov[\widehat{\beta}_\kappa]). \tag{12.106}$$

(e) Show that when $\kappa = 0$, we have the least squares estimator, and that $\text{MSE}_0 = \sigma^2\sum\delta_i^{-2}$. Thus if any of the δ_i's are near 0, the MSE can be very large. (f) Show that the $\|\text{Bias}_\kappa\|^2$ is increasing in κ, and the $\text{trace}(Cov[\widehat{\beta}_\kappa])$ is decreasing in κ, for $\kappa \geq 0$. Also, show that

$$\frac{\partial}{\partial\kappa}\text{MSE}_\kappa\Big|_{\kappa=0} = -2\sigma^2\sum\delta_i^{-2}. \tag{12.107}$$

Argue that for small enough κ, the ridge estimator has a lower MSE than least squares. Note that the smaller the δ_i's, the more advantage the ridge estimator has. (This result is due to Hoerl and Kennard (1970).)

Exercise 12.7.18. Consider the regression model with just one x and no intercept, so that $y_i = \beta x_i + e_i$, $i = 1,\ldots,n$. (a) Show that $\|\mathbf{y} - \mathbf{x}b\|^2 = SS_e + (b - \widehat{\beta})^2\sum x_i^2$, where $\widehat{\beta}$ is the least squares estimator $\sum x_iy_i/\sum x_i^2$. (b) Show that the b that minimizes the lasso objective function in (12.64) also minimizes $h(b) = (b - \widehat{\beta})^2 + \lambda^*|b|$ as in (12.65), where $\lambda^* \geq 0$. (c) Show that for $b \neq 0$ the derivative of h exists and

$$h'(b) = 2(b - \widehat{\beta}) + \lambda^*\,\text{Sign}(b). \tag{12.108}$$

(d) Show that if $h'(b) = 0$, then $b = \widehat{\beta} - (\lambda^*/2)\,\text{Sign}(b)$. Also, show that such a b exists if and only if $|\widehat{\beta}| \geq \lambda^*/2$, in which case b and $\widehat{\beta}$ have the same sign. [Hint: Look at the signs of the two sides of the equation depending on whether $\widehat{\beta}$ is bigger or smaller than $\lambda^*/2$.]

Chapter 13

Likelihood, Sufficiency, and MLEs

13.1 Likelihood function

If we know θ, then the density tells us what $\mathbf{X}$ is likely to be. In statistics, we do not know θ, but we do observe $\mathbf{X} = \mathbf{x}$, and wish to know what values θ is likely to be. The analog to the density for the statistical problem is the **likelihood function**, which is the same as the density, but considered as a function of θ for fixed $\mathbf{x}$. The function is not itself a density (usually), because there is no particular reason to believe that the integral over θ for fixed $\mathbf{x}$ is 1. Rather, the likelihood function gives the relative likelihood of various values of θ. It is fundamental to Bayesian inference, and extremely useful in frequentist inference. It encapsulates the relationship between θ and the data.

Definition 13.1. *Suppose $\mathbf{X}$ has density $f(\mathbf{x} \mid \theta)$ for $\theta \in \mathcal{T}$. Then a **likelihood function** for observation $\mathbf{X} = \mathbf{x}$ is*

$$L(\theta\,; \mathbf{x}) = c_{\mathbf{x}} f(\mathbf{x} \mid \theta), \quad \theta \in \mathcal{T}, \tag{13.1}$$

where $c_{\mathbf{x}}$ is any positive constant.

Likelihoods are to be interpreted in only relative fashion, that is, to say the likelihood of a particular θ_1 is $L(\theta_1; \mathbf{y})$ does not mean anything by itself. Rather, meaning is attributed to saying that the relative likelihood of θ_1 to θ_2 (in light of the data $\mathbf{y}$) is $L(\theta_1; \mathbf{y})/L(\theta_2; \mathbf{y})$. There is a great deal of controversy over *what* exactly the relative likelihood is. We do not have to worry about that particularly, since we are just using likelihood as a means to an end. The general idea, though, is that the data supports θ's with relatively high likelihood.

For example, suppose $X \sim \text{Binomial}(n, \theta)$, $\theta \in (0, 1)$, $n = 5$. The pmf is

$$f(x \mid \theta) = \binom{n}{x} \theta^x (1 - \theta)^{n-x}, \quad x = 0, \dots, n. \tag{13.2}$$

The likelihood is

$$L(\theta\,; x) = c_x\, \theta^x (1 - \theta)^{n-x}, \quad 0 < \theta < 1. \tag{13.3}$$

See Figure 13.1 for graphs of the two functions.

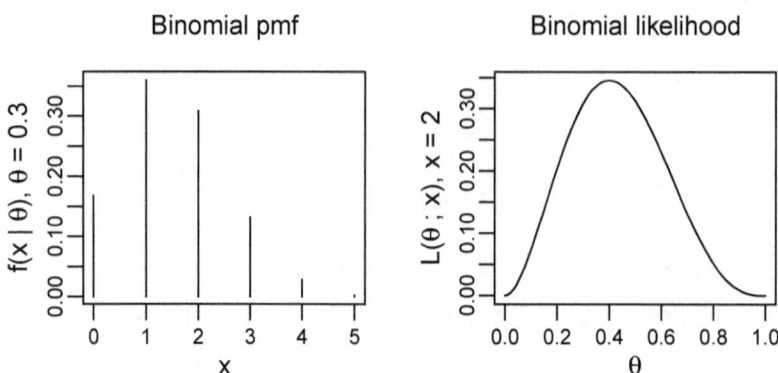

Figure 13.1: Binomial pmf and likelihood, where $X \sim \text{Binomial}(5, \theta)$.

Here, c_x is a constant that may depend on x but *not* on θ. Likelihood is not probability, in particular because θ is not necessarily random. Even if θ is random, the likelihood is not the pdf, but rather the part of the pdf that depends on $\mathbf{x}$. That is, suppose π is the prior pdf of θ. Then the posterior pdf can be written

$$f(\theta \mid \mathbf{x}) = \frac{f(\mathbf{x} \mid \theta)\pi(\theta)}{\int_{\mathcal{T}} f(\mathbf{x} \mid \theta^*)\pi(\theta^*)d\theta^*} = L(\theta; \mathbf{x})\pi(\theta). \tag{13.4}$$

(Note: These densities should have subscripts, e.g., $f(\mathbf{x} \mid \theta)$ should be $f_{\mathbf{X} \mid \theta}(\mathbf{x} \mid \theta)$, but I hope leaving them off is not too confusing.) Here, the c_x is the inverse of the integral in the denominator of (13.4) (the marginal density, which does not depend on θ as it is integrated away). Thus though the likelihood is not a density, it does tell us how to update the prior to obtain the posterior.

13.2 Likelihood principle

As we will see in this and later chapters, likelihood functions are very useful in inference, whether from a Bayesian or frequentist point of view. Going beyond the utility of likelihood, the **likelihood principle** is a fairly strong rule that purports to judge inferences. It basically says that if two experiments yield the same likelihood, then they should yield the same inference. We first define what it means for two outcomes to have the same likelihood, then briefly illustrate the principle. Berger and Wolpert (1988) goes into much more depth.

Definition 13.2. *Suppose we have two models, $\mathbf{X}$ with density $f(\mathbf{x} \mid \theta)$ and $\mathbf{Y}$ with density $g(\mathbf{y} \mid \theta)$, that depend on the same parameter θ with space $\mathcal{T}$. Then $\mathbf{x}$ and $\mathbf{y}$ have the **same likelihood** if for some positive constants $c_{\mathbf{x}}$ and $c_{\mathbf{y}}$ in (13.1),*

$$L(\theta; \mathbf{x}) = L^*(\theta; \mathbf{y}) \text{ for all } \theta \in \mathcal{T}, \tag{13.5}$$

where L and L^* are their respective likelihoods.

The two models in Definition 13.2 could very well be the same, in which case **x** and **y** are two possible elements of $\mathcal{X}$. As an example, suppose the model has $\mathbf{X} = (X_1, X_2)$, where the elements are iid $N(\mu, 1)$, and $\mu \in \mathcal{T} = \mathbb{R}$. Then the pdf is

$$f(\mathbf{x} \mid \mu) = \frac{1}{2\pi} e^{-\frac{1}{2}\Sigma(x_i - \mu)^2} = \frac{1}{2\pi} e^{-\frac{1}{2}(x_1^2 + x_2^2) + (x_1 + x_2)\mu - \mu^2}. \tag{13.6}$$

Consider two possible observed vectors, $\mathbf{x} = (1, 2)$ and $\mathbf{y} = (-3, 6)$. These observations have quite different pdf values. Their ratio is

$$\frac{f(1, 2 \mid \mu)}{f(-3, 6 \mid \mu)} = 485165195. \tag{13.7}$$

This ratio is interesting in two ways: It shows the probability of being near **x** is almost half a billion times larger than the probability of being near **y**, and it shows the ratio does not depend on μ. That is, their likelihoods are the same:

$$L(\mu; \mathbf{x}) = c_{(1,3)} \, e^{3\mu - \mu^2} \quad \text{and} \quad L(\mu; \mathbf{y}) = c_{(-3,6)} \, e^{3\mu - \mu^2}. \tag{13.8}$$

It does not matter what the constants are. We could just take $c_{(1,3)} = c_{(-3,6)} = 1$, but the important aspect is that the sums of the two observations are both 3, and the sum is the only part of the data that hits μ.

13.2.1 Binomial and negative binomial

The models can be different, but they do need to share the same parameter. For example, suppose we have a coin with probability of heads being $\theta \in (0, 1)$, and we intend to flip it a number of times independently. Here are two possible experiments:

- **Binomial.** Flip the coin $n = 10$ times, and count X, the number of heads, so that $X \sim \text{Binomial}(10, \theta)$.

- **Negative binomial.** Flip the coin until there are 4 heads, and count Y, the number of tails obtained. This Y is Negative Binomial$(4, \theta)$.

The space of the Negative Binomial(r, θ) is $\mathcal{Y} = \{0, 1, 2, \dots\}$, and the pmf is given in Table 1.2 on page 9:

$$g(y \mid \theta) = \binom{K - 1 + y}{K - 1} \theta^K (1 - \theta)^y. \tag{13.9}$$

Next, suppose we perform the binomial experiment and obtain $X = 4$ heads out of 10 flips. The likelihood is the usual binomial one:

$$L(\theta; 4) = \binom{10}{4} \theta^4 (1 - \theta)^6. \tag{13.10}$$

Also, suppose we perform the negative binomial experiment and happen to see $Y = 6$ tails before the $K = 4^{th}$ head. The likelihood here is

$$L^*(\theta; 6) = \binom{9}{3} \theta^4 (1 - \theta)^6. \tag{13.11}$$

The likelihoods are the same. I left the constants there to illustrate that the pmfs are definitely different, but erasing the constants leaves the same $\theta^4(1-\theta)^6$. These two likelihoods are based on different random variables. The binomial has a fixed number of flips but could have any number of heads (between 0 and 10), while the negative binomial has a fixed number of heads but could have any number of flips (over 4). In particular, either experiment with the given outcome would yield the same posterior for θ.

The likelihood principle says that if two outcomes (whether they are from the same experiment or not) have the same likelihood, then any inference made about θ based on the outcomes must be the same. Any inference that is not the same under the two scenarios are said to "violate the likelihood principle." Bayesian inference does not violate the likelihood principle, nor does maximum likelihood estimation, as long as your inference is just "Here is the estimate ..."

Unbiased estimation **does** violate the likelihood principle. Keeping with the above example, we know that X/n is an unbiased estimator of θ for the binomial. For $Y \sim \text{Negative Binomial}(K, \theta)$, the unbiased estimator is found by ignoring the last flip, because that we know is always heads, so would bias the estimate if used. That is,

$$E[\widehat{\theta}_U^*] = \theta, \quad \widehat{\theta}_U^* = \frac{K-1}{Y+K-1}. \tag{13.12}$$

See Exercise 4.4.15.

Now we test out the inference: "The unbiased estimate of θ is ...":

- Binomial$(10, \theta)$, with outcome $x = 4$. "The unbiased estimate of θ is $\widehat{\theta}_U = 4/10 = 2/5$."

- Negative Binomial$(4, \theta)$, with outcome $y = 6$. "The unbiased estimate of θ is $\widehat{\theta}_U^* = 3/9 = 1/3$."

Those two situations have the same likelihood, but different estimates, thus violating the likelihood principle!

The problem with unbiasedness is that it depends on the entire density, i.e., on outcomes not observed, so different densities would give different expected values. For that reason, any inference that involves the operating characteristics of the procedure violates the likelihood principle.

Whether one decides to fully accept the likelihood principle or not, it provides an important guide for any kind of inference, as we shall see.

13.3   Sufficiency

Consider again (13.6), or more generally, $X_1, \ldots, X_n$ iid $N(\mu, 1)$. The likelihood is

$$L(\mu ; x_1, \ldots, x_n) = e^{\mu \sum x_i - \frac{n}{2}\mu^2}, \tag{13.13}$$

where the $\exp(-\sum x_i^2/2)$ part can be dropped as it does not depend on μ. Note that this function depends on the x_i's only through their sum, that is, as in (13.8), if $\mathbf{x}$ and $\mathbf{x}^*$ have the same sum, they have the same likelihood. Thus the likelihood principle says that all we need to know is the sum of the x_i's to make an inference about μ. This sum is a **sufficient statistic**.

Definition 13.3. *Consider the model with space $\mathcal{X}$ and parameter space $\mathcal{T}$. A function*

$$s : \mathcal{X} \longrightarrow \mathcal{S} \qquad (13.14)$$

*is a **sufficient statistic** if for some function b,*

$$b : \mathcal{S} \times \mathcal{T} \longrightarrow [0, \infty), \qquad (13.15)$$

the constant in the likelihood can be chosen so that

$$L(\boldsymbol{\theta}; \mathbf{x}) = b(s(\mathbf{x}), \boldsymbol{\theta}). \qquad (13.16)$$

Thus $\mathbf{S} = s(\mathbf{X})$ is a sufficient statistic (it may be a vector) if by knowing $\mathbf{S}$, you know the likelihood, i.e., it is sufficient for performing any inference. That is handy, because you can reduce your data set, which may be large, to possibly just a few statistics without losing any information. More importantly, it turns out that the best inferences depend on just the sufficient statistics.

We next look at some examples. First, we note that for any model, the data $\mathbf{x}$ is itself sufficient, because the likelihood depends on $\mathbf{x}$ through $\mathbf{x}$.

13.3.1 IID

If $X_1, \ldots, X_n$ are iid with density $f(x_i \mid \boldsymbol{\theta})$, then no matter what the model, the order statistics (see Section 5.5) are sufficient. To see this fact, write

$$L(\boldsymbol{\theta}; \mathbf{x}) = f(x_1 \mid \boldsymbol{\theta}) \cdots f(x_n \mid \boldsymbol{\theta}) = f(x_{(1)} \mid \boldsymbol{\theta}) \cdots f(x_{(n)} \mid \boldsymbol{\theta}) = b((x_{(1)}, \ldots, x_{(n)}), \boldsymbol{\theta}),$$
$$(13.17)$$

because the order statistics are just the x_i's in a particular order.

13.3.2 Normal distribution

If $X_1, \ldots, X_n$ are iid $N(\mu, 1)$, $\mu \in \mathcal{T} = \mathbb{R}$, then we have several candidates for sufficient statistic:

$$
\begin{aligned}
\text{The data itself}: \quad & s_1(\mathbf{x}) = \mathbf{x}; \\
\text{The order statistics}: \quad & s_2(\mathbf{x}) = (x_{(1)}, \ldots, x_{(n)}); \\
\text{The sum}: \quad & s_3(\mathbf{x}) = \sum x_i; \\
\text{The mean}: \quad & s_4(\mathbf{x}) = \bar{x}; \\
\text{Partial sums}: \quad & s_5(\mathbf{x}) = (x_1 + x_2, x_3 + x_4 + x_5, x_6) \quad (\text{if } n = 6). \qquad (13.18)
\end{aligned}
$$

An important fact is that any one-to-one function of a sufficient statistic is also sufficient, because knowing one means you know the other, hence you know the likelihood. For example, the mean and sum are one-to-one. Also, note that the dimension of the sufficient statistics in (13.18) are different ($n, n, 1, 1,$ and 3, respectively). Generally, one prefers the most compact one, in this case either the mean or sum. Each of those are functions of the others. In fact, they are **minimal sufficient**.

Definition 13.4. *A statistic $s(\mathbf{x})$ is **minimal sufficient** if it is sufficient, and given any other sufficient statistic $t(\mathbf{x})$, there is a function h such that $s(\mathbf{x}) = h(t(\mathbf{x}))$.*

This concept is important, but we will later focus on the more restrictive notion of completeness.

Which statistics are sufficient depend crucially on the parameter space. In the above, we assumed the variance known. But suppose $X_1, \ldots, X_N$ are iid $N(\mu, \sigma^2)$, with $(\mu, \sigma^2) \in \mathcal{T} \equiv \mathbb{R} \times (0, \infty)$. Then the likelihood is

$$L(\mu, \sigma^2 ; \mathbf{x}) = \frac{1}{\sigma^n} e^{-\frac{1}{2\sigma^2} \Sigma(x_i - \mu)^2} = \frac{1}{\sigma^n} e^{-\frac{1}{2\sigma^2} \Sigma x_i^2 + \frac{1}{\sigma^2} \mu \Sigma x_i - \frac{n}{2\sigma^2} \mu^2}. \tag{13.19}$$

Now we cannot eliminate the $\sum x_i^2$ part, because it involves σ^2. Here the sufficient statistic is two-dimensional:

$$s(\mathbf{x}) = (s_1(\mathbf{x}), s_2(\mathbf{x})) = \left(\sum x_i, \sum x_i^2 \right), \tag{13.20}$$

and the b function is

$$b(s_1, s_2) = \frac{1}{\sigma^n} e^{-\frac{1}{2\sigma^2} s_2 + \frac{1}{\sigma^2} \mu s_1 - \frac{n}{2\sigma^2} \mu^2}. \tag{13.21}$$

13.3.3 Uniform distribution

Suppose $X_1, \ldots, X_n$ are iid Uniform$(0, \theta)$, $\theta \in (0, \infty)$. The likelihood is

$$L(\theta ; \mathbf{x}) = \prod \frac{1}{\theta} I[0 < x_i < \theta] = \begin{cases} \frac{1}{\theta^n} & \text{if } 0 < x_i < \theta \text{ for all } x_i \\ 0 & \text{if } \quad \text{not} \end{cases}. \tag{13.22}$$

All the x_i's are less than θ if and only if the largest one is, so that we can write

$$L(\theta ; \mathbf{x}) = \begin{cases} \frac{1}{\theta^n} & \text{if } 0 < x_{(n)} < \theta \\ 0 & \text{if } \quad \text{not} \end{cases}. \tag{13.23}$$

Thus the likelihood depends on $\mathbf{x}$ only through the maximum, hence $x_{(n)}$ is sufficient.

13.3.4 Laplace distribution

Now suppose $X_1, \ldots, X_n$ are iid Laplace(θ), $\theta \in \mathbb{R}$, which has pdf $\exp(-|x_i - \theta|)/2$, so that the likelihood is

$$L(\theta ; \mathbf{x}) = e^{-\Sigma |x_i - \theta|}. \tag{13.24}$$

Because the data are iid, the order statistics are sufficient, but unfortunately, there is not another sufficient statistic with smaller dimension. The absolute value bars cannot be removed. Similar models based on the Cauchy, logistic, and others have the same problem.

13.3.5 Exponential families

In some cases, the sufficient statistic does reduce the dimensionality of the data significantly, such as in the iid normal case, where no matter how large n is, the sufficient statistic is two-dimensional. In other cases, such as the Laplace above, there is no dimensionality reduction, so one must still carry around n values. Exponential families are special families in which there is substantial reduction for iid variables or vectors. Because the likelihood of an iid sample is the product of individual likelihoods, statistics "add up" when they are in an exponent.

The vector $\mathbf{X}$ has an **exponential family** distribution if its density (pdf or pmf) depends on a $p \times 1$ vector $\boldsymbol{\theta}$ and can be written

$$f(\mathbf{x} \mid \boldsymbol{\theta}) = a(\mathbf{x})e^{\theta_1 t_1(\mathbf{x}) + \cdots + \theta_p t_p(\mathbf{x}) - \psi(\boldsymbol{\theta})} \qquad (13.25)$$

for some functions $t_1(\mathbf{x}), \ldots, t_p(\mathbf{x})$, $a(\mathbf{x})$, and $\psi(\boldsymbol{\theta})$. The $\boldsymbol{\theta}$ is called the **natural parameter** and the $t(\mathbf{x}) = (t_1(\mathbf{x}), \ldots, t_p(\mathbf{x}))'$ is the vector of **natural sufficient statistics**. Now if $\mathbf{X}_1, \ldots, \mathbf{X}_n$ are iid vectors with density (13.25), then the joint density is

$$f(\mathbf{x}_1, \ldots, \mathbf{x}_n \mid \boldsymbol{\theta}) = \prod f(\mathbf{x}_i \mid \boldsymbol{\theta}) = [\prod a(\mathbf{x}_i)]e^{\theta_1 \Sigma_i t_1(\mathbf{x}_i) + \cdots + \theta_p \Sigma_i t_p(\mathbf{x}_i) - n\psi(\boldsymbol{\theta})}, \qquad (13.26)$$

hence has sufficient statistic

$$s(\mathbf{x}_1, \ldots, \mathbf{x}_n) = \begin{pmatrix} \Sigma_i t_1(\mathbf{x}_i) \\ \vdots \\ \Sigma_i t_p(\mathbf{x}_i) \end{pmatrix} = \sum_i t(\mathbf{x}_i), \qquad (13.27)$$

which has dimension p no matter how large n.

The natural parameters and sufficient statistics are not necessarily the most "natural" to us. For example, in the normal case of (13.19), the natural parameters can be taken to be

$$\theta_1 = \frac{\mu}{\sigma^2} \quad \text{and} \quad \theta_2 = -\frac{1}{2\sigma^2}. \qquad (13.28)$$

The corresponding statistics are

$$t(x_i) = \begin{pmatrix} t_1(x_i) \\ t_2(x_i) \end{pmatrix} = \begin{pmatrix} x_i \\ x_i^2 \end{pmatrix}. \qquad (13.29)$$

There are other choices, e.g., we could switch the negative sign from θ_2 to t_2.

Other exponential families include the Poisson, binomial, gamma, beta, multivariate normal, and multinomial.

13.4 Conditioning on a sufficient statistic

The intuitive meaning of a sufficient statistic is that once you know the statistic, nothing else about the data helps in inference about the parameter. For example, in the iid case, once you know the values of the x_i's, it does not matter in what order they are listed. This notion can be formalized by finding the conditional distribution of the data given the sufficient statistic, and showing that it does not depend on the parameter.

First, a lemma that makes it easy to find the conditional density, if there is one.

Lemma 13.5. *Suppose $\mathbf{X}$ has space $\mathcal{X}$ and density $f_{\mathbf{X}}$, and $\mathbf{S} = s(\mathbf{X})$ is a function of $\mathbf{X}$ with space $\mathcal{S}$ and density $f_{\mathbf{S}}$. Then the conditional density of $\mathbf{X}$ given $\mathbf{S}$, if it exists, is*

$$f_{\mathbf{X} \mid \mathbf{S}}(\mathbf{x} \mid \mathbf{s}) = \frac{f_{\mathbf{X}}(\mathbf{x})}{f_{\mathbf{S}}(\mathbf{s})} \quad \text{for } \mathbf{x} \in \mathcal{X}_{\mathbf{s}} \equiv \{\mathbf{x} \in \mathcal{X} \mid s(\mathbf{x}) = \mathbf{s}\}. \qquad (13.30)$$

The caveat "if it exists" in the lemma is unnecessary in the discrete case, because the conditional pmf always will exist. But in continuous or mixed cases, the resulting conditional distribution may not have a density with respect to Lebesgue measure. It will have a density with respect to some measure, though, which one would see in a measure theory course.

Proof. (*Discrete case*) Suppose $\mathbf{X}$ is discrete. Then by Bayes theorem,

$$f_{\mathbf{X}|S}(\mathbf{x}\,|\,s) = \frac{f_{S|\mathbf{X}}(s\,|\,\mathbf{x})\, f_{\mathbf{X}}(\mathbf{x})}{f_S(s)}, \quad \mathbf{x} \in \mathcal{X}_s. \tag{13.31}$$

Because S is a function of $\mathbf{X}$,

$$f_{S|\mathbf{X}}(s\,|\,\mathbf{x}) = P[s(\mathbf{X}) = s\,|\,\mathbf{X} = \mathbf{x}] = \left\{ \begin{array}{ll} 1 & \text{if}\quad s(\mathbf{x}) = s \\ 0 & \text{if}\quad s(\mathbf{x}) \neq s \end{array} \right\} = \left\{ \begin{array}{ll} 1 & \text{if}\quad \mathbf{x} \in \mathcal{X}_s \\ 0 & \text{if}\quad \mathbf{x} \notin \mathcal{X}_s \end{array} \right. .$$
$$\tag{13.32}$$

Thus $f_{S|\mathbf{X}}(s\,|\,\mathbf{x}) = 1$ in (13.31), because $\mathbf{x} \in \mathcal{X}_s$, so we can erase it, yielding (13.30). □

Here is the main result.

Lemma 13.6. *Suppose $s(\mathbf{x})$ is a sufficient statistic for a model with data $\mathbf{X}$ and parameter space $\mathcal{T}$. Then the conditional distribution $\mathbf{X}\,|\,s(\mathbf{X}) = s$ does not depend on θ.*

Before giving a proof, consider the example with $X_1, \ldots, X_n$ iid Poisson(θ), $\theta \in (0, \infty)$. The likelihood is

$$L(\theta\,;\mathbf{x}) = \prod e^{-\theta}\theta^{x_i} = e^{-n\theta}\theta^{\sum x_i}. \tag{13.33}$$

We see that $s(\mathbf{x}) = \sum x_i$ is sufficient. We know that $S = \sum X_i$ is Poisson($n\theta$), hence

$$f_S(s) = e^{-n\theta}\frac{(n\theta)^s}{s!}. \tag{13.34}$$

Then by Lemma 13.5, the conditional pmf of $\mathbf{X}$ given S is

$$\begin{aligned} f_{\mathbf{X}|S}(\mathbf{x}\,|\,s) &= \frac{f_{\mathbf{X}}(\mathbf{x})}{f_S(s)} \\ &= \frac{e^{-n\theta}\theta^{\sum x_i}/\prod x_i!}{e^{-n\theta}(n\theta)^s/s!} \\ &= \frac{s!}{\prod x_i!}\frac{1}{n^s}, \quad \mathbf{x} \in \mathcal{X}_s = \{\mathbf{x}\,|\,\textstyle\sum x_i = s\}. \end{aligned} \tag{13.35}$$

That is a multinomial distribution:

$$\mathbf{X}\,|\,\textstyle\sum X_i = s \sim \text{Multinomial}_n(s, (\tfrac{1}{n}, \ldots, \tfrac{1}{n})). \tag{13.36}$$

But the main point of Lemma 13.6 is that this distribution is independent of θ. Thus knowing the sum means the exact values of the x_i's do not reveal anything extra about θ.

The key to the result lies in (13.33) and (13.34), where we see that the likelihoods of $\mathbf{X}$ and $s(\mathbf{X})$ are the same, if $\sum x_i = s$. This fact is a general one.

Lemma 13.7. *Suppose $s(\mathbf{X})$ is a sufficient statistic for the model with data $\mathbf{X}$, and consider the model for S where $S =^{\mathcal{D}} s(\mathbf{X})$. Then the likelihoods for $\mathbf{X} = \mathbf{x}$ and $S = s$ are the same if $s = s(\mathbf{x})$.*

This result should make sense, because it is saying that the sufficient statistic contains the same information about θ that the full data do. One consequence is that instead of having to work with the full model, one can work with the sufficient statistic's model. For example, instead of working with $X_1, \ldots, X_n$ iid $N(\mu, \sigma)$, you can work with just the two independent random variables $\overline{X} \sim N(\mu, \sigma^2/n)$ and $S^2 \sim \sigma^2 \chi^2_{n-1}/n$ without losing any information.

We prove the lemma in the discrete case.

Proof. Let $f_X(x \mid \theta)$ be the pmf of X, and $f_S(s \mid \theta)$ be that of S. Because $s(X)$ is sufficient, the likelihood for X can be written

$$L(\theta; x) = c_x \, f_X(x \mid \theta) = b(s(x), \theta) \tag{13.37}$$

by Definition 13.3, for some c_x and $b(s, \theta)$. The pmf of S is

$$\begin{aligned} f_S(s \mid \theta) = P_\theta[s(X) = s] &= \sum_{x \in \mathcal{X}_s} f_X(x \mid \theta) \quad \text{(where } \mathcal{X}_s = \{x \in \mathcal{X} \mid s(x) = s\}) \\ &= \sum_{x \in \mathcal{X}_s} b(s(x), \theta)/c_x \\ &= b(s, \theta) \sum_{x \in \mathcal{X}_s} 1/c_x \quad \text{(because in the summation, } s(x) = s) \\ &= b(s, \theta) d_s, \end{aligned} \tag{13.38}$$

where d_s is that sum of $1/c_x$'s, which does not depend on θ. Thus the likelihood of S can be written

$$L^*(\theta; s) = b(s, \theta), \tag{13.39}$$

the same as L in (13.37). □

A formal proof for the continuous case proceeds by introducing appropriate extra variables Y so that X and (S, Y) are one-to-one, then using Jacobians, then integrating out the Y. We will not do that, but special cases can be done easily, e.g., if $X_1, \ldots, X_n$ are iid $N(\mu, 1)$, one can directly show that X and $s(X) = \sum X_i \sim N(n\mu, n)$ have the same likelihood.

Now for the proof of Lemma 13.6, again in the discrete case. Basically, we just repeat the calculations for the Poisson.

Proof. As in the proof of Lemma 13.7, the pmfs of X and S can be written, respectively,

$$f_X(x \mid \theta) = L(\theta; x)/c_x \quad \text{and} \quad f_S(s \mid \theta) = L^*(\theta; s)d_s, \tag{13.40}$$

where the likelihoods are equal in the sense that

$$L(\theta; x) = L^*(\theta; s) \quad \text{if } x \in \mathcal{X}_s. \tag{13.41}$$

Then by Lemma 13.5,

$$\begin{aligned} f_{X \mid S}(x \mid s, \theta) &= \frac{f_X(x \mid \theta)}{f_S(s \mid \theta)} \quad \text{for } x \in \mathcal{X}_s \\ &= \frac{L(\theta; x)/c_x}{L^*(\theta; s)d_s} \quad \text{for } x \in \mathcal{X}_s \\ &= \frac{1}{c_x d_s} \quad \text{for } x \in \mathcal{X}_s \end{aligned} \tag{13.42}$$

by (13.41), which does not depend on θ. □

Many statisticians would switch Definition 13.3 and Lemma 13.6 for sufficiency. That is, a statistic $s(\mathbf{X})$ is defined to be sufficient if the conditional distribution $\mathbf{X} \mid s(\mathbf{X}) = s$ does not depend on the parameter. Then one can show that the likelihood depends on $\mathbf{x}$ only through $s(\mathbf{x})$, that result being **Fisher's factorization theorem**.

We end this section with a couple of additional examples that derive the conditional distribution of $\mathbf{X}$ given $s(\mathbf{X})$.

13.4.1 IID

Suppose $X_1, \ldots, X_n$ are iid continuous random variables with pdf $f(x_i)$. No matter what the model, we know that the order statistics are sufficient, so we will suppress the θ. We can proceed as in the proof of Lemma 13.6. Letting $s(\mathbf{x}) = (x_{(1)}, \ldots, x_{(n)})$, we know that the joint pdfs of $\mathbf{X}$ and $\mathbf{S} \equiv s(\mathbf{X})$ are, respectively,

$$f_{\mathbf{X}}(\mathbf{x}) = \prod f(x_i) \quad \text{and} \quad f_S(\mathbf{s}) = n! \prod f(x_{(i)}). \tag{13.43}$$

The products of the pdfs are the same, just written in different orders. Thus

$$P[\mathbf{X} = \mathbf{x} \mid s(\mathbf{X}) = \mathbf{s}] = \frac{1}{n!} \text{ for } \mathbf{x} \in \mathcal{X}_{\mathbf{s}}. \tag{13.44}$$

The $\mathbf{s}$ is a particular set of ordered values, and $\mathcal{X}_{\mathbf{s}}$ is the set of all $\mathbf{x}$ that have the same values as $\mathbf{s}$, but in any order. To illustrate, suppose that $n = 3$ and $s(\mathbf{X}) = \mathbf{s} = (1, 2, 7)$. Then $\mathbf{X}$ has a conditional chance of $1/6$ of being any $\mathbf{x}$ with order statistics $(1, 2, 7)$:

$$\mathcal{X}_{(1,2,7)} = \{(1,2,7), (1,7,2), (2,1,7), (2,7,1), (7,1,2), (7,2,1)\}. \tag{13.45}$$

The discrete case works as well, although the counting is a little more complicated when there are ties. For example, if $s(\mathbf{x}) = (1, 3, 3, 4)$, there are $4!/2! = 12$ different orderings.

13.4.2 Normal mean

Suppose $X_1, \ldots, X_n$ are iid $N(\mu, 1)$, so that $\overline{X}$ is sufficient (as is $\sum X_i$). We wish to find the conditional distribution of $\mathbf{X}$ given $\overline{X} = \overline{x}$. Because we have normality and linear functions, everything is straightforward. First we need the joint distribution of $\mathbf{W} = (\overline{X}, \mathbf{X}')'$, which is a linear transformation of $\mathbf{X} \sim N(\mu \mathbf{1}_n, \mathbf{I}_n)$, hence multivariate normal. We could figure out the matrix for the transformation, but all we really need are the mean and covariance matrix of $\mathbf{W}$. The mean and covariance of the $\mathbf{X}$ part we know, and the mean and variance of $\overline{X}$ are μ and $1/n$, respectively. All that is left is the covariance of the X_i's with $\overline{X}$, which are all the same, and can be found directly:

$$Cov[X_i, \overline{X}] = \frac{1}{n} \sum_{j=1}^{n} Cov[X_i, X_j] = \frac{1}{n}, \tag{13.46}$$

because X_i is independent of all the X_j's except for X_i itself, with which its covariance is 1. Thus

$$\mathbf{W} = \begin{pmatrix} \overline{X} \\ \mathbf{X} \end{pmatrix} \sim N \left(\mu \mathbf{1}_{n+1}, \begin{pmatrix} \frac{1}{n} & \frac{1}{n} \mathbf{1}'_n \\ \frac{1}{n} \mathbf{1}_n & \mathbf{I}_n \end{pmatrix} \right). \tag{13.47}$$

For the conditional distribution, we use Lemma 7.8. The X there is $\overline{X}$ here, and the $\mathbf{Y}$ there is the $\mathbf{X}$ here. Thus

$$\Sigma_{XX} = \frac{1}{n}, \ \Sigma_{YX} = \frac{1}{n}\mathbf{1}_n, \ \Sigma_{YY} = \mathbf{I}_n, \ \mu_X = \mu, \ \text{and} \ \mu_Y = \mu\mathbf{1}_n, \tag{13.48}$$

hence

$$\beta = \frac{1}{n}\mathbf{1}_n\frac{1}{n^{-1}} = \mathbf{1}_n \ \text{and} \ \alpha = \mu\mathbf{1}_n - \beta\mu = \mu\mathbf{1}_n - \mathbf{1}_n\mu = \mathbf{0}_n. \tag{13.49}$$

Thus the conditional mean is

$$E[\mathbf{X}\mid\overline{X}=\overline{x}] = \alpha + \beta\,\overline{x} = \overline{x}\,\mathbf{1}_n. \tag{13.50}$$

That result should not be surprising. It says that if you know the sample mean is $\overline{x}$, you expect on average the observations to be $\overline{x}$. The conditional covariance is

$$\Sigma_{YY} - \Sigma_{YX}\Sigma_{XX}^{-1}\Sigma_{XY} = \mathbf{I}_n - \frac{1}{n}\mathbf{1}_n\frac{1}{n^{-1}}\frac{1}{n}\mathbf{1}_n' = \mathbf{I}_n - \frac{1}{n}\mathbf{1}_n\mathbf{1}_n' = \mathbf{H}_n, \tag{13.51}$$

the centering matrix from (7.38). Putting it all together:

$$\mathbf{X}\mid\overline{X}=\overline{x} \sim N(\overline{x}\,\mathbf{1}_n, \mathbf{H}_n). \tag{13.52}$$

We can pause and reflect that this distribution is free of μ, as it had better be. Note also that if we subtract the conditional mean, we have

$$\mathbf{X} - \overline{X}\,\mathbf{1}_n\mid\overline{X}=\overline{x} \sim N(\mathbf{0}_n, \mathbf{H}_n). \tag{13.53}$$

That conditional distribution is free of $\overline{x}$, meaning the vector is independent of $\overline{X}$. But this is the vector of deviations, and we already knew it is independent of $\overline{X}$ from (7.43).

13.4.3 Sufficiency in Bayesian analysis

Since the Bayesian posterior distribution depends on the data only through the likelihood function for the data, and the likelihood for the sufficient statistic is the same as that for the data (Lemma 13.7), one need deal with just the sufficient statistic when finding the posterior. See Exercise 13.8.7.

13.5 Rao-Blackwell: Improving an estimator

We see that sufficient statistics are nice in that we do not lose anything by restricting to them. In fact, they are more than just convenient — if you base an estimate on more of the data than the sufficient statistic, then you do lose something. For example, suppose $X_1, \ldots, X_n$ are iid $N(\mu, 1)$, $\mu \in \mathbb{R}$, and we wish to estimate

$$g(\mu) = P_\mu[X_i \leq 10] = \Phi(10 - \mu), \tag{13.54}$$

Φ being the distribution function of the standard normal. An unbiased estimator is

$$\delta(\mathbf{x}) = \frac{\#\{x_i\mid x_i \leq 10\}}{n} = \frac{1}{n}\sum I[x_i \leq 10]. \tag{13.55}$$

Note that this estimator is *not* a function of just $\bar{x}$, the sufficient statistic. The claim is that there is another unbiased estimator that is a function of just $\bar{x}$ and is better, i.e., has lower variance.

Fortunately, there is a way to find such an estimator besides guessing. We use conditioning as in the previous section, and the results on conditional means and variances in Section 6.4.3. Start with an estimator $\delta(\mathbf{x})$ of $g(\theta)$, and a sufficient statistic $\mathbf{S} = s(\mathbf{X})$. Then consider the conditional expected value of δ:

$$\delta^*(\mathbf{s}) = E[\delta(\mathbf{X}) \mid s(\mathbf{X}) = \mathbf{s}]. \tag{13.56}$$

First, we need to make sure δ^* *is* an estimator, which means that it does not depend on the unknown θ. But by Lemma 13.6, we know that the conditional distribution of $\mathbf{X}$ given $\mathbf{S}$ does not depend on θ, and δ does not depend on θ because it is an estimator, hence δ^* is an estimator. If we condition on something not sufficient, then we may not end up with an estimator.

Is δ^* a good estimator? From (6.38), we know it has the same expected value as δ:

$$E_{\boldsymbol{\theta}}[\delta^*(\mathbf{S})] = E_{\boldsymbol{\theta}}[\delta(\mathbf{X})], \tag{13.57}$$

so that

$$\text{Bias}_{\boldsymbol{\theta}}[\delta^*] = \text{Bias}_{\boldsymbol{\theta}}[\delta]. \tag{13.58}$$

Thus we haven't done worse in terms of bias, and in particular if δ is unbiased, so is δ^*.

Turning to variance, we have the "variance-conditional variance" equation (6.43), which translated to our situation here is

$$Var_{\boldsymbol{\theta}}[\delta(\mathbf{X})] = E_{\boldsymbol{\theta}}[v(\mathbf{S})] + Var_{\boldsymbol{\theta}}[\delta^*(\mathbf{S})], \tag{13.59}$$

where

$$v(\mathbf{s}) = Var_{\boldsymbol{\theta}}[\delta(\mathbf{X}) \mid s(\mathbf{X}) = \mathbf{s}]. \tag{13.60}$$

Whatever v is, it is not negative, hence

$$Var_{\boldsymbol{\theta}}[\delta^*(\mathbf{S})] \leq Var_{\boldsymbol{\theta}}[\delta(\mathbf{X})]. \tag{13.61}$$

Thus variance-wise, the δ^* is no worse than δ. In fact, δ^* is strictly better unless $v(\mathbf{S})$ is zero with probability one. But in that case, δ and δ^* are the same, so that δ itself is a function of just $\mathbf{S}$ already.

Finally, if the bias is the same and the variance of δ^* is lower, then the mean squared error of δ^* is better. To summarize:

Theorem 13.8. Rao-Blackwell. *If δ is an estimator of $g(\theta)$, and $s(\mathbf{X})$ is sufficient, then $\delta^*(\mathbf{s})$ given in (13.56) has the same bias as δ, and smaller variance and MSE, unless δ is a function of just $s(\mathbf{X})$, in which case δ and δ^* are the same.*

13.5.1 Normal probability

Consider the estimator $\delta(\mathbf{x})$ in (13.55) for $g(\mu) = \Phi(10 - \mu)$ in (13.54) in the normal case. With $\bar{X}$ being the sufficient statistic, we can find the conditional expected value of δ. We start by finding the conditional expected value of just one of the $I[x_i \leq 10]$'s. It turns out that the conditional expectation is the same for each i, hence the conditional expected value of δ is the same as the condition expected value of just one. So we are interested in finding

$$\delta^*(\bar{x}) = E[I[X_i \le 10] \,|\, \overline{X} = \bar{x}] = P[X_i \le 10 \,|\, \overline{X} = \bar{x}]. \tag{13.62}$$

From (13.52) we have that

$$X_i \,|\, \overline{X} = \bar{x} \sim N\left(\bar{x}, 1 - \frac{1}{n}\right), \tag{13.63}$$

hence

$$
\begin{aligned}
\delta^*(\bar{x}) &= P[X_i \le 10 \,|\, \overline{X} = \bar{x}] \\
&= P[N(\bar{x}, 1 - 1/n) \le 10] \\
&= \Phi\left(\frac{10 - \bar{x}}{\sqrt{1 - 1/n}}\right).
\end{aligned} \tag{13.64}
$$

This estimator is then guaranteed to be unbiased, and have a lower variance that δ. It would have been difficult to come up with this estimator directly, or even show that it is unbiased, but the original δ is quite straightforward, as is the conditional calculation.

13.5.2 IID

In Section 13.4.1, we saw that when the observations are iid, the conditional distribution of $\mathbf{X}$ given the order statistics is uniform over all the permutations of the observations. One consequence is that any estimator must be invariant under permutations, or else it can be improved. For a simple example, consider estimating μ, the mean, with the simple estimator $\delta(\mathbf{X}) = X_1$. Then with $s(\mathbf{x})$ being the order statistics,

$$\delta^*(\mathbf{s}) = E[X_1 \,|\, s(\mathbf{X}) = \mathbf{s}] = \frac{1}{n} \sum s_i = \bar{s} \tag{13.65}$$

because X_1 is conditionally equally likely to be any of the order statistics. Of course, the mean of the order statistics is the same as the $\bar{x}$, so we have that $\overline{X}$ is a better estimate than X_1, which we knew. The procedure applied to any weighted average will also end up with the mean, e.g.,

$$
\begin{aligned}
E\left[\frac{1}{2}X_1 + \frac{1}{3}X_2 + \frac{1}{6}X_4 \,\middle|\, s(\mathbf{X}) = \mathbf{s}\right] &= \frac{1}{2}E[X_1 \,|\, s(\mathbf{X}) = \mathbf{s}] + \frac{1}{3}E[X_2 \,|\, s(\mathbf{X}) = \mathbf{s}] \\
&\quad + \frac{1}{6}E[X_4 \,|\, s(\mathbf{X}) = \mathbf{s}] \\
&= \frac{1}{2}\bar{s} + \frac{1}{3}\bar{s} + \frac{1}{6}\bar{s} \\
&= \bar{s}.
\end{aligned} \tag{13.66}
$$

Turning to the variance σ^2, because $X_1 - X_2$ has mean 0 and variance $2\sigma^2$, $\delta(\mathbf{x}) = (x_1 - x_2)^2/2$ is an unbiased estimator of σ^2. Conditioning on the order statistics, we obtain the mean of all the $(x_i - x_j)^2/2$'s with $i \ne j$:

$$\delta^*(\mathbf{s}) = E\left[\frac{(X_1 - X_2)^2}{2} \,\middle|\, s(\mathbf{X}) = \mathbf{s}\right] = \frac{1}{n(n-1)} \sum\sum_{i \ne j} \frac{(x_i - x_j)^2}{2}. \tag{13.67}$$

After some algebra, we can write

$$\delta^*(s(\mathbf{x})) = \frac{\sum(x_i - \bar{x})^2}{n-1}, \tag{13.68}$$

the usual unbiased estimate of σ^2.

The above estimators are special cases of **U-statistics**, for which there are many nice asymptotic results. A U-statistic is based on a **kernel** $h(x_1, \ldots, x_d)$, a function of a subset of the observations. The corresponding U-statistic is the symmetrized version of the kernel, i.e., the conditional expected value,

$$u(\mathbf{x}) = E[h(X_1, \ldots, X_d) \mid s(\mathbf{X}) = s(\mathbf{x})]$$

$$= \frac{1}{n(n-1)\cdots(n-d+1)} \sum \cdots \sum_{i_1, \ldots, i_d, \ distinct} h(x_{i_1}, \ldots, x_{i_d}). \tag{13.69}$$

See Serfling (1980) for more on these statistics.

13.6   Maximum likelihood estimates

If $L(\boldsymbol{\theta}; \mathbf{x})$ reveals how likely $\boldsymbol{\theta}$ is in light of the data $\mathbf{x}$, it seems reasonable that the most likely $\boldsymbol{\theta}$ would be a decent estimate of $\boldsymbol{\theta}$. In fact, it is reasonable, and the resulting estimator is quite popular.

Definition 13.9. *Given the model with likelihood $L(\boldsymbol{\theta}; \mathbf{x})$ for $\boldsymbol{\theta} \in \mathcal{T}$, the **maximum likelihood estimate (MLE)** at observation $\mathbf{x}$ is the unique value of $\boldsymbol{\theta}$ that maximizes $L(\boldsymbol{\theta}; \mathbf{x})$ over $\boldsymbol{\theta} \in \mathcal{T}$, if such unique value exists. Otherwise, the MLE does not exist at $\mathbf{x}$.*

By convention, the MLE of any function of the parameter is the function of the MLE:

$$\widehat{g(\boldsymbol{\theta})} = g(\widehat{\boldsymbol{\theta}}), \tag{13.70}$$

a plug-in estimator. See Exercises 13.8.10 through 13.8.12 for some justification of the convention.

There are times when the likelihood does not technically have a maximum, but there is an obvious limit of the θ's that approach the supremum. For example, suppose $X \sim \text{Uniform}(0, \theta)$. Then the likelihood at $x > 0$ is

$$L(\theta; x) = \frac{1}{\theta} I_{\{x < \theta\}}(\theta) = \begin{cases} \frac{1}{\theta} & \text{if } \theta > x \\ 0 & \text{if } \theta \leq x \end{cases}. \tag{13.71}$$

Figure 13.2 exhibits the likelihood function when $x = 4$. The highest point occurs at $\theta = 4$, almost. To be precise, there is no maximum, because the graph is not continuous at $\theta = 4$. But we still take the MLE to be $\widehat{\theta} = 4$ in this case, because we could switch the filled-in dot from $(4, 0)$ to $(4, 1/4)$ by taking $X \sim \text{Uniform}(0, \theta]$, so that the pdf at $x = \theta$ is $1/\theta$, not 0.

Often, the MLE is found by differentiating the likelihood, or the log of the likelihood (called the **loglikelihood**). Because the log function is strictly increasing, the same value of θ maximizes the likelihood and the loglikelihood. For example, in the binomial example, from (13.3) the loglikelihood is (dropping the c_x)

$$l(\theta; x) = \log(L(\theta; x)) = x \log \theta + (n - x) \log(1 - \theta). \tag{13.72}$$

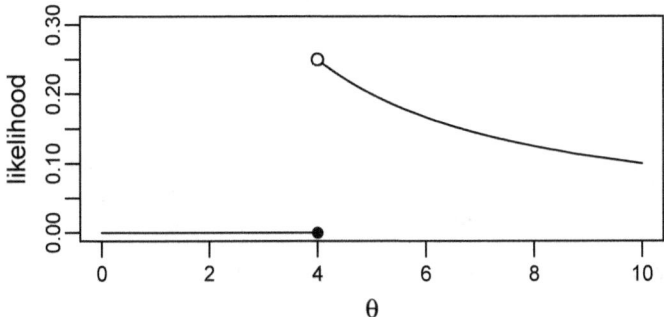

Figure 13.2: Uniform$(0,\theta)$ likelihood when $x = 4$.

Then

$$l'(\theta\,;x) = \frac{x}{\theta} - \frac{n-x}{1-\theta}. \tag{13.73}$$

Set that expression to 0 and solve for θ to obtain

$$\widehat{\theta} = \frac{x}{n}. \tag{13.74}$$

From Figure 13.1, one can see that the maximum is at $x/n = 0.4$.

If $\boldsymbol{\theta}$ is $p \times 1$, then one must do a p-dimensional maximization. For example, suppose $X_1, \ldots, X_n$ are iid $N(\mu, \sigma^2)$, $(\mu, \sigma^2) \in \mathbb{R} \times (0, \infty)$. Then the loglikelihood is (dropping the $\sqrt{2\pi}$'s),

$$\log(L(\mu, \sigma^2\,;x_1, \ldots, x_n)) = -\frac{1}{2\sigma^2} \sum(x_i - \mu)^2 - \frac{n}{2} \log(\sigma^2). \tag{13.75}$$

We could go ahead and differentiate with respect to μ and σ^2, obtaining two equations. But notice that μ appears only in the sum of squares part, and we have already, in Exercise 12.7.1, found that $\widehat{\mu} = \overline{x}$. (It is not hard to prove by differentiation.) Then for the variance, we need to maximize

$$\log(L(\overline{x}, \sigma^2\,;x_1, \ldots, x_n)) = -\frac{1}{2\sigma^2} \sum(x_i - \overline{x})^2 - \frac{n}{2} \log(\sigma^2). \tag{13.76}$$

Differentiating with respect to σ^2,

$$\frac{\partial}{\partial(\sigma^2)} \log(L(\overline{x}, \sigma^2\,;x_1, \ldots, x_n)) = \frac{1}{2\sigma^4} \sum(x_i - \overline{x})^2 - \frac{n}{2} \frac{1}{\sigma^2}, \tag{13.77}$$

and setting to 0 leads to

$$\widehat{\sigma}^2 = \frac{\sum(x_i - \overline{x})^2}{n} = s^2. \tag{13.78}$$

So the MLE of (μ, σ^2) is $(\overline{x}, s^2)$. It is then easy to find the MLEs of functions of (μ, σ^2), e.g., the MLE of the coefficient of variation σ/μ is $s/\overline{x}$ as in (13.70).

Chapter 14 delves more deeply into likelihood estimation. In particular, under conditions, we can obtain an automatic estimate of the standard error based on the Fisher information.

13.7   Functions of estimators

If $\widehat{\theta}$ is an estimator of θ, then $g(\widehat{\theta})$ is an estimator of $g(\theta)$. Do the properties of $\widehat{\theta}$ transfer to $g(\widehat{\theta})$? Maybe, maybe not. Although there are exceptions (such as when g is linear for the first two statements), generally

If $\widehat{\theta}_U$ is an unbiased estimator of θ	then $g(\widehat{\theta}_U)$	is **not**	an unbiased estimator of $g(\theta)$;
If $\widehat{\theta}_B$ is the Bayes posterior mean of θ	then $g(\widehat{\theta}_B)$	is **not**	the Bayes posterior mean of $g(\theta)$;
If $\widehat{\theta}_{mle}$ is the MLE of θ	then $g(\widehat{\theta}_{mle})$	**is**	the MLE of $g(\theta)$.

The basic reason for the first two statements is the following:

Theorem 13.10. *If Y is a random variable with finite mean, and g is a function of y, then*

$$E[g(Y)] \neq g(E[Y]) \tag{13.79}$$

unless

- *g is linear: $g(y) = a + b\,y$ for constants a and b;*

- *Y is essentially constant: $P[Y = \mu] = 1$ for some μ;*

- *You are lucky.*

A simple example is $g(x) = x^2$. If $E[X^2] = E[X]^2$, then X has variance 0. As long as X is not a constant, $E[X^2] > E[X]^2$.

13.7.1   Poisson distribution

Suppose $X \sim \text{Poisson}(\theta)$, $\theta \in (0, \infty)$. Then for estimating θ, $\widehat{\theta}_U = X$ is unbiased, and happens to be the MLE as well. For a Bayes estimate, take the prior $\Theta \sim \text{Exponential}(1)$. The likelihood is $L(\theta\,;x) = \exp(-\theta)\theta^x$, hence the posterior is

$$\pi(\theta\,|\,x) = cL(\theta\,;x)\pi(\theta) = ce^{-\theta}\theta^x e^{-\theta} = c\theta^x e^{-2\theta}, \tag{13.80}$$

which is Gamma($x + 1, 2$). Thus from Table 1.1, the posterior mean with respect to this prior is

$$\widehat{\theta}_B = E[\theta\,|\,X = x] = \frac{x+1}{2}. \tag{13.81}$$

Now consider estimating $g(\theta) = \exp(-\theta)$, which is the $P_\theta[X = 0]$. (So if θ is the average number of telephone calls coming in an hour, $\exp(-\theta)$ is the chance there are 0 calls in the next hour.) Is $g(\widehat{\theta}_U)$ unbiased? No:

$$E[g(\widehat{\theta}_U)] = E[e^{-X}] = e^{-\theta} \sum_{x=0}^{\infty} e^{-x} \frac{\theta^x}{x!}$$

$$= e^{-\theta} \sum_{x=0}^{\infty} \frac{(e^{-1}\theta)^x}{x!}$$

$$= e^{-\theta} e^{e^{-1}\theta}$$

$$= e^{\theta(e^{-1}-1)} \neq e^{-\theta}. \tag{13.82}$$

There is an unbiased estimator for g, namely $I[X = 0]$.

Turn to the posterior mean of $g(\theta)$, which is

$$\widehat{g(\theta)}_B = E[g(\theta) \mid X = x] = \int_0^{\infty} g(\theta)\pi(\theta \mid x)d\theta$$

$$= \frac{2^{x+1}}{\Gamma(x+1)} \int_0^{\infty} e^{-\theta}\theta^x\, e^{-2\theta}d\theta$$

$$= \frac{2^{x+1}}{\Gamma(x+1)} \int_0^{\infty} \theta^x\, e^{-3\theta}d\theta$$

$$= \frac{2^{x+1}}{\Gamma(x+1)} \frac{\Gamma(x+1)}{3^{x+1}}$$

$$= \left(\frac{2}{3}\right)^{x+1} \neq e^{-\widehat{\theta}_B} = e^{-\frac{x+1}{2}}. \tag{13.83}$$

See Exercise 13.8.10 for the MLE.

13.8   Exercises

Exercise 13.8.1. Imagine a particular phone in a call center, and consider the time between calls. Let X_1 be the waiting time in minutes for the first call, X_2 the waiting time for the second call after the first, etc. Assume that $X_1, X_2, \ldots, X_n$ are iid Exponential(θ). There are two devices that may be used to record the waiting times between the calls. The old mechanical one can measure each waiting time up to only an hour, while the new electronic device can measure the waiting time with no limit. Thus there are two possible experiments:

Old: Using the old device, one observes $Y_1, \ldots, Y_n$, where $Y_i = \min\{X_i, 60\}$, that is, if the true waiting time is over 60 minutes, the device records 60.

New: Using the new device, one observes the true $X_1, \ldots, X_n$.

(a) Using the old device, what is $P_\theta[Y_i = 60]$? (b) Using the old device, find $E_\theta[Y_i]$. Is $\overline{Y}_n$ an unbiased estimate of $1/\theta$? (c) Using the new device, find $E_\theta[X_i]$. Is $\overline{X}_n$ an unbiased estimate of $1/\theta$?

In what follows, suppose the actual waiting times are 10, 12, 25, 35, 38 (so $n = 5$). (d) What is the likelihood for these data when using the old device? (e) What is the likelihood for these data using the new device? Is it the same as for the old device?

(f) Let $\mu = 1/\theta$. What is the MLE of μ using the old device for these data? The new device? Are they the same? (g) Let the prior on θ be Exponential(25). What is the posterior mean of μ for these data using the old device? The new device? Are they the same? (h) Look at the answers to parts (b), (c), and (f). What is odd about the situation?

Exercise 13.8.2. In this question X and Y are independent, with

$$X \sim \text{Poisson}(2\theta), \ Y \sim \text{Poisson}(2(1-\theta)), \ \theta \in (0,1). \tag{13.84}$$

(a) Which of the following are sufficient statistics for this model? (i) (X,Y); (ii) $X+Y$; (iii) $X-Y$; (iv) $X/(X+Y)$; (v) $(X+Y,X-Y)$. (b) If $X=Y=0$, which of the following are versions of the likelihood? (i) $\exp(-2)\exp(4\theta)$; (ii) $\exp(-2)$; (iii) 1; (iv) $\exp(-2\theta)$. (c) What value(s) of θ maximize the likelihood when $X=Y=0$? (d) What is the MLE for θ when $X+Y > 0$? (e) Suppose θ has the Beta(a,b) prior. What is the posterior mean of θ? (f) Which of the following estimators of θ are unbiased? (i) $\delta_1(x,y) = x/2$; (ii) $\delta_2(x,y) = 1 - y/2$; (iii) $\delta_3(x,y) = (x-y)/4 + 1/2$. (g) Find the MSE for each of the three estimators in part (f). Also, find the maximum MSE for each. Which has the lowest maximum? Which is best for θ near 0? Which is best for θ near 1?

Exercise 13.8.3. Suppose X and Y are independent, with $X \sim \text{Binomial}(n,\theta)$ and $Y \sim \text{Binomial}(m,\theta)$, $\theta \in (0,1)$, and let $T = X+Y$. (a) Does the conditional distribution of $X \mid T = t$ depend on θ? (b) Find the conditional distribution from part (a) for $n=6, m=3, t=4$. (c) What is $E[X \mid T = t]$ for $n=6, m=3, t=4$? (d) Now suppose X and Y are independent, but with $X \sim \text{Binomial}(n,\theta_1)$ and $Y \sim \text{Binomial}(m,\theta_2)$, $\theta = (\theta_1,\theta_2) \in (0,1) \times (0,1)$. Does the conditional distribution $X \mid T = t$ depend on θ?

Exercise 13.8.4. Suppose X_1 and X_2 are independent $N(0,\sigma^2)$'s, $\sigma^2 \in (0,\infty)$, and let $R = \sqrt{X_1^2 + X_2^2}$. (a) Find the conditional space of $\mathbf{X}$ given $R = r$, $\mathcal{X}_r$. (b) Find the pdf of R. (c) Find the "density" of $\mathbf{X} \mid R = r$. (It is not the density with respect to Lebesgue measure on $\mathbb{R}^2$, but still is a density.) Does it depend on σ^2? Does it depend on r? Does it depend on (x_1,x_2) other than through r? How does it relate to the conditional space? (d) What do you think the conditional distribution of $\mathbf{X} \mid R = r$ is?

Exercise 13.8.5. For each model, indicate which statistics are sufficient. (They do not need to be minimal sufficient. There may be several correct answers for each model.) The models are each based on $X_1, \ldots, X_n$ iid with some distribution. (Assume that $n > 3$.) Here are the distributions and parameter spaces: (a) $N(\mu,1)$, $\mu \in \mathbb{R}$. (b) $N(0,\sigma^2)$, $\sigma^2 > 0$. (c) $N(\mu,\sigma^2)$, $(\mu,\sigma^2) \in \mathbb{R} \times (0,\infty)$. (d) Uniform$(\theta,1+\theta)$, $\theta \in \mathbb{R}$. (e) Uniform$(0,\theta)$, $\theta > 0$. (f) Cauchy(θ), $\theta \in \mathbb{R}$ (so the pdf is $1/(\pi(1-(x-\theta)^2))$. (g) Gamma(α,λ), $(\alpha,\lambda) \in (0,\infty) \times (0,\infty)$. (h) Beta$(\alpha,\beta)$, $(\alpha,\beta) \in (0,\infty) \times (0,\infty)$. (i) Logistic$(\theta)$, $\theta \in \mathbb{R}$ (so the pdf is $\exp(x-\theta)/(1+\exp(x-\theta))^2)$. (j) The "shifted exponential (α,λ)", which has pdf given in (13.87), where $(\alpha,\lambda) \in \mathbb{R} \times (0,\infty)$. (k) The model has the single distribution Uniform$(0,1)$.

The choices of sufficient statistics are below. For each model, decide which of the following are sufficient for that model. (1) $(X_1, \ldots, X_n)$, (2) $(X_{(1)}, \ldots, X_{(n)})$, (3) $\sum_{i=1}^n X_i$, (4) $\sum_{i=1}^n X_i^2$, (5) $(\sum_{i=1}^n X_i, \sum_{i=1}^n X_i^2)$, (6) $X_{(1)}$, (7) $X_{(n)}$, (8) $(X_{(1)}, X_{(n)})$ (9) $(X_{(1)}, \overline{X})$, (10) $(\overline{X}, S^2)$, (11) $\prod_{i=1}^n X_i$, (12) $(\prod_{i=1}^n X_i, \sum_{i=1}^n X_i)$, (13) $(\prod_{i=1}^n X_i, \prod_{i=1}^n (1-X_i))$, (14) 0.

Exercise 13.8.6. Suppose $X_1, X_2, X_3, \ldots$ are iid Laplace(μ, σ), so that the pdf of x_i is $exp(-|x_i - \mu|/\sigma)/(2\sigma)$. (a) Find a minimal sufficient statistic if $-\infty < \mu < \infty$ and $\sigma > 0$. (b) Find a minimal sufficient statistic if $\mu = 0$ (known) and $\sigma > 0$.

Exercise 13.8.7. Suppose that $\mathbf{X} \mid \Theta = \theta$ has density $f(\mathbf{x} \mid \theta)$, and $\mathbf{S} = s(\mathbf{x})$ is a sufficient statistic. Show that for any prior density π on Θ, the posterior density for $\Theta \mid \mathbf{X} = \mathbf{x}$ is the same as that for $\Theta \mid \mathbf{S} = \mathbf{s}$ if $\mathbf{s} = s(\mathbf{x})$. [Hint: Use Lemma 13.7 and (13.16) to show that both posterior densities equal

$$\frac{b(\mathbf{s}, \theta)\pi(\theta)}{\int b(\mathbf{s}, \theta^*)\pi(\theta^*)d\theta^*} . \quad] \tag{13.85}$$

Exercise 13.8.8. Show that each of the following models is an exponential family model. Give the natural parameters and statistics. In each case, the data are $X_1, \ldots, X_n$ with the given distribution. (a) Poisson(λ) where $\lambda > 0$. (b) Exponential(λ) where $\lambda > 0$. (c) Gamma(α, λ), where $\alpha > 0$ and $\lambda > 0$. (d) Beta(α, β), where $\alpha > 0$ and $\beta > 0$.

Exercise 13.8.9. Show that each of the following models is an exponential family model. Give the natural parameters and statistics. (a) $X \sim$ Binomial(n, p), where $p \in (0, 1)$. (b) $\mathbf{X} \sim$ Multinomial$(n, \mathbf{p})$, where $\mathbf{p} = (p_1, \ldots, p_k)$, $p_k > 0$ for all k, and $\sum p_k = 1$. Take the natural sufficient statistic to be $\mathbf{X}$. (c) Take $\mathbf{X}$ as in part (b), but take the natural sufficient statistic to be $(X_1, \ldots, X_{K-1})$. [Hint: Set $X_K = n - X_1 - \cdots - X_{K-1}$.]

Exercise 13.8.10. Suppose $X \sim$ Poisson(θ), $\theta \in (0, \infty)$, so that the MLE of θ is $\widehat{\theta} = X$. Reparameterize to $\tau = g(\theta) = \exp(-\theta)$, so that the parameter space of τ is $(0,1)$. (a) Find the pmf for X in terms of τ, $f^*(x \mid \tau)$. (b) Find the MLE of τ for the pmf in part (a). Does it equal $\exp(-\widehat{\theta})$?

Exercise 13.8.11. Consider the statistical model with densities $f(\mathbf{x} \mid \theta)$ for $\theta \in \mathcal{T}$. Suppose the function $g : \mathcal{T} \to \mathcal{O}$ is one-to-one and onto, so that a reparameterization of the model has densities $f^*(\mathbf{x} \mid \omega)$ for $\omega \in \Omega$, where $f^*(\mathbf{x} \mid \omega) = f(\mathbf{x} \mid g^{-1}(\omega))$. (a) Show that $\widehat{\theta}$ uniquely maximizes $f(\mathbf{x} \mid \theta)$ over θ if and only if $\widehat{\omega} \equiv g(\widehat{\theta})$ uniquely maximizes $f^*(\mathbf{x} \mid \omega)$ over ω. [Hint: Show that $f(\mathbf{x} \mid \widehat{\theta}) > f(\mathbf{x} \mid \theta)$ for all $\theta \neq \widehat{\theta}$ implies $f^*(\mathbf{x} \mid \widehat{\omega}) > f^*(\mathbf{x} \mid \omega)$ for all $\omega \neq \widehat{\omega}$, and *vice versa*.] (b) Argue that if $\widehat{\theta}$ is the MLE of θ, then $g(\widehat{\theta})$ is the MLE of ω.

Exercise 13.8.12. Again consider the model in Exercise 13.8.11, but now suppose $g : \mathcal{T} \to \mathcal{O}$ is just onto, not one-to-one. Let g^* be any function of θ such that the joint function $h(\theta) = (g(\theta), g^*(\theta))$, $h : \mathcal{T} \to \mathcal{L}$, is one-to-one and onto, and set the reparameterized density as $f^*(\mathbf{x} \mid \lambda) = f(\mathbf{x} \mid h^{-1}(\lambda))$, $\lambda \in \mathcal{L}$. Exercise 13.8.11 shows that if $\widehat{\theta}$ uniquely maximizes $f(\mathbf{x} \mid \theta)$ over $\mathcal{T}$, then $\widehat{\lambda} = h(\widehat{\theta})$ uniquely maximizes $f^*(\mathbf{x} \mid \lambda)$ over $\mathcal{L}$. Argue that if $\widehat{\theta}$ is the MLE of θ, that it is legitimate to define $g(\widehat{\theta})$ to be the MLE of $\omega = g(\theta)$.

Exercise 13.8.13. Recall the fruit fly example in Section 6.4.4. Equation (6.114) has the pmf for one observation (Y_1, Y_2). The data consist of n iid such observations, $(Y_{i1}, Y_{i2}), i = 1, \ldots, n$. Let n_{ij} be the number of pairs (Y_{i1}, Y_{i2}) that equal (i, j) for

$i = 0, 1$ and $j = 0, 1$. (a) Show that the loglikelihood can be written

$$l_n(\theta) = (n_{00} + n_{01} + n_{10}) \log(1 - \theta) + n_{00} \log(2 - \theta)$$
$$+ (n_{01} + n_{10} + n_{11}) \log(\theta) + n_{11} \log(1 + \theta). \quad (13.86)$$

(b) The data can be found in (6.54). Each Y_{ij} is the number of CUs in its genotype, so that $(TL, TL) \Rightarrow 0$ and $(TL, CU) \Rightarrow 1$. Find $n_{00}, n_{01} + n_{10}$, and n_{11}, and fill the values into the loglikelihood. (c) Sketch the likelihood. Does there appear to be a unique maximum? If so, what is it approximately?

Exercise 13.8.14. Suppose $X_1, \ldots, X_n$ are iid with shifted exponential pdf,

$$f(x_i \mid \alpha, \lambda) = \lambda e^{-\lambda(x_i - \alpha)} I[x_i \geq \alpha], \quad (13.87)$$

where $(\alpha, \lambda) \in \mathbb{R} \times (0, \infty)$. Find the MLE of (α, λ) when $n = 4$ and the data are 10,7,12,15. [Hint: First find the MLE of α for fixed λ, and note that it does not depend λ.]

Exercise 13.8.15. Let $X_1, \ldots, X_n$ be iid $N(\mu, \sigma^2)$, $-\infty < \mu < \infty$ and $\sigma^2 > 0$. Consider estimates of σ^2 of the form $\widehat{\sigma}_c^2 = c \sum_{i=1}^n (X_i - \overline{X})^2$ for some $c > 0$. (a) Find the expected value, bias, and variance of $\widehat{\sigma}_c^2$. (b) For which value of c is the estimator unbiased? For which value is it the MLE? (c) Find the expected mean square error (MSE) of the estimator. For which value of c is the MSE mimimized? (d) Is the MLE unbiased? Does the MLE minimize the MSE? Does the unbiased estimator minimize the MSE?

Exercise 13.8.16. Suppose $U_1, \ldots, U_n$ are iid Uniform$(\mu - 1, \mu + 1)$. The likelihood does not have a unique maximum. Let $u_{(1)}$ be the minimum and $u_{(n)}$ be the maximum of the data. (a) The likelihood is maximized for any μ in what interval? (b) Recall the midrange $u_{mr} = (u_{(1)} + u_{(n)})/2$ from Exercise 5.6.15. Is u_{mr} one of the maxima of the likelihood?

Exercise 13.8.17. Let $X_1, \ldots, X_n$ be a sample from the Cauchy(θ) distribution (which has pdf $1/(\pi(1 + (x - \theta)^2)))$, where $\theta \in \mathbb{R}$. (a) If $n = 1$, show that the MLE of θ is X_1. (b) Suppose $n = 7$ and the observations are 10,2,4,2,5,7,1. Plot the loglikelihood and likelihood equations. Is the MLE of θ unique? Does the likelihood equation have a unique root?

Exercise 13.8.18. Consider the simple linear regression model, where $Y_1, \ldots, Y_n$ are independent, and $Y_i \sim N(\alpha + \beta x_i, \sigma^2)$ for $i = 1, \ldots, n$. The x_i's are fixed, and assume they are not all equal. (a) Find the likelihood of $\mathbf{Y} = (Y_1, \ldots, Y_n)'$. Write down the log of the likelihood, $l(\alpha, \beta, \sigma^2 ; \mathbf{y})$. (b) Fix σ^2. Why is the MLE of (α, β) the same as the least squares estimate of (α, β)? (c) Let $\widehat{\alpha}$ and $\widehat{\beta}$ be the MLEs, and let $SS_e = \sum(y_i - \widehat{\alpha} - \widehat{\beta} x_i)^2$ be the residual sum of squares. Write $l(\widehat{\alpha}, \widehat{\beta}, \sigma^2 ; \mathbf{y})$ as a function of SS_e and σ^2 (and n). (d) Find the MLE of σ^2. Is it unbiased?

Exercise 13.8.19. Now look at the multiple linear model, where $\mathbf{Y} \sim N(\mathbf{x}\boldsymbol{\beta}, \sigma^2 \mathbf{I}_n)$ as in (12.10), and assume that $\mathbf{x}'\mathbf{x}$ is invertible. (a) Show that the pdf of $\mathbf{Y}$ is

$$f(\mathbf{y} \mid \boldsymbol{\beta}, \sigma^2) = \frac{1}{(\sqrt{2\pi}\sigma)^n} e^{-\frac{1}{2\sigma^2} \|\mathbf{y} - \mathbf{x}\boldsymbol{\beta}\|^2}. \quad (13.88)$$

[Hint: Use (7.28).] (b) Use (12.17) and (12.18) to write $\|y - x\beta\|^2 = SS_e + (\widehat{\beta}_{LS} - \beta)'x'x(\widehat{\beta}_{LS} - \beta)$, where $\widehat{\beta}_{LS}$ is the least squares estimate of β, and SS_e is the sum of squared errors. (c) Use parts (a) and (b) to write the likelihood as

$$L(\beta, \sigma^2 ; y) = \frac{1}{\sigma^2} e^{-\frac{1}{2\sigma^2}(SS_e + (\widehat{\beta}_{LS} - \beta)'x'x(\widehat{\beta}_{LS} - \beta))}. \tag{13.89}$$

(d) From (13.89), argue that the sufficient statistic is $(\widehat{\beta}_{LS}, SS_e)$.

Exercise 13.8.20. Continue with the multiple regression model in Exercise 13.8.19. (a) Find $\widehat{\beta}$, the MLE of β. (You can keep σ^2 fixed here.) Is it the same as the least squares estimate? (b) Find $\widehat{\sigma}^2$, the MLE of σ^2. Is it unbiased? (c) Find the value of the loglikelihood at the MLE, $l(\widehat{\beta}, \widehat{\sigma}^2 ; y)$.

Exercise 13.8.21. Suppose that for some power λ,

$$\frac{Y^\lambda - 1}{\lambda} \sim N(\mu, \sigma^2). \tag{13.90}$$

Generally, λ will be in the range from -2 to 2. We are assuming that the parameters are such that the chance of Y being non-positive is essentially 0, so that λ can be a fraction (i.e., tranforms like $\sqrt{Y}$ are real). (a) What is the limit as $\lambda \to 0$ of $(y^\lambda - 1)/\lambda$ for $y > 0$? (b) Find the pdf of Y. (Don't forget the Jacobian. Note that if $W \sim N(\mu, \sigma^2)$, then $y = g(w) = (\lambda w + 1)^{1/\lambda}$. So $g^{-1}(y)$ is $(y^\lambda - 1)/\lambda$.)
Now suppose $Y_1, \ldots, Y_n$ are independent, and $x_1, \ldots, x_n$ are fixed, with

$$\frac{Y_i^\lambda - 1}{\lambda} \sim N(\alpha + \beta x_i, \sigma^2). \tag{13.91}$$

In regression, one often takes transformations of the variables to get a better fitting model. Taking logs or square roots of the y_i's (and maybe the x_i's) can often be effective. The goal here is to find the best transformation by finding the MLE of λ, as well as of α and β and σ^2. This λ represents a power transformation of the Y_i's, called the **Box-Cox transformation**.
The loglikelihood of the parameters based on the y_i's depends on α, β, σ^2 and λ. For fixed λ, it can be maximized using the usual least-squares theory. So let

$$RSS_\lambda = \sum \left(\frac{y_i^\lambda - 1}{\lambda} - \widehat{\alpha}_\lambda - \widehat{\beta}_\lambda x_i \right)^2 \tag{13.92}$$

be the residual sum of squares, considering λ fixed and the $(y_i^\lambda - 1)/\lambda$'s as the dependent variable's observations. (c) Show that the loglikelihood can be written as the function of λ,

$$h(\lambda) \equiv l_n(\widehat{\alpha}_\lambda, \widehat{\beta}_\lambda, \widehat{\sigma}_\lambda^2, \lambda ; y_i's) = -\frac{n}{2} \log(RSS_\lambda) + \lambda \sum \log(y_i). \tag{13.93}$$

Then to maximize this over λ, one tries a number of values for λ, each time performing a new regression on the $(y_i^\lambda - 1)/\lambda$'s, and takes the λ that maximizes the loglikelihood. (d) As discussed in Section 12.5.2, Jung et al. (2014) collected data on the $n = 94$ most dangerous hurricanes in the US since 1950. Let Y_i be the estimate of damage by hurricane i in millions of 2014 dollars (plus 1, to avoid taking log of 0), and x_i be the minimum atmospheric pressure in the storm. Lower pressure leads to more severe storms. These two variables can be downloaded directly into R using the command

source("http://istics.net/r/hurricanes.R")

Apply the results on the Box-Cox transformation to these data. Find the $h(\lambda)$ for a grid of values of λ from -2 to 2. What value of λ maximizes the loglikelihood, and what is that maximum value? (e) Usually one takes a more understandable power near the MLE. Which of the following transformations has loglikelihood closest to the MLE's: $1/y^2$, $1/y$, $1/\sqrt{y}$, $\log(y)$, $\sqrt{y}$, y, or y^2? (f) Plot x versus $(y^{\hat{\lambda}} - 1)/\hat{\lambda}$, where $\hat{\lambda}$ is the MLE. Does it look like the usual assumptions for linear regression in (12.10) are reasonable?

Exercise 13.8.22. Suppose $\mathbf{X} \sim$ Multinomial$(n, \mathbf{p})$. Let the parameter be $(p_1, \ldots, p_{K-1})$, so that $p_K = 1 - p_1 - \cdots - p_{K-1}$. The parameter space is then $\{(p_1, \ldots, p_{K-1}) \mid 0 < p_i$ for each i, and $p_1 + \cdots + p_{K-1} < 1\}$. Show that the MLE of $\mathbf{p}$ is $\mathbf{X}/n$.

Exercise 13.8.23. Take $(X_1, X_2, X_3, X_4) \sim$ Multinomial$(n, (p_1, p_2, p_3, p_4))$. Put the p_i's in a table:

$$
\begin{array}{cc|c}
p_1 & p_2 & \alpha \\
p_3 & p_4 & 1 - \alpha \\
\hline
\beta & 1 - \beta & 1
\end{array}
\qquad (13.94)
$$

Here, $\alpha = p_1 + p_2$ and $\beta = p_1 + p_3$. Assume the model that the rows and columns are independent, meaning

$$p_1 = \alpha\beta, p_2 = \alpha(1 - \beta), p_3 = (1 - \alpha)\beta, p_4 = (1 - \alpha)(1 - \beta). \qquad (13.95)$$

(a) Write the loglikelihood as a function of α and β (not the p_i's). (b) Find the MLEs of α and β. (c) What are the MLEs of the p_i's?

Exercise 13.8.24. Suppose $X_1, \ldots, X_n$ are iid $N(\mu, 1)$, $\mu \in \mathbb{R}$, so that with $\mathbf{X} = (X_1, \ldots, X_n)'$, we have from (13.53) the conditional distribution

$$\mathbf{X} \mid \overline{X} = \overline{x} \sim N(\overline{x} \, \mathbf{1}_n, \mathbf{H}_n), \qquad (13.96)$$

where $\mathbf{H}_n = \mathbf{I}_n - (1/n) \, \mathbf{1}_n\mathbf{1}_n'$ is the centering matrix. Assume $n \geq 2$. (a) Find $E[X_1 \mid \overline{X} = \overline{x}]$. (b) Find $E[X_1^2 \mid \overline{X} = \overline{x}]$. (c) Find $E[X_1 X_2 \mid \overline{X} = \overline{x}]$. (d) Consider estimating $g(\mu) = 0$. (i) Is $\delta(\mathbf{X}) = X_1 - X_2$ an unbiased estimator of 0? (ii) What is the variance of δ? (iii) Find $\delta^*(\overline{x}) \equiv E[\delta(\mathbf{X}) \mid \overline{X} = \overline{x}]$. Is it unbiased? (iv) What is the variance of δ^*? (v) Which estimator has a lower variance? (e) Consider estimating $g(\mu) = \mu^2$. (i) The estimator $\delta(\mathbf{X}) = X_1^2 - a$ is unbiased for what a? (ii) What is the variance of δ? (iii) Find $\delta^*(\overline{x}) \equiv E[\delta(\mathbf{X}) \mid \overline{X} = \overline{x}]$. Is it unbiased? (iv) What is the variance of δ^*? (v) Which estimator has a lower variance? (f) Continue estimating $g(\mu) = \mu^2$. (i) The estimator $\delta(\mathbf{X}) = X_1 X_2 - a$ is unbiased for what a? (ii) What is the variance of δ? (iii) Find $\delta^*(\overline{x}) \equiv E[\delta(\mathbf{X}) \mid \overline{X} = \overline{x}]$. Is it unbiased? (iv) What is the variance of δ^*? (v) Which estimator has a lower variance? (g) Compare the estimators δ^* in (e)(iii) and (f)(iii).

Chapter 14

More on Maximum Likelihood Estimation

In the previous chapter we have seen some situations where maximum likelihood has yielded fairly reasonable estimators. In fact, under certain conditions, MLEs are consistent, asymptotically normal as $n \to \infty$, and have optimal asymptotic standard errors. We will focus on the iid case, but the results have much wider applicability.

The first three sections of this chapter show how to use maximum likelihood to find estimates and their asymptotic standard errors and confidence intervals. Sections 14.4 on present the technical conditions and proofs for the results. Most of the presentation presumes a one-dimensional parameter. Section 14.8 extends the results to multidimensional parameters.

14.1 Score function

Suppose $\mathbf{X}_1, \ldots, \mathbf{X}_n$ are iid, each with space $\mathcal{X}$ and density $f(\mathbf{x} \mid \theta)$, where $\theta \in \mathcal{T} \subset \mathbb{R}$, so that θ is one-dimensional. There are a number of technical conditions that need to be satisfied for what follows, which will be presented in Section 14.4. Here, we note that we do need to have continuous first, second, and third derivatives with respect to the parameters, and

$$f(\mathbf{x} \mid \theta) > 0 \quad \text{for all} \ \mathbf{x} \in \mathcal{X}, \ \theta \in \mathcal{T}. \tag{14.1}$$

In particular, (14.1) rules out the Uniform$(0, \theta)$, since the sample space would depend on the parameters. Which is not to say that the MLE is bad in this case, but that the asymptotic normality, etc., does not hold.

By independence, the overall likelihood is $\prod f(\mathbf{x}_i \mid \theta)$, hence the loglikelihood is

$$l_n(\theta; \mathbf{x}_1, \ldots, \mathbf{x}_n) = \sum_{i=1}^{n} \log(f(\mathbf{x}_i \mid \theta)) = \sum_{i=1}^{n} l_1(\theta; \mathbf{x}_i), \tag{14.2}$$

where $l_1(\theta; \mathbf{x}) = \log(f(\mathbf{x} \mid \theta))$ is the loglikelihood for one observation. The MLE is found by differentiating the loglikelihood, which is the sum of the derivatives of the individual loglikelihoods, or **score** functions. For one observation, the score is $l_1'(\theta; \mathbf{x}_i)$. The score for the entire set of data is $l_n'(\theta; \mathbf{x}_1, \ldots, \mathbf{x}_n)$, the sum of the individual scores. The MLE $\widehat{\theta}_n$ then satisfies

$$l'_n(\widehat{\theta}_n ; x_1, \ldots, x_n) = \sum_{i=1}^{n} l'_1(\widehat{\theta}_n ; x_i) = 0. \tag{14.3}$$

We are assuming that there is a unique solution, and that it does indeed maximize the loglikelihood.

If one is lucky, there is a closed-form solution. Mostly there will not be, so that some iterative procedure will be necessary. The Newton-Raphson method is a popular approach, which can be quite quick if it works. The idea is to expand l'_n in a one-step Taylor series around an initial guess of the solution, $\theta^{(0)}$, then solve for $\widehat{\theta}$ to obtain the next guess $\theta^{(1)}$. Given the jth guess $\theta^{(j)}$, we have (dropping the x_i's for simplicity)

$$l'_n(\widehat{\theta}) \approx l'_n(\theta^{(j)}) + (\widehat{\theta} - \theta^{(j)})l''_n(\theta^{(j)}). \tag{14.4}$$

We know that $l'_n(\widehat{\theta}) = 0$, so we can solve approximately for $\widehat{\theta}$:

$$\widehat{\theta} \approx \theta^{(j)} - \frac{l'_n(\theta^{(j)})}{l''_n(\theta^{(j)})} \equiv \theta^{(j+1)}. \tag{14.5}$$

This $\theta^{(j+1)}$ is our next guess for the MLE. We iterate until the process converges, if it does, in which case we have our $\widehat{\theta}$.

14.1.1 Fruit flies

Go back to the fruit fly example in Section 6.4.4. Equation (6.114) has the pmf of each observation, and (6.54) contains the data. Exercise 13.8.13 shows that the loglikelihood can be written

$$l_n(\theta) = 7\log(1 - \theta) + 5\log(2 - \theta) + 5\log(\theta) + 3\log(1 + \theta). \tag{14.6}$$

Starting with the guess $\theta^{(0)} = 0.5$, the iterations for Newton-Raphson proceed as follows:

j	$\theta^{(j)}$	$l'_n(\theta^{(j)})$	$l''_n(\theta^{(j)})$
0	0.500000	−5.333333	−51.55556
1	0.396552	0.038564	−54.50161
2	0.397259	0.000024	−54.43377
3	0.397260	0	−54.43372

(14.7)

The process has converged sufficiently, so we have $\widehat{\theta}_{MLE} = 0.3973$. Note this estimate is very close to the Dobzhansky estimate of 0.4 found in (6.65) and therebelow.

14.2 Fisher information

The likelihood, or loglikelihood, is supposed to reflect the relative support for various values of the parameter given by the data. The MLE is the value with the most support, but we would also like to know which other values have almost as much support. One way to assess the range of highly-supported values is to look at the likelihood near the MLE. If it falls off quickly as we move from the MLE, then we

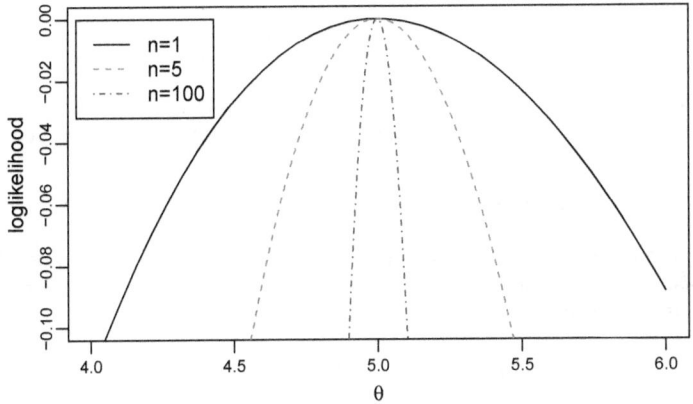

Figure 14.1: Loglikelihood for the Poisson with $\bar{x} = 5$ and $n = 1, 5,$ and 100.

have more confidence that the true parameter is near the MLE than if the likelihood falls way slowly. For example, consider Figure 14.1. The data are iid Poisson(θ)'s with sample mean being 5, and the three curves are the loglikelihoods for $n = 1, 5,$ and 100. In each case the maximum is at $\hat{\theta} = 5$. The flattest loglikelihood is that for $n = 1$, and the one with the narrowest curve is that for $n = 100$. Note that for $n = 1$, there are many values of θ that have about the same likelihood as the MLE. Thus they are almost as likely. By contrast, for $n = 100$, there is a distinct drop off from the maximum as one moves away from the MLE. Thus there is more information about the parameter. The $n = 5$ case is in between the other two. Of course, we expect more information with larger n. One way to quantify the information is to look at the second derivative of the loglikelihood at the MLE: The more negative, the more informative.

In general, the negative second derivative of the loglikelihood is called the **observed Fisher information** in the data. It can be written

$$\widehat{\mathcal{I}}_n(\theta; \mathbf{x}_1, \ldots, \mathbf{x}_n) = -l_n''(\theta; \mathbf{x}_1, \ldots, \mathbf{x}_n) = \sum_{i=1}^{n} \widehat{\mathcal{I}}_1(\theta; \mathbf{x}_i), \qquad (14.8)$$

where $\widehat{\mathcal{I}}_1(\theta; \mathbf{x}_i) = -l_1''(\theta; \mathbf{x}_i)$ is the observed Fisher information in the single observation $\mathbf{x}_i$. The idea is that the larger the information, the more we know about θ. In the Poisson example above, the observed information is $\sum x_i / \theta^2$, hence at the MLE $\hat{\theta} = \bar{x}_n$ it is $n / \bar{x}_n$. Thus the information is directly proportional to n (for fixed $\bar{x}_n$).

The (expected) **Fisher information** in one observation, $\mathcal{I}_1(\theta)$, is the expected value of the observed Fisher information:

$$\mathcal{I}_1(\theta) = E[\widehat{\mathcal{I}}_1(\theta; \mathbf{X}_i)]. \qquad (14.9)$$

The Fisher information in the entire iid sample is

$$\mathcal{I}_n(\theta) = E[\widehat{\mathcal{I}}_n(\theta ; \mathbf{X}_1, \ldots, \mathbf{X}_n)] = E\left[\sum_{i=1}^{n} \widehat{\mathcal{I}}_1(\theta ; \mathbf{X}_i)\right] = n\mathcal{I}_1(\theta). \tag{14.10}$$

In the Poisson case, $\mathcal{I}_1(\theta) = E[X_i/\theta^2] = 1/\theta$, hence $\mathcal{I}_n(\theta) = n/\theta$.
    For multidimensional parameters, the Fisher information is a matrix. See (14.80).

14.3 Asymptotic normality

One of the more amazing properties of the MLE is that, under very general conditions, it is asymptotically normal, with variance the inverse of the Fisher information. In the one-parameter iid case,

$$\sqrt{n}(\widehat{\theta}_n - \theta) \longrightarrow^{\mathcal{D}} N\left(0, \frac{1}{\mathcal{I}_1(\theta)}\right). \tag{14.11}$$

To eliminate the dependence on θ in the normal, we can use Slutsky to obtain

$$\sqrt{n\mathcal{I}_1(\widehat{\theta}_n)} \, (\widehat{\theta}_n - \theta) \longrightarrow^{\mathcal{D}} N(0,1). \tag{14.12}$$

It turns out that we can also use the observed Fisher information in place of the $n\mathcal{I}_1$:

$$\sqrt{\widehat{\mathcal{I}}_n(\widehat{\theta}_n)} \, (\widehat{\theta}_n - \theta) \longrightarrow^{\mathcal{D}} N(0,1). \tag{14.13}$$

    Consider the fruit fly example in Section 14.1.1. Since the score function is the first derivative of the loglikelihood, minus the first derivative of the score function is the observed Fisher information. Thus the Newton-Raphson process in (14.7) automatically presents us with $\widehat{\mathcal{I}}_n(\widehat{\theta}_n) = -l_n''(\widehat{\theta}_n) = 54.4338$, hence an approximate 95% confidence interval for θ is

$$\left(0.3973 \pm \frac{2}{\sqrt{54.4338}}\right) = (0.3973 \pm 2 \times 0.1355) = (0.1263, 0.6683). \tag{14.14}$$

A rather wide interval.

14.3.1 Sketch of the proof

The regularity conditions and statement and proof of the main asymptotic results require a substantial amount of careful analysis, which we present in Sections 14.4 to 14.6. Here we give the basic idea behind the asymptotic normality in (14.11). Starting with the Taylor series as in the Newton-Raphson algorithm, write

$$l_n'(\widehat{\theta}_n) \approx l_n'(\theta) + (\widehat{\theta}_n - \theta)l_n''(\theta), \tag{14.15}$$

where θ is the true value of the parameter, and the dependence on the x_i's is suppressed. Rearranging and inserting n's in the appropriate places, we obtain

$$\sqrt{n}(\widehat{\theta}_n - \theta) \approx \sqrt{n}\frac{l_n'(\theta)}{l_n''(\theta)} = \frac{\sqrt{n}\frac{1}{n}\sum l_1'(\theta ; X_i)}{\frac{1}{n}\sum l_1''(\theta ; X_i)}$$

$$= \frac{\sqrt{n}\frac{1}{n}\sum l_1'(\theta ; X_i)}{-\frac{1}{n}\sum \widehat{\mathcal{I}}_1(\theta ; X_i)}, \tag{14.16}$$

since $\widehat{\mathcal{I}}_1(\theta; X_i) = -l_1''(\theta; X_i)$.

We will see in Lemma 14.1 that

$$E_\theta[l_1'(\theta; X_i)] = 0 \quad \text{and} \quad Var_\theta[l_1'(\theta; X_i)] = \mathcal{I}_1(\theta). \tag{14.17}$$

Thus the central limit theorem shows that

$$\sqrt{n}\,\frac{1}{n}\sum l_1'(\theta; X_i) \longrightarrow^{\mathcal{D}} N(0, \mathcal{I}_1(\theta)). \tag{14.18}$$

Since $E[\widehat{\mathcal{I}}_1(\theta; X_i)] = \mathcal{I}_1(\theta)$ by definition, the law of large numbers shows that

$$\frac{1}{n}\sum \widehat{\mathcal{I}}_1(\theta; X_i) \longrightarrow^{P} \mathcal{I}_1(\theta). \tag{14.19}$$

Finally, Slutsky shows that

$$\frac{\sqrt{n}\,\frac{1}{n}\sum l_1'(\theta; X_i)}{-\frac{1}{n}\sum \widehat{\mathcal{I}}_1(\theta; X_i)} \longrightarrow^{\mathcal{D}} \frac{N(0, \mathcal{I}_1(\theta))}{-\mathcal{I}_1(\theta)} = N\left(0, \frac{1}{\mathcal{I}_1(\theta)}\right), \tag{14.20}$$

as desired. Theorem 14.6 below deals more carefully with the approximation in (14.16) to justify (14.11).

If the justification in this section is satisfactory, you may want to skip to Section 14.7 on asymptotic efficiency, or Section 14.8 on the multiparameter case.

14.4 Cramér's conditions

Cramér (1999) was instrumental in applying rigorous mathematics to the study of statistics. In particular, he provided technical conditions under which the likelihood results are valid. The conditions easily hold in exponential families, but for other densities they may or may not be easy to verify. We start with $X_1, \ldots, X_n$ iid, each having space $\mathcal{X}$ and pdf $f(x \mid \theta)$, where $\theta \in \mathcal{T} = (a, b)$ for fixed $-\infty \leq a < b \leq \infty$. First, we need that the space of X_i is the same for each θ, which is satisfied if

$$f(x \mid \theta) > 0 \quad \text{for all } x \in \mathcal{X}, \theta \in \mathcal{T}. \tag{14.21}$$

We also need that

$$\frac{\partial f(x \mid \theta)}{\partial \theta}, \frac{\partial^2 f(x \mid \theta)}{\partial \theta^2}, \frac{\partial^3 f(x \mid \theta)}{\partial \theta^3} \quad \text{exist for all } x \in \mathcal{X}, \theta \in \mathcal{T}. \tag{14.22}$$

In order for the score and information functions to exist and behave correctly, assume that for any $\theta \in \mathcal{T}$,

$$\int_\mathcal{X} \frac{\partial f(x \mid \theta)}{\partial \theta} dx = \frac{\partial}{\partial \theta} \int_\mathcal{X} f(x \mid \theta) dx \quad (= 0)$$

$$\text{and} \quad \int_\mathcal{X} \frac{\partial^2 f(x \mid \theta)}{\partial \theta^2} dx = \frac{\partial^2}{\partial \theta^2} \int_\mathcal{X} f(x \mid \theta) dx \quad (= 0). \tag{14.23}$$

(Replace the integrals with sums for the discrete case.)

Recall the Fisher information in one observation from (14.8) and (14.9) is given by

$$\mathcal{I}_1(\theta) = -E_\theta[l_1''(\theta; x)]. \tag{14.24}$$

Assume that

$$0 < \mathcal{I}_1(\theta) < \infty \quad \text{for all } \theta \in \mathcal{T}. \tag{14.25}$$

We have the following.

Lemma 14.1. *If (14.21), (14.22), and (14.23) hold, then*

$$E_\theta[l_1'(\theta\,;X)] = 0 \quad and \quad Var_\theta[l_1'(\theta\,;X)] = \mathcal{I}_1(\theta). \tag{14.26}$$

Proof. First, since $l_1(\theta\,;x) = \log(f(x\,|\,\theta))$,

$$
\begin{aligned}
E_\theta[l_1'(\theta\,;X)] &= E_\theta\left[\frac{\partial}{\partial\theta}\log(f(x\,|\,\theta))\right] \\
&= \int_\mathcal{X} \frac{\partial f(x\,|\,\theta)/\partial\theta}{f(x\,|\,\theta)}\,f(x\,|\,\theta)dx \\
&= \int_\mathcal{X} \frac{\partial f(x\,|\,\theta)}{\partial\theta}\,dx \\
&= 0
\end{aligned}
\tag{14.27}
$$

by (14.23). Next, write

$$
\begin{aligned}
\mathcal{I}_1(\theta) = -E_\theta[l_1''(\theta\,;X)] &= -E_\theta\left[\frac{\partial^2}{\partial\theta^2}\log(f(X|\theta))\right] \\
&= -E_\theta\left[\frac{\partial^2 f(X|\theta)/\partial\theta^2}{f(X|\theta)} - \left(\frac{\partial f(X|\theta)/\partial\theta}{f(X|\theta)}\right)^2\right] \\
&= -\int_\mathcal{X} \frac{\partial^2 f(x|\theta)/\partial\theta^2}{f(x|\theta)}f(x|\theta)dx + E_\theta[l_1'(\theta\,;X)^2] \\
&= -\frac{\partial^2}{\partial\theta^2}\int_\mathcal{X} f(x|\theta)dx + E_\theta[l_1'(\theta\,;X)^2] \\
&= E_\theta[l_1'(\theta\,;X)^2]
\end{aligned}
\tag{14.28}
$$

again by (14.23), which with $E_\theta[l_1'(\theta\,;X)] = 0$ proves (14.26).   □

One more technical assumption we need is that for each $\theta \in \mathcal{T}$ (which will take the role of the "true" value of the parameter), there exists an $\epsilon > 0$ and a function $M(x)$ such that

$$|l_1'''(t\,;x)| \le M(x) \text{ for } \theta - \epsilon < t < \theta + \epsilon, \text{ and } E_\theta[M(X)] < \infty. \tag{14.29}$$

14.5   Consistency

First we address the question of whether the MLE is a consistent estimator of θ. The short answer is "Yes," although things can get sticky if there are multiple maxima. But before we get to the results, there are some mathematical prerequisites to deal with.

14.5.1   Convexity and Jensen's inequality

Definition 14.2. *Convexity. A function $g : \mathcal{X} \to \mathbb{R}$, $\mathcal{X} \subset \mathbb{R}$, is **convex** if for each $x_0 \in \mathcal{X}$, there exist α_0 and β_0 such that*

$$g(x_0) = \alpha_0 + \beta_0 x_0, \tag{14.30}$$

and

$$g(x) \geq \alpha_0 + \beta_0 x \quad \text{for all } x \in \mathcal{X}. \tag{14.31}$$

The function is **strictly convex** *if (14.30) holds, and*

$$g(x) > \alpha_0 + \beta_0 x \quad \text{for all } x \in \mathcal{X}, \ x \neq x_0. \tag{14.32}$$

The definition basically means that the tangent to g at any point lies below the function. If $g''(x)$ exists for all x, then g is convex if and only if $g''(x) \geq 0$ for all x, and it is strictly convex if and only if $g''(x) > 0$ for all x. Notice that the line defined by a_0 and b_0 need not be unique. For example, $g(x) = |x|$ is convex, but when $x_0 = 0$, any line through $(0,0)$ with slope between ± 1 will lie below g.

By the same token, any line segment connecting two points on the curve lies above the curve, as in the next lemma.

Lemma 14.3. *If g is convex, $x, y \in \mathcal{X}$, and $0 < \epsilon < 1$, then*

$$\epsilon g(x) + (1 - \epsilon)g(y) \geq g(\epsilon x + (1 - \epsilon)y). \tag{14.33}$$

If g is strictly convex, then

$$\epsilon g(x) + (1 - \epsilon)g(y) > g(\epsilon x + (1 - \epsilon)y) \quad \text{for } x \neq y. \tag{14.34}$$

Rather than prove this lemma, we will prove the more general result for random variables.

Lemma 14.4. *Jensen's inequality. Suppose that X is a random variable with space $\mathcal{X}$, and that $E[X]$ exists. If the function g is convex, then*

$$E[g(X)] \geq g(E[X]), \tag{14.35}$$

where $E[g(X)]$ may be $+\infty$. Furthermore, if g is strictly convex and X is not constant,

$$E[g(X)] > g(E[X]). \tag{14.36}$$

Proof. We'll prove it just in the strictly convex case, when X is not constant. The other case is easier. Apply Definition 14.2 with $x_0 = E[X]$, so that

$$g(E[X]) = \alpha_0 + \beta_0 E[X], \quad \text{and} \quad g(x) > \alpha_0 + \beta_0 x \ \text{for all} \ x \neq E[X]. \tag{14.37}$$

But then

$$E[g(X)] > E[\alpha_0 + \beta_0 X] = \alpha_0 + \beta_0 E[X] = g(E[X]), \tag{14.38}$$

because there is a positive probability $X \neq E[X]$. $\qquad\square$

Why does Lemma 14.4 imply Lemma 14.3? [Take X to be the random variable with $P[X = x] = \epsilon$ and $P[X = y] = 1 - \epsilon$.]

A mnemonic device for which way the inequality goes is to think of the convex function $g(x) = x^2$. Jensen implies that

$$E[X^2] \geq E[X]^2, \tag{14.39}$$

but that is the same as saying that $Var[X] \geq 0$. Also, $Var[X] > 0$ unless X is a constant.

Convexity is also defined for $\mathbf{x}$ being a $p \times 1$ vector, so that $\mathcal{X} \subset \mathbb{R}^p$, in which case the line $\alpha_0 + \beta_0 x$ in Definition 14.2 becomes a hyperplane $\alpha_0 + \boldsymbol{\beta}_0' \mathbf{x}$. Jensen's inequality follows as well, where we just turn X into a vector.

14.5.2   A consistent sequence of roots

For now, we assume that the space does not depend on θ (14.21), and the first derivative of the loglikelihood in (14.22) is continuous. We need **identifiability**, which
means that if $\theta_1 \neq \theta_2$, then the distributions of X_i under θ_1 and θ_2 are different. Also,
for each n and $x_1, \ldots, x_n$, there exists a unique solution to $l'_n(\theta; x_1, \ldots, x_n) = 0$:

$$l'_n(\widehat{\theta}_n; x_1, \ldots, x_n) = 0, \ \ \widehat{\theta}_n \in \mathcal{T}. \tag{14.40}$$

Note that this $\widehat{\theta}_n$ is a function of $x_1, \ldots, x_n$. It is also generally the maximum likelihood estimate, although it is possible it is a local minimum or an inflection point
rather than the maximum.

   Now suppose θ is the true parameter, and take $\epsilon > 0$. Look at the difference,
divided by n, of the likelihoods at θ and $\theta + \epsilon$:

$$\frac{1}{n}\left(l_n(\theta; x_1, \ldots, x_n) - l_n(\theta + \epsilon; x_1, \ldots, x_n)\right) = \frac{1}{n}\sum_{i=1}^{n}\log\left(\frac{f(x_i \mid \theta)}{f(x_i \mid \theta + \epsilon)}\right)$$

$$= \frac{1}{n}\sum_{i=1}^{n} -\log\left(\frac{f(x_i \mid \theta + \epsilon)}{f(x_i \mid \theta)}\right). \tag{14.41}$$

The final expression is the mean of iid random variables, hence by the WLLN it
converges in probability to the expected value of the summand (and dropping the
x_i's in the notation for convenience):

$$\frac{1}{n}\left(l_n(\theta) - l_n(\theta + \epsilon)\right) \longrightarrow^{\mathcal{P}} E_\theta\left[-\log\left(\frac{f(X \mid \theta + \epsilon)}{f(X \mid \theta)}\right)\right]. \tag{14.42}$$

   Now apply Jensen's inequality, Lemma 14.4, to that expected value, with $g(x) = -\log(x)$, and the random variable being $f(X \mid \theta + \epsilon)/f(X \mid \theta)$. This g is strictly convex, and the random variable is not constant because the parameters are different
(identifiability), hence

$$E_\theta\left[-\log\left(\frac{f(X \mid \theta + \epsilon)}{f(X \mid \theta)}\right)\right] > -\log\left(E_\theta\left[\frac{f(X \mid \theta + \epsilon)}{f(X \mid \theta)}\right]\right)$$

$$= -\log\left(\int_{\mathcal{X}}\frac{f(x \mid \theta + \epsilon)}{f(x \mid \theta)}f(x \mid \theta)dx\right)$$

$$= -\log\left(\int_{\mathcal{X}}f(x \mid \theta + \epsilon)dx\right)$$

$$= -\log(1)$$

$$= 0. \tag{14.43}$$

The same result holds for $\theta - \epsilon$, hence

$$\frac{1}{n}\left(l_n(\theta) - l_n(\theta + \epsilon)\right) \longrightarrow^{\mathcal{P}} c > 0 \text{ and}$$

$$\frac{1}{n}\left(l_n(\theta) - l_n(\theta - \epsilon)\right) \longrightarrow^{\mathcal{P}} d > 0. \tag{14.44}$$

These equations mean that eventually, the likelihood at θ is higher than that at $\theta \pm \epsilon$.
Precisely,

$$P_\theta[l_n(\theta) > l_n(\theta + \epsilon) \text{ and } l_n(\theta) > l_n(\theta - \epsilon)] \longrightarrow 1. \tag{14.45}$$

Note that if $l_n(\theta) > l_n(\theta + \epsilon)$ and $l_n(\theta) > l_n(\theta - \epsilon)$, then between $\theta - \epsilon$ and $\theta + \epsilon$, the likelihood goes up then comes down again. Because the derivative is continuous, somewhere between $\theta \pm \epsilon$ the derivative must be 0. By assumption, that point is the unique root $\hat{\theta}_n$. It is also the maximum. Which means that

$$l_n(\theta) > l_n(\theta + \epsilon) \text{ and } l_n(\theta) > l_n(\theta - \epsilon) \Longrightarrow \theta - \epsilon < \hat{\theta}_n < \theta + \epsilon. \tag{14.46}$$

By (14.45), the left hand side of (14.46) has probability going to 1, hence

$$P[|\hat{\theta}_n - \theta| < \epsilon] \to 1 \Longrightarrow \hat{\theta}_n \longrightarrow^P \theta, \tag{14.47}$$

and the MLE is consistent.

The requirement that there is a unique root (14.40) for all n and set of x_i's is too strong. The main problem is that sometimes the maximum of the likelihood does not exist over $\mathcal{T} = (a, b)$, but at a or b. For example, in the binomial case, if the number of successes is 0, then the MLE of p would be 0, which is not in $(0,1)$. Thus in the next theorem, we need only that probably there is a unique root.

Theorem 14.5. *Suppose that*

$$P_\theta[l_n'(t; X_1, \ldots, X_n) \text{ has a unique root } \hat{\theta}_n \in \mathcal{T}] \longrightarrow 1. \tag{14.48}$$

Then

$$\hat{\theta}_n \longrightarrow^P \theta. \tag{14.49}$$

Technically, if there is not a unique root, you can choose $\hat{\theta}_n$ to be whatever you want, but typically it would be either one of a number of roots, or one of the limiting values a and b. Equation (14.48) does not always hold. For example, in the Cauchy location-family case, the number of roots goes in distribution to $1 + \text{Poisson}(1/\pi)$ (Reeds, 1985), so there is always a good chance of two or more roots. But it will be true that if you pick the right root, e.g., the one closest to the median, it will be consistent.

14.6 Proof of asymptotic normality

To find the asymptotic distribution of the MLE, we first expand the derivative of the likelihood around $\theta = \hat{\theta}_n$:

$$l_n'(\hat{\theta}_n) = l_n'(\theta) + (\hat{\theta}_n - \theta) l_n''(\theta) + \tfrac{1}{2}(\hat{\theta}_n - \theta)^2 l_n'''(\theta_n^*), \quad \theta_n^* \text{ between } \theta \text{ and } \hat{\theta}_n. \tag{14.50}$$

(Recall that these functions depend on the x_i's.) If $\hat{\theta}_n$ is a root of l_n' as in (14.40), then

$$0 = l_n'(\theta) + (\hat{\theta}_n - \theta) l_n''(\theta) + \tfrac{1}{2}(\hat{\theta}_n - \theta)^2 l_n'''(\theta_n^*)$$

$$\Longrightarrow (\hat{\theta}_n - \theta)(l_n''(\theta) + \tfrac{1}{2}(\hat{\theta}_n - \theta) l_n'''(\theta_n^*)) = -l_n'(\theta)$$

$$\Longrightarrow \sqrt{n}\,(\hat{\theta}_n - \theta) = -\frac{\sqrt{n}\,\tfrac{1}{n} l_n'(\theta)}{\tfrac{1}{n} l_n''(\theta) + (\hat{\theta}_n - \theta) \tfrac{1}{2n} l_n'''(\theta_n^*)}. \tag{14.51}$$

The task is then to find the limits of the three terms on the right: the numerator and the two summands in the denominator.

Theorem 14.6. *Cramér. Suppose that the assumptions in Section 14.4 hold, i.e., (14.21), (14.22), (14.23), (14.25), and (14.29). Also, suppose that $\widehat{\theta}_n$ is a consistent sequence of roots of (14.40), that is, $l'_n(\widehat{\theta}_n) = 0$ and $\widehat{\theta}_n \to^P \theta$, where θ is the true parameter. Then*

$$\sqrt{n}\,(\widehat{\theta}_n - \theta) \longrightarrow^D N\left(0, \frac{1}{\mathcal{I}_1(\theta)}\right). \tag{14.52}$$

Proof. From the sketch of the proof in Section 14.3.1, (14.18) gives us

$$\sqrt{n}\,\frac{1}{n}\,l'_n(\theta) \longrightarrow^D N(0, \mathcal{I}_1(\theta)), \tag{14.53}$$

and (14.19) gives us

$$\frac{1}{n}\,l''_n(\theta) \longrightarrow^P -\mathcal{I}_1(\theta). \tag{14.54}$$

Consider the $M(x_i)$ from assumption (14.29). By the WLLN,

$$\frac{1}{n}\sum_{i=1}^{n} M(X_i) \longrightarrow^P E_\theta[M(X)] < \infty, \tag{14.55}$$

and we have assumed that $\widehat{\theta}_n \to^P \theta$, hence

$$(\widehat{\theta}_n - \theta)\frac{1}{n}\sum_{i=1}^{n} M(X_i) \longrightarrow^P 0. \tag{14.56}$$

Thus for any $\delta > 0$,

$$P[|\widehat{\theta}_n - \theta| < \delta \ \text{ and } \ |(\widehat{\theta}_n - \theta)\frac{1}{n}\sum_{i=1}^{n} M(X_i)| < \delta] \longrightarrow 1. \tag{14.57}$$

Now take the $\delta < \epsilon$, where ϵ is from the assumption (14.29). Then

$$
\begin{aligned}
|\widehat{\theta}_n - \theta| < \delta &\Longrightarrow |\theta_n^* - \theta| < \delta \\
&\Longrightarrow |l'''_1(\theta_n^*; x_i)| \leq M(x_i) \quad \text{by (14.29)} \\
&\Longrightarrow |\frac{1}{n}\,l'''_n(\theta_n^*)| \leq \frac{1}{n}\sum_{i=1}^{n} M(x_i).
\end{aligned} \tag{14.58}
$$

Thus

$$|\widehat{\theta}_n - \theta| < \delta \ \text{ and } \ |(\widehat{\theta}_n - \theta)\frac{1}{n}\sum_{i=1}^{n} M(X_i)| < \delta \Longrightarrow |(\widehat{\theta}_n - \theta)\frac{1}{n}\,l'''_n(\theta_n^*)| < \delta, \tag{14.59}$$

and (14.57) shows that

$$P[|(\widehat{\theta}_n - \theta)\frac{1}{n}\,l'''_n(\theta_n^*)| < \delta] \longrightarrow 1. \tag{14.60}$$

That is,

$$(\widehat{\theta}_n - \theta)\frac{1}{n}\,l'''_n(\theta_n^*) \longrightarrow^P 0. \tag{14.61}$$

Putting together (14.53), (14.54), and (14.61),

$$\sqrt{n}\,(\widehat{\theta}_n - \theta) = -\frac{\sqrt{n}\,\frac{1}{n}\,l'_n(\theta)}{\frac{1}{n}\,l''_n(\theta) + (\widehat{\theta}_n - \theta)\,\frac{1}{2n}\,l'''_n(\theta_n^*)}$$

$$\xrightarrow{\mathcal{D}} -\frac{N(0, \mathcal{I}_1(\theta))}{-\mathcal{I}_1(\theta) + 0}$$

$$= N\left(0, \frac{1}{\mathcal{I}_1(\theta)}\right), \tag{14.62}$$

which proves the theorem (14.52). $\qquad\qquad\square$

Note. The assumption that we have a consistent sequence of roots can be relaxed to the condition (14.48), that is, $\widehat{\theta}_n$ has to be a root of l'_n only with high probability:

$$P[l'_n(\widehat{\theta}_n) = 0 \text{ and } \widehat{\theta}_n \in \mathcal{T}] \longrightarrow 1. \tag{14.63}$$

If $\mathcal{I}_1(\theta)$ is continuous, $\mathcal{I}_n(\widehat{\theta}_n)/n \xrightarrow{\mathcal{P}} \mathcal{I}_1(\theta)$, so that (14.12) holds here, too. It may be that $\mathcal{I}_1(\theta)$ is annoying to calculate. One can instead use the observed Fisher Information as in (14.13), $\mathcal{I}_n(\widehat{\theta}_n)$. The advantage is that the second derivative itself is used, and the expected value of it does not need to be calculated. Using $\widehat{\theta}_n$ yields a consistent estimate of $I_1(\theta)$:

$$\frac{1}{n}\widehat{\mathcal{I}}_n(\widehat{\theta}_n) = -\frac{1}{n}l''_n(\theta) - (\widehat{\theta}_n - \theta)\frac{1}{n}l'''_n(\theta_n^*)$$

$$\xrightarrow{\mathcal{P}} \mathcal{I}_1(\theta) + 0, \tag{14.64}$$

by (14.54) and (14.61). It is thus legitimate to use, for large n, either of the following as approximate 95% confidence intervals:

$$\widehat{\theta}_n \pm 2\frac{1}{\sqrt{n\,\mathcal{I}_1(\widehat{\theta}_n)}} \tag{14.65}$$

or

$$\widehat{\theta}_n \pm 2\frac{1}{\sqrt{\widehat{\mathcal{I}}_n(\widehat{\theta}_n)}} \quad \text{or, equivalently,} \quad \widehat{\theta}_n \pm 2\frac{1}{\sqrt{-l''_n(\widehat{\theta}_n)}}. \tag{14.66}$$

14.7 Asymptotic efficiency

We do not expected the MLE to be unbiased. In fact, it may be that the mean or variance of the MLE does not exist. For example, the MLE for $1/\lambda$ in the Poisson case is $1/\overline{X}_n$, which does not have a finite mean because there is a positive probability that $\overline{X}_n = 0$. But under the given conditions, if n is large, the MLE is close in distribution to a random variable that is unbiased and has optimal (in a sense given below) asymptotic variance.

A sequence δ_n is a *consistent and asymptotically normal* sequence of estimators of $g(\theta)$ if

$$\delta_n \xrightarrow{\mathcal{P}} g(\theta) \quad \text{and} \quad \sqrt{n}\,(\delta_n - g(\theta)) \xrightarrow{\mathcal{D}} N(0, \sigma_g^2(\theta)) \tag{14.67}$$

for some $\sigma_g^2(\theta)$. That is, it is consistent and asymptotically normal. The asymptotic normality implies the consistency, because $g(\theta)$ is subtracted from the estimator in the second convergence.

Theorem 14.7. *Suppose the conditions in Section 14.4 hold. If the sequence δ_n is a consistent and asymptotically normal estimator of $g(\theta)$, and g' is continuous, then*

$$\sigma_g^2(\theta) \geq \frac{g'(\theta)^2}{\mathcal{I}_1(\theta)} \tag{14.68}$$

for all $\theta \in \mathcal{T}$ except perhaps for a set of Lebesgue measure 0.

See Bahadur (1964) for a proof. That coda about "Lebesgue measure 0" is there because it is possible to trick up the estimator so that it is "superefficient" at a few points. If $\sigma_g^2(\theta)$ is continuous in θ, then you can ignore that bit. Also, the conditions need not be quite as strict as in Section 14.4 in that the part about the third derivative in (14.22) can be dropped, and (14.29) can be changed to be about the second derivative.

Definition 14.8. *If the conditions above hold, then the **asymptotic efficiency** of the sequence δ_n is*

$$AE_\theta(\delta_n) = \frac{g'(\theta)^2}{\mathcal{I}_1(\theta)\sigma_g^2(\theta)}. \tag{14.69}$$

*If the asymptotic efficiency is 1, then the sequence is said to be **asymptotically efficient**.*

A couple of immediate consequences follow, presuming the conditions hold.

1. The maximum likelihood estimator of θ is asymptotically efficient, because $\sigma^2(\theta) = 1/\mathcal{I}_1(\theta)$ and $g'(\theta) = 1$.

2. If $\widehat{\theta}_n$ is an asymptotically efficient estimator of θ, then $g(\widehat{\theta}_n)$ is an asymptotically efficient estimator of $g(\theta)$ by the Δ-method.

Recall that in Section 9.2.1 we introduced the asymptotic relative efficiency of two estimators. Here, we see that the asymptotic efficiency of an estimator is its asymptotic relative efficiency to the MLE.

14.7.1   Mean and median

Recall Section 9.2.1, where we compared the median and mean as estimators of the sample median θ in some location families. Here we look at the asymptotic efficiencies. For given base pdf f, the densities we consider are $f(x - \mu)$ for $\mu \in \mathbb{R}$. In order to satisfy condition (14.21), we need that $f(x) > 0$ for all $x \in \mathbb{R}$, which rules out the uniform.

We first need to find the Fisher information. Since $l_1(\mu; x_i) = \log(f(x_i - \mu))$, $l_1'(\mu; x_i) = -f'(x_i - \mu)/f(x_i - \mu)$. Using (14.28), we have that

$$\mathcal{I}_1(\mu) = E[l_1'(\mu; X_i)^2] = \int_{-\infty}^{\infty} \left(\frac{f'(x-\mu)}{f(x-\mu)}\right)^2 f(x-\mu)dx$$

$$= \int_{-\infty}^{\infty} \frac{f'(x)^2}{f(x)}dx. \tag{14.70}$$

Note that the information does not depend on μ.

For example, consider the logistic, which has

$$f(x) = \frac{e^x}{(1+e^x)^2}. \tag{14.71}$$

Then

$$\frac{f'(x)}{f(x)} = \frac{\partial}{\partial x} \log(f(x)) = \frac{\partial}{\partial x}(x - 2\log(1 + e^x)) = 1 - 2\frac{e^x}{1 + e^x}, \tag{14.72}$$

and

$$
\begin{aligned}
\mathcal{I}_1(0) &= \int_{-\infty}^{\infty} \left(1 - 2\frac{e^x}{1 + e^x}\right)^2 \frac{e^x}{(1+e^x)^2} dx \\
&= \int_0^1 (1 - 2(1-u))^2 u(1-u)\frac{du}{u(1-u)} \\
&= 4\int_0^1 (u - \tfrac{1}{2})^2 du = \frac{1}{3},
\end{aligned}
\tag{14.73}
$$

where we make the change of variables $u = 1/(1+e^x)$.

Exercise 14.9.6 finds the Fisher information for the normal, Cauchy, and Laplace. The Laplace does not satisfy the conditions, since its pdf $f(x)$ is not differentiable at $x = 0$, but the results still hold as long as we take $Var[l_1'(\theta; X_i)]$ as $\mathcal{I}_1(0)$. The next table exhibits the Fisher information, and asymptotic efficiencies of the mean and median, for these distributions. The σ^2 is the variance of X_i, and the τ^2 is the variance in the asymptotic distribution of $\sqrt{n}(\text{Median}_n - \mu)$, found earlier in (9.32).

Base distribution	σ^2	τ^2	$\mathcal{I}_1(\mu)$	AE(Mean)	AE(Median)
Normal$(0,1)$	1	$\pi/2$	1	1	$2/\pi \approx 0.6366$
Cauchy	∞	$\pi^2/4$	$1/2$	0	$8/\pi^2 \approx 0.8106$
Laplace	2	1	1	$1/2$	1
Logistic	$\pi^2/3$	4	$1/3$	$9/\pi^2 \approx 0.9119$	$3/4$

$$\tag{14.74}$$

For these cases, the MLE is asymptotically efficient; in the normal case the MLE is the mean, and in the Laplace case the MLE is the median. If you had to choose between the mean and the median, but weren't sure which of the distributions is in effect, the median would be the safer choice. Its efficiency ranges from about 64% to 100%, while the mean's efficiency can be 50% or even 0.

Lehmann (1991) (an earlier edition of Lehmann and Casella (2003)) in Table 4.4 has more calculations for the asymptotic efficiencies of some trimmed means. The α trimmed mean for a sample of n observations is the mean of the remaining observations after removing the smallest and largest floor$(n\alpha)$ observations, where floor(x) is the largest integer less than or equal to x. The regular mean has $\alpha = 0$, and the median has $\alpha = 1/2$ (or slightly lower than $1/2$ if n is even). Here are some of the asymptotic efficiencies:

$f \downarrow; \alpha \rightarrow$	0	1/8	1/4	3/8	1/2
Normal	1.00	0.94	0.84	0.74	0.64
Cauchy	0.00	0.50	0.79	0.88	0.81
t_3	0.50	0.96	0.98	0.92	0.81
t_5	0.80	0.99	0.96	0.88	0.77
Laplace	0.50	0.70	0.82	0.91	1.00
Logistic	0.91	0.99	0.95	0.86	0.75

$$(14.75)$$

This table can help you choose what trimming amount you would want to use, depending on what you think your f might be. You can see that between the mean and a small amount of trimming (1/8), the efficiencies of most distributions go up substantially, while the normal's goes down only a small amount. With 25% trimming, all have at least a 79% efficiency.

14.8   Multivariate parameters

The work so far assumed that θ was one-dimensional (although the data could be multidimensional). Everything follows for multidimensional parameters θ, with some extended definitions. Now assume that $\mathcal{T} \subset \mathbb{R}^K$, and that $\mathcal{T}$ is open. The score function, the derivative of the loglikelihood, is now K-dimensional:

$$\nabla l_n(\theta) = \nabla l_n(\theta; x_1, \ldots, x_n) = \sum_{i=1}^{n} \nabla l_1(\theta; x_i), \qquad (14.76)$$

where

$$\nabla l_1(\theta; x_i) = \begin{pmatrix} \frac{\partial l_1(\theta; x_i)}{\partial \theta_1} \\ \vdots \\ \frac{\partial l_1(\theta; x_i)}{\partial \theta_K} \end{pmatrix}. \qquad (14.77)$$

The MLE then satisfies the equations

$$\nabla l_n(\widehat{\theta}_n) = 0. \qquad (14.78)$$

As in Lemma 14.1,

$$E[\nabla l_1(\theta; \mathbf{X}_i)] = 0. \qquad (14.79)$$

Also, the Fisher information in one observation is a $K \times K$ matrix,

$$\mathcal{I}_1(\theta) = \operatorname{Cov}_\theta[\nabla l_1(\theta; \mathbf{X}_i)] = E_\theta[\widehat{\mathcal{I}}_1(\theta; \mathbf{X}_i)], \qquad (14.80)$$

where $\widehat{\mathcal{I}}_1$ is the observed Fisher information matrix in one observation defined by

$$\widehat{\mathcal{I}}_1(\theta; x_i) = \begin{bmatrix} \frac{\partial^2 l_1(\theta; x_i)}{\partial \theta_1^2} & \frac{\partial^2 l_1(\theta; x_i)}{\partial \theta_1 \partial \theta_2} & \cdots & \frac{\partial^2 l_1(\theta; x_i)}{\partial \theta_1 \partial \theta_K} \\ \frac{\partial^2 l_1(\theta; x_i)}{\partial \theta_2 \partial \theta_1} & \frac{\partial^2 l_1(\theta; x_i)}{\partial \theta_2^2} & \cdots & \frac{\partial^2 l_1(\theta; x_i)}{\partial \theta_2 \partial \theta_K} \\ \vdots & \vdots & \ddots & \vdots \\ \frac{\partial^2 l_1(\theta; x_i)}{\partial \theta_K \partial \theta_1} & \frac{\partial^2 l_1(\theta; x_i)}{\partial \theta_K \partial \theta_2} & \cdots & \frac{\partial^2 l_1(\theta; x_i)}{\partial \theta_K^2} \end{bmatrix}. \qquad (14.81)$$

I won't detail all the assumptions, but they are basically the same as before, except that they apply to all the partial and mixed partial derivatives. The equation (14.25),

$$0 < \mathcal{I}_1(\boldsymbol{\theta}) < \infty \text{ for all } \boldsymbol{\theta} \in \mathcal{T}, \tag{14.82}$$

means that $\mathcal{I}_1(\boldsymbol{\theta})$ is positive definite, and all its elements are finite. The two main results are next.

1. If $\widehat{\boldsymbol{\theta}}_n$ is a consistent sequence of roots of the derivative of the loglikelihood, then

$$\sqrt{n} \, (\widehat{\boldsymbol{\theta}}_n - \boldsymbol{\theta}) \longrightarrow^{\mathcal{D}} N(0, \mathcal{I}_1^{-1}(\boldsymbol{\theta})). \tag{14.83}$$

2. If δ_n is a consistent and asymptotically normal sequence of estimators of $g(\boldsymbol{\theta})$, where the partial derivatives of g are continuous, and

$$\sqrt{n} \, (\delta_n - g(\boldsymbol{\theta})) \longrightarrow^{\mathcal{D}} N(0, \sigma_g^2(\boldsymbol{\theta})), \tag{14.84}$$

then

$$\sigma_g^2(\boldsymbol{\theta}) \geq \mathbf{D}_g(\boldsymbol{\theta}) \mathcal{I}_1^{-1}(\boldsymbol{\theta}) \mathbf{D}_g(\boldsymbol{\theta})' \tag{14.85}$$

for all $\boldsymbol{\theta} \in \mathcal{T}$ (except possibly for a few), where $\mathbf{D}_g$ is the $1 \times K$ vector of partial derivatives $\partial g(\boldsymbol{\theta})/\partial \theta_i$ as in the multivariate Δ-method in (9.48).

If $\widehat{\boldsymbol{\theta}}_n$ is the MLE of $\boldsymbol{\theta}$, then the lower bound in (14.85) is the variance in the asymptotic distribution of $\sqrt{n}(g(\widehat{\boldsymbol{\theta}}_n) - g(\boldsymbol{\theta}))$. Which is to say that the MLE of $g(\boldsymbol{\theta})$ is asymptotically efficient.

14.8.1 Non-IID models

Often the observations under consideration are not iid, as in the regression model (12.3) where the Y_i's are independent but have different means depending on their x_i's. Under suitable conditions, the asymptotic results will still hold for the MLE. In such case, the asymptotic distributions would use the Fisher's information (observed or expected) on the left-hand side:

$$\begin{aligned}\widehat{\mathcal{I}}_n^{1/2}(\widehat{\boldsymbol{\theta}}_n)(\widehat{\boldsymbol{\theta}}_n - \boldsymbol{\theta}) &\longrightarrow^{\mathcal{D}} N(0, \mathbf{I}_K) \text{ or}\\ \mathcal{I}_n^{1/2}(\widehat{\boldsymbol{\theta}}_n)(\widehat{\boldsymbol{\theta}}_n - \boldsymbol{\theta}) &\longrightarrow^{\mathcal{D}} N(0, \mathbf{I}_K).\end{aligned} \tag{14.86}$$

Of course, these two convergences hold in the iid case as well.

14.8.2 Common mean

Suppose $X_1, \ldots, X_n$ and $Y_1, \ldots, Y_n$ are all independent, and

$$X_i's \text{ are } N(\mu, \theta_X), \quad Y_i's \text{ are } N(\mu, \theta_Y). \tag{14.87}$$

That is, the X_i's and Y_i's have the same means, but possibly different variances. Such data may arise when two unbiased measuring devices with possibly different precisions are used. We can use the likelihood results we have seen so far by pairing up the X_i's with the Y_i's, so that we have $(X_1, Y_1), \ldots, (X_n, Y_n)$ as iid vectors. In fact, similar results will still hold if $n \neq m$, as long as the ratio n/m has a limit strictly between 0 and 1.

Exercise 14.9.3 shows that the score function in one observation is

$$
\nabla l_1(\mu, \theta_X, \theta_Y; x_i, y_i) = \begin{pmatrix} \frac{x_i - \mu}{\theta_X} + \frac{y_i - \mu}{\theta_Y} \\[2mm] -\frac{1}{2}\frac{1}{\theta_X} + \frac{1}{2}\frac{(x_i - \mu)^2}{\theta_X^2} \\[2mm] -\frac{1}{2}\frac{1}{\theta_Y} + \frac{1}{2}\frac{(y_i - \mu)^2}{\theta_Y^2} \end{pmatrix}, \tag{14.88}
$$

and the Fisher information in one observation is

$$
\mathcal{I}_1(\mu, \theta_X, \theta_Y) = \begin{pmatrix} \frac{1}{\theta_X} + \frac{1}{\theta_Y} & 0 & 0 \\[2mm] 0 & \frac{1}{2\theta_X^2} & 0 \\[2mm] 0 & 0 & \frac{1}{2\theta_Y^2} \end{pmatrix}. \tag{14.89}
$$

A multivariate version of the Newton-Raphson algorithm in (14.5) replaces the observed Fisher information with its expectation. Specifically, letting $\theta = (\mu, \theta_X, \theta_Y)'$ be the parameter vector, we obtain the jth guess from the $(j-1)^{st}$ one via

$$
\theta^{(j)} = \theta^{(j-1)} + \mathcal{I}_n^{-1}(\theta^{(j-1)}) \nabla l_n(\theta^{(j-1)}). \tag{14.90}
$$

We could use the observed Fisher information, but it is not diagonal, so the expected Fisher information is easier to invert. A bit of algebra shows that the updating reduces to the following:

$$
\mu^{(j)} = \frac{\bar{x}_n \theta_Y^{(j-1)} + \bar{y}_n \theta_X^{(j-1)}}{\theta_X^{(j-1)} + \theta_Y^{(j-1)}},
$$

$$
\theta_X^{(j)} = s_X^2 + (\bar{x}_n - \mu^{(j-1)})^2, \quad \text{and}
$$

$$
\theta_Y^{(j)} = s_Y^2 + (\bar{y}_n - \mu^{(j-1)})^2. \tag{14.91}
$$

Here, $s_X^2 = \sum (x_i - \bar{x}_n)^2/n$, and similarly for s_Y^2.

Denoting the MLE of μ by $\hat{\mu}_n$, we have that

$$
\sqrt{n}(\hat{\mu}_n - \mu) \xrightarrow{D} N\left(0, \frac{\theta_X \theta_Y}{\theta_X + \theta_Y}\right). \tag{14.92}
$$

14.8.3   Logistic regression

In Chapter 12, we looked at linear regression, where the mean of the Y_i's is assumed to be a linear function of some x_i's. If the Y_i's are Bernoulli(p_i), so take on only the values 0 and 1, then a linear model on $E[Y_i] = p_i$ may not be appropriate as the $\alpha + \beta x_i$ could easily fall outside of the [0,1] range. A common solution is to model the logit of the p_i's, which we saw way back in Exercise 1.7.15. The logit of a probability p is the log odds of the probability:

$$
\text{logit}(p) = \log\left(\frac{p}{1-p}\right). \tag{14.93}
$$

This transformation has range $\mathbb{R}$. A simple logistic regression model is based on $(x_1, Y_1), \ldots, (x_n, Y_n)$ independent observations, where for each i, x_i is fixed, and $Y_i \sim$ Bernoulli(p_i) with

$$\text{logit}(p_i) = \beta_0 + \beta_1 x_i, \tag{14.94}$$

β_0 and β_1 being the parameters. Multiple logistic regression has several x-variables, so that the model is

$$\text{logit}(p_i) = \beta_0 + \beta_1 x_{i1} + \cdots + \beta_K x_{iK}. \tag{14.95}$$

Analogous to the notation in (12.9) for linear regression, we write

$$\text{logit}(\mathbf{p}) = \mathbf{x}\boldsymbol{\beta} = \begin{pmatrix} 1 & x_{11} & x_{12} & \cdots & x_{1K} \\ 1 & x_{21} & x_{22} & \cdots & x_{2K} \\ \vdots & \vdots & \vdots & \cdots & \vdots \\ 1 & x_{n1} & x_{n2} & \cdots & x_{nK} \end{pmatrix} \begin{pmatrix} \beta_0 \\ \beta_1 \\ \beta_2 \\ \vdots \\ \beta_K \end{pmatrix}, \tag{14.96}$$

where $p = (p_1, \ldots, p_n)'$ and by logit$(\mathbf{p})$ we mean $(\text{logit}(p_1), \ldots, \text{logit}(p_n))'$.

To use maximum likelihood, we first have to find the likelihood as a function of β. The inverse function of $z = \text{logit}(p)$ is $p = e^z/(1+e^z)$, so that since the likelihood of $Y \sim$ Bernoulli(p) is $p^y(1-p)^{1-y} = (p/(1-p))^y(1-p)$, we have that the likelihood for the data $\mathbf{Y} = (Y_1, \ldots, Y_n)'$ is

$$\begin{aligned} L_n(\boldsymbol{\beta}; \mathbf{y}) &= \prod_{i=1}^n \left(\frac{p_i}{1-p_i}\right)^{y_i} (1-p_i) \\ &= \prod_{i=1}^n (e^{\mathbf{x}_i\boldsymbol{\beta}})^{y_i} (1+e^{\mathbf{x}_i\boldsymbol{\beta}})^{-1} \\ &= e^{\mathbf{y}'\mathbf{x}\boldsymbol{\beta}} \prod_{i=1}^n (1+e^{\mathbf{x}_i\boldsymbol{\beta}})^{-1}, \end{aligned} \tag{14.97}$$

where $\mathbf{x}_i$ is the ith row of $\mathbf{x}$. Note that we have an exponential family.

Since the observations do not have the same distribution (the distribution of Y_i depends on $\mathbf{x}_i$), we deal with the score and Fisher information of the entire sample. The score function can be written as

$$\nabla l(\boldsymbol{\beta}; \mathbf{y}) = \mathbf{x}'(\mathbf{y} - \mathbf{p}), \tag{14.98}$$

keeping in mind that the p_i's are functions of $\mathbf{x}_i\boldsymbol{\beta}$. The Fisher information is the same as the observed Fisher information, which can be written as

$$\mathcal{I}_n(\boldsymbol{\beta}) = \text{Cov}[\nabla l(\boldsymbol{\beta}; \mathbf{y})] = \mathbf{x}' \text{diag}(p_1(1-p_1), \ldots, p_n(1-p_n))\mathbf{x}, \tag{14.99}$$

where $\text{diag}(a_1, \ldots, a_n)$ is the diagonal matrix with the a_i's along the diagonal. The MLE then can be found much as in (14.90), though using software such as R is easier.

Fahrmeir and Kaufmann (1985) show that the asymptotic normality is valid here, even though we do not have iid observations, under some conditions: The minimum eigenvalue of $\mathcal{I}_n(\boldsymbol{\beta})$ goes to ∞, and $\mathbf{x}_n'\mathcal{I}_n^{-1}(\boldsymbol{\beta})\mathbf{x}_n \to 0$, as $n \to \infty$. The latter follows from the former if the $\mathbf{x}_i$'s are bounded.

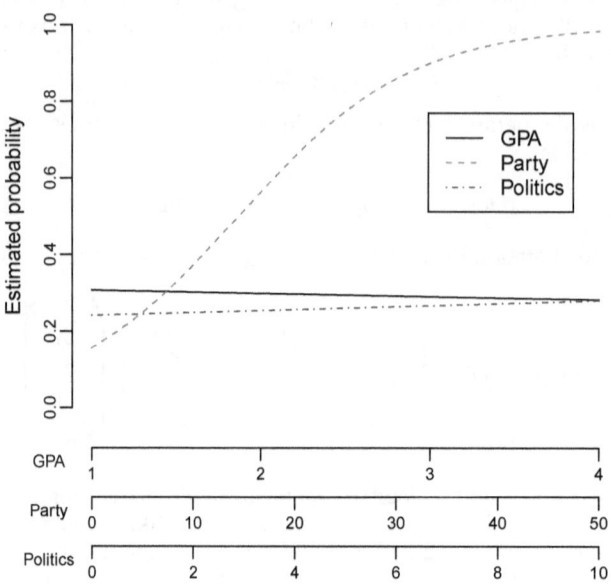

Figure 14.2: The estimated probabilities of being Greek as a function of the variables GPA, party, and politics, each with the other two being held at their average value.

When there are multiple observations with the same x_i value, the model can be equivalently but more compactly represented as binomials. That is, the data are $Y_1, \ldots, Y_q$, independent, where $Y_i \sim \text{Binomial}(n_i, p_i)$, and $\text{logit}(p_i) = x_i \beta$ as in (14.95). Now $n = \sum_{i=1}^{q} n_i$, $\mathbf{p}$ is $q \times 1$, and $\mathbf{x}$ is $q \times (K+1)$ in (14.96). The likelihood can be written

$$L_n(\boldsymbol{\beta}; \mathbf{y}) = e^{\mathbf{y}'\mathbf{x}\boldsymbol{\beta}} \prod_{i=1}^{q} (1 + e^{x_i\boldsymbol{\beta}})^{-n_i}, \tag{14.100}$$

so the score is

$$\nabla l(\boldsymbol{\beta}; \mathbf{y}) = \mathbf{x}'(\mathbf{y} - (n_1 p_1, \ldots, n_q p_q)') \tag{14.101}$$

and the Fisher information is

$$\mathcal{I}_n(\boldsymbol{\beta}) = \text{Cov}[\nabla l(\boldsymbol{\beta}; \mathbf{y})] = \mathbf{x}' \text{diag}(n_1 p_1(1 - p_1), \ldots, n_q p_q(1 - p_q))\mathbf{x}. \tag{14.102}$$

Being Greek

Here we will use a survey of $n = 788$ students to look at some factors that are related to people being Greek in the sense of being a member of a fraternity or sorority. The Y_i's are then 1 if that person is Greek, and 0 if not. The x-variables we will consider are gender (0 = male, 1 = female), GPA (grade point average), political views (from 0 (most liberal) to 10 (most conservative)), and average number of hours per week spent partying. Next are snippets of the 788×1 vector **y** and the 788×4 matrix **x**:

$$
\mathbf{y} = \begin{pmatrix} 0 \\ 0 \\ 0 \\ 1 \\ \vdots \\ 0 \end{pmatrix} \quad \text{and} \quad \mathbf{x} = \begin{pmatrix} 1 & 0 & 2.8 & 16 & 3 \\ 1 & 1 & 3.9 & 0 & 8 \\ 1 & 0 & 3.1 & 4 & 3 \\ 1 & 0 & 3.7 & 10 & 6 \\ \vdots & \vdots & \vdots & \vdots & \vdots \\ 1 & 1 & 3.8 & 4 & 5 \end{pmatrix}. \tag{14.103}
$$

To use R for finding the MLE, let y be the **y** vector, and x be the **x** matrix without the first column of 1's. The following will calculate and display the results:

```
lreg <- glm(y~x,family="binomial")
summary(lreg)
```

The estimate of β is

$$
\widehat{\beta} = \begin{pmatrix} -2.2133 \\ 0.5023 \\ -0.0384 \\ 0.1166 \\ 0.0673 \end{pmatrix}. \tag{14.104}
$$

The standard errors can be found by first estimating the $\text{logit}(p_i)$'s, turning them into $\widehat{p}_i$'s, inserting these into the formula (14.99) for Fisher's information matrix, then inverting the matrix, to obtain an estimate of the asymptotic covariance:

$$
\widehat{Cov}(\widehat{\beta}) = \mathcal{I}_n^{-1}(\widehat{\beta}) = \frac{1}{100} \begin{pmatrix} 38.1772 & -1.8705 & -9.5683 & -0.1392 & -0.7459 \\ -1.8705 & 3.5100 & -0.3490 & 0.0336 & 0.0505 \\ -9.5683 & -0.3490 & 2.9939 & -0.0125 & 0.0032 \\ -0.1392 & 0.0336 & -0.0125 & 0.0196 & 0.0022 \\ -0.7459 & 0.0505 & 0.0032 & 0.0022 & 0.1466 \end{pmatrix}.
$$
$$\tag{14.105}$$

The estimated standard errors of the coefficients are the square roots of the diagonals. The next table has the estimates and standard errors (ignoring the intercept), plus the approximate 95% confidence intervals.

	$\widehat{\beta}_i$	$se(\widehat{\beta}_i)$	Confidence interval
Gender	0.5023	0.1874	(0.1276, 0.8770)
GPA	-0.0384	0.1730	(-0.3845, 0.3076)
Party	0.1166	0.0140	(0.0886, 0.1446)
Politics	0.0673	0.0383	(-0.0092, 0.1439)

(14.106)

Thus we see that gender and hours partying have strong positive association with being Greek, GPA does not have much association at all, and politics is maybe mildly positive (the more conservative, the more likely Greek).

Interpreting the coefficients is a little tricky. The ith coefficient estimates the average additional log odds of being Greek associated with a one unit increase in the ith variable, holding the other variables constant. For example, one more hour partying is associated with 0.1166 increase in log odds. Translating to probabilities is non-linear, so depends on the original level. Here we will set all variables except for the ith at their average, and vary the ith, to see the effect on the estimate probabilities.

For gender, we set that variable at 0 and 1, yielding 22.27% and 32.13%, respectively. Thus women are about 10% more likely to be Greek, at the average of the other variables. Figure 14.2 plots the probabilities versus the other three variables. It is clear that partying has the most drastic relationship with being Greek. GPA and politics are both fairly flat.

14.9   Exercises

Exercise 14.9.1. Find the loglikelihood, score function, and Fisher's information in one observation for the following distributions: (a) Bernoulli$(p), p \in (0, 1)$. (b) Exponential$(\lambda), \lambda > 0$. (c) $N(0, \lambda), \lambda > 0$. (d) $N(\mu, \mu^2), \mu > 0$ (so that the coefficient of variation is always 1).

Exercise 14.9.2. Continue with Exercise 13.8.13 on the fruit fly data. From (6.114) we have that the data $(N_{00}, N_{01}, N_{10}, N_{11})$ is Multinomial$(n, \mathbf{p}(\theta))$, where $N_{ab} = \#\{(Y_{i1}, Y_{i2}) = (a, b)\}$, and

$$\mathbf{p}(\theta) = (\tfrac{1}{2}(1-\theta)(2-\theta), \tfrac{1}{2}\theta(1-\theta), \tfrac{1}{2}\theta(1-\theta), \tfrac{1}{2}\theta(1+\theta)). \tag{14.107}$$

Thus as in (13.86), the loglikelihood is

$$l_n(\theta) = (n_{00} + n_{01} + n_{10})\log(1-\theta) + n_{00}\log(2-\theta)$$
$$+ (n_{01} + n_{10} + n_{11})\log(\theta) + n_{11}\log(1+\theta). \tag{14.108}$$

(a) The observed Fisher information for the n observations is then

$$\widehat{\mathcal{I}}_n(\theta; n_{00}, n_{01}, n_{10}, n_{11}) = \frac{1}{(1-\theta)^2} A + \frac{1}{(2-\theta)^2} B + \frac{1}{\theta^2} C + \frac{1}{(1+\theta)^2} D. \tag{14.109}$$

Find A, B, C, D as functions of the n_{ij}'s. (b) Show that the expected Fisher information is

$$\mathcal{I}_n(\theta) = \frac{n}{2}\left(\frac{2+\theta}{1-\theta} + \frac{1-\theta}{2-\theta} + \frac{3-\theta}{\theta} + \frac{\theta}{1+\theta}\right) = \frac{3n}{(1-\theta)(2-\theta)\theta(1+\theta)}. \tag{14.110}$$

(c) The Dobzhansky estimator of θ is given in (6.65) to be $\widehat{\theta}_D = \sum\sum y_{ij}/(2n)$. Exercise 6.8.16 shows that its variance is $3\theta(1-\theta)/(4n)$. Find the asymptotic efficiency of $\widehat{\theta}_D$. Graph the efficiency it as a function of θ. What is its minimum? Maximum? For what values of θ, if any, is the Dobzhansky estimator fully efficient (AE=1)?

Exercise 14.9.3. Consider the common mean problem from Section 14.8.2, so that $X_1, \ldots, X_n$ are iid $N(\mu, \theta_X)$, $Y_1, \ldots, Y_n$ are iid $N(\mu, \theta_Y)$, and the X_i's are independent of the Y_i's. (a) Show that the score function in one observation is

$$\nabla l_1(\mu, \theta_X, \theta_Y; x_i, y_i) = \begin{pmatrix} \frac{x_i - \mu}{\theta_X} + \frac{y_i - \mu}{\theta_Y} \\ -\frac{1}{2}\frac{1}{\theta_X} + \frac{1}{2}\frac{(x_i - \mu)^2}{\theta_X^2} \\ -\frac{1}{2}\frac{1}{\theta_Y} + \frac{1}{2}\frac{(y_i - \mu)^2}{\theta_Y^2} \end{pmatrix}. \tag{14.111}$$

What is the expected value of the score? (b) The observed Fisher information matrix in one observation is

$$\widehat{\mathcal{I}}_1(\mu, \theta_X, \theta_Y; x_i, y_i) = \begin{pmatrix} \frac{1}{\theta_X} + \frac{1}{\theta_Y} & \frac{x_i - \mu}{\theta_X^2} & * \\ \frac{x_i - \mu}{\theta_X^2} & -\frac{1}{2\theta_X^2} + \frac{(x_i - \mu)^2}{\theta_X^3} & * \\ * & * & * \end{pmatrix}. \tag{14.112}$$

Find the missing elements. (c) Show that the Fisher information in one observation, $\mathcal{I}_1(\mu, \theta_X, \theta_Y)$, is as in (14.89). (d) Verify the asymptotic variance in (14.92).

Exercise 14.9.4. Continue with the data in Exercise 14.9.3. Instead of the MLE, consider estimators of μ of the form $\widehat{\mu}_a = a\bar{X}_n + (1 - a)\bar{Y}_n$ for some $a \in [0, 1]$. (a) Find $E[\widehat{\mu}_a]$ and $Var[\widehat{\mu}_a]$. Is $\widehat{\mu}_a$ unbiased? (b) Show that the variance is minimized for a equalling $\alpha = \theta_Y/(\theta_X + \theta_Y)$. Is $\widehat{\mu}_\alpha$ an unbiased estimator of μ? (c) Let $\widehat{\alpha}_n = S_Y^2/(S_X^2 + S_Y^2)$. Does $\widehat{\alpha}_n \to^P \alpha$? (d) Consider the estimator $\widehat{\mu}_{\widehat{\alpha}_n}$. Show that

$$\sqrt{n}(\widehat{\mu}_{\widehat{\alpha}_n} - \mu) \longrightarrow^{\mathcal{D}} N\left(0, \frac{\theta_X \theta_Y}{\theta_X + \theta_Y}\right). \tag{14.113}$$

[Hint: First show that $\sqrt{n}(\widehat{\mu}_\alpha - \mu) \sim N(0, \theta_X \theta_Y/(\theta_X + \theta_Y))$. Then show that

$$\sqrt{n}(\widehat{\mu}_{\widehat{\alpha}_n} - \mu) - \sqrt{n}(\widehat{\mu}_\alpha - \mu) = \sqrt{n}(\bar{X}_n - \bar{Y}_n)(\widehat{\alpha}_n - \alpha) \longrightarrow^P 0.] \tag{14.114}$$

(e) What is the asymptotic efficiency of $\widehat{\mu}_{\widehat{\alpha}_n}$?

Exercise 14.9.5. Suppose X is from an exponential family model with pdf $f(x \mid \theta) = a(x) \exp(\theta x - \psi(\theta))$ and parameter space $\theta \in (b, d)$. (It could be that $b = -\infty$ and/or $d = \infty$.) (a) Show that the cumulant generating function is $c(t) = \psi(t + \theta) - \psi(\theta)$. For which values of t is it finite? Is it finite for t in a neighborhood of 0? (b) Let $\mu(\theta) = E_\theta[X]$ and $\sigma^2(\theta) = Var_\theta[X]$. Show that $\mu(\theta) = \psi'(\theta)$ and $\sigma^2(\theta) = \psi''(\theta)$. (c) Show that the score function for one observation is $l_1'(\theta; x) = x - \mu(\theta)$. (d) Show that the observed Fisher information and expected Fisher information in one observation are both $\mathcal{I}_1(\theta) = \sigma^2(\theta)$.

Now suppose $X_1, \ldots, X_n$ are iid from $f(x \mid \theta)$. (e) Show that the MLE based on the n observations is $\widehat{\theta}_n = \mu^{-1}(\bar{x}_n)$. (f) Show that $d\mu^{-1}(w)/dw = 1/\sigma^2(\mu^{-1}(w))$. (g) Use the Δ-method to show that

$$\sqrt{n}(\widehat{\theta}_n - \theta) \longrightarrow^{\mathcal{D}} N\left(0, \frac{1}{\sigma^2(\theta)}\right), \tag{14.115}$$

which proves (14.11) directly for one-dimensional exponential families.

Exercise 14.9.6. This exercise is based on the location family model with pdfs $f(x - \mu)$ for $\mu \in \mathbb{R}$. For each part, verify the Fisher information in one observation for f being the pdf of the given distribution. (a) Normal(0,1), $\mathcal{I}_1(0) = 1$. (b) Laplace. In this case, the first derivative of $\log(f(x))$ is not differentiable at $x = 0$, but because the distribution is continuous, you can ignore that point when calculating $\mathcal{I}_1(\mu) = Var_\mu[l_1'(\mu; X)] = 1$. It will not work to use the second derivative to find the Fisher information. (c) Cauchy, $\mathcal{I}_1(0) = 1/2$. [Hint: Start by showing that

$$\mathcal{I}_1(0) = \frac{4}{\pi} \int_{-\infty}^{\infty} \frac{x^2}{(1 + x^2)^3} dx = \frac{8}{\pi} \int_0^{\infty} \frac{x^2}{(1 + x^2)^3} dx = \frac{4}{\pi} \int_0^1 \sqrt{u}\sqrt{1 - u} \, du, \tag{14.116}$$

using the change of variables $u = 1/(1 + x^2)$. Then note that the integral looks like part of a beta density.]

Exercise 14.9.7. Agresti (2013), Table 4.2, summarizes data on the relationship between snoring and heart disease for $n = 2484$ adults. Observation (Y_i, x_i) indicates whether person i had heart disease ($Y_i = 1$) or did not have heart disease ($Y_i = 0$), and the amount the person snored, in four categories. The table summarizes the data:

Heart disease? $\rightarrow$		Yes	No
Frequency of snoring $\downarrow$	x_i		
Never	-3	24	1355
Occasionally	-1	35	603
Nearly every night	1	21	192
Every night	3	30	224

$$(14.117)$$

(So that there are 24 people who never snore and have $Y_i = 1$, and 224 people who snore every night and have $Y_i = 0$.) The model is the linear logistic one, with logit$(p_i) = \alpha + \beta x_i$, $i = 1, \ldots, n$. The x_i's are categorical, but in order of snoring frequency, so we will code them $x_i = -3, -1, 1, 3$, as in the table. The MLEs are $\widehat{\alpha} = -2.79558$, $\widehat{\beta} = 0.32726$. (a) Find the Fisher information in the entire sample evaluated at the MLE, $\mathcal{I}_n(\widehat{\alpha}, \widehat{\beta})$. (b) Find $\mathcal{I}_n^{-1}(\widehat{\alpha}, \widehat{\beta})$. (c) Find the standard errors of $\widehat{\alpha}$ and $\widehat{\beta}$. Does the slope appear significantly different than 0? (d) For an individual i with $x_i = 3$, find the MLE of logit(p_i) and its standard error. (e) For the person in part (d), find the MLE of p_i and its standard error. [Hint: Use the Δ-method.]

Exercise 14.9.8. Consider a set of teams. The chance team A beats team B in a single game is p_{AB}. If these two teams do not play often, or at all, one cannot get a very good estimate of p_{AB} by just looking at those games. But we often do have information of how they did against other teams, good and bad, which should help in estimating p_{AB}.

Suppose p_A is the chance team A beats the "typical" team, and p_B the chance that team B beats the typical team. Then even if A and B have never played each other, one can use the following idea to come up with a p_{AB}: Both teams flip a coin independently, where the chance of heads is p_A for team A's coin and p_B for team B's. Then if both are heads or both are tails, they flip again. Otherwise, whoever got the heads wins. They keep flipping until someone wins. (a) What is the probability team A beats team B, p_{AB}, in this scenario? (As a function of p_A, p_B.) If $p_A = p_B$, what is p_{AB}? If $p_B = 0.5$ (so that team B is typical), what is p_{AB}? If $p_A = 0.6$ and $p_B = 0.4$, what is p_{AB}? If both are very good: $p_A = 0.9999$ and $p_B = 0.999$, what is p_{AB}? (b) Now let o_A, o_B be their odds of beating a typical team, (odds $= p/(1-p)$). Find o_{AB}, the odds of team A beating team B, as a function of the individual odds (so the answer is in terms of o_A, o_B). (c) Let γ_A and γ_B be the corresponding logits (log odds), so that $\gamma_i = $ logit(p_i). Find $\gamma_{AB} = \log(o_{AB})$ as a function of the γ's. (d) Now suppose there are 4 teams, $i = 1, 2, 3, 4$, and γ_i is the logit for team i beating the typical team. Then the logits for team i beating team j, the γ_{ij}'s, can be written as a linear transformation of the γ_i's:

$$\begin{pmatrix} \gamma_{12} \\ \gamma_{13} \\ \gamma_{14} \\ \gamma_{23} \\ \gamma_{24} \\ \gamma_{34} \end{pmatrix} = \mathbf{x} \begin{pmatrix} \gamma_1 \\ \gamma_2 \\ \gamma_3 \\ \gamma_4 \end{pmatrix}. \qquad (14.118)$$

What is the matrix $\mathbf{x}$? This model for the logits is called the **Bradley-Terry model** (Bradley and Terry, 1952).

Exercise 14.9.9. Continue Exercise 14.9.8 on the Bradley-Terry model. We look at the numbers of times each team in the National League (baseball) beat the other teams in 2015. The original data can be found at http://espn.go.com/mlb/standings/grid/_/year/2015. There are 15 teams. Each row is a paired comparison of a pair of teams. The first two columns of row ij contain the y_{ij}'s and $(n_{ij} - y_{ij})$'s, where y_{ij} is the number of times team i beat team j, and n_{ij} is the number of games they played. The rest of the matrix is the $\mathbf{x}$ matrix for the logistic regression. The model is that $Y_{ij} \sim \text{Binomial}(n_{ij}, p_{ij})$, where $\text{logit}(p_{ij}) = \gamma_i - \gamma_j$. So we have a logistic regression model, where the $\mathbf{x}$ matrix is that in the previous problem, expanded to 15 teams. Because the sum of each row is 0, the matrix is not full rank, so we drop the last column, which is equivalent to setting $\gamma_{15} = 0$, which means that team #15, the Nationals, are the "typical" team. That is ok, since the logits depend only on the differences of the γ_i's. The file http://istics.net/r/nl2015.R contains the data in an R matrix.

Here are the data for just the three pairings among the Cubs, Cardinals, and Brewers:

	W	L
Cubs vs. Brewers	14	5
Cubs vs. Cardinals	8	11
Brewers vs. Cardinals	6	13

$$(14.119)$$

Thus the Cubs and Brewers played 19 times, the Cubs winning 14 and the Brewers winning 5. We found the MLEs and Fisher information for this model using all the teams. The estimated coefficients for these three teams are

$$\widehat{\gamma}_{\text{Cubs}} = 0.4525, \quad \widehat{\gamma}_{\text{Brewers}} = -0.2477, \quad \widehat{\gamma}_{\text{Cardinals}} = 0.5052. \qquad (14.120)$$

The part of the inverse of the Fisher information at the MLE pertaining to these three teams is

	Cubs	Brewers	Cardinals
Cubs	0.05892	0.03171	0.03256
Brewers	0.03171	0.05794	0.03190
Cardinals	0.03256	0.03190	0.05971

$$(14.121)$$

(a) For each of the three pairings of the three above teams, find the estimate of $\text{logit}(p_{ij})$ and its standard error. For which pair, if any, does the logit appear significantly different from 0? (I.e., 0 is not in the approximate 95% confidence interval.) (b) For each pair, find the estimated p_{ij} and its standard error. (c) Find the estimated expected number of wins and losses for each matchup. That is, find $(n_{ij}\widehat{p}_{ij}, n_{ij}(1 - \widehat{p}_{ij}))$ for each pair. Compare these to the actual results in (14.119). Are the estimates close to the actual wins and losses?

Exercise 14.9.10. Suppose

$$\begin{pmatrix} X_1 \\ Y_1 \end{pmatrix}, \dots, \begin{pmatrix} X_n \\ Y_n \end{pmatrix} \text{ are iid } N\left(\begin{pmatrix} 0 \\ 0 \end{pmatrix}, \begin{pmatrix} 1 & \rho \\ \rho & 1 \end{pmatrix}\right) \tag{14.122}$$

for $\rho \in (-1,1)$. This question will consider the following estimators of ρ:

$$R_{1n} = \frac{1}{n} \sum_{i=1}^{n} X_i Y_i;$$

$$R_{2n} = \frac{\sum_{i=1}^{n} X_i Y_i}{\sqrt{\sum_{i=1}^{n} X_i^2 \; \sum_{i=1}^{n} Y_i^2}};$$

$$R_{3n} = \text{the MLE.} \tag{14.123}$$

(a) Find $\mathcal{I}_1(\rho)$, Fisher's information in one observation. (b) What is the asymptotic variance for R_{1n}, that is, what is the $\sigma_1^2(\rho)$ in

$$\sqrt{n} \, (R_{1n} - \rho) \longrightarrow^{\mathcal{D}} N(0, \sigma_1^2(\rho))? \tag{14.124}$$

(c) Data on $n = 107$ students, where the X_i's are the scores on the midterms, and Y_i are the scores on the final, has

$$\sum x_i y_i = 73.31, \quad \sum x_i^2 = 108.34, \quad \sum y_i^2 = 142.80. \tag{14.125}$$

(Imagine the scores are normalized so that the population means are 0 and population variances are 1.) Find the values of R_{1n}, R_{2n}, and R_{3n} for these data. Calculating the MLE requires a numerical method like Newton-Raphson. Are these estimates roughly the same? (d) Find the asymptotic efficiency of the three estimators. Which one (among these three) has the best asymptotic efficiency? (See (9.71) for the asymptotic variance of R_{2n}.) Is there an obvious worst one?

Chapter 15

Hypothesis Testing

Estimation addresses the question, "What is θ?" Hypothesis testing addresses questions like, "Is $\theta = 0$?" Confidence intervals do both. It will give a range of plausible values, and if you wonder whether $\theta = 0$ is plausible, you just check whether 0 is in the interval. But hypothesis testing also addresses broader questions in which confidence intervals may be clumsy. Some types of questions for hypothesis testing:

- Is a particular drug more effective than a placebo?

- Are cancer and smoking related?

- Is the relationship between amount of fertilizer and yield linear?

- Is the distribution of income the same among men and women?

- In a regression setting, are the errors independent? Normal? Homoscedastic?

The main feature of hypothesis testing problems is that there are two competing models under consideration, the **null hypothesis** model and the **alternative hypothesis** model. The random variable (vector) $\mathbf{X}$ and space $\mathcal{X}$ are the same in both models, but the sets of distributions are different, being denoted $\mathcal{P}_0$ and $\mathcal{P}_A$ for the null and alternative, respectively, where $\mathcal{P}_0 \cap \mathcal{P}_A = \emptyset$. If P is the probability distribution for X, then the hypotheses are written

$$H_0: P \in \mathcal{P}_0 \text{ versus } H_A: P \subset \mathcal{P}_A. \tag{15.1}$$

Often both models will be parametric:

$$\mathcal{P}_0 = \{P_{\boldsymbol{\theta}} \,|\, \boldsymbol{\theta} \in \mathcal{T}_0\} \text{ and } \mathcal{P}_A = \{P_{\boldsymbol{\theta}} \,|\, \boldsymbol{\theta} \in \mathcal{T}_A\}, \text{ with } \mathcal{T}_0, \mathcal{T}_A \subset \mathcal{T}, \mathcal{T}_0 \cap \mathcal{T}_A = \emptyset, \tag{15.2}$$

for some overall parameter space $\mathcal{T}$. It is not unusual, but also not required, that $\mathcal{T}_A = \mathcal{T} - \mathcal{T}_0$. In a parametric setting, the hypotheses are written

$$H_0: \boldsymbol{\theta} \in \mathcal{T}_0 \text{ versus } H_A: \boldsymbol{\theta} \in \mathcal{T}_A. \tag{15.3}$$

Mathematically, there is no particular reason to designate one of the hypotheses null and the other alternative. In practice, the null hypothesis tends to be the one that represents the status quo, or that nothing unusual is happening, or that everything is ok, or that the new isn't any better than the old, or that the defendant is innocent. For example, in the Salk polio vaccine study (Exercise 6.8.9), the null hypothesis would be that the vaccine has no effect. In simple linear regression, the typical null hypothesis would be that the slope is 0, i.e., the distributions of the Y_i's do not depend on the x_i's. One may also wish to test the assumptions in regression: The null hypothesis would be that the residuals are iid Normal$(0, \sigma_e^2)$'s.

Section 16.4 considers model selection, in which there are a number of models, and we wish to choose the best in some sense. Hypothesis testing could be thought of as a special case of model selection, where there are just two models, but it is more useful to keep the notions separate. In model selection, the models have the same status, while in hypothesis testing the null hypothesis is special in representing a status quo. (Though hybrid model selection/hypothesis testing situations could be imagined.)

We will look at two primary approaches to hypothesis testing. The **accept/reject** or **fixed α** or **Neyman-Pearson** approach is frequentist and action-oriented: Based on the data **x**, you either accept or reject the null hypothesis. The evaluation of any procedure is based on the chance of making the wrong decision. The **Bayesian** approach starts with a prior distribution (on the parameters, as well as on the truth of the two hypotheses), and produces the posterior probability that the null hypothesis is true. In the latter case, you can decide to accept or reject the null based on a cutoff for its probability. We will also discuss **p-values**, which arise in the frequentist paradigm, and are often misinterpreted as the posterior probabilities of the null.

15.1   Accept/Reject

There is a great deal of terminology associated with the accept/reject paradigm, but the basics are fairly simple. Start with a **test statistic** $T(\mathbf{x})$, which is a function $T : \mathcal{X} \to \mathbb{R}$ that measures in some sense the difference between the data **x** and the null hypothesis. To illustrate, let $X_1, \ldots, X_n$ be iid $N(\mu, \sigma_0^2)$, where σ_0^2 is known, and test the hypotheses

$$H_0 : \mu = \mu_0 \text{ versus } H_A : \mu \neq \mu_0. \tag{15.4}$$

The usual test statistic is based on the z statistic:

$$T(x_1, \ldots, x_n) = |z|, \text{ where } z = \frac{\bar{x} - \mu_0}{\sigma_0 / \sqrt{n}}. \tag{15.5}$$

The larger T, the more one would doubt the null hypothesis. Next, choose a cutoff point c that represents how large the test statistic can be before rejecting the null hypothesis. That is,

$$\text{The test} \begin{cases} \text{Rejects the null} & \text{if} \quad T(\mathbf{x}) > c \\ \text{Accepts the null} & \text{if} \quad T(\mathbf{x}) \leq c \end{cases}. \tag{15.6}$$

Or it may reject when $T(\mathbf{x}) \geq c$ and accept when $T(\mathbf{x}) < c$.

In choosing c, there are two types of error to balance called, rather colorlessly,

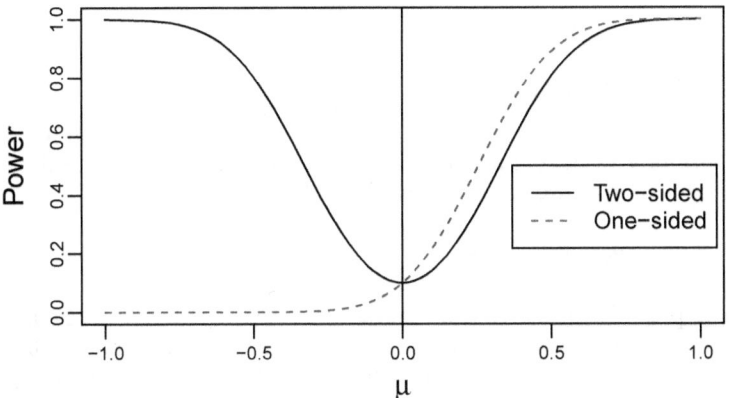

Figure 15.1: The power function for the z test when $\alpha = 0.10$, $\mu_0 = 0$, $n = 25$, and $\sigma_0^2 = 1$.

Type I and Type II errors:

Truth $\downarrow$	Action	
	Accept H_0	Reject H_0
H_0	OK	Type I error (false positive)
H_A	Type II error (false negative)	OK

$$(15.7)$$

It would have been better if the terminology had been in line with medical and other usage, where a *false positive* is rejecting the null when it is true, e.g., saying you have cancer when you don't, and a *false negative* is accepting the null when it is false, e.g., saying everything is ok when it is not. In any case, the larger c, the smaller chance of a false positive, but the greater chance of a false negative.

Common practice is to set a fairly low limit (such as 5% or 1%), the **level**, on the chance of a Type I error:

Definition 15.1. *A hypothesis test (15.7) has level α if*

$$P_\theta[T(\mathbf{X}) > c] \leq \alpha \text{ for all } \theta \in \mathcal{T}_0. \tag{15.8}$$

Note that a test with level 0.05 also has level 0.10. A related concept is the **size** of a test, which is the smallest α for which it is level α:

$$\text{Size} = \sup_{\theta \in \mathcal{T}_0} P_\theta[T(\mathbf{X}) > c]. \tag{15.9}$$

Usually the size and level are the same, or close, and rarely is the distinction between the two made.

Traditionally, more emphasis is on the *power* than the probability of Type II error, where $\text{Power}_\theta = 1 - P_\theta[\text{Type II Error}]$ when $\theta \in \mathcal{T}_A$. Power is good. Designing a good study involves making sure that one has a large enough sample size that the power is high enough.

Under the null hypothesis, the z statistic in (15.5) is distributed $N(0,1)$. Thus the size as a function of c is $2(1 - \Phi(c))$, where Φ is the $N(0,1)$ distribution function. To obtain a size of α we would take $c = z_{\alpha/2}$, the $(1 - \alpha/2)^{nd}$ quantile of the normal. The power function is also straightforward to calculate:

$$\text{Power}_\mu = 1 - \Phi\left(c - \frac{\sqrt{n}\mu}{\sigma_0}\right) + \Phi\left(-c - \frac{\sqrt{n}\mu}{\sigma_0}\right). \tag{15.10}$$

Figure 15.1 plots the power function for $\alpha = 0.10$ (so $c = z_{0.10} = 1.645$), $\mu_0 = 0$, $n = 25$, and $\sigma_0^2 = 1$, denoting it "two-sided" since the alternative contains both sides of μ_0. Note that the power function is continuous and crosses the null hypothesis at the level 0.10. Thus we cannot decrease the size without decreasing the power, or increase the power without increasing the size.

A one-sided version of the testing problem in (15.4) would have the alternative being just one side of the null, for example,

$$H_0: \mu \leq \mu_0 \quad \textit{versus} \quad H_A: \mu > \mu_0. \tag{15.11}$$

The test would reject when $z > c'$, where now c' is the $(1 - \alpha)^{th}$ quantile of a standard normal. With the same level $\alpha = 0.10$, the c' here is 1.282. The power of this test is the "one-sided" curve in Figure 15.1. We can see that though it has the same size as the two-sided test, its power is better for the μ's in its alternative. Since the two-sided test has to guard against both sides, its power is somewhat lower.

There are a number of approaches to finding reasonable test statistics. Section 15.2 leverages results we have for estimation. Section 15.3 and Chapter 16 develop tests based on the likelihood. Section 15.4 presents Bayes tests. Chapter 17 looks at randomization tests, and Chapter 18 applies randomization to nonparametric tests, many of which are based on ranks. Chapters 21 and 22 compare tests decision-theoretically.

15.1.1 Interpretation

In practice, one usually doesn't want to reject the null unless there is substantial evidence against it. The situation is similar to the courts in a criminal trial. The defendant is "presumed innocent until proven guilty." That is, the jury imagines the defendant is innocent, then considers the evidence, and only if the evidence piles up so that the jury believes the defendant is "guilty beyond reasonable doubt" does it convict. The accept/reject approach to hypothesis testing parallels the courts with the following connections:

Courts	Testing	
Defendant innocent	Null hypothesis true	
Evidence	Data	(15.12)
Defendant declared guilty	Reject the null	
Defendant declared not guilty	Accept the null	

Notice that the jury does not say that the defendant is innocent, but either guilty or not guilty. "Not guilty" is a way to say that there is not enough evidence to convict, *not* to say the jury is confident the defendant is innocent. Similarly, in hypothesis testing accepting the hypothesis does *not* mean one is confident it is true. Rather, it may be true, or just that there is not enough evidence to reject it. In fact, one may prefer to replace the choice "accept the null" with "fail to reject the null." "Reasonable doubt" is quantified by level.

15.2 Tests based on estimators

If the null hypothesis sets a parameter or set of parameters equal to a fixed constant, that is, $H_0: \boldsymbol{\theta} = \boldsymbol{\theta}_0$ for known $\boldsymbol{\theta}_0$, then the methods developed for estimating $\boldsymbol{\theta}$ can be applied to testing. In the one-parameter hypothesis, we could find a $100(1 - \alpha)\%$ confidence interval for θ, then reject the null hypothesis if the θ_0 is not in the interval. Such a test has level α. Bootstrapped confidence intervals can be used for approximate level α tests. **Randomization tests** (Chapter 17) provide another resampling-based approach.

In normal-based models, z tests and t tests are often available. In (15.4) through (15.6) we saw the z test for testing $\mu = \mu_0$ when σ^2 is known. If we have the same iid $N(\mu, \sigma^2)$ situation, but do not know σ^2, then the (two-sided) hypotheses become

$$H_0: \mu = \mu_0, \sigma^2 > 0 \ \textit{versus} \ H_A: \mu \neq \mu_0, \sigma^2 > 0. \tag{15.13}$$

Here we use the t statistic:

Reject H_0 when $|T(x_1, \ldots, x_n)| > t_{n-1, \alpha/2}$, where $T(x_1, \ldots, x_n) = \dfrac{\bar{x} - \mu_0}{s_* / \sqrt{n}}$, (15.14)

$s_*^2 = \sum (x_i - \bar{x})^2 / (n - 1)$, and $t_{n-1, \alpha/2}$ is the $(1 - \alpha/2)^{nd}$ quantile of a Student's t_{n-1}.

In Exercise 7.8.12, we looked at a confidence interval for the difference in means in a two-sample model, where $X_1, \ldots, X_n$ are iid $N(\mu, \sigma^2)$, $Y_1, \ldots, Y_m$ are iid $N(\gamma, \sigma^2)$, and the X_i's and Y_i's are independent. Here we test

$$H_0: \mu = \gamma, \sigma^2 > 0 \ \textit{versus} \ H_A: \mu \neq \gamma, \sigma^2 > 0. \tag{15.15}$$

We can again use a t test, where we reject the null when $|T| > t_{n+m-2, \alpha/2}$, where

$$T = \frac{\bar{x} - \bar{y}}{s_{pooled} \sqrt{\frac{1}{n} + \frac{1}{m}}}, \quad s_{pooled}^2 = \frac{\sum (x_i - \bar{x})^2 + \sum (y_i - \bar{y})^2}{n + m - 2}. \tag{15.16}$$

Or for normal linear regression, testing $\beta_i = 0$ uses $T = \hat{\beta}_i / se(\hat{\beta}_i)$ and rejects when $|T| > t_{n-p, \alpha/2}$.

More generally, we often have the asymptotic normal result that if $\theta = \theta_0$,

$$Z = \frac{\hat{\theta} - \theta_0}{se(\hat{\theta})} \longrightarrow^{\mathcal{D}} N(0, 1), \tag{15.17}$$

so that an approximate z test rejects the null when $|Z| > z_{\alpha/2}$. If $\boldsymbol{\theta}$ is $K \times 1$, and for some $\hat{\mathbf{C}}$ we have

$$\hat{\mathbf{C}}^{-1/2} (\hat{\boldsymbol{\theta}} - \boldsymbol{\theta}_0) \longrightarrow^{\mathcal{D}} N(\mathbf{0}, \mathbf{I}_K) \tag{15.18}$$

as in (14.86) for MLEs, then an approximate χ^2 test would reject $H_0: \boldsymbol{\theta} = \boldsymbol{\theta}_0$ when

$$(\hat{\boldsymbol{\theta}} - \boldsymbol{\theta}_0)' \hat{\mathbf{C}}^{-1} (\hat{\boldsymbol{\theta}} - \boldsymbol{\theta}_0) > \chi_{K, \alpha}^2, \tag{15.19}$$

$\chi_{K, \alpha}^2$ being the $(1 - \alpha)^{th}$ quantile of a χ_K^2.

15.2.1   Linear regression

Let $\mathbf{Y} \sim N(\mathbf{x}\boldsymbol{\beta}, \sigma^2 \mathbf{I}_n)$, where $\boldsymbol{\beta}$ is $p \times 1$, and $\mathbf{x}'\mathbf{x}$ is invertible. We saw above that we can use a t test to test a single $\beta_i = 0$. We can also test whether a set of β_i's is zero, which often arises in analysis of variance models, and any time one has a set of related x-variables. Partition the $\boldsymbol{\beta}$ and its least squares estimator into the first p_1 and last p_2 components, $p = p_1 + p_2$:

$$\boldsymbol{\beta} = \begin{pmatrix} \boldsymbol{\beta}_1 \\ \boldsymbol{\beta}_2 \end{pmatrix} \quad \text{and} \quad \widehat{\boldsymbol{\beta}} = \begin{pmatrix} \widehat{\boldsymbol{\beta}}_1 \\ \widehat{\boldsymbol{\beta}}_2 \end{pmatrix}. \tag{15.20}$$

Then $\boldsymbol{\beta}_1$ and $\widehat{\boldsymbol{\beta}}_1$ are $p_1 \times 1$, and $\boldsymbol{\beta}_2$ and $\widehat{\boldsymbol{\beta}}_2$ are $p_2 \times 1$. We want to test

$$H_0 : \boldsymbol{\beta}_2 = \mathbf{0} \quad \textit{versus} \quad H_A : \boldsymbol{\beta}_2 \neq \mathbf{0}. \tag{15.21}$$

Theorem 12.1 shows that $\widehat{\boldsymbol{\beta}} \sim N(\boldsymbol{\beta}, \sigma^2 \mathbf{C})$ where $\mathbf{C} = (\mathbf{x}'\mathbf{x})^{-1}$. If we partition $\mathbf{C}$ in accordance with $\boldsymbol{\beta}$, i.e.,

$$\mathbf{C} = \begin{pmatrix} \mathbf{C}_{11} & \mathbf{C}_{12} \\ \mathbf{C}_{21} & \mathbf{C}_{22} \end{pmatrix}, \quad \mathbf{C}_{11} \text{ is } p_1 \times p_1 \text{ and } \mathbf{C}_{22} \text{ is } p_2 \times p_2, \tag{15.22}$$

then we have $\widehat{\boldsymbol{\beta}}_2 \sim N(\boldsymbol{\beta}, \sigma^2 \mathbf{C}_{22})$. Similar to (15.19), if $\boldsymbol{\beta}_2 = \mathbf{0}$,

$$U \equiv \frac{1}{\sigma^2} \widehat{\boldsymbol{\beta}}_2' \mathbf{C}_{22}^{-1} \widehat{\boldsymbol{\beta}}_2 \sim \chi^2_{p_2}. \tag{15.23}$$

We cannot use U directly, since σ^2 is unknown, but we can estimate it with $\widehat{\sigma}^2 = SS_e/(n-p)$ from (12.25), where $SS_e = \|\mathbf{y} - \mathbf{x}\widehat{\boldsymbol{\beta}}\|^2$, the residual sum of squares from the original model. Theorem 12.1 also shows that SS_e is independent of $\widehat{\boldsymbol{\beta}}$, hence of U, and $V \equiv SS_e/\sigma^2 \sim \chi^2_{n-p}$. We thus have the ingredients for an F random variable, defined in Exercise 7.8.18. That is, under the null,

$$F \equiv \frac{U/p_2}{V/(n-p)} = \frac{\widehat{\boldsymbol{\beta}}_2' \mathbf{C}_{22}^{-1} \widehat{\boldsymbol{\beta}}_2}{p_2 \widehat{\sigma}^2} \sim F_{p_2, n-p}. \tag{15.24}$$

The F test rejects the null when $F > F_{p_2, n-p, \alpha}$, where $F_{p_2, n-p, \alpha}$ is the $(1-\alpha)^{th}$ quantile of an $F_{p_2, n-p}$.

15.3   Likelihood ratio test

We will start simple, where each hypothesis has exactly one distribution. Let f be the density of the data $\mathbf{X}$, and consider the hypotheses

$$H_0 : f = f_0 \quad \textit{versus} \quad H_A : f = f_A, \tag{15.25}$$

where f_0 and f_A are the null and alternative densities, respectively, under consideration. The densities could be from the same family with different parameter values, or densities from distinct families, e.g., $f_0 = N(0, 1)$ and $f_A =$ Cauchy.

Recalling the meaning of likelihood, it would make sense to reject f_0 in favor of f_A if f_A is sufficiently more likely than f_0. Consider basing the test on the **likelihood ratio**,

$$LR(\mathbf{x}) = \frac{f_A(\mathbf{x})}{f_0(\mathbf{x})}. \tag{15.26}$$

We will see in Section 21.3 that the Neyman-Pearson lemma guarantees that such a test is best in the sense that it has the highest power among tests with its size.

For example, suppose $X \sim \text{Binomial}(n, p)$, and the hypotheses are

$$H_0: p = \frac{1}{2} \quad versus \quad H_A: p = \frac{3}{4}. \tag{15.27}$$

Then

$$LR(x) = \frac{(3/4)^x (1/4)^{n-x}}{(1/2)^n} = \frac{3^x}{2^n}. \tag{15.28}$$

We then reject the null hypothesis if $LR(x) > c$, where we choose c to give us the desired level. But since $LR(x)$ is strictly increasing in x, there exists a c' such that $LR(x) > c$ if and only if $x > c'$. For example, if $n = 10$, taking $c' = 7.5$ yields a level $\alpha = 0.05468 (= P[X \in \{8, 9, 10\} \mid p = \frac{1}{2}])$. We could go back and figure out what c is, but there is no need. What we really want is the test, which is to reject $H_0: p = 1/2$ when $x > 7.5$. Its power is 0.5256, which is the best you can do with the given size.

When one or both of the hypotheses are not simple (**composite** is the terminology for not simple), it is not so obvious how to proceed, because the likelihood ratio $f(\mathbf{x} \mid \theta_A)/f(\mathbf{x} \mid \theta_0)$ will depend on which $\theta_0 \in \mathcal{T}_0$ and/or $\theta_A \in \mathcal{T}_A$. Two possible solutions are to average or to maximize over the parameter spaces. That is, possible test statistics are

$$\frac{\int_{\mathcal{T}_A} f(\mathbf{x} \mid \theta_A) \rho_A(\theta_A) d\theta_A}{\int_{\mathcal{T}_0} f(\mathbf{x} \mid \theta_0) \rho_0(\theta_0) d\theta_0} \quad \text{and} \quad \frac{\sup_{\theta_A \in \mathcal{T}_A} f(\mathbf{x} \mid \theta_A)}{\sup_{\theta_0 \in \mathcal{T}_0} f(\mathbf{x} \mid \theta_0)} = \frac{f(\mathbf{x} \mid \widehat{\theta}_A)}{f(\mathbf{x} \mid \widehat{\theta}_0)}, \tag{15.29}$$

where ρ_0 and ρ_A are prior probability densities over $\mathcal{T}_0$ and $\mathcal{T}_A$, respectively, and $\widehat{\theta}_0$ and $\widehat{\theta}_A$ are the respective MLEs for θ over the two parameter spaces. The latter ratio is the (maximum) likelihood ratio statistic, which is discussed in Section 16.1. Score tests, which are often simpler than the likelihood ratio tests, are presented in Section 16.3. The former ratio in (15.29) is the statistic in what is called a **Bayes test**, which is key to the Bayesian testing presented next.

15.4 Bayesian testing

The Neyman-Pearson approach is all about action: Either accept the null or reject it. As in the courts, it does not suggest the degree to which the null is plausible or not. By contrast, the Bayes approach produces the probabilities the null and alternative are true, given the data and a prior distribution. Start with the simple versus simple case as in (15.25), where the null hypothesis is that the density is f_0, and the alternative that the density is f_A. The prior π is given by

$$P[H_0] (= P[H_0 \text{ is true}]) = \pi_0, \quad P[H_A] = \pi_A, \tag{15.30}$$

where $\pi_0 + \pi_A = 1$. Where do these probabilities come from? Presumably, from a reasoned consideration of all that is known prior to seeing the data. Or, one may try to be fair and take $\pi_0 = \pi_A = 1/2$. The densities are then the conditional distributions of X given the hypotheses are true:

$$\mathbf{X} \mid H_0 \sim f_0 \text{ and } \mathbf{X} \mid H_A \sim f_A. \tag{15.31}$$

Bayes theorem (Theorem 6.3 on page 94) gives the posterior probabilities:

$$P[H_0 \mid \mathbf{X} = \mathbf{x}] = \frac{\pi_0 f_0(\mathbf{x})}{\pi_0 f_0(\mathbf{x}) + \pi_A f_A(\mathbf{x})} = \frac{\pi_0}{\pi_0 + \pi_A LR(\mathbf{x})},$$

$$P[H_A \mid \mathbf{X} = \mathbf{x}] = \frac{\pi_A f_A(\mathbf{x})}{\pi_0 f_0(\mathbf{x}) + \pi_A f_A(\mathbf{x})} = \frac{\pi_A LR(\mathbf{x})}{\pi_0 + \pi_A LR(\mathbf{x})}, \tag{15.32}$$

where $LR(\mathbf{x}) = f_A(\mathbf{x})/f_0(\mathbf{x})$ is the likelihood ratio from (15.26). In dividing numerator and denominator by $f_0(\mathbf{x})$, we are assuming it is not zero. Thus these posteriors depend on the data only through the likelihood ratio. That is, the posterior does not violate the likelihood principle. Hypothesis tests do violate the likelihood principle: Whether they reject depends on the c, which is calculated from f_0, not the likelihood.

Odds are actually more convenient here, where the odds of an event B are

$$Odds[B] = \frac{P[B]}{1 - P[B]}, \text{ hence } P[B] = \frac{Odds[B]}{1 + Odds[B]}. \tag{15.33}$$

The prior odds in favor of H_A are then π_A/π_0, and the posterior odds are

$$\begin{aligned} Odds[H_A \mid \mathbf{X} = \mathbf{x}] &= \frac{P[H_A \mid \mathbf{X} = \mathbf{x}]}{P[H_0 \mid \mathbf{X} = \mathbf{x}]} \\ &= \frac{\pi_A}{\pi_0} LR(\mathbf{x}) \\ &= Odds[H_A] \times LR(\mathbf{x}). \end{aligned} \tag{15.34}$$

That is,

$$\text{Posterior odds} = (\text{Prior odds}) \times (\text{Likelihood ratio}), \tag{15.35}$$

which neatly separates the contribution to the posterior of the prior and the data. If a decision is needed, one would choose a cutoff point k, say, and reject the null if the posterior odds exceed k. But this test is the same as the Neyman-Pearson test based on (15.26) with cutoff point c. The difference is that in the present case, the cutoff would not be chosen to achieve a certain level, but rather on an assessment of what probability of H_0 is too low to accept the null.

Take the example in (15.27) with $n = 10$, so that $X \sim \text{Binomial}(10, p)$, and the hypotheses are

$$H_0: p = \tfrac{1}{2} \text{ versus } H_A: p = \tfrac{3}{4}. \tag{15.36}$$

If the prior odds are even, i.e., $\pi_0 = \pi_A = 1/2$, so that the prior odds are 1, then the posterior odds are equal to the likelihood ratio, giving the following:

x	Odds$[H_A \mid X = x] = LR(x)$	$100 \times P[H_0 \mid X = x]$
0	0.0010	99.90
1	0.0029	99.71
2	0.0088	99.13
3	0.0264	97.43
4	0.0791	92.67
5	0.2373	80.82
6	0.7119	58.41
7	2.1357	31.89
8	6.4072	13.50
9	19.2217	4.95
10	57.6650	1.70

$$(15.37)$$

Thus if you see $X = 2$ heads, the posterior probability that $p = 1/2$ is about 99%, and if $X = 9$, it is about 5%. Note that using the accept/reject test as in Section 15.3, $X = 8$ would lead to rejecting the null with $\alpha \approx 5.5\%$, whereas here the posterior probability of the null is 13.5%.

In the composite situation, it is common to take a stagewise prior. That is, as in (15.29), we have distributions over the two parameter spaces, as well as the prior marginal probabilities of the hypotheses. Thus the prior π is specified by

$$\Theta \mid H_0 \sim \rho_0, \quad \Theta \mid H_A \sim \rho_A, \quad P[H_0] = \pi_0, \quad \text{and} \quad P[H_1] = \pi_A, \tag{15.38}$$

where ρ_0 is a probability distribution on $\mathcal{T}_0$, and ρ_A is one on $\mathcal{T}_A$. Conditioning on one hypothesis, we can find the marginal distribution of the $\mathbf{X}$ by integrating its density (in the pdf case) with respect to the conditional density on $\boldsymbol{\theta}$. That is, for the null,

$$f(\mathbf{x} \mid H_0) = \int_{\mathcal{T}_0} f(\mathbf{x}, \boldsymbol{\theta} \mid H_0) d\boldsymbol{\theta} = \int_{\mathcal{T}_0} f(\mathbf{x} \mid \boldsymbol{\theta} \& H_0) f(\boldsymbol{\theta} \mid H_0) d\boldsymbol{\theta}$$

$$= \int_{\mathcal{T}_0} f(\mathbf{x} \mid \boldsymbol{\theta}) \rho_0(\boldsymbol{\theta}) d\boldsymbol{\theta}. \tag{15.39}$$

The alternative is similar. Now we are in the same situation as (15.31), where the ratio has the integrated densities, i.e.,

$$B_{A0}(\mathbf{x}) = \frac{f(\mathbf{x} \mid H_A)}{f(\mathbf{x} \mid H_0)} = \frac{\int_{\mathcal{T}_A} f(\mathbf{x} \mid \boldsymbol{\theta}) \rho_A(\boldsymbol{\theta}) d\boldsymbol{\theta}}{\int_{\mathcal{T}_0} f(\mathbf{x} \mid \boldsymbol{\theta}) \rho_0(\boldsymbol{\theta}) d\boldsymbol{\theta}} = \frac{E_{\rho_A}[f(\mathbf{x} \mid \Theta)]}{E_{\rho_0}[f(\mathbf{x} \mid \Theta)]}. \tag{15.40}$$

(The final ratio is applicable when the ρ_0 and ρ_A do not have pdfs.) This ratio is called the **Bayes factor** for H_A versus H_0, which is where the "B_{A0}" notation arises. It is often inverted, so that $B_{0A} = 1/B_{A0}$ is the Bayes factor for H_0 versus H_A. See Jeffreys (1961).

For example, consider testing a normal mean is 0. That is, $X_1, \ldots, X_n$ are iid $N(\mu, \sigma^2)$, with $\sigma^2 > 0$ known, and we test

$$H_0 : \mu = 0 \quad versus \quad H_A : \mu \neq 0. \tag{15.41}$$

Take the prior probabilities $\pi_0 = \pi_A = 1/2$. Under the null, there is only $\mu = 0$, so $\rho_0[M = 0] = 1$. For the alternative, take a normal centered at 0 as the prior ρ_A:

$$M \mid H_A \sim N(0, \sigma_0^2), \tag{15.42}$$

where σ_0^2 is known. (Technically, we should remove the value 0 from the distribution, but it has 0 probability anyway.)

We know the sufficient statistic is $\overline{X}_n$, hence as in Section 13.4.3, we can base the analysis on $\overline{X}_n \sim N(\mu, \sigma^2/n)$. The denominator in the likelihood ratio in (15.40) is the $N(0, \sigma^2/n)$ density at $\overline{x}_n$. The numerator is the marginal density under the alternative, which using (7.106) can be shown to be $N(0, \sigma_0^2 + \sigma^2/n)$. Thus the Bayes factor is

$$B_{A0}(\overline{x}_n) = \frac{\phi(\overline{x}_n \mid 0, \sigma^2/n)}{\phi(\overline{x}_n \mid 0, \sigma_0^2 + \sigma^2/n)}, \tag{15.43}$$

where $\phi(z \mid \mu, \sigma^2)$ is the $N(\mu, \sigma^2)$ pdf. Exercise 15.7.5 rewrites the Bayes factor as

$$B_{A0}(\mathbf{x}) = \frac{1}{\sqrt{1+\tau}} e^{\frac{1}{2}z^2 \tau/(1+\tau)}, \quad \text{where } z = \sqrt{n}\, \frac{\overline{x}_n}{\sigma} \text{ and } \tau = \frac{n\sigma_0^2}{\sigma^2}. \tag{15.44}$$

The z is the usual z statistic, and τ is the ratio of the prior variance to $Var[\overline{X}_n \mid M = \mu]$, similar to the quantities in (11.50). Then with $\pi_0 = \pi_A$, we have that

$$P[H_0 \mid \overline{X}_n = \overline{x}_n] = \frac{1}{1 + B_{A0}(\overline{x}_n)}. \tag{15.45}$$

Even if one does not have a good idea from the context of the data what σ_0^2 should be, a value (or values) needs to be chosen. Berger and Sellke (1987) consider this question extensively. Here we give some highlights. Figure 15.2 plots the posterior probability of the null hypothesis as a function of τ for values of the test statistic $z = 1, 2, 3$. A small value of τ indicates a tight prior around 0 under the alternative, which does not sufficiently distinguish the alternative from the null, and leads to a posterior probability of the null towards $1/2$. At least for the larger values of z, as τ increases, the posterior probability of the null quickly decreases, then bottoms out and slowly increases. In fact, for any $z \neq 0$, the posterior probability of the null approaches 1 as $\tau \to \infty$. Contrast this behavior with that for probability intervals (Section 11.6.1), which stabilize fairly quickly as the posterior variance increases to infinity.

A possibly reasonable choice of τ would be one where the posterior probability is fairly stable. For $|z| > 1$, the minimum is achieved at $\tau = z^2 - 1$, which may be reasonable though most favorable to the alternative. Choosing $\tau = 1$ means the prior variance and variance of the sample mean given μ are equal. Choosing $\tau = n$ equates the prior variance and variance of one X_i given μ. This is the choice Berger and Sellke (1987) use as one close to the Cauchy proposed by Jeffreys (1961), and deemed reasonable by Kass and Wasserman (1995). It also is approximated by the Bayes information criterion (BIC) presented in Section 16.5. Some value within that range is defensible. The table below shows the posterior probability of the null as a percentage for various values of z and relationship between the prior and sample variances.

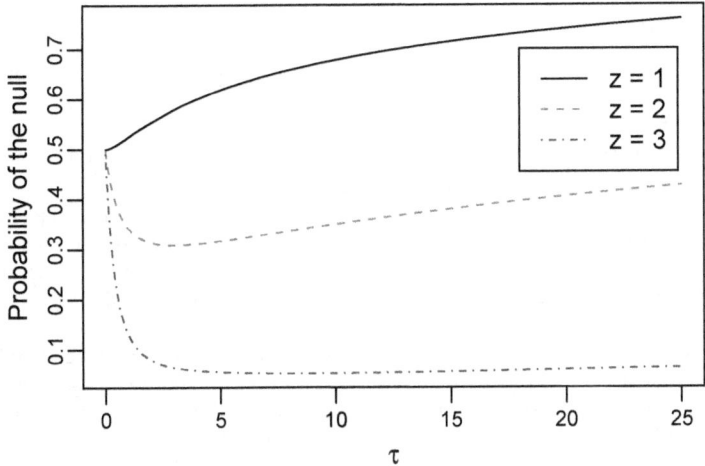

Figure 15.2: The posterior probability of the null hypothesis as a function of τ for values of $z = 1$ (upper curve), $z = 2$ (middle curve), and $z = 3$ (lower curve). See (15.44).

z	$\sigma_0^2 = \sigma^2/n$	Minimum	$\sigma_0^2 = \sigma^2$ n=100	n=1000
0	58.58	50.00	90.95	96.94
0.5	57.05	50.00	89.88	96.54
1.0	52.41	50.00	85.97	95.05
1.5	44.62	44.53	76.74	91.14
2.0	34.22	30.86	58.11	81.10
2.5	22.87	15.33	31.29	58.24
3.0	12.97	5.21	10.45	26.09
3.5	6.20	1.25	2.28	6.51
4.0	2.52	0.22	0.36	1.06
4.5	0.89	0.03	0.04	0.13
5.0	0.27	0	0	0.01

$$(15.46)$$

It is hard to make any sweeping recommendations, but one fact may surprise you. A z of 2 is usually considered substantial evidence against the null. Looking at the table, the prior that most favors the alternative still gives the null a posterior probability of over 30%, as high as 81% for the Jeffreys-type prior with $n = 1000$. When $z = 3$, the lowest posterior probability is 5.21%, but other reasonable priors give values twice that. It seems that z must be at least 3.5 or 4 to have real doubts about the null, at least according to this analysis.

Kass and Raftery (1995) discuss many aspects of Bayesian testing and computation of Bayes factors, and Lecture 2 of Berger and Bayarri (2012) has a good overview of current approaches to Bayesian testing.

15.5 P-values

In the accept/reject setup, the outcome of the test is simply an action: Accept the null or reject it. The size α is the (maximum) chance of rejecting the null given that it is true. But just knowing the action does not reveal how strongly the null is rejected or accepted. For example, when testing the null $\mu = 0$ versus $\mu \neq 0$ based on $X_1, \ldots, X_n$ iid $N(\mu, 1)$, an $\alpha = 0.05$ level test rejects the null when $Z = \sqrt{n}|\overline{X}_n| > 1.96$. The test then would reject the null at the 5% level if $z = 2$. It would also reject the null at the 5% level if $z = 10$. But intuitively, $z = 10$ provides much stronger evidence against the null than does $z = 2$. The Bayesian posterior probability of the null is a very comprehensible assessment of evidence for or against the null, but does require a prior.

A frequentist measure of evidence often used is the **p-value**, which can be thought of as the smallest size such that the test of that size rejects the null. Equivalently, it is the (maximum) chance under the null of seeing something as or more extreme than the observed test statistic. That is, if $T(\mathbf{X})$ is the test statistic,

$$\text{p-value}(\mathbf{x}) = \sup_{\theta \in \mathcal{T}_0} P[T(\mathbf{X}) \geq T(\mathbf{x}) \,|\, \theta]. \tag{15.47}$$

In the normal mean case with $T(x_1, \ldots, x_n) = |z|$ where $z = \sqrt{n}\,\overline{x}_n$, the p-value is

$$\text{p-value}(x_1, \ldots, x_n) = P[|\sqrt{n}\,\overline{X}_n| \geq |z| \,|\, \mu = 0] = P[|N(0,1)| \geq |z|]. \tag{15.48}$$

If $z = 2$, the p-value is 4.45%, if $z = 3$ the p-value is 0.27%, and if $z = 10$, the p-value is essentially 0 ($\approx 10^{-21}\%$). Thus the lower the p-value, the more evidence against the null.

The next lemma shows that if one reports the p-value instead of the accept/reject decision, the reader can decide on the level to use, and easily perform the test.

Lemma 15.2. *Consider the hypothesis testing problem based on $\mathbf{X}$, where the null hypothesis is $H_0: \theta \in \mathcal{T}_0$. For test statistic $T(\mathbf{x})$, define the p-value as in (15.47). Then for $\alpha \in (0,1)$,*

$$\sup_{\theta \in \mathcal{T}_0} P[\text{p-value}(\mathbf{X}) \leq \alpha \,|\, \theta] \leq \alpha, \tag{15.49}$$

hence the test that rejects the null when p-value$(\mathbf{x}) \leq \alpha$ has level α.

Proof. Take a $\theta_0 \in \mathcal{T}_0$. If we let F be the distribution function of $-T(\mathbf{X})$ when $\theta = \theta_0$, then the p-value is $F(-t(\mathbf{x}))$. Equation (4.31) shows that $P[F(-T(\mathbf{X})) \leq \alpha] \leq \alpha$, i.e.,

$$P[\text{p-value}(\mathbf{X}) \leq \alpha \,|\, \theta_0] \leq \alpha, \tag{15.50}$$

and (15.49) follows by taking the supremum over θ_0 on the left-hand side. $\qquad \square$

It is not necessary that p-values be defined through a test statistic. We could alternately define a p-value to be any function p-value$(\mathbf{x})$ that satisfies (15.50) for all $\alpha \in (0,1)$. Thus the p-value is the test statistic, where small values lead to rejection.

A major stumbling block to using the p-value as a measure of evidence is to know how to interpret the p-value. A common mistake is to conflate the p-value with the probability that the null is true given the data:

$$P[T(\mathbf{X}) \geq T(\mathbf{x}) \,|\, \mu = 0] \quad \overset{?}{\approx} \quad P[M = 0 \,|\, T(\mathbf{X}) = T(\mathbf{x})]. \tag{15.51}$$

There are a couple of obvious problems with removing the question marks in (15.51). First, the conditioning is reversed, i.e., in general $P[A \mid B] \neq P[B \mid A]$. Second, the right-hand side is taking the test statistic exactly equal to the observed one, while the left-hand side is looking at being greater than or equal to the observed.

More convincing is to calculate the two quantities in (15.51). The right-hand side is found in (15.46). Even taking the minimum posterior probabilities, we have the following:

$z \rightarrow$	1	1.5	2	2.5	3	3.5	4	4.5	5
p-value	31.73	13.36	4.55	1.24	0.27	0.047	0.006	0.0007	0.0001
$P[H_0 \mid Z = z]$	50.00	44.53	30.86	15.33	5.21	1.247	0.221	0.0297	0.0031
$\widehat{P}[H_0 \mid Z = z]$	49.75	42.23	27.65	12.90	4.16	0.961	0.166	0.0220	0.0022

$$(15.52)$$

The p-values far overstate the evidence against the null compared to the posterior probabilities. The ratio of p-value to posterior probability is 6.8 for $z = 2$, 19.3 for $z = 3$, and 53.6 for $z = 5$. In fact, as Berger and Sellke (1987) note, the p-value for $Z = z$ is similar to the posterior probability for $Z = z - 1$. Sellke, Bayarri, and Berger (2001) provide a simple formula that adjusts the p-value to approximate a reasonable lower bound for the posterior probability of the null. Let p be the p-value, and suppose that $p < 1/e$. Then for a simple null and a general class of alternatives (those with decreasing failure rate, though we won't go into that),

$$\widehat{P}[H_0 \mid T(\mathbf{X}) = t(\mathbf{x})] \geq \frac{B_{0A}^*}{1 + B_{0A}^*} \quad \text{where } B_{0A}^* = -e\, p \log(p). \quad (15.53)$$

The bound is $1/2$ if $p > 1/e$. See Exercise 15.7.8 for an illustration. The third row in (15.52) contains these values, which are indeed very close to the actual probabilities.

15.6 Confidence intervals from tests

Hypothesis tests can often be "inverted" to find confidence intervals for parameters in certain models. At the beginning of this chapter, we noted that one could test the null hypothesis that $\eta = \eta_0$ by first finding a confidence interval for η, and rejecting the null hypothesis if η_0 is not in that interval. We can reverse this procedure, taking as a confidence interval all η_0 that are not rejected by the hypothesis test.

Formally, we have $\mathbf{X} \sim P_{\boldsymbol{\theta}}$ and a parameter of interest, say $\eta = \eta(\boldsymbol{\theta})$, with space $\mathcal{H}$. (We will work with one-dimensional η, but it could be multidimensional.) Then for fixed $\eta_0 \in \mathcal{H}$, consider testing

$$H_0: \eta = \eta_0 \quad (\text{i.e., } \boldsymbol{\theta} \in \mathcal{T}_{\eta_0} = \{\boldsymbol{\theta} \in \mathcal{T} \mid \eta(\boldsymbol{\theta}) = \eta_0\}), \quad (15.54)$$

where we can use whatever alternative fits the situation. For given α, we need a function $T(\mathbf{x}; \eta_0)$ such that for each $\eta_0 \in \mathcal{H}$, $T(\mathbf{x}; \eta_0)$ is a test statistic for the null in (15.54). That is, there is a cutoff point c_{η_0} for each η_0 such that

$$\sup_{\boldsymbol{\theta} \in \mathcal{T}_{\eta_0}} P_{\boldsymbol{\theta}}[T(\mathbf{X}; \eta_0) > c_{\eta_0}] \leq \alpha. \quad (15.55)$$

Now we use this T as a pivotal quantity as in (11.9). Consider the set defined for each $\mathbf{x}$ via

$$C(\mathbf{x}) = \{\eta_0 \in \mathcal{H} \mid T(\mathbf{x}; \eta_0) \leq c_{\eta_0}\}. \quad (15.56)$$

This set is a $100(1 - \alpha)\%$ confidence region for η, since for each θ,

$$P_\theta[\eta(\theta) \in C(\mathbf{X})] = 1 - P_\theta[T(\mathbf{X}; \eta(\theta)) > c_{\eta(\theta)}] \geq 1 - \alpha \qquad (15.57)$$

by (15.55). We call it a "confidence region" since in general it may not be an interval.

For a simple example, suppose $X_1, \ldots, X_n$ are iid $N(\mu, \sigma^2)$, and we wish a confidence interval for μ. To test $H_0 : \mu = \mu_0$ with level α, we can use Student's t, $T(\mathbf{x}; \mu_0) = |\bar{x} - \mu_0| / (s_* / \sqrt{n})$. The cutoff point is $t_{n-1,\alpha/2}$ for any μ_0. Thus

$$C(\mathbf{x}) = \left\{ \mu_0 \ \middle| \ \frac{|\bar{x} - \mu_0|}{s_* / \sqrt{n}} \leq t_{n-1,\alpha/2} \right\} = \left[\bar{x} - t_{n-1,\alpha/2} \frac{s_*}{\sqrt{n}}, \bar{x} + t_{n-1,\alpha/2} \frac{s_*}{\sqrt{n}} \right], \qquad (15.58)$$

as before, except the interval is closed instead of open.

A more interesting situation is $X \sim \text{Binomial}(n, p)$ for n small enough that the normal approximation may not work. (Generally, people are comfortable with the approximation if $np \geq 5$ and $n(1 - p) \geq 5$.) The **Clopper-Pearson** interval inverts the two-sided binomial test to obtain an exact confidence interval. It is exact in the sense that it is guaranteed to have level at least the nominal one. It tends to be very conservative, so may not be the best choice. See Brown, Cai, and DasGupta (2001). Consider the two-sided testing problem

$$H_0 : p = p_0 \quad \textit{versus} \quad H_A : p \neq p_0 \qquad (15.59)$$

for fixed $p_0 \in (0, 1)$. Given level α, the test we use has (approximately) equal tails. It rejects the null hypothesis if

$$x \leq k(p_0) \ \text{ or } \ x \geq l(p_0), \ \text{ where}$$
$$k(p_0) = \max\{\text{integer } k \mid P_{p_0}[X \leq k] \leq \alpha/2\} \ \text{ and}$$
$$l(p_0) = \min\{\text{integer } l \mid P_{p_0}[X \geq l] \leq \alpha/2\}. \qquad (15.60)$$

For data x, the confidence interval consists of all p_0 that are not rejected:

$$C(x) = \{p_0 \mid k(p_0) + 1 \leq x \leq l(p_0) - 1\}. \qquad (15.61)$$

Let b_x and a_x be the values of p defined by

$$P_{b_x}[X \leq x] = \frac{\alpha}{2} = P_{a_x}[X \geq x]. \qquad (15.62)$$

Both $k(p_0)$ and $l(p_0)$ are nondecreasing in p_0, hence as in Exercise 15.7.12,

$$k(p_0) + 1 \leq x \quad \Leftrightarrow \quad \begin{cases} p_0 < b_x & \text{if} \quad 0 \leq x \leq n - 1 \\ p_0 < 1 & \text{if} \qquad x = n \end{cases}. \qquad (15.63)$$

Similarly,

$$l(p_0) - 1 \geq x \quad \Leftrightarrow \quad \begin{cases} p_0 > 0 & \text{if} \qquad x = 0 \\ p_0 > a_x & \text{if} \quad 1 \leq x \leq n \end{cases}. \qquad (15.64)$$

So the confidence interval in (15.61) is given by

$$C(0) = (0, b_0); \ \ C(x) = (a_x, b_x), 1 \leq x \leq n - 1; \ \ C(n) = (a_n, 1). \qquad (15.65)$$

We can use Exercise 5.6.17 to more easily solve for a_x ($x \leq n - 1$) and b_x ($x \geq 0$) in (15.62):

$$a_x = q(\alpha/2; x, n - x + 1) \ \text{ and } \ b_x = q(1 - \alpha/2; x + 1, n - x), \qquad (15.66)$$

where $q(\gamma; a, b)$ is the γth quantile of a $\text{Beta}(a, b)$.

15.7   Exercises

Exercise 15.7.1. Suppose $X \sim$ Exponential(λ), and we wish to test the hypotheses $H_0: \lambda = 1$ versus $H_A: \lambda \neq 1$. Consider the test that rejects the null hypothesis when $|X - 1| > 3/4$. (a) Find the size α of this test. (b) Find the power of this test as a function of λ. (c) For which λ is the power at its minimum? What is the power at this value? Is it less than the size? Is that a problem?

Exercise 15.7.2. Suppose $X_1, \ldots, X_n$ are iid Uniform$(0, \theta)$, and we test $H_0: \theta \leq 1/2$ versus $H_A: \theta > 1/2$. Take the test statistic to be $T = \max\{X_1, \ldots, X_n\}$, so we reject when $T > c_\alpha$. Find c_α so that the size of the test is $\alpha = 0.05$. Calculate c_α for $n = 10$.

Exercise 15.7.3. Suppose $X \mid P = p \sim$ Binomial$(5, p)$, and we wish to test $H_0: p = 1/2$ versus $H_A: p = 3/5$. Consider the test that rejects the null when $X = 5$. (a) Find the size α and power of this test. Is the test very powerful? (b) Find $LR(x)$ and $P[H_0 \mid X = x]$ as functions of x when the prior is that $P[H_0] = P[H_A] = 1/2$. When $x = 5$, is the posterior probability of the null hypothesis close to α?

Exercise 15.7.4. Take $X \mid P = p \sim$ Binomial$(5, p)$, and now test $H_0: p = 1/2$ versus $H_A: p \neq 1/2$. Consider the test that rejects the null hypothesis when $X \in \{0, 5\}$. (a) Find the size α of this test. (b) Consider the prior distribution where $P[H_0] = P[H_A] = 1/2$, and $P \mid H_A \sim$ Uniform$(0, 1)$ (where $p = 1/2$ is removed from the uniform, with no ill effects). Find $P[X = x \mid H_0]$, $P[X = x \mid H_A]$, and the Bayes factor, B_{A0}. (c) Find $P[H_0 \mid X = x]$. When $x = 0$ or 5, is the posterior probability of the null hypothesis close to α?

Exercise 15.7.5. Let $\phi(z \mid \mu, \sigma^2)$ denote the $N(\mu, \sigma^2)$ pdf. (a) Show that the Bayes factor in (15.43) can be written as

$$B_{A0}(\overline{x}_n) = \frac{\phi(\overline{x}_n \mid 0, \sigma^2/n)}{\phi(\overline{x}_n \mid 0, \sigma_0^2 + \sigma^2/n)} = \frac{1}{\sqrt{1+\tau}} e^{\frac{1}{2}z^2 \tau/(1+\tau)}, \tag{15.67}$$

where $z = \sqrt{n}\,\overline{x}_n/\sigma$ and $\tau = n\sigma_0^2/\sigma^2$, as in (15.44). (b) Show that as $\tau \to \infty$ with z fixed (so the prior variance goes to infinity), the Bayes factor goes to 0, hence the posterior probability of the null goes to 1. (c) For fixed z, show that the minimum of $B_{A0}(\overline{x}_n)$ over τ is acheived when $\tau = z^2 - 1$.

Exercise 15.7.6. Let $X \sim$ Binomial(n, p), and test $H_0: p = 1/2$ versus $H_A: p \neq 1/2$. (a) Suppose $n = 25$, and let the test statistic be $T(x) = |x - 12.5|$, so that the null is rejected if $T(x) > c$ for some constant c. Find the sizes for the tests with (i) $c = 5$; (ii) $c = 6$; (iii) $c = 7$.

For the rest of the question, consider a beta prior for the alternative, i.e., $P \mid H_A \sim$ Beta(γ, γ). (b) What is the prior mean given the alternative? Find the marginal pmf for X given the alternative. What distribution does it represent? [Hint: Recall Definition 6.2 on page 86.] (c) Show that for $X = x$, the Bayes factor is

$$B_{A0}(x) = 2^n \frac{\Gamma(2\gamma)}{\Gamma(\gamma)^2} \frac{\Gamma(x+\gamma)\Gamma(n-x+\gamma)}{\Gamma(n+2\gamma)}. \tag{15.68}$$

(d) Suppose $n = 25$ and $x = 7$. Show that the p-value for the test statistic in part (a) is 0.0433. Find the Bayes factor in (15.68), as well as the posterior probability of the null assuming $\pi_0 = \pi_A = 1/2$, for γ equal to various values from 0.1 to 5. What is the minimum posterior probability (approximately)? Does it come close to the p-value? For what values of α is the posterior probability relatively stable?

Exercise 15.7.7. Consider testing a null hypothesis using a test statistic T. Assume that the distribution of T under the null is the same for all parameters in the null hypothesis, and is continuous with distribution function $F_T(t)$. Show that under the null, the p-value has a Uniform(0,1) distribution. [Hint: First note that the p-value equals $1 - F_T(t)$, then use (4.30).]

Exercise 15.7.8. In Exercise 15.7.7, we saw that in many situations, the p-value has a Uniform(0,1) distribution under the null. In addition, one would usually expect the p-value to have a distribution "smaller" than the uniform when the alternative holds. A simple parametric abstraction of this notion is to have $X \sim \text{Beta}(\gamma, 1)$, and test $H_0: \gamma = 1$ versus $H_A: 0 < \gamma < 1$. Then X itself is the p-value. (a) Show that under the null, $X \sim \text{Uniform}(0,1)$. (b) Show that for α in the alternative, $P[X < x \mid \gamma] > P[X < x \mid H_0]$. (What is the distribution function?). (c) Show that the Bayes factor for a prior density $\pi(\gamma)$ on the alternative space is

$$B_{A0} = \int_0^1 \gamma x^{\gamma-1} \pi(\gamma) d\gamma. \tag{15.69}$$

(d) We wish to find an upper bound for B_{A0}. Argue that B_{A0} is less than or equal to the supremum of the integrand, $\gamma x^{\gamma-1}$, over $0 < \gamma < 1$. (e) Show that

$$\sup_{0<\gamma<1} \gamma x^{\gamma-1} = \begin{cases} -\dfrac{1}{e \, x \log(x)} & \text{if } x \le \frac{1}{e} \\ 1 & \text{if } x > \frac{1}{e} \end{cases}. \tag{15.70}$$

(f) We have that an upper bound on the Bayes factor B_{A0} is given in (15.70). Show that the lower bound on the posterior probability of the null is as given in (15.53), where the p there is the x here. [Recall that $B_{0A} = 1/B_{A0}$.]

Exercise 15.7.9. Meta-analysis is a general term for combining results from many different studies. It is very common in health studies. One aspect of meta-analysis is combining independent tests, where several studies test the same null hypothesis, and one wishes to combine the hypothesis tests into one overall test. The simplest form of the problem assumes p independent p-values, $U_1, \dots, U_p$, one from each study, where the parameter for the ith study is θ_i. If $\theta_i = 0$ then $U_i \sim \text{Uniform}(0,1)$, and if $\theta_i \ne 0$, U_i is smaller than a Uniform(0,1) in some sense. Then we test the combined null,

$$H_0: \boldsymbol{\theta} = \mathbf{0} \quad versus \quad H_A: \boldsymbol{\theta} \ne \mathbf{0}, \tag{15.71}$$

where $\boldsymbol{\theta} = (\theta_1, \dots, \theta_p)$. There have been many omnibus methods proposed, where "omnibus" means they can be used no matter what the actual distributions of the original test statistics are. The most famous is Fisher's procedure, which is based on the product of the p-values. Other tests include those based on the sum, minimum, or maximum of the p-values. In any case, the resulting statistic is not itself a p-value. We need to find the null distribution of the combined statistic, and from that find the overall p-value. (a) The Fisher, or Fisher-Pearson, statistic is usually written as a function of the product, $T_P(\mathbf{U}) = -2\log(\prod U_i) = -2\sum \log(U_i)$. Show that under the null, the $-2\log(U_i)$'s are iid Exponential(1/2), which is the same as χ_2^2. What is the null distribution of $T_P(\mathbf{U})$? (b) Tippett's test uses the statistic $T_{Min}(\mathbf{U}) = \min\{U_i\}$, the value of the most significant p-value. For given α, find the c_α such that $P[T_{Min}(\mathbf{U}) \le c_\alpha \mid H_0] = \alpha$. (c) The maximum test is based on $T_{Max}(\mathbf{U}) = \max\{U_i\}$, the least significant of the p-values. Find the c_α such that $P[T_{Max}(\mathbf{U}) \le c_\alpha \mid H_0] = \alpha$.

(d) The Edgington test uses the sum of the p-values, $T_S(\mathbf{U}) = \sum U_i$. Find c_α such that $P[T_S(\mathbf{U}) \leq c_\alpha \mid H_0]$ for $p = 2$ and $\alpha \in (0, .5)$. [See Exercise 1.7.6.] (e) The Liptak-Stouffer statistic transforms each U_i into a normal under the null. The statistic is $T_N(\mathbf{U}) = -\sum \Phi^{-1}(U_i)$, where Φ^{-1} is the inverse of the $N(0,1)$ distribution function. What is the null distribution of each $-\Phi^{-1}(U_i)$? Of $T_N(\mathbf{U})$?

Exercise 15.7.10. When the null and alternative hypotheses are the same dimensionality, it is possible for p-values and Bayesian posterior probabilities of the null to be similar. For example, suppose $X \sim N(\mu, 1)$, and we test the one-sided hypotheses $H_0: \mu \leq 0$ versus $H_A: \mu > 0$. The test statistic is X itself. (a) Show that the p-value for $X = x$ is $1 - \Phi(x)$, where Φ is the $N(0,1)$ distribution function. (b) We do not have to treat the two hypotheses separately to develop the prior in this case. Take the overall prior to be $M \sim N(0, \sigma_0^2)$. Show that under this prior, $P[H_0] = P[H_A] = 1/2$. (c) Using the prior in part (b), find the posterior distribution of $M \mid X = x$, and then $P[H_0 \mid X = x]$. (d) What is the limit of $P[H_0 \mid X = x]$ as $\sigma_0^2 \to \infty$? How does it compare to the p-value?

Exercise 15.7.11. Consider the polio vaccine example from Exercise 6.8.9, where X_V is the number of subjects in the vaccine group that contracted polio, and X_C is the number in the control group. We model X_V and X_C as independent, where

$$X_V \sim \text{Poisson}(c_V \theta_V) \quad \text{and} \quad X_C \sim \text{Poisson}(c_C \theta_C). \tag{15.72}$$

Here θ_V and θ_C are the population rates of polio cases per 100,000 subjects for the two groups. The sample sizes for the two groups are $n_V = 200,745$ and $n_C = 201,229$, so that $c_V = 2.00745$ and $c_C = 2.01229$. We wish to test

$$H_0: \theta_V = \theta_C \quad \text{versus} \quad H_A: \theta_V \neq \theta_C. \tag{15.73}$$

Here we will perform a Bayesian test. (a) Let $X \mid \Theta = \theta \sim \text{Poisson}(c\theta)$ and $\Theta \sim \text{Gamma}(\alpha, \lambda)$ for given $\alpha > 0, \lambda > 0$. Show that the marginal density of X is

$$f(x \mid \alpha, \lambda) = \frac{\Gamma(x + \alpha)}{x! \Gamma(\alpha)} \frac{c^x \lambda^\alpha}{(c + \lambda)^{x + \alpha}}. \tag{15.74}$$

(If α is a positive integer, then this f is the negative binomial pmf. The density is in fact a generalization of the negative binomial to real-valued $\alpha > 0$.) (b) Take X_V and X_C in (15.72) and the hypotheses in (15.73). Suppose the prior given the alternative hypothesis has Θ_C and Θ_V independent, both $\text{Gamma}(\alpha, \lambda)$. Show that the marginal for (X_C, X_V) under the alternative is $f(x_V \mid \alpha, \lambda) f(x_C \mid \alpha, \lambda)$. (c) Under the null, let $\Theta_V = \Theta_C = \Theta$, their common value, and set the prior $\Theta \sim \text{Gamma}(\alpha, \lambda)$. Show that the marginal joint pmf of (X_V, X_C) is

$$f(x_V, x_C \mid H_A) = \frac{\Gamma(x_V + x_C + \alpha)}{x_V! x_C! \Gamma(\alpha)} \frac{c_V^{x_V} c_C^{x_C} \lambda^\alpha}{(\lambda + c_V + c_C)^{x_V + x_C + \alpha}}. \tag{15.75}$$

[This distribution is the negative multinomial.] (d) For the vaccine group, there were $x_V = 57$ cases of polio, and for the control group, there were $x_C = 142$ cases. Find (numerically) the Bayes factor $B_{A0}(x_V, x_C)$, the ratio of integrated likelihoods as in (15.40), when $\alpha = 1, \lambda = 1/25$. [Hint: You may wish to find the logs of the various quantities first.] (e) What is the posterior probability of the null? What do you conclude about the effectiveness of the polio vaccine based on this analysis? (f) [Extra credit: Try some other values of α and λ. Does it change the conclusion much?]

Exercise 15.7.12. This exercise verifies the results for the Clopper-Pearson confidence interval procedure in (15.65) and (15.66). Let $X \sim \text{Binomial}(n, p)$ for $0 < p < 1$, and fix $\alpha \in (0, 1)$. As in (15.60), for $p_0 \in (0, 1)$, define $k(p_0) = \max\{\text{integer } k \mid P_{p_0}[X \le k] \le \alpha/2\}$, and for given x, suppose that $k(p_0) \le x - 1$. (a) Argue that for any $p \in (0, 1)$, $k(p) \le n - 1$. Thus if $x = n$, $k(p_0) \le x - 1$ implies that $p < 1$. (b) Now suppose $x \in \{0, \ldots, n-1\}$, and as in (15.62) define b_x to satisfy $P_{b_x}[X \le x] = \alpha/2$, so that $k(b_x) = x$. Show that if $p < b_x$, then $P_p[X \le x] > \alpha/2$. Argue that therefore, $k(p) \le x - 1$ if and only if $p < b_x$. (c) Conclude that (15.63) holds. (d) Exercise 5.6.17 shows that $P_p[X \le x] = P[\text{Beta}(x+1, n-x) > p]$. Use this fact to prove (15.66).

Exercise 15.7.13. Continue with the setup in Exercise 15.7.12. (a) Suppose $x = 0$. Show that the confidence interval is $C(0) = (0, 1 - \sqrt[n]{\alpha/2})$. (b) Show that $C(n) = (\sqrt[n]{\alpha/2}, 1)$.

Exercise 15.7.14. Suppose $X_1, \ldots, X_n$ are iid $N(\mu_X, \sigma_X^2)$ and $Y_1, \ldots, Y_m$ are iid $N(\mu_Y, \sigma_Y^2)$, and the X_i's are independent of the Y_i's. The goal is a confidence interval for the ratio of means, μ_X/μ_Y. We could use the Δ-method on $\bar{x}/\bar{y}$, or bootstrap as in Exercise 11.7.11. Here we use an idea of Fieller (1932), who allowed correlation between X_i and Y_i. Assume σ_X^2 and σ_Y^2 are known. Consider the null hypothesis $H_0: \mu_X/\mu_Y = \gamma_0$ for some fixed γ_0. Write the null hypothesis as $H_0: \mu_X - \gamma_0\mu_Y = 0$. Let $T(\mathbf{x}, \mathbf{y}, \gamma_0)$ be the z-statistic (15.5) based on $\bar{x} - \gamma_0\bar{y}$, so that the two-sided level α test rejects the null when $T(\mathbf{x}, \mathbf{y}, \gamma_0)^2 \ge z_{\alpha/2}^2$. We invert this test as in (15.56) to find a $100(1 - \alpha)\%$ confidence region for γ. (a) Show that the confidence region can be written

$$C(\mathbf{x}, \mathbf{y}) = \{\gamma_0 \mid a_y(\gamma_0 - \bar{x}\bar{y}/a_y)^2 < \bar{x}^2\bar{y}^2/a_y - a_x\}, \qquad (15.76)$$

where $a_x = \bar{x}^2 - z_{\alpha/2}^2\sigma_X^2/n$ and $a_y = \bar{y}^2 - z_{\alpha/2}^2\sigma_Y^2/m$. (b) Let $c = \bar{x}^2\bar{y}^2/a_y - a_x$. Show that if $a_y > 0$ and $c > 0$, the confidence interval is $\bar{x}\bar{y}/a_y \pm d$. What is d? (c) What is the confidence interval if $a_y > 0$ but $c < 0$? (d) What is the confidence interval if $a_y < 0$ and $c > 0$? Is that reasonable? [Hint: Note that $a_y < 0$ means that μ_Y is not significantly different than 0.] (e) Finally, suppose $a_y < 0$ and $c < 0$. Show that the confidence region is $(-\infty, u) \cup (v, \infty)$. What are u and v?

Chapter 16

Likelihood Testing and Model Selection

16.1 Likelihood ratio test

The **likelihood ratio test (LRT)** (or **maximum likelihood ratio test**, though LRT is more common), uses the likelihood ratio statistic, but substitutes the MLE for the parameter value in each density as in the second ratio in (15.29). The statistic is

$$\Lambda(\mathbf{x}) = \frac{\sup_{\boldsymbol{\theta}_A \in T_A} f(\mathbf{x} \mid \boldsymbol{\theta}_A)}{\sup_{\boldsymbol{\theta}_0 \in T_0} f(\mathbf{x} \mid \boldsymbol{\theta}_0)} = \frac{f(\mathbf{x} \mid \widehat{\boldsymbol{\theta}}_A)}{f(\mathbf{x} \mid \widehat{\boldsymbol{\theta}}_0)}, \tag{16.1}$$

where $\widehat{\boldsymbol{\theta}}_0$ is the MLE of $\boldsymbol{\theta}$ under the null model, and $\widehat{\boldsymbol{\theta}}_A$ is the MLE of $\boldsymbol{\theta}$ under the alternative model. Notice that in the simple versus simple situation, $\Lambda(\mathbf{x}) = LR(\mathbf{x})$. In many situations this statistic leads to a reasonable test, in the same way that the MLE is often a good estimator. Also, under appropriate conditions, it is easy to find the cutoff point to obtain the approximate level α:

$$\text{Under } H_0, \ 2\log(\Lambda(\mathbf{X})) \longrightarrow^{\mathcal{D}} \chi^2_\nu, \ \nu = \dim(T_A) - \dim(T_0), \tag{16.2}$$

the *dims* being the number of free parameters in the two parameter spaces, which may or not may not be easy to determine.

First we will show some examples, then formalize the above result, at least for simple null hypotheses.

16.1.1 Normal mean

Suppose $X_1, \ldots, X_n$ are iid $N(\mu, \sigma^2)$, and we wish to test whether $\mu = 0$, with σ^2 unknown:

$$H_0 \colon \mu = 0, \sigma^2 > 0 \ \textit{versus} \ H_A \colon \mu \neq 0, \sigma^2 > 0. \tag{16.3}$$

Here, $\boldsymbol{\theta} = (\mu, \sigma^2)$, so we need to find the MLE under the two models. Start with the null. Here, $T_0 = \{(0, \sigma^2) \mid \sigma^2 > 0\}$. Thus the MLE of μ is $\widehat{\mu}_0 = 0$, since it is the only possibility. For σ^2, we then maximize

$$\frac{1}{\sigma^n} e^{-\Sigma x_i^2 / (2\sigma^2)}, \tag{16.4}$$

which by the usual calculations yields

$$\widehat{\sigma}_0^2 = \sum x_i^2 / n, \quad \text{hence} \quad \widehat{\theta}_0 = (0, \sum x_i^2 / n). \tag{16.5}$$

The alternative has space $\mathcal{T}_A = \{(\mu, \sigma^2) \,|\, \mu \neq 0, \sigma^2 > 0\}$, which from Section 13.6 yields MLEs

$$\widehat{\theta}_A = (\overline{x}, s^2), \quad \text{where} \quad s^2 = \widehat{\sigma}_A^2 = \frac{\sum (x_i - \overline{x})^2}{n}. \tag{16.6}$$

Notice that not only is the MLE of μ different in the two models, but so is the MLE of σ^2. (Which should be reasonable, because if you know the mean is 0, you do not have to use $\overline{x}$ in the variance.) Next, stick those estimates into the likelihood ratio, and see what happens (the $\sqrt{2\pi}$'s cancel):

$$
\begin{aligned}
\Lambda(\mathbf{x}) &= \frac{f(\mathbf{x} \,|\, \overline{x}, s^2)}{f(\mathbf{x} \,|\, 0, \sum x_i^2 / n)} \\[2mm]
&= \frac{\frac{1}{s^n} \, e^{-\Sigma(x_i - \overline{x})^2 / (2s^2)}}{\frac{1}{(\sum x_i^2 / n)^{n/2}} \, e^{-\Sigma x_i^2 / (2\Sigma x_i^2 / n)}} \\[2mm]
&= \left(\frac{\sum x_i^2 / n}{s^2} \right)^{n/2} \frac{e^{-n/2}}{e^{-n/2}} \\[2mm]
&= \left(\frac{\sum x_i^2}{\sum (x_i - \overline{x})^2} \right)^{n/2}.
\end{aligned}
\tag{16.7}
$$

Using (16.2), we have that the LRT

$$\text{Rejects } H_0 \text{ when } \quad 2\log(\Lambda(\mathbf{x})) > \chi_{\nu, \alpha}^2, \tag{16.8}$$

which has a size of approximately α. To find the degrees of freedom, we count up the free parameters in the alternative space, which is two (μ and σ^2), and the free parameters in the null space, which is just one (σ^2). Thus the difference is $\nu = 1$.

In this case, we can also find an exact test. We start by rewriting the LRT:

$$
\begin{aligned}
\Lambda(\mathbf{x}) > c &\iff \frac{\sum (x_i - \overline{x})^2 + n\overline{x}^2}{\sum (x_i - \overline{x})^2} > c^* = c^{2/n} \\[2mm]
&\iff \frac{n\overline{x}^2}{\sum (x_i - \overline{x})^2} > c^{**} = c^* - 1 \\[2mm]
&\iff \frac{(\sqrt{n}\overline{x})^2}{\sum (x_i - \overline{x})^2 / (n-1)} > c^{***} = (n-1)c^{**} \\[2mm]
&\iff |T_n(\mathbf{x})| > c^{****} = \sqrt{c^{***}},
\end{aligned}
\tag{16.9}
$$

where T_n is the t statistic,

$$T_n(\mathbf{x}) = \frac{\sqrt{n}\overline{x}}{\sqrt{\sum (x_i - \overline{x})^2 / (n-1)}}. \tag{16.10}$$

Thus the cutoff is $c^{****} = t_{n-1, \alpha/2}$. Which is to say, the LRT in this case is the usual two-sided t-test. We could reverse the steps to find the original cutoff c in (16.8), but it is not necessary since we can base the test on the T_n.

16.1.2 Linear regression

Here we again look at the multiple regression testing problem from Section 15.2.1. The model (12.9) is

$$\mathbf{Y} \sim N(\mathbf{x}\boldsymbol{\beta}, \sigma^2 \mathbf{I}_n), \tag{16.11}$$

where $\boldsymbol{\beta}$ is $p \times 1$, $\mathbf{x}$ is $n \times p$, and we will assume that $\mathbf{x}'\mathbf{x}$ is invertible. In simple linear regression, it is common to test whether the slope is zero, leaving the intercept to be arbitrary. More generally, we test whether some components of $\boldsymbol{\beta}$ are zero. Partition $\boldsymbol{\beta}$ and $\mathbf{x}$ into two parts,

$$\boldsymbol{\beta} = \begin{pmatrix} \boldsymbol{\beta}_1 \\ \boldsymbol{\beta}_2 \end{pmatrix}, \quad \mathbf{x} = (\mathbf{x}_1, \mathbf{x}_2), \tag{16.12}$$

where $\boldsymbol{\beta}_1$ is $p_1 \times 1$, $\boldsymbol{\beta}_2$ is $p_2 \times 1$, $\mathbf{x}_1$ is $n \times p_1$, $\mathbf{x}_2$ is $n \times p_2$, and $p = p_1 + p_2$. We test

$$H_0: \boldsymbol{\beta}_2 = \mathbf{0} \;\; versus \;\; H_A: \boldsymbol{\beta}_2 \neq \mathbf{0}, \tag{16.13}$$

so that $\boldsymbol{\beta}_1$ is unspecified. Using (13.88), we can take the loglikelihood to be

$$l(\boldsymbol{\beta}, \sigma^2; \mathbf{y}) = -\frac{1}{2\sigma^2}\|\mathbf{y} - \mathbf{x}\boldsymbol{\beta}\|^2 - \frac{n}{2}\log(\sigma^2). \tag{16.14}$$

Under the alternative, the MLE of $\boldsymbol{\beta}$ is the least squares estimate for the unrestricted model, and the MLE of σ^2 is the residual sum of squares over n. (See Exercise 13.8.20.) Using the projections as in (12.14), we have

$$\widehat{\sigma}_A^2 = \frac{1}{n}\|\mathbf{y} - \widehat{\boldsymbol{\beta}}_A\|^2 = \frac{1}{n}\mathbf{y}'\mathbf{Q}_{\mathbf{x}}\mathbf{y}, \quad \mathbf{Q}_{\mathbf{x}} = \mathbf{I}_n - \mathbf{P}_{\mathbf{x}}, \quad \mathbf{P}_{\mathbf{x}} = \mathbf{x}(\mathbf{x}'\mathbf{x})^{-1}\mathbf{x}'. \tag{16.15}$$

Under the null, the model is $\mathbf{Y} \sim N(\mathbf{x}_1\boldsymbol{\beta}_1, \sigma^2\mathbf{I}_n)$, hence

$$\widehat{\boldsymbol{\beta}}_0 = \begin{pmatrix} \widehat{\boldsymbol{\beta}}_{01} \\ \mathbf{0} \end{pmatrix}, \quad \text{and} \quad \widehat{\sigma}_0^2 = \frac{1}{n}\|\mathbf{y} - \mathbf{x}_1\widehat{\boldsymbol{\beta}}_{01}\|^2 = \frac{1}{n}\mathbf{y}'\mathbf{Q}_{\mathbf{x}_1}\mathbf{y}, \tag{16.16}$$

where $\widehat{\boldsymbol{\beta}}_{01} = (\mathbf{x}_1'\mathbf{x}_1)^{-1}\mathbf{x}_1'\mathbf{y}$. The maximal loglikelihoods are

$$l_i(\widehat{\boldsymbol{\beta}}_i, \widehat{\sigma}_i^2; \mathbf{y}) = -\frac{n}{2}\log(\widehat{\sigma}_i^2) - \frac{n}{2}, \quad i = 0, A, \tag{16.17}$$

so that

$$2\log(\Lambda(\mathbf{y})) = n\log\left(\frac{\mathbf{y}'\mathbf{Q}_{\mathbf{x}_1}\mathbf{y}}{\mathbf{y}'\mathbf{Q}_{\mathbf{x}}\mathbf{y}}\right). \tag{16.18}$$

We can find the exact distribution of the statistic under the null, but it is easier to derive after rewriting the model bit so that the two submatrices of $\mathbf{x}$ are orthogonal. Let

$$\mathbf{x}^* = \mathbf{x}\mathbf{A} \text{ and } \boldsymbol{\beta}^* = \mathbf{A}^{-1}\boldsymbol{\beta} \text{ where } \mathbf{A} = \begin{pmatrix} \mathbf{I}_{p_1} & -(\mathbf{x}_1'\mathbf{x}_1)^{-1}\mathbf{x}_1'\mathbf{x}_2 \\ \mathbf{0} & \mathbf{I}_{p_2} \end{pmatrix}. \tag{16.19}$$

Thus $\mathbf{x}\boldsymbol{\beta} = \mathbf{x}^*\boldsymbol{\beta}^*$. Exercise 16.7.3 shows that

$$\boldsymbol{\beta}^* = \begin{pmatrix} \boldsymbol{\beta}_1^* \\ \boldsymbol{\beta}_2 \end{pmatrix} \text{ and } \mathbf{x}^* = (\mathbf{x}_1, \mathbf{x}_2^*), \text{ where } \mathbf{x}_1'\mathbf{x}_2^* = 0. \tag{16.20}$$

Now x_1 has not changed, nor has β_2, hence the hypotheses in (16.13) remain the same. The exercise also shows that, because $x_1'x_2^* = 0$,

$$\mathbf{P_x} = \mathbf{P_{x^*}} = \mathbf{P_{x_1}} + \mathbf{P_{x_2^*}} \implies \mathbf{Q_x} = \mathbf{Q_{x^*}} \text{ and } \mathbf{Q_{x_1}} = \mathbf{Q_x} + \mathbf{P_{x_2^*}}. \tag{16.21}$$

We can then write the ratio in the log in (16.18) as

$$\frac{y'Q_{x_1}y}{y'Q_xy} = \frac{y'Q_xy + y'P_{x_2^*}y}{y'Q_xy} = 1 + \frac{y'P_{x_2^*}y}{y'Q_xy}. \tag{16.22}$$

It can further be shown that

$$Y'P_{x_2^*}Y \sim \sigma^2\chi_{p_2}^2 \text{ and is independent of } Y'Q_xY \sim \sigma^2\chi_{n-p}^2, \tag{16.23}$$

and from Exercise 7.8.18 on the F distribution,

$$\frac{y'P_{x_2^*}y/p_2}{y'Qy/(n-p)} = \frac{y'P_{x_2^*}y}{p_2\widehat{\sigma}^2} \sim F_{p_2,n-p}, \tag{16.24}$$

where $\widehat{\sigma}^2 = y'Qy/(n-p)$, the unbiased estimator of σ^2. Thus the LRT is equivalent to the F test. What might not be obvious, but is true, is that this F statistic is the same as the one we found in (15.24). See Exercise 16.7.4.

16.1.3 Independence in a 2 × 2 table

Consider the setup in Exercise 13.8.23, where $\mathbf{X} \sim$ Multinomial$(n, \mathbf{p})$, and $\mathbf{p} = (p_1, p_2, p_3, p_4)$, arranged as a 2 × 2 contingency table:

$$\begin{array}{cc|c} p_1 & p_2 & \alpha \\ p_3 & p_4 & 1-\alpha \\ \hline \beta & 1-\beta & 1 \end{array} \tag{16.25}$$

The $\alpha = p_1 + p_2$ and $\beta = p_3 + p_4$. The null hypothesis we want to test is that rows and columns are independent, that is,

$$H_0: p_1 = \alpha\beta, \quad p_2 = \alpha(1-\beta), \quad p_3 = (1-\alpha)\beta, \quad p_4 = (1-\alpha)(1-\beta). \tag{16.26}$$

The alternative is that the $\mathbf{p}$ is unrestricted:

$$H_A: \mathbf{p} \in \{\boldsymbol{\theta} \in \mathbb{R}^4 \,|\, \theta_i > 0 \text{ and } \theta_1 + \cdots + \theta_4 = 1\}. \tag{16.27}$$

Technically, we should exclude the null from the alternative space.

Now for the LRT. The likelihood is

$$L(\mathbf{p}; \mathbf{x}) = p_1^{x_1} p_2^{x_2} p_3^{x_3} p_4^{x_4}. \tag{16.28}$$

Denoting the MLE of p_i under the null by $\widehat{p}_{0i}$ and under alternative by $\widehat{p}_{Ai}$ for each i, we have that LRT statistic can be written

$$2\log(\Lambda(\mathbf{x})) = 2\sum_{i=1}^{4} x_i \log\left(\frac{n\widehat{p}_{Ai}}{n\widehat{p}_{0i}}\right). \tag{16.29}$$

The n's in the logs are unnecessary, but it is convention in contingency tables to write this statistic in terms of the expected counts in the cells, $E[X_i] = np_i$, so that

$$2\log(\Lambda(\mathbf{x})) = 2 \sum_{i=1}^{4} Obs_i \log\left(\frac{Exp_{Ai}}{Exp_{0i}}\right), \tag{16.30}$$

where Obs_i is the observed count x_i, and the Exp's are the expected counts under the two hypotheses.

Exercise 13.8.23 shows that the MLEs under the null in (16.26) of α and β are $\widehat{\alpha} = (x_1 + x_2)/n$ and $\widehat{\beta} = (x_1 + x_3)/n$, hence

$$\widehat{p}_{01} = \widehat{\alpha}\widehat{\beta}, \ \ \widehat{p}_{02} = \widehat{\alpha}(1 - \widehat{\beta}), \ \ \widehat{p}_{03} = (1 - \widehat{\alpha})\widehat{\beta}, \ \ \widehat{p}_{04} = (1 - \widehat{\alpha})(1 - \widehat{\beta}). \tag{16.31}$$

Under the alternative, from Exercise 13.8.22 we know that $\widehat{p}_{Ai} = x_i/n$ for each i. In this case, $Exp_{Ai} = Obs_i$. The statistic is then

$$2\log(\Lambda(\mathbf{x})) = 2\left(x_1 \log\left(\frac{x_1}{n\widehat{\alpha}\widehat{\beta}}\right) + x_2 \log\left(\frac{x_2}{n\widehat{\alpha}(1 - \widehat{\beta})}\right)\right.$$
$$\left. + x_3 \log\left(\frac{x_3}{n(1 - \widehat{\alpha})\widehat{\beta}}\right) + x_4 \log\left(\frac{x_4}{n(1 - \widehat{\alpha})(1 - \widehat{\beta})}\right)\right). \tag{16.32}$$

The alternative space (16.27) has only three free parameters, since one is free to choose three p_i's (within bounds), but then the fourth is set since they must sum to 1. The null space is the set of $\mathbf{p}$ that satisfy the parametrization given in (16.26) for $(\alpha, \beta) \in (0, 1) \times (0, 1)$, yielding two free parameters. Thus the difference in dimensions is 1, hence the $2\log(\Lambda(\mathbf{x}))$ is asymptotically χ_1^2.

16.1.4 Checking the dimension

For many hypotheses, it is straightforward to count the number of free parameters in the parameter space. In some more complicated models, it may not be so obvious. I don't know of a universal approach to counting the number, but if you do have what you think is a set of free parameters, you can check its validity by checking the Cramér conditions for the model when written in terms of those parameters. In particular, if there are K parameters, the parameter space must be open in $\mathbb{R}^K$, and the Fisher information matrix must be finite and positive definite.

16.2 Asymptotic null distribution of the LRT statistic

Similar to the way that, under conditions, the MLE is asymptotically (multivariate) normal, the $2\log(\Lambda(\mathbf{X}))$ is asymptotically χ^2 under the null as given in (16.2). We will not present the proof of the general result, but sketch it when the null is simple. We have $\mathbf{X}_1, \ldots, \mathbf{X}_n$ iid, each with density $f(\mathbf{x} \mid \boldsymbol{\theta})$, $\boldsymbol{\theta} \in \mathcal{T} \subset \mathbb{R}^K$, and make the assumptions in Section 14.4 used for likelihood estimation. Consider testing

$$H_0: \boldsymbol{\theta} = \boldsymbol{\theta}_0 \ \ versus \ \ H_A: \boldsymbol{\theta} \in \mathcal{T} - \{\boldsymbol{\theta}_0\} \tag{16.33}$$

for fixed $\boldsymbol{\theta}_0 \in \mathcal{T}$. One of the assumptions is that $\mathcal{T}$ is an open set, so that $\boldsymbol{\theta}_0$ is in the interior of $\mathcal{T}$. In particular, this assumption rules out one-sided tests. The MLE under the null is thus the fixed $\widehat{\boldsymbol{\theta}}_0 = \boldsymbol{\theta}_0$, and the MLE under the alternative, $\widehat{\boldsymbol{\theta}}_A$, is the usual MLE.

From (16.1), since the $\mathbf{x}_i$'s are iid,

$$l_n(\boldsymbol{\theta}; \mathbf{x}_1, \ldots, \mathbf{x}_n) = l_n(\boldsymbol{\theta}) = \sum_{i=1}^{n} l_1(\boldsymbol{\theta}; \mathbf{x}_i), \tag{16.34}$$

where $l_1(\boldsymbol{\theta}; \mathbf{x}_i) = \log(f(\mathbf{x}_i \mid \boldsymbol{\theta}))$ is the loglikelihood in one observation, and we have dropped the $\mathbf{x}_i$'s from the notation of l_n for simplicity, as in Section 14.1. Thus the log of the likelihood ratio is

$$2\log(\Lambda(\mathbf{x})) = 2\left(l_n(\widehat{\boldsymbol{\theta}}_A) - l_n(\boldsymbol{\theta}_0)\right). \tag{16.35}$$

Expand $l_n(\boldsymbol{\theta}_0)$ around $\widehat{\boldsymbol{\theta}}_A$ in a Taylor series:

$$l_n(\boldsymbol{\theta}_0) \approx l_n(\widehat{\boldsymbol{\theta}}_A) + (\boldsymbol{\theta}_0 - \widehat{\boldsymbol{\theta}}_A)'\nabla l_n(\widehat{\boldsymbol{\theta}}_A) - \tfrac{1}{2}(\boldsymbol{\theta}_0 - \widehat{\boldsymbol{\theta}}_A)'\widehat{\boldsymbol{\mathcal{I}}}_n(\widehat{\boldsymbol{\theta}}_A)(\boldsymbol{\theta}_0 - \widehat{\boldsymbol{\theta}}_A), \tag{16.36}$$

where the score function ∇l_n is given in (14.77) and the observed Fisher information matrix $\widehat{\boldsymbol{\mathcal{I}}}_n$ is given in (14.81).

As in (14.78), $\nabla l_n(\widehat{\boldsymbol{\theta}}_A) = \mathbf{0}$ because the score is zero at the MLE. Thus by (16.35),

$$2\log(\Lambda(\mathbf{x})) \approx (\boldsymbol{\theta}_0 - \widehat{\boldsymbol{\theta}}_A)'\widehat{\boldsymbol{\mathcal{I}}}_n(\widehat{\boldsymbol{\theta}}_A)(\boldsymbol{\theta}_0 - \widehat{\boldsymbol{\theta}}_A). \tag{16.37}$$

Now suppose H_0 is true, so that $\boldsymbol{\theta}_0$ is the true value of the parameter. As in (14.86), we have

$$\widehat{\boldsymbol{\mathcal{I}}}_n^{1/2}(\widehat{\boldsymbol{\theta}}_A)(\widehat{\boldsymbol{\theta}}_A - \boldsymbol{\theta}_0) \longrightarrow^{\mathcal{D}} \mathbf{Z} = N(\mathbf{0}, \mathbf{I}_K). \tag{16.38}$$

Then using the mapping Lemma 9.1 on page 139,

$$2\log(\Lambda(\mathbf{x})) \longrightarrow^{\mathcal{D}} \mathbf{Z}'\mathbf{Z} \sim \chi_K^2. \tag{16.39}$$

16.2.1 Composite null

Again with $\mathcal{T} \subset \mathbb{R}^K$, where $\mathcal{T}$ is open, we consider the null hypothesis that sets part of $\boldsymbol{\theta}$ to zero. That is, partition

$$\boldsymbol{\theta} = \begin{pmatrix} \boldsymbol{\theta}^{(1)} \\ \boldsymbol{\theta}^{(2)} \end{pmatrix}, \quad \boldsymbol{\theta}^{(1)} \text{ is } K_1 \times 1, \quad \boldsymbol{\theta}^{(2)} \text{ is } K_2 \times 1, \quad K_1 + K_2 = K. \tag{16.40}$$

The problem is to test

$$H_0: \boldsymbol{\theta}^{(1)} = \mathbf{0} \ \text{ versus } \ H_A: \boldsymbol{\theta}^{(1)} \neq \mathbf{0}, \tag{16.41}$$

with $\boldsymbol{\theta}^{(2)}$ unspecified. More precisely,

$$H_0: \boldsymbol{\theta} \in \mathcal{T}_0 = \{\boldsymbol{\theta} \in \mathcal{T} \mid \boldsymbol{\theta}^{(1)} = \mathbf{0}\} \ \text{ versus } \ H_A: \boldsymbol{\theta} \in \mathcal{T}_A = \mathcal{T} - \mathcal{T}_0. \tag{16.42}$$

The parameters in $\boldsymbol{\theta}^{(2)}$ are called **nuisance parameters**, because they are not of primary interest, but still need to be dealt with. Without them, we would have a nice simple null. The main result follows. See Theorem 7.7.4 in Lehmann (2004) for proof.

Theorem 16.1. *If the Cramér assumptions in Section 14.4 hold, then under the null, the LRT statistic* Λ *for problem (16.42) has*

$$2 \log(\Lambda(\mathbf{X})) \longrightarrow^{\mathcal{D}} \chi^2_{K_1}. \tag{16.43}$$

This theorem takes care of the simple null case as well, where $K_2 = 0$.

Setting some θ_i's to zero may seem to be an overly restrictive type of null, but many testing problems can be reparameterized into that form. For example, suppose $X_1, \ldots, X_n$ are iid $N(\mu_X, \sigma^2)$, and $Y_1, \ldots, Y_n$ are iid $N(\mu_Y, \sigma^2)$, and the X_i's and Y_i's are independent. We wish to test $\mu_X = \mu_Y$, with σ^2 unknown. Then the hypotheses are

$$H_0 : \mu_X = \mu_Y, \sigma^2 > 0 \;\; versus \;\; H_A : \mu_X \neq \mu_Y, \sigma^2 > 0. \tag{16.44}$$

To put these in the form (16.42), take a one-to-one reparameterizations

$$\boldsymbol{\theta} = \begin{pmatrix} \mu_X - \mu_Y \\ \mu_X + \mu_Y \\ \sigma^2 \end{pmatrix}; \;\; \boldsymbol{\theta}^{(1)} = \mu_X - \mu_Y \text{ and } \boldsymbol{\theta}^{(2)} = \begin{pmatrix} \mu_X + \mu_Y \\ \sigma^2 \end{pmatrix}. \tag{16.45}$$

Then

$$\mathcal{T}_0 = \{0\} \times \mathbb{R} \times (0, \infty) \text{ and } \mathcal{T} = \mathbb{R}^2 \times (0, \infty). \tag{16.46}$$

Here, $K_1 = 1$, and there are $K_2 = 2$ nuisance parameters, $\mu_X + \mu_Y$ and σ^2. Thus the asymptotic χ^2 has 1 degree of freedom.

16.3   Score tests

When the null hypothesis is simple, tests based directly on the score function can often be simpler to implement than the LRT, since we do not need to find the MLE under the alternative. Start with the iid model having a one-dimensional parameter, so that $X_1, \ldots, X_n$ are iid with density $f(x_i \mid \theta)$, $\theta \in \mathbb{R}$. Consider the one-sided testing problem,

$$H_0 : \theta = \theta_0 \;\; versus \;\; H_A : \theta > \theta_0. \tag{16.47}$$

The best test for a simple alternative $\theta_A > \theta_0$ has test statistic $\prod f(x_i \mid \theta_A) / \prod f(x_i \mid \theta_0)$. Here we take the log of that ratio, and expand it in a Taylor series in θ_A around θ_0, so that

$$\sum_{i=1}^{n} \log \left(\frac{f(x_i \mid \theta_A)}{f(x_i \mid \theta_0)} \right) = l_n(\theta_A) - l_n(\theta_0)$$

$$\approx (\theta_A - \theta_0)' l_n'(\theta_0), \tag{16.48}$$

where $l_n'(\theta)$ is the score function in n observations as in (14.3). The test statistic in (16.48) is approximately the best statistic for alternative θ_A when θ_A is very close to θ_0.

For fixed $\theta_A > \theta_0$, the test that rejects the null when $(\theta_A - \theta_0) l_n'(\theta_0) > c$ is then the same that rejects when $l_n'(\theta_0) > c^*$. Since we are in the iid case, $l_n'(\theta) = \sum_{i=1}^{n} l_1'(\theta; x_i)$, where l_1' is the score for one observation. Under the null hypothesis,

$$E_{\theta_0}[l_1'(\theta_0; X_i)] = 0 \text{ and } Var_{\theta_0}[l_1'(\theta_0; X_i)] = \mathcal{I}_1(\theta_0) \tag{16.49}$$

as in Lemma 14.1 on page 226, hence by the central limit theorem,

$$T_n(\mathbf{x}) \equiv \frac{l'_n(\theta_0; \mathbf{X})}{\sqrt{n \, \mathcal{I}_1(\theta_0)}} \longrightarrow^{\mathcal{D}} N(0,1) \quad \text{under the null hypothesis.} \tag{16.50}$$

Then the one-sided score test

$$\text{Rejects the null hypothesis when } T_n(\mathbf{x}) > z_\alpha, \tag{16.51}$$

which is approximately level α. Note that we did not need the MLE under the alternative.

For example, suppose the X_i's are iid under the Cauchy location family, so that

$$f(x_i \,|\, \theta) = \frac{1}{\pi} \frac{1}{1 + (x_i - \theta)^2}. \tag{16.52}$$

We wish to test

$$H_0: \theta = 0 \quad versus \quad H_A: \theta > 0, \tag{16.53}$$

The score at $\theta = \theta_0 = 0$ and information in one observation are, respectively,

$$l'_1(0; x_i) = \frac{2x_i}{1 + x_i^2} \quad \text{and} \quad \mathcal{I}_1(0) = \frac{1}{2}. \tag{16.54}$$

See (14.74) for $\mathcal{I}_1$. Then the test statistic is

$$T_n(\mathbf{x}) = \frac{l'_n(0; \mathbf{x})}{\sqrt{n \, \mathcal{I}_1(0)}} = \frac{\sum_{i=1}^{n} \frac{2x_i}{1+x_i^2}}{\sqrt{n/2}} = \frac{2\sqrt{2}}{\sqrt{n}} \sum_{i=1}^{n} \frac{x_i}{1 + x_i^2}, \tag{16.55}$$

and for an approximate 0.05 level, the cutoff point is $z_{0.05} = 1.645$. If the information is difficult to calculate, which it is not in this example, then you can use the observed information at θ_0 instead. This test has relatively good power for small θ. However, note that as θ gets large, so do the x_i's, hence $T_n(\mathbf{x})$ becomes small. Thus the power at large θ's is poor, even below the level α.

16.3.1 Many-sided

Now consider a more general problem, where $\theta \in \mathcal{T} \subset \mathbb{R}^K$, and the null, θ_0, is in the interior of $\mathcal{T}$. The testing problem is

$$H_0: \theta = \theta_0 \quad versus \quad H_A: \theta \in \mathcal{T} - \{\theta_0\}. \tag{16.56}$$

Here the score is the vector of partial derivatives of the loglikelihood, $\nabla l_1(\theta; \mathbf{X})$, as in (14.77). Equations (14.79) and (14.80) show that, under the null,

$$E[\nabla l_1(\theta; \mathbf{X}_i)] = 0 \quad \text{and} \quad Cov[\nabla l_1(\theta; \mathbf{X}_i)] = \mathcal{I}_1(\theta_0). \tag{16.57}$$

Thus, again in the iid case, the multivariate central limit theorem shows that

$$\frac{1}{\sqrt{n}} \nabla l_n(\theta_0) \longrightarrow^{\mathcal{D}} N(0, \mathcal{I}_1(\theta_0)). \tag{16.58}$$

Then the mapping theorem and (7.53) on the χ^2 shows that under the null

$$S_n^2 \equiv \frac{1}{n} \nabla l_n(\boldsymbol{\theta}_0)' \boldsymbol{I}_1^{-1}(\boldsymbol{\theta}_0) \nabla l_n(\boldsymbol{\theta}_0) \longrightarrow^{\mathcal{D}} \chi_K^2. \tag{16.59}$$

The S_n is the statistic for the score test, and the approximate level α score test rejects the null hypothesis if

$$S_n^2 > \chi_{K,\alpha}^2. \tag{16.60}$$

Multinomial distribution

Suppose $\mathbf{X} \sim \text{Multinomial}(n, (p_1, p_2, p_3))$, and we wish to test that the probabilities are equal:

$$H_0: p_1 = p_2 = \tfrac{1}{3} \text{ versus } H_A: (p_1, p_2) \neq (\tfrac{1}{3}, \tfrac{1}{3}). \tag{16.61}$$

We've left out p_3 because it is a function of the other two. Leaving it in will violate the openness of the parameter space in $\mathbb{R}^3$.

The loglikelihood is

$$l_n(p_1, p_2 ; \mathbf{x}) = x_1 \log(p_1) + x_2 \log(p_2) + x_3 \log(1 - p_1 - p_2). \tag{16.62}$$

Exercise 16.7.12 shows that the score at the null is

$$\nabla l_n(\tfrac{1}{3}, \tfrac{1}{3}) = 3 \begin{pmatrix} x_1 - x_3 \\ x_2 - x_3 \end{pmatrix}, \tag{16.63}$$

and the Fisher information matrix is

$$\boldsymbol{I}_n(\tfrac{1}{3}, \tfrac{1}{3}) = 3n \begin{pmatrix} 2 & 1 \\ 1 & 2 \end{pmatrix}. \tag{16.64}$$

Thus since $\boldsymbol{I}_n = n\boldsymbol{I}_1$, after some manipulation, the score statistic is

$$\begin{aligned} S_n^2 &= \nabla l_n(\boldsymbol{\theta}_0)' \boldsymbol{I}_n^{-1}(\boldsymbol{\theta}_0) \nabla l_n(\boldsymbol{\theta}_0) \\ &= \frac{3}{n} \begin{pmatrix} X_1 - X_3 \\ X_2 - X_3 \end{pmatrix}' \begin{pmatrix} 2 & 1 \\ 1 & 2 \end{pmatrix}^{-1} \begin{pmatrix} X_1 - X_3 \\ X_2 - X_3 \end{pmatrix} \\ &= \frac{1}{n} ((X_1 - X_3)^2 + (X_2 - X_3)^2 + (X_1 - X_2)^2). \end{aligned} \tag{16.65}$$

The cutoff point is $\chi_{2,\alpha}^2$, because there are $K = 2$ parameters. The statistic looks reasonable, because the X_i's would tend to be different if their p_i's were. Also, it may not look like it, but this S_n^2 is the same as the **Pearson χ^2 statistic** for these hypotheses, which is

$$X^2 = \sum_{i=1}^3 \frac{(X_i - n/3)^2}{n/3} = \sum_{i=1}^3 \frac{(Obs_i - Exp_i)^2}{Exp_i}. \tag{16.66}$$

Here, Obs_i is the observed count x_i as above, and Exp_i is the expected count under the null, which here is $n/3$ for each i.

16.4   Model selection: AIC and BIC

We often have a number of models we wish to consider, rather than just two as in hypothesis testing. (Note also that hypothesis testing may not be appropriate even when choosing between two models, e.g., when there is no obvious allocation to "null" and "alternative" models.) For example, in the regression or logistic regression model, each subset of explanatory variables defines a different model. Here, we assume there are K models under consideration, labeled $M_1, M_2, \ldots, M_K$. Each model is based on the same data, $\mathbf{Y}$, but has its own density and parameter space:

$$\text{Model } M_k \;\Rightarrow\; \mathbf{Y} \sim f_k(\mathbf{y} \mid \boldsymbol{\theta}_k), \; \boldsymbol{\theta}_k \in \mathcal{T}_k. \tag{16.67}$$

The densities need not have anything to do with each other, i.e., one could be normal, another uniform, another logistic, etc., although often they will be of the same family. It is possible that the models will overlap, so that several models might be correct at once, e.g., when there are nested models.

Let

$$l_k(\boldsymbol{\theta}_k\,;\,\mathbf{y}) = \log(L_k(\boldsymbol{\theta}_k\,;\,\mathbf{y})) = \log(f_k(\mathbf{y} \mid \boldsymbol{\theta}_k)) + C(\mathbf{y}), \; k = 1, \ldots, K, \tag{16.68}$$

be the loglikelihoods for the models. The constant $C(\mathbf{y})$ is arbitrary, being the log of the constant multiplier in the likelihood from Definition 13.1 on page 199. As long as it is the same for each k, it will not affect the outcome of the following procedures. Define the **deviance** of the model M_k at parameter value $\boldsymbol{\theta}_k$ by

$$\text{Deviance}(M_k(\boldsymbol{\theta}_k)\,;\,\mathbf{y}) = -2\, l_k(\boldsymbol{\theta}_k\,;\,\mathbf{y}). \tag{16.69}$$

It is a measure of fit of the model to the data; the smaller the deviance, the better the fit. The MLE of $\boldsymbol{\theta}_k$ for model M_k minimizes this deviance, giving us the **observed deviance**,

$$\text{Deviance}(M_k(\widehat{\boldsymbol{\theta}}_k)\,;\,\mathbf{y}) = -2\, l_k(\widehat{\boldsymbol{\theta}}_k\,;\,\mathbf{y}) = -2 \max_{\boldsymbol{\theta}_k \in \mathcal{T}_k} l_k(\boldsymbol{\theta}_k\,;\,\mathbf{y}). \tag{16.70}$$

Note that the likelihood ratio statistic in (16.2) is just the difference in observed deviance of the two hypothesized models:

$$2\log(\Lambda(\mathbf{y})) = \text{Deviance}(H_0(\widehat{\boldsymbol{\theta}}_0)\,;\,\mathbf{y}) - \text{Deviance}(H_A(\widehat{\boldsymbol{\theta}}_A)\,;\,\mathbf{y}). \tag{16.71}$$

At first blush one might decide the best model is the one with the smallest observed deviance. The problem with that approach is that because the deviances are based on minus the maximum of the likelihoods, the model with the best observed deviance will be the largest model, i.e., one with highest dimension. Instead, we add a penalty depending on the dimension of the parameter space, as for Mallows' C_p in (12.61). The two most popular likelihood-based procedures are the Bayes information criterion (BIC) of Schwarz (1978) and the Akaike information criterion (AIC) of Akaike (1974) (who actually meant for the "A" to stand for "An"):

$$\text{BIC}(M_k\,;\,\mathbf{y}) = \text{Deviance}(M_k(\widehat{\boldsymbol{\theta}}_k)\,;\,\mathbf{y}) + \log(n)d_k, \text{ and}$$
$$\text{AIC}(M_k\,;\,\mathbf{y}) = \text{Deviance}(M_k(\widehat{\boldsymbol{\theta}}_k)\,;\,\mathbf{y}) + 2d_k, \tag{16.72}$$

where

$$d_k = \dim(\mathcal{T}_k). \tag{16.73}$$

Whichever criterion is used, it is implemented by finding the value for each model, then choosing the model with the smallest value of the criterion, or looking at the models with the smallest values.

Note that the only difference between AIC and BIC is the factor multiplying the dimension in the penalty component. The BIC penalizes each dimension more heavily than does the AIC, at least if $n > 7$, so tends to choose more parsimonious models. In more complex situations than we deal with here, the *deviance information criterion* is useful, which uses more general definitions of the deviance. See Spiegelhalter, Best, Carlin, and van der Linde (2002).

The AIC and BIC have somewhat different motivations. The BIC, as hinted at by the "Bayes" in the name, is an attempt to estimate the Bayes posterior probability of the models. More specifically, if the prior probability that model M_k is the true one is π_k, then the BIC-based estimate of the posterior probability is

$$\widehat{P}^{\text{BIC}}[M_k \mid \mathbf{y}] = \frac{e^{-\frac{1}{2}\,\text{BIC}(M_k;\mathbf{y})}\pi_k}{e^{-\frac{1}{2}\,\text{BIC}(M_1;\mathbf{y})}\pi_1 + \cdots + e^{-\frac{1}{2}\,\text{BIC}(M_K;\mathbf{y})}\pi_K}. \tag{16.74}$$

If the prior probabilities are taken to be equal, then because each posterior probability has the same denominator, the model that has the highest posterior probability is indeed the model with the smallest value of BIC. The advantage of the posterior probability form is that it is easy to assess which models are nearly as good as the best, if there are any.

The next two sections present some further details on the two criteria.

16.5 BIC: Motivation

To see where the approximation in (16.74) arises, we first need a prior on the parameter space. As we did in Section 15.4 for hypothesis testing, we decompose the overall prior into conditional ones for each model. The marginal probability of each model is the prior probability:

$$\pi_k = P[M_k]. \tag{16.75}$$

For a model M, where the parameter is d-dimensional, let the prior be

$$\boldsymbol{\theta} \mid M \sim N_d(\boldsymbol{\theta}_0, \boldsymbol{\Sigma}_0). \tag{16.76}$$

Then the density of $\mathbf{Y}$ in (16.67), conditioning on the model, is

$$g(\mathbf{y} \mid M) = \int_{\mathcal{T}} f(\mathbf{y} \mid \boldsymbol{\theta})\phi(\boldsymbol{\theta} \mid \boldsymbol{\theta}_0, \boldsymbol{\Sigma}_0)d\boldsymbol{\theta}, \tag{16.77}$$

where ϕ is the multivariate normal pdf.

We will use the Laplace approximation, as in Schwarz (1978), to approximate this density. The following requires a number of regularity assumptions, not all of which we will detail, including the Cramér conditions in Section 14.4. In particular, we assume $\mathbf{Y}$ consists of n iid observations, where n is large. Since $l_n(\boldsymbol{\theta}) \equiv l_n(\boldsymbol{\theta};\mathbf{y}) = \log(f(\mathbf{y} \mid \boldsymbol{\theta}))$,

$$g(\mathbf{y} \mid M) = \int_{\mathcal{T}} e^{l_n(\boldsymbol{\theta})}\phi(\boldsymbol{\theta} \mid \boldsymbol{\theta}_0, \boldsymbol{\Sigma}_0)d\boldsymbol{\theta}. \tag{16.78}$$

The Laplace approximation expands $l_n(\theta)$ around its maximum, the maximum occurring at the maximum likelihood estimator $\hat{\theta}$. Then as in (16.36),

$$l_n(\theta) \approx l_n(\hat{\theta}) + (\theta - \hat{\theta})' \nabla l_n(\hat{\theta}) - \tfrac{1}{2}(\theta - \hat{\theta})' \hat{\mathcal{I}}_n(\hat{\theta})(\theta - \hat{\theta})$$
$$= l_n(\hat{\theta}) - \tfrac{1}{2}(\theta - \hat{\theta})' \hat{\mathcal{I}}_n(\hat{\theta})(\theta - \hat{\theta}), \qquad (16.79)$$

where the score function at the MLE is zero: $\nabla l_n(\hat{\theta}) = \mathbf{0}$, and $\hat{\mathcal{I}}_n$ is the $d \times d$ observed Fisher information matrix. Then (16.78) and (16.79) combine to show that

$$g(\mathbf{y} \mid M) \approx e^{l_n(\hat{\theta})} \int_{\mathcal{T}} e^{-\frac{1}{2}(\theta - \hat{\theta})' \hat{\mathcal{I}}_n(\hat{\theta})(\theta - \hat{\theta})} \phi(\theta \mid \theta_0, \Sigma_0) d\theta. \qquad (16.80)$$

Kass and Wasserman (1995) give precise details on approximating this g. We will be more heuristic. To whit, the first term in the integrand in (16.80) looks like a $N(\theta, \hat{\mathcal{I}}_n^{-1}(\hat{\theta}))$ pdf for $\hat{\theta}$ without the constant, which constant would be $\sqrt{|\hat{\mathcal{I}}_n(\hat{\theta})|} / \sqrt{2\pi}^d$. Putting in and taking out the constant yields

$$g(\mathbf{y} \mid M) \approx e^{l_n(\hat{\theta})} \frac{\sqrt{2\pi}^d}{\sqrt{|\hat{\mathcal{I}}_n(\hat{\theta})|}} \int_{\mathbb{R}^d} \phi(\hat{\theta} \mid \theta, \hat{\mathcal{I}}_n^{-1}(\hat{\theta})) \phi(\theta \mid \theta_0, \Sigma_0) d\theta. \qquad (16.81)$$

Mathematically, this integral is the marginal pdf of $\hat{\theta}$ when its conditional distribution given θ is $N(\theta, \hat{\mathcal{I}}_n^{-1}(\hat{\theta}))$ and the prior distribution of θ given the model is $N(\theta_0, \Sigma_0)$ as in (16.76). Exercise 7.8.15(e) shows that this marginal is then $N(\theta_0, \Sigma_0 + \hat{\mathcal{I}}_n^{-1}(\hat{\theta}))$. Using this marginal pdf yields

$$g(\mathbf{y} \mid M) \approx e^{l_n(\hat{\theta})} \frac{\sqrt{2\pi}^d}{\sqrt{|\hat{\mathcal{I}}_n(\hat{\theta})|}} \frac{1}{\sqrt{2\pi}^d \sqrt{|\Sigma_0 + \hat{\mathcal{I}}_n^{-1}(\hat{\theta})|}} e^{-\frac{1}{2}(\hat{\theta} - \theta_0)'(\Sigma_0 + \hat{\mathcal{I}}_n^{-1}(\hat{\theta}))^{-1}(\hat{\theta} - \theta_0)}$$
$$= e^{l_n(\hat{\theta})} \frac{1}{\sqrt{|\hat{\mathcal{I}}_n(\hat{\theta})\Sigma_0 + \mathbf{I}_d|}} e^{-\frac{1}{2}(\hat{\theta} - \theta_0)'(\Sigma_0 + \hat{\mathcal{I}}_n^{-1}(\hat{\theta}))^{-1}(\hat{\theta} - \theta_0)}, \qquad (16.82)$$

where the $\mathbf{I}_d$ is the $d \times d$ identity matrix.

In Section 15.4 on Bayesian testing, we saw some justification for taking a prior that has about as much information as does one observation. In this case, the information in the n observations is $\hat{\mathcal{I}}_n(\hat{\theta})$, so it would be reasonable to take Σ_0^{-1} to be $\hat{\mathcal{I}}_n(\hat{\theta})/n$, giving us

$$g(\mathbf{y} \mid M) \approx e^{l_n(\hat{\theta})} \frac{1}{\sqrt{|n\mathbf{I}_d + \mathbf{I}_d|}} e^{-\frac{1}{2}(\hat{\theta} - \theta_0)'((n+1)\hat{\mathcal{I}}_n^{-1}(\hat{\theta}))^{-1}(\hat{\theta} - \theta_0)}$$
$$= e^{l_n(\hat{\theta})} \frac{1}{\sqrt{n+1}^d} e^{-\frac{1}{2}(\hat{\theta} - \theta_0)'((n+1)\hat{\mathcal{I}}_n^{-1}(\hat{\theta}))^{-1}(\hat{\theta} - \theta_0)}. \qquad (16.83)$$

The final approximation in the BIC works on the logs:

$$\log(g(\mathbf{y} \mid M)) \approx l_n(\hat{\theta}) - \frac{d}{2} \log(n + 1) - \frac{1}{2}(\hat{\theta} - \theta_0)'((n + 1)\hat{\mathcal{I}}_n^{-1}(\hat{\theta}))^{-1}(\hat{\theta} - \theta_0)$$
$$\approx l_n(\hat{\theta}) - \frac{d}{2} \log(n). \qquad (16.84)$$

The last step shows two further approximations. For large n, replacing $n + 1$ with n in the log is very minor. The justification for erasing the final quadratic term is that the first term on the right, $l_n(\widehat{\theta})$, is of order n, and the second term is of order $\log(n)$, while the final term is of constant order since $\widehat{\boldsymbol{\mathcal{I}}}_n(\widehat{\theta})/(n+1) \approx \boldsymbol{\mathcal{I}}_1(\theta)$. Thus for large n it can be dropped. There are a number of approximations and heuristics in this derivation, and indeed the resulting approximation may not be especially good. See Berger, Ghosh, and Mukhopadhyay (2003), for example. A nice property is that under conditions, if one of the considered models is the correct one, then the BIC chooses the correct model as $n \to \infty$.

The final expression in (16.84) is the BIC approximation to the log of the marginal density. The BIC statistic itself is based on the deviance, that is, for model M,

$$BIC(M\,;\mathbf{y}) = -2\log(\widehat{g}(\mathbf{y}\,|\,M)) = -2\log(l_n(\widehat{\theta})) + d\log(n)$$
$$= \text{Deviance}(M(\widehat{\theta})\,;\mathbf{y}) + \log(n)d, \tag{16.85}$$

as in (16.72). Given a number of models, $M_1, \ldots, M_K$, each with its own marginal prior probability π_k and conditional marginal density $g_k(\mathbf{y}\,|\,M_k)$, the posterior probability of the model is

$$P[M_k\,|\,\mathbf{y}] = \frac{g_k(\mathbf{y}\,|\,M_k)\pi_k}{g_1(\mathbf{y}\,|\,M_1)\pi_1 + \cdots + g_K(\mathbf{y}\,|\,M_K)\pi_K}. \tag{16.86}$$

Thus from (16.85), we have the BIC-based estimate of g_k,

$$\widehat{g}_k(\mathbf{y}\,|\,M_k) = e^{-\frac{1}{2}BIC(M_k\,;y)}, \tag{16.87}$$

hence replacing the g_k's in (16.86) with their estimates yields the estimated posterior given in (16.74).

16.6 AIC: Motivation

The Akaike information criterion can be thought of as a generalization of Mallows' C_p from Section 12.5.3, based on deviance rather than error sum of squares. To evaluate model M_k as in (16.67), we imagine fitting the model based on the data $\mathbf{Y}$, then testing it out on a new (unobserved) variable, $\mathbf{Y}^{New}$, which has the same distribution as and is independent of $\mathbf{Y}$. The measure of discrepancy between the model and the new variable is the deviance in (16.69), where the parameter is estimated using $\mathbf{Y}$. We then take the expected value, yielding the expected predictive deviance,

$$EPredDev_k = E[\text{Deviance}(M_k(\widehat{\theta}_k)\,;\mathbf{Y}^{New})]. \tag{16.88}$$

The expected value is over $\widehat{\theta}_k$, which depends on only $\mathbf{Y}$, and $\mathbf{Y}^{New}$.

As for Mallows' C_p, we estimate the expected predictive deviance using the observed deviance, then add a term to ameliorate the bias. Akaike (1974) argues that for large n, if M_k is the true model,

$$\delta = EPredDev_k - E[\text{Deviance}(M_k(\widehat{\theta}_k)\,;\mathbf{Y})] \approx 2d_k, \tag{16.89}$$

where d_k is the dimension of the model as in (16.73), from which the estimate AIC in (16.72) arises. A good model is then one with a small AIC.

Note also that by adjusting the priors $\pi_k = P[M_k]$ in (16.74), one can work it so that the model with the lowest AIC has the highest posterior probability. See Exercise 16.7.18.

Akaike's original motivation was information-theoretic, based on the **Kullback-Leibler** divergence from density f to density g. This divergence is defined as

$$\text{KL}(f \,||\, g) = -\int g(\mathbf{w}) \log\left(\frac{f(\mathbf{w})}{g(\mathbf{w})}\right) d\mathbf{w}. \tag{16.90}$$

For fixed g, the Kullback-Leibler divergence is positive unless $g = f$, in which case it is zero. For the Akaike information criterion, g is the true density of $\mathbf{Y}$ and $\mathbf{Y}^{New}$, and for model k, f is the density estimated using the maximum likelihood estimate of the parameter, $f_k(\mathbf{w} \,|\, \widehat{\boldsymbol{\theta}})$, where $\widehat{\boldsymbol{\theta}}$ is based on $\mathbf{Y}$. Write

$$\text{KL}(f_k(\mathbf{w} \,|\, \widehat{\boldsymbol{\theta}}_k) \,||\, g) = -\int g(\mathbf{w}) \log(f_k(\mathbf{w} \,|\, \widehat{\boldsymbol{\theta}})) d\mathbf{w} + \int g(\mathbf{w}) \log(g(\mathbf{w})) d\mathbf{w}$$

$$= \tfrac{1}{2} E[\text{Deviance}(M_k(\widehat{\boldsymbol{\theta}}_k) ; \mathbf{Y}^{New}) \,|\, \mathbf{Y} = \mathbf{y}] - \text{Entropy}(g). \tag{16.91}$$

(The $\mathbf{w}$ is representing the $\mathbf{Y}^{New}$, and the dependence on the observed $\mathbf{y}$ is through only $\widehat{\boldsymbol{\theta}}_k$.) Here the g, the true density of $\mathbf{Y}$, does not depend on the model M_k, hence neither does its entropy, defined by $-\int g(\mathbf{w}) \log(g(\mathbf{w})) d\mathbf{w}$. Thus $EPredDev_k$ from (16.88) is equivalent to (16.91) upon taking the further expectation over $\mathbf{Y}$.

One slight logical glitch in the development is that while the theoretical criterion (16.88) is defined assuming $\mathbf{Y}$ and $\mathbf{Y}^*$ have the true distribution, the approximation in (16.89) assumes the true distribution is contained in the model M_k. Thus it appears that the approximation is valid for all models under consideration only if the true distribution is contained in all the models. Even so, the AIC is a legitimate method for model selection. See the book Burnham and Anderson (2003) for more information.

Rather than justify the result in full generality, we will follow Hurvich and Tsai (1989) and derive the exact value for Δ in multiple regression.

16.6.1 Multiple regression

The multiple regression model (12.9) is

$$\text{Model } M : \mathbf{Y} \sim N_n(\mathbf{x}\boldsymbol{\beta}, \sigma^2 \mathbf{I}_n), \quad \boldsymbol{\beta} \in \mathbb{R}^p, \tag{16.92}$$

where $\mathbf{x}$ is $n \times p$. Now from (16.14),

$$l(\boldsymbol{\beta}, \sigma^2 ; \mathbf{y}) = -\frac{n}{2} \log(\sigma^2) - \frac{1}{2} \frac{\|\mathbf{y} - \mathbf{x}\boldsymbol{\beta}\|^2}{\sigma^2}. \tag{16.93}$$

Exercise 13.8.20 shows that MLEs are

$$\widehat{\boldsymbol{\beta}} = (\mathbf{x}'\mathbf{x})^{-1}\mathbf{x}'\mathbf{y} \text{ and } \widehat{\sigma}^2 = \frac{1}{n} \|\mathbf{y} - \mathbf{x}\widehat{\boldsymbol{\beta}}\|^2. \tag{16.94}$$

Using (16.69), we see that the deviances evaluated at the data $\mathbf{Y}$ and the unobserved $\mathbf{Y}^{New}$ are, respectively,

$$\text{Deviance}(M(\widehat{\boldsymbol{\beta}}, \widehat{\sigma}^2) ; \mathbf{Y}) = n \log(\widehat{\sigma}^2) + n, \text{ and}$$

$$\text{Deviance}(M(\widehat{\boldsymbol{\beta}}, \widehat{\sigma}^2) ; \mathbf{Y}^{New}) = n \log(\widehat{\sigma}^2) + \frac{\|\mathbf{Y}^{New} - \mathbf{x}\widehat{\boldsymbol{\beta}}\|^2}{\widehat{\sigma}^2}. \tag{16.95}$$

The first terms on the right-hand sides in (16.95) are the same, hence the difference in (16.89) is

$$\delta = E[\|\mathbf{U}\|^2/\widehat{\sigma}^2] - n, \quad \text{where} \quad \mathbf{U} = \mathbf{Y}^{New} - \mathbf{x}\widehat{\beta} = \mathbf{Y}^{New} - \mathbf{P}_x\mathbf{Y}. \tag{16.96}$$

From Theorem 12.1 on page 183, we know that $\widehat{\beta}$ and $\widehat{\sigma}^2$ are independent, and further both are independent of $\mathbf{Y}^{New}$, hence we have

$$E[\|\mathbf{U}\|^2/\widehat{\sigma}^2] = E[\|\mathbf{U}\|^2]E[1/\widehat{\sigma}^2]. \tag{16.97}$$

Exercise 16.7.16 shows that

$$\delta = \frac{n}{n-p-2}\, 2(p+1), \tag{16.98}$$

where in the "$(p+1)$" term, the "p" is the number of β_i's and the "1" is for the σ^2. Then from (16.89), the estimate of $EPredDev$ is

$$\text{AICc}(M\,;\mathbf{y}) = \text{Deviance}(M(\widehat{\beta}, \widehat{\sigma}^2)\,;\mathbf{y}) + \frac{n}{n-p-2}\, 2(p+1). \tag{16.99}$$

The lower case "c" stands for "corrected." For large n, $\Delta \approx 2(p+1)$.

16.7 Exercises

Exercise 16.7.1. Continue with the polio example from Exercise 15.7.11, where here we look at the LRT. Thus we have X_V and X_C independent,

$$X_V \sim \text{Poisson}(c_V\theta_V) \quad \text{and} \quad X_C \sim \text{Poisson}(c_C\theta_C), \tag{16.100}$$

where c_V and c_V are known constants, and $\theta_V > 0$ and $\theta_C > 0$. We wish to test

$$H_0 : \theta_V = \theta_C \quad \text{versus} \quad H_A : \theta_V \neq \theta_C. \tag{16.101}$$

(a) Under the alternative, find the MLEs of θ_V and θ_C. (b) Under the null, find the common MLE of θ_V and θ_C. (c) Find the $2\log(\Lambda)$ version of the LRT statistic. What are the degrees of freedom in the asymptotic χ^2? (d) Now look at the polio data presented in Exercises 15.7.11, where $x_V = 57$, $x_C = 142$, $c_V = 2.00745$, and $c_C = 2.01229$. What are the values of the MLEs for these data? What is the value of the $2\log(\Lambda)$? Do you reject the null hypothesis?

Exercise 16.7.2. Suppose $X_1,\ldots,X_n$ are iid $N(\mu_X,\sigma^2)$, and $Y_1,\ldots,Y_m$ are iid $N(\mu_Y, \sigma^2)$, and the X_i's and Y_i's are independent. We wish to test $\mu_X = \mu_Y$, with σ^2 unknown. Then the hypotheses are

$$H_0 : \mu_X = \mu_Y, \sigma^2 > 0 \;\;\textit{versus}\;\; H_A : \mu_X \neq \mu_Y, \sigma^2 > 0. \tag{16.102}$$

(a) Find the MLEs of the parameters under the null hypothesis. (b) Find the MLEs of the parameters under the alternative hypothesis. (c) Find the LRT statistic. (d) Letting $\widehat{\sigma}_0^2$ and $\widehat{\sigma}_A^2$ be the MLEs for σ^2 under the null and alternative, respectively, show that

$$(n+m)(\widehat{\sigma}_0^2 - \widehat{\sigma}_A^2) = k_{n,m}\,(\bar{x} - \bar{y})^2 \tag{16.103}$$

for some constant $k_{n,m}$ that depends on n and m. Find the $k_{n,m}$. (e) Let T be the two-sample t-statistic,

$$T = \frac{\bar{x} - \bar{y}}{s_{pooled}\sqrt{\frac{1}{n} + \frac{1}{m}}}, \tag{16.104}$$

where s_{pooled}^2 is the pooled variance estimate,

$$s_{pooled}^2 = \frac{\sum(x_i - \bar{x})^2 + \sum(y_i - \bar{y})^2}{n + m - 2}. \tag{16.105}$$

What is the distribution of T under the null hypothesis? (You don't have to prove it; just say what the distribution is.) (f) Show that the LRT statistic is an increasing function of T^2.

Exercise 16.7.3. Consider the regression problem in Section 16.1.2, where $\mathbf{Y} \sim N(\mathbf{x}\boldsymbol{\beta}, \sigma^2 \mathbf{I}_n)$, $\mathbf{x}'\mathbf{x}$ is invertible, $\boldsymbol{\beta} = (\boldsymbol{\beta}_1', \boldsymbol{\beta}_2')'$, $\mathbf{x} = (\mathbf{x}_1, \mathbf{x}_2)$, and we test

$$H_0 : \boldsymbol{\beta}_2 = \mathbf{0} \ versus \ H_A : \boldsymbol{\beta}_2 \neq \mathbf{0}. \tag{16.106}$$

(a) Let $\mathbf{A}$ be an invertible $p \times p$ matrix, and rewrite the model with $\mathbf{x}\boldsymbol{\beta}$ replaced by $\mathbf{x}^*\boldsymbol{\beta}^*$, where $\mathbf{x}^* = \mathbf{x}\mathbf{A}$ and $\boldsymbol{\beta}^* = \mathbf{A}^{-1}\boldsymbol{\beta}$, so that $\mathbf{x}\boldsymbol{\beta} = \mathbf{x}^*\boldsymbol{\beta}^*$. Show that the projection matrices for $\mathbf{x}$ and $\mathbf{x}^*$ are the same, $\mathbf{P}_{\mathbf{x}} = \mathbf{P}_{\mathbf{x}^*}$ (hence $\mathbf{Q}_{\mathbf{x}} = \mathbf{Q}_{\mathbf{x}^*}$). (b) Now take $\mathbf{A}$ as in (16.19):

$$\mathbf{A} = \begin{pmatrix} \mathbf{I}_{p_1} & -(\mathbf{x}_1'\mathbf{x}_1)^{-1}\mathbf{x}_1'\mathbf{x}_2 \\ \mathbf{0} & \mathbf{I}_{p_2} \end{pmatrix}. \tag{16.107}$$

Show that with this $\mathbf{A}$, $\mathbf{x}^* = (\mathbf{x}_1, \mathbf{x}_2^*)$ and $\boldsymbol{\beta}^* = (\boldsymbol{\beta}_1^{*\prime}, \boldsymbol{\beta}_2')'$, where $\mathbf{x}_2^* = \mathbf{Q}_{\mathbf{x}_1}\mathbf{x}_2$. Give $\boldsymbol{\beta}_1^*$ explicitly. We use this $\mathbf{x}^*$ and $\boldsymbol{\beta}^*$ for the remainder of this exercise. (c) Show that $\mathbf{x}_1'\mathbf{x}_2^* = \mathbf{0}$. (d) Show that

$$\mathbf{P}_{\mathbf{x}} = \mathbf{P}_{\mathbf{x}_1} + \mathbf{P}_{\mathbf{x}_2^*}. \tag{16.108}$$

(e) Writing $\mathbf{Y} \sim N(\mathbf{x}^*\boldsymbol{\beta}^*, \sigma^2 \mathbf{I}_n)$, show that the joint distribution of $\mathbf{Q}_{\mathbf{x}}\mathbf{Y}$ and $\mathbf{P}_{\mathbf{x}_2^*}\mathbf{Y}$ is

$$\begin{pmatrix} \mathbf{Q}_{\mathbf{x}}\mathbf{Y} \\ \mathbf{P}_{\mathbf{x}_2^*}\mathbf{Y} \end{pmatrix} \sim N\left(\begin{pmatrix} \mathbf{0} \\ \mathbf{x}_2^*\boldsymbol{\beta}_2 \end{pmatrix}, \sigma^2 \begin{pmatrix} \mathbf{Q}_{\mathbf{x}} & \mathbf{0} \\ \mathbf{0} & \mathbf{P}_{\mathbf{x}_2^*} \end{pmatrix} \right). \tag{16.109}$$

(f) Finally, argue that $\mathbf{Y}'\mathbf{P}_{\mathbf{x}_2^*}\mathbf{Y}$ is independent of $\mathbf{Y}'\mathbf{Q}_{\mathbf{x}}\mathbf{Y}$, and when $\boldsymbol{\beta}_2 = \mathbf{0}$, $\mathbf{Y}'\mathbf{P}_{\mathbf{x}_2^*}\mathbf{Y} \sim \sigma^2\chi_{p_2}^2$. Thus the statistic in (16.24) is $F_{p_2, n-p}$ under the null hypothesis.

Exercise 16.7.4. Continue with the model in Section 16.1.2 and Exercise 16.7.3. (a) With $\mathbf{x}^* = \mathbf{x}\mathbf{A}$ for $\mathbf{A}$ invertible, show that $\hat{\boldsymbol{\beta}}$ is the least squares estimate in the original model if and only if $\hat{\boldsymbol{\beta}}^*$ is the least squares estimate in the model with $\mathbf{Y} \sim N(\mathbf{x}^*\boldsymbol{\beta}^*, \sigma^2\mathbf{I}_n)$. [Hint: Start by noting that if $\hat{\boldsymbol{\beta}}$ minimizes $\|\mathbf{y} - \mathbf{x}\boldsymbol{\beta}\|^2$ over $\boldsymbol{\beta}$, then it must minimize $\|\mathbf{y} - \mathbf{x}^*\mathbf{A}^{-1}\boldsymbol{\beta}\|^2$ over $\boldsymbol{\beta}$.] (b) Apply part (a) to show that with the $\mathbf{A}$ in (16.107), the least squares estimate of $\boldsymbol{\beta}_2$ is the same whether using the model with $\mathbf{x}\boldsymbol{\beta}$ or $\mathbf{x}^*\boldsymbol{\beta}^*$. (c) We know that using the model with $\mathbf{x}\boldsymbol{\beta}$, the least squares estimator $\hat{\boldsymbol{\beta}} \sim N(\boldsymbol{\beta}, \sigma^2\mathbf{C})$ where $\mathbf{C} = (\mathbf{x}'\mathbf{x})^{-1}$, hence $Cov[\hat{\boldsymbol{\beta}}_2] = \mathbf{C}_{22}$, the lower-right $p_2 \times p_2$ block of $\mathbf{C}$. Show that for $\boldsymbol{\beta}_2$ using $\mathbf{x}^*\boldsymbol{\beta}^*$,

$$\hat{\boldsymbol{\beta}}_2 = (\mathbf{x}_2^{*\prime}\mathbf{x}_2^*)^{-1}\mathbf{x}_2^{*\prime}\mathbf{Y} \sim N(\boldsymbol{\beta}_2, \sigma^2(\mathbf{x}_2^{*\prime}\mathbf{x}_2^*)^{-1}), \tag{16.110}$$

hence $\mathbf{C}_{22}^{-1} = (\mathbf{x}_2^{*\prime}\mathbf{x}_2^*)^{-1}$. (d) Show that

$$\widehat{\boldsymbol{\beta}}_2' \mathbf{C}_{22}^{-1} \widehat{\boldsymbol{\beta}}_2 = \mathbf{y}' \mathbf{P}_{\mathbf{x}_2^*} \mathbf{y}. \tag{16.111}$$

(e) Argue then that the F statistic in (16.24) is the same as that in (15.24).

Exercise 16.7.5. Refer back to the snoring and heart disease data in Exercise 14.9.7. The data consists of $(Y_1, x_1), \dots, (Y_n, x_n)$ independent observations, where each $Y_i \sim$ Bernoulli(p_i) indicates whether person i has heart disease. The p_i's follow a logistic regression model, $\text{logit}(p_i) = \alpha + \beta x_i$, where x_i is the extent to which the person snores. Here are the data again:

Heart disease? $\rightarrow$		Yes	No	
Frequency of snoring $\downarrow$	x_i			
Never	-3	24	1355	(16.112)
Occasionally	-1	35	603	
Nearly every night	1	21	192	
Every night	3	30	224	

The MLEs are $\widehat{\alpha} = -2.79558$, $\widehat{\beta} = 0.32726$. Consider testing $H_0 : \beta = 0$ versus $H_A : \beta \neq 0$. We could perform an approximate z-test, but here find the LRT. The loglikelihood is $\sum_{i=1}^n (y_i \log(p_i) + (1 - y_i) \log(1 - p_i))$. (a) Find the value of the loglikelihood under the alternative, $l_n(\widehat{\alpha}, \widehat{\beta})$. (b) The null hypothesis implies that the p_i's are all equal. What is their common MLE under H_0? What is the value of the loglikelihood? (c) Find the $2\log(\Lambda(\mathbf{y}, \mathbf{x}))$ statistic. (d) What are the dimensions of the two parameter spaces? What are the degrees of freedom in the asymptotic χ^2? (e) Test the null hypothesis with $\alpha = 0.05$. What do you conclude?

Exercise 16.7.6. Continue with the snoring and heart disease data from Exercise 16.7.5. The model fit was a simple linear logistic regression. It is possible there is a more complicated relationship between snoring and heart disease. This exercise tests the "goodness-of-fit" of this model. Here, the linear logistic regression model is the null hypothesis. The alternative is the "saturated" model where p_i depends on x_i in an arbitrary way. That is, the alternative hypothesis is that there are four probabilities corresponding to the four possible x_i's: $q_{-3}, q_{-1}, q_1,$ and q_3. Then for person i, $p_i = q_{x_i}$. The hypotheses are

$$H_0 : \text{logit}(p_i) = \alpha + \beta x_i, (\alpha, \beta) \in \mathbb{R}^2 \text{ versus } H_A : p_i = q_{x_i}, (q_{-3}, q_{-1}, q_1, q_3) \in (0, 1)^4. \tag{16.113}$$

(a) Find the MLEs of the four q_j's under the alternative, and the value of the loglikelihood. (b) Find the $2\log(\Lambda(\mathbf{y}, \mathbf{x}))$ statistic. (The loglikelihood for the null here is the same as that for the alternative in the previous exercise.) (c) Find the dimensions of the two parameter spaces, and the degrees of freedom in the χ^2. (d) Do you reject the null for $\alpha = 0.05$? What do you conclude?

Exercise 16.7.7. Lazarsfeld, Berelson, and Gaudet (1968) collected some data to determine the relationship between level of education and intention to vote in an election. The variables of interest were

- X = Education: 0 = Some high school, 1 = No high school;

- Y = Interest: 0 = Great political interest, 1 = Moderate political interest, 2 = No political interest;

- Z = Vote: 0 = Intends to vote, 1 = Does not intend to vote.

Here is the table of counts N_{ijk}:

	Y = 0		Y = 1		Y = 2	
	Z = 0	Z = 1	Z = 0	Z = 1	Z = 0	Z = 1
X = 0	490	5	917	69	74	58
X = 1	279	6	602	67	145	100

(16.114)

That is, $N_{000} = 490$ people had $X = Y = Z = 0$, etc. You would expect X and Z to be dependent, that is, people with more education are more likely to vote. That's not the question. The question is whether education and voting are conditionally independent given interest, that is, once you know someone's level of political interest, knowing their educational level does not help you predict whether they vote.

The model is that the vector $\mathbf{N}$ of counts is Multinomial$(n, \mathbf{p})$, with $n = 2812$ and $K = 12$ categories. Under the alternative hypothesis, there is no restriction on the parameter $\mathbf{p}$, where

$$p_{ijk} = P[X = i,\ Y = j,\ Z = k], \qquad (16.115)$$

and $i = 0, 1$; $j = 0, 1, 2$; $k = 0, 1$. The null hypothesis is that X and Z are conditionally independent given Y. Define the following parameters:

$$r_{ij} = P[X = i \mid Y = j], \quad s_{kj} = P[Z = k \mid Y = j], \quad \text{and} \quad t_j = P[Y = j]. \qquad (16.116)$$

(a) Under the null, p_{ijk} is what function of the r_{ij}, s_{kj}, t_j? (b) Under the alternative hypothesis, what are the MLEs of the p_{ijk}'s? (Give the numerical answers.) (c) Under the null hypothesis, what are the MLEs of the r_{ij}'s, s_{kj}'s, and t_j's? What are the MLEs of the p_{ijk}'s? (d) Find the loglikelihoods under the null and alternatives. What is the value of $2\log(\Lambda(\mathbf{n}))$ for testing the null vs. the alternative? (e) How many free parameters are there for the alternative hypothesis? How many free parameters are there for the null hypothesis among the r_{ij}'s? How many free parameters are there for the null hypothesis among the s_{kj}'s? How many free parameters are there for the null hypothesis among the t_j's? How many free parameters are there for the null hypothesis total? (f) What are the degrees of freedom for the asymptotic χ^2 distribution under the null? What is the p-value? What do you conclude? (Use level 0.05.)

Exercise 16.7.8. Suppose

$$\begin{pmatrix} X_1 \\ Y_1 \end{pmatrix}, \cdots, \begin{pmatrix} X_n \\ Y_n \end{pmatrix} \quad \text{are iid} \quad N\left(\begin{pmatrix} 0 \\ 0 \end{pmatrix}, \sigma^2 \begin{pmatrix} 1 & \rho \\ \rho & 1 \end{pmatrix} \right). \qquad (16.117)$$

for $-1 < \rho < 1$ and $\sigma^2 > 0$. The problem is to find the likelihood ratio test of

$$H_0 : \rho = 0, \sigma^2 > 0 \quad \text{versus} \quad H_A : \rho \neq 0, \sigma^2 > 0. \qquad (16.118)$$

Set $T_1 = \sum(X_i^2 + Y_i^2)$ and $T_2 = \sum X_i Y_i$, the sufficient statistics. (a) Show that the MLE of σ^2 is $T_1/(2n)$ under H_0. (b) Under H_A, let $U_i = (X_i + Y_i)/\sqrt{2}$ and $V_i = (X_i - Y_i)/\sqrt{2}$. Show that U_i and V_i are independent, $U_i \sim N(0, \theta_1)$ and $V_i \sim N(0, \theta_2)$, where $\theta_1 = \sigma^2(1 + \rho)$ and $\theta_2 = \sigma^2(1 - \rho)$. (c) Find the MLEs of θ_1 and θ_2 in terms of the U_i's and V_i's. (d) Find the MLEs of σ^2 and ρ in terms of T_1 and T_2. (e) Use parts (a) and (d) to derive the form of the likelihood ratio test. Show that it is equivalent to rejecting H_0 when $2(T_2/T_1)^2 > c$.

Exercise 16.7.9. Find the score test based on $X_1, \ldots, X_n$, iid with the Laplace location family density $(1/2)\exp(-|x_i - \mu|)$, for testing $H_0 : \mu = 0$ versus $H_A : \mu > 0$. Recall from (14.74) that the Fisher information here is $\mathcal{I}_1(\mu) = 1$, even though the assumptions don't all hold. (This test is called the sign test. See Section 18.1.)

Exercise 16.7.10. In this question, $X_1, \ldots, X_n$ are iid with some location family distribution, density $f(x - \mu)$. The hypotheses to test are

$$H_0 : \mu = 0 \ \textit{versus} \ H_A : \mu > 0. \tag{16.119}$$

For each situation, find the statistic for the score test expressed so that the statistic is asymptotically $N(0,1)$ under the null. In each case, the score statistic will be

$$\frac{c \sum_{i=1}^n h(x_i)}{\sqrt{n}} \tag{16.120}$$

for some function h and constant c. The f's: (a) $f \sim N(0,1)$. (b) $f \sim$ Laplace. (See Exercise 16.7.9.) (c) $f \sim$ Logistic.

Other questions: (d) For which (if any) of the above distributions is the score statistic *exactly* $N(0,1)$? (e) Which distribution(s) (if any) has corresponding score statistic that has the same distribution under the null for any of the above distributions?

Exercise 16.7.11. Suppose $X_1, \ldots, X_n$ are iid Poisson(λ). Find the approximate level $\alpha = 0.05$ score test for testing $H_0 : \lambda = 1$ versus $H_A : \lambda > 1$.

Exercise 16.7.12. Consider the testing problem in (16.61), where $\mathbf{X} \sim$ Multinomial $(n, (p_1, p_2, p_3))$ and we test the null that $p_1 = p_2 = 1/3$. (a) Show that the score function and Fisher information matrix at the null are as given in (16.63) and (16.64). (b) Verify the step from the second to third lines in (16.65) that shows that the score test function is $((X_1 - X_3)^2 + (X_2 - X_3)^2 + (X_1 - X_2)^2)/n$. (c) Find the $2\log(\Lambda(\mathbf{x}))$ version of the LRT statistic. Show that it can be written as $2\sum_{i=1}^3 Obs_i \log(Obs_i / Exp_i)$.

Exercise 16.7.13. Here we refer back to the snoring and heart disease data in Exercises 16.7.5 and 16.7.6. Consider four models:

- M_0: The p_i's are all equal (the null in Exercise 16.7.5);
- M_1: The linear logistic model: $\text{logit}(p_i) = \alpha + \beta x_i$ (the alternative in Exercise 16.7.5 and the null in Exercise 16.7.6);
- M_2: The quadratic logistic model: $\text{logit}(p_i) = \alpha + \beta x_i + \gamma z_i$;
- M_3: The saturated model: There is no restriction on the p_i's (the alternative in Exercise 16.7.6).

The quadratic model M_2 fits a parabola to the logits, rather than just a straight line as in M_1. We could take $z_i = x_i^2$, but an equivalent model uses the more numerically convenient "orthogonal polynomials" with $\mathbf{z} = (1, -1, -1, 1)'$. The MLEs of the parameters in the quadratic model are $\hat{\alpha} = -2.7733$, $\hat{\beta} = 0.3352$, $\hat{\gamma} = -0.2484$. (a)

For each model, find the numerical value of the maximum loglikelihood (the form $\sum(y_i \log(\widehat{p}_i) + (1 - y_i) \log(1 - \widehat{p}_i)))$. (b) Find the dimensions and BICs for the four models. Which has the best (lowest) BIC? (c) Find the BIC-based estimates of the posterior probabilities $P[M_k \mid Y = y]$. What do you conclude? (d) Now focus on just models M_1 and M_3, the linear logistic model and saturated model. In Exercise 16.7.6, we (just) rejected M_1 in favor of M_3 at the $\alpha = 0.05$ level. Find the posterior probability of M_1 among just these two models. Is it close to 5%? What do you conclude about the fit of the linear model?

Exercise 16.7.14. This questions uses data on diabetes patients. The data can be found at `http://www-stat.stanford.edu/~hastie/Papers/LARS/`. There are $n = 442$ patients, and 10 baseline measurements, which are the predictors. The dependent variable is a measure of the progress of the disease one year after the baseline measurements were taken. The ten predictors include age, sex, BMI, blood pressure, and six blood measurements (hdl, ldl, glucose, etc.) denoted S1, ..., S6. The prediction problem is to predict the progress of the disease for the next year based on these measurements. Here are the results for some selected subsets:

Name	Subset $\mathcal{A}$	$q_{\mathcal{A}}$	$RSS_{\mathcal{A}}/n$	
A	$\{1,4\}$	2	3890.457	
B	$\{1,4,10\}$	3	3205.190	
C	$\{1,4,5,10\}$	4	3083.052	
D	$\{1,4,5,6,10\}$	5	3012.289	
E	$\{1,3,4,5,8,10\}$	6	2913.759	(16.121)
F	$\{1,3,4,5,6,10\}$	6	2965.772	
G	$\{1,3,4,5,6,7,10\}$	7	2876.684	
H	$\{1,3,4,5,6,9,10\}$	7	2885.248	
I	$\{1,3,4,5,6,7,9,10\}$	8	2868.344	
J	$\{1,3,4,5,6,7,9,10,11\}$	9	2861.347	
K	$\{1,2,3,4,5,6,7,8,9,10,11\}$	11	2859.697	

The $q_{\mathcal{A}}$ is the number of predictors included in the model, and $RSS_{\mathcal{A}}$ is the residual sum of squares for the model. (a) Find the AIC, BIC, and Mallows' C_p's for these models. Find the BIC-based posterior probabilities of them. (b) Which are the top two models for each criterion? (c) What do you see?

Exercise 16.7.15. This exercise uses more of the hurricane data from Exercise 13.8.21 originally analyzed by Jung et al. (2014). The model is a normal linear regression model, where Y is the log of the number of deaths (plus one). The three explanatory variables are minimum atmospheric pressure, gender of the hurricane's name (1=female, 0=male), and square root of damage costs (in millions of dollars). There are $n = 94$ observations. The next table has the residual sums of squared errors, SS_e, for each regression model obtained by using a subset of the explanatory variables:

MinPressure	Gender	$\sqrt{\text{Damage}}$	SS_e	
0	0	0	220.16	
0	0	1	100.29	
0	1	0	218.48	
0	1	1	99.66	(16.122)
1	0	0	137.69	
1	0	1	97.69	
1	1	0	137.14	
1	1	1	97.16	

For each model, the included variables are indicated with a "1." For your information, we have the least squares estimates and their standard errors for the model with all three explanatory variables:

	Estimate	Std. Error	
Intercept	12.8777	7.9033	
MinPressure	−0.0123	0.0081	(16.123)
Gender	0.1611	0.2303	
Damage	0.0151	0.0025	

(a) Find the dimensions and BICs for the models. Which one has the best BIC? (b) Find the BIC-based estimates of the posterior probabilities of the models. Which ones have essentially zero probability? Which ones have the highest probability? (c) For each of the three variables, find the probability it is in the true model. Is the gender variable very likely to be in the model?

Exercise 16.7.16. Consider the linear model in (16.92), $Y \sim N_n(x\beta, \sigma^2 I_n)$, where β is $p \times 1$ and $x'x$ is invertible. Let $\widehat{\sigma}^2 = \|y - x\widehat{\beta}\|^2/n$ denote the MLE of σ^2, where $\widehat{\beta}$ is the MLE of β. (a) Show that $E[1/\widehat{\sigma}^2] = n/(\sigma^2(n - p - 2))$ if $n > p + 2$. (b) Suppose Y^{New} is independent of Y and has the same distribution as Y, and let $U = Y^{New} - P_x Y$ as in (16.96), where $P_x = x(x'x)^{-1}x'$ is given in (12.14). Show that $E[U] = 0$ and $Cov[U] = I_n + P_x$. (c) Argue that $E[\|U\|^2] = \sigma^2 \text{trace}(I_n + P_x) = \sigma^2(n + p)$. (d) Show that then, $\delta = E[\|U\|^2/\widehat{\sigma}^2] - n = 2n(p + 1)/(n - p - 2)$.

Exercise 16.7.17. Consider the subset selection model in regression, but where σ^2 is known. In this case, Mallows' C_p can be given as

$$C_p(\mathcal{A}) - \frac{RSS_\mathcal{A}}{\sigma^2} + 2q_\mathcal{A} - n, \tag{16.124}$$

where, as in Exercise 16.7.14, $\mathcal{A}$ is the set of predictors included in the model, $q_\mathcal{A}$ is the number of predictors, and $RSS_\mathcal{A}$ is the residual sum of squares. Show that the $AIC(\mathcal{A})$ is a monotone function of $C_p(\mathcal{A})$. (Here, n and σ^2 are known constants.)

Exercise 16.7.18. Show that in (16.74), if we take the prior probabilities as

$$\pi_k \propto \left(\frac{\sqrt{n}}{e}\right)^{d_k}, \tag{16.125}$$

where d_k is the dimension of Model k, then the model that maximizes the estimated posterior probability is the model with the lowest AIC. Note that except for very small n, this prior places relatively more weight on higher-dimensional models.

Chapter 17

Randomization Testing

Up to now, we have been using the **sampling model** for inference. That is, we have assumed the the the data arose by sampling from a (usually infinite) population. E. g., in the two-sample means problem, the model assumes independent random samples from the two populations, and the goal is to infer something about the difference in population means. By contrast, many good studies, especially in agriculture, psychology, and medicine, proceed by first obtaining a group of subjects (farm plots, rats, people), then randomly assigning some of the subjects to one treatment and the rest to another treatment, often a placebo. For example, in the polio study (Exercise 6.8.9), in selected school districts, second graders whose parents volunteered became the subjects of the experiment. About half were randomly assigned to receive the polio vaccine and the rest were assigned a placebo. Such a design yields a **randomization model**: the set of subjects is the population, and the statistical randomness arises from the randomization within this small population, rather than the sampling from one or two larger populations. Inference is then on the subjects at hand, where we may want to estimate what the means would be if every subject had one treatment, or everyone had the placebo. The distribution of a test statistic depends not on sampling new subjects, but on how the subjects are randomly allocated to treatment.

A key aspect of the randomization model is that under an appropriate null, the distribution of the test statistic can often be found exactly by calculating it under all possible randomizations, of which there are (usually) only a finite number. If the number is too large, sampling a number of randomizations, or asymptotic approximations, are used. Sections 17.1 and 17.2 illustrate the two-treatment randomization model, for numerical and categorical data. Section 17.3 considers a randomization model using the sample correlation as test statistic.

Interestingly, the tests developed for randomization models can also be used in many sampling models. By conditioning on an appropriate statistic, under the null, the **randomization distribution** of the test statistic is the same as it would be under the randomization model. The resulting tests, or p-values, are again exact, and importantly have desired properties under the unconditional sampling distribution. See Section 17.4.

In Chapter 18 we look at some traditional **nonparametric** testing procedures. These are based on using the signs and/or ranks of the original data, and have null sampling distributions that follow directly from the randomization distributions found in the previous sections. They are again exact, and often very robust.

17.1   Randomization model: Two treatments

To illustrate a randomization model, we look at a study by Zelazo, Zelazo, and Kolb (1972) on whether walking exercises helped infants learn to walk. The researchers took 24 one-week-old male infants, and randomly assigned them to one of four treatment groups, so that there were six in each group. We will focus on just two groups: The **walking exercise group**, who were given exercise specifically developed to teach walking, and the **regular exercise group**, who were given the same amount of exercise, but without the specific walking exercises. The outcome measured was the age in months when the infant first walked. The question is whether the walking exercise helped the infant walk sooner. The data are

$$
\begin{array}{c|c}
\text{Walking group} & \text{Regular group} \\
\hline
9 \;\; 9.5 \;\; 9.75 \;\; 10 \;\; 13 \;\; 9.5 & 11 \;\; 10 \;\; 10 \;\; 11.75 \;\; 10.5 \;\; 15
\end{array}
\tag{17.1}
$$

The randomization model takes the $N = 12$ infants as the population, wherein each observation has two values attached to it: x_i is the age first walked if given the walking exercises, and y_i is the age first walked if given the regular exercises. Thus the population is

$$
\mathcal{P} = \{(x_1, y_1), \ldots, (x_N, y_N)\},
\tag{17.2}
$$

but we observe x_i for only $n = 6$ of the infants, and observe y_i only for the other $m = 6$. Conceptually, the randomization could be accomplished by randomly permuting the twelve observations, then assigning the first six to the walking group, and the rest to the regular group.

There are two popular null hypotheses to consider. The **exact** null says that the walking treatment has no effect, which means $x_i = y_i$ for all i. The **average** null states that the averages over the twelve infants of the walking and regular outcomes are equal. We will deal with the exact null one here. The alternative could be general, i.e., $x_i \neq y_i$, but we will take the specific one-sided alternative that the walking exercise is superior on average for these twelve:

$$
H_0 : x_i = y_i, \; i = 1, \ldots, N \; versus \; H_A : \frac{1}{N} \sum_{i=1}^{N} x_i < \frac{1}{N} \sum_{i=1}^{N} y_i.
\tag{17.3}
$$

A reasonable test statistic is the difference of means for the two observed groups:

$$
t_{obs} = \frac{1}{n} \sum_{i \in \mathcal{W}} x_i - \frac{1}{m} \sum_{i \in \mathcal{R}} y_i,
\tag{17.4}
$$

where "$\mathcal{W}$" indicates those assigned to the walking group, and "$\mathcal{R}$" to the regular group. From (17.1) we can calculate the observed $T = -1.25$.

We will find the p-value, noting that small values of T favor the alternative. The x_i's and y_i's are not random here. What is random according to the design of the experiment is which observations are assigned to which treatment. Under the null, we actually do observe all the (x_i, y_i), because $x_i = y_i$. Thus the null distribution of the statistic is based on randomly assigning n of the values to the walking group, and m to the regular group. One way to represent this randomization is to use permutations of the vector of values $\mathbf{z} = (9, 9.5, 9.75, \ldots, 10.5, 15)'$ from (17.1). An $N \times N$ permutation matrix $\mathbf{p}$ has exactly one 1 in each row and one 1 in each column, and the rest of the elements are 0, so that $\mathbf{pz}$ just permutes the elements of $\mathbf{z}$. For example, if $N = 4$,

$$
\mathbf{p} = \begin{pmatrix} 0 & 1 & 0 & 0 \\ 0 & 0 & 0 & 1 \\ 0 & 0 & 1 & 0 \\ 1 & 0 & 0 & 0 \end{pmatrix} \tag{17.5}
$$

is a 4×4 permutation matrix, and

$$
\mathbf{pz} = \begin{pmatrix} 0 & 1 & 0 & 0 \\ 0 & 0 & 0 & 1 \\ 0 & 0 & 1 & 0 \\ 1 & 0 & 0 & 0 \end{pmatrix} \begin{pmatrix} z_1 \\ z_2 \\ z_3 \\ z_4 \end{pmatrix} = \begin{pmatrix} z_2 \\ z_4 \\ z_3 \\ z_1 \end{pmatrix}. \tag{17.6}
$$

Let $\mathcal{S}_N$ be the set of all $N \times N$ permutation matrices. It is called the **symmetric group**.

The difference in means in (17.4) can be represented as a linear function of the data: $t_{obs} = \mathbf{a}'\mathbf{z}$ where

$$
\mathbf{a}' = \left(\frac{1}{n}, \ldots, \frac{1}{n}, -\frac{1}{m}, \ldots, -\frac{1}{m} \right) = \left(\frac{1}{n} \mathbf{1}'_n, -\frac{1}{m} \mathbf{1}'_m \right), \tag{17.7}
$$

i.e., there are n of the $1/n$'s and m of the $-1/m$'s. The **randomization distribution** of the statistic is then given by

$$
T(\mathbf{Pz}) = \mathbf{a}'\mathbf{Pz}, \quad \mathbf{P} \sim \text{Uniform}(\mathcal{S}_N). \tag{17.8}
$$

The p-value is the chance of being no larger than $t_{obs} = T(\mathbf{z})$:

$$
\text{p-value}(t_{obs}) = P[T(\mathbf{Pz}) \leq t_{obs}] = \frac{1}{\#\mathcal{S}_N} \#\{\mathbf{p} \in \mathcal{S}_N \mid T(\mathbf{pz}) \leq t_{obs}\}. \tag{17.9}
$$

There are $12! \approx 480$ million such permutations, but since we are looking at just the averages of batches of six, the order of the observations within each group is irrelevant. Thus there are really only $\binom{12}{6} = 924$ allocations to the two groups we need to find. It is not hard (using R, especially the function combn) to calculate all the possibilities. The next table exhibits some of the permutations:

Walking group	Regular group	$T(\mathbf{pz})$
9 9.5 9.75 10 13 9.5	11 10 10 11.75 10.5 15	-1.250
9 9.5 10 13 10 10.5	9.75 9.5 11 10 11.75 15	-0.833
9 9.5 11 10 10.5 15	9.75 10 13 9.5 10 11.75	0.167
$\vdots$	$\vdots$	$\vdots$
9.5 9.5 11 10 11.75 15	9 9.75 10 13 10 10.5	0.750
9.75 13 9.5 10 11.75 10.5	9 9.5 10 11 10 15	0.000

(17.10)

Figure 17.1 graphs the distribution of these $T(\mathbf{pz})$'s. The p-value then is the proportion of them less than or equal to the observed -1.25, which is $123/924 = 0.133$. Thus we do not reject the null hypothesis that there is no treatment effect.

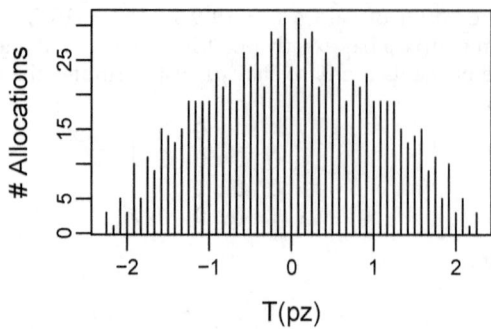

Figure 17.1: The number of allocations corresponding to each value of $T(\mathbf{pz})$.

Note that any statistic could be equally easily used. For example, using the difference of medians, we calculate the p-value to be 0.067, smaller but still not significant. In Section 18.2.2 we will see that the Mann-Whitney/Wilcoxon statistic yields a significant p-value of 0.028. When the sample sizes are larger, it becomes impossible to enumerate all the possibilities. In such cases, we can either simulate a number of randomizations, or use asymptotic considerations as in Section 17.5.

Other randomization models lead to corresponding randomization p-values. For example, it may be that the observations are paired up, and in each pair one observation is randomly given the treatment and the other a placebo. Then the randomization p-value would look at the statistic for all possible interchanges within pairs.

17.2   Fisher's exact test

The idea in Section 17.1 can be extended to cases where the outcomes are binary. For example, Li, Harmer, Fisher, McAuley, Chaumeton, Eckstrom, and Wilson (2005) report on a study to assess whether tai chi, a Chinese martial art, can improve balance in the elderly. A group of 188 people over 70 years old were randomly assigned to two groups, each of which had the same number and length of exercise sessions over a six-month period, but one group practiced tai chi and the other stretching. (There were actually 256 people to begin, but some dropped out before the treatments started, and others did not fully report on their outcomes.) One outcome reported was the number of falls during the six-month duration of the study. Here is the observed table, where the two outcomes are 1 = "no falls" and 0 = "one or more falls."

Group	No falls	One or more falls	Total
Tai chi	68	27	95
Stretching	50	43	93
Total	118	70	188

(17.11)

The tai chi group did have fewer falls with about 72% of the people experiencing no falls, while about 54% of the control group had no falls. To test whether this difference is statistically significant, we proceed as for the walking exercises example.

Take the population as in (17.2) to consist of $(x_i, y_i), i = 1, \ldots, N = 188$, where x_i indicates whether person i would have had no falls if in the tai chi group, and y_i indicates whether the person would have had no falls if in the stretching group. The random element is the subset of people assigned to tai chi. Similar to (17.3), we take the null hypothesis that the specific exercise has no effect, and the alternative that tai chi is better:

$$H_0 : x_i = y_i, \ i = 1, \ldots, N \ \textit{versus} \ H_A : \#\{i \,|\, x_i = 1\} > \#\{i \,|\, y_i = 1\}. \tag{17.12}$$

The statistic we'll use is the number of people in the tai chi group who had no falls:

$$t_{obs} = \{i \in \text{tai chi group} \,|\, x_i = 1\}, \tag{17.13}$$

and now large values of the statistic support the alternative. Here, $t_{obs} = 68$. Since the total numbers in the margins of (17.11) are known, any one of the other entries in the table could also be used.

As in the walking example, we take $\mathbf{z}$ to be the observed vector under the null (so that $x_i = y_i$), where the observations in the tai chi group are listed first, and here $\mathbf{a}$ sums up the tai chi values. Then the randomization distribution of the test statistic is given by

$$T(\mathbf{Pz}) = \mathbf{a}'\mathbf{Pz}, \ \text{where} \ \mathbf{a} = \begin{pmatrix} \mathbf{1}_{95} \\ \mathbf{0}_{93} \end{pmatrix} \ \text{and} \ \mathbf{z} = \begin{pmatrix} \mathbf{1}_{68} \\ \mathbf{0}_{27} \\ \mathbf{1}_{50} \\ \mathbf{0}_{43} \end{pmatrix}, \tag{17.14}$$

with $\mathbf{P} \sim \text{Uniform}(\mathcal{S}_N)$ again.

We can find the exact distribution of $T(\mathbf{Pz})$. The probability that $T(\mathbf{Pz}) = t$ is the probability that the first 95 elements of $\mathbf{Pz}$ have t ones and $95 - t$ zeroes. There are 118 ones in the $\mathbf{z}$ vector, so that there are $\binom{118}{t}$ ways to choose the t ones, and $\binom{70}{95-t}$ ways to choose the zeroes. Since there are $\binom{188}{95}$ ways to choose first 95 without regard to outcome, we have

$$P[T(\mathbf{Pz}) = t] = \frac{\binom{118}{t}\binom{70}{95-t}}{\binom{188}{95}}, \ \ 25 \le t \le 95. \tag{17.15}$$

This distribution is the Hypergeometric(118,70,95) distribution, where the pmf of the Hypergeometric(k, l, n) is

$$f(t \,|\, k, l, n) = \frac{\binom{k}{t}\binom{l}{n-t}}{\binom{N}{n}}, \ \ \max\{0, n-l\} \le t \le \min\{k, n\}, \tag{17.16}$$

and k, l, n are nonnegative integers, $N = k + l$. (There are several common parameterizations of the hypergeometric. We use the one in R.) A generic 2×2 table corresponding to (17.11) is

	Success	Failure	Total
Treatment 1	t	$n - t$	n
Treatment 2	$k - t$	$l - n + t$	m
Total	k	l	N

(17.17)

The p-value for our data is then

$$P[T(\mathbf{Pz}) \geq t_{obs}] = P[\text{Hypergeometric}(118,70,95) \geq 68] = 0.00863. \qquad (17.18)$$

Thus we would reject the null hypothesis that the type of exercise has no effect, i.e., the observed superiority of tai chi is statistically significant.

The test here is called **Fisher's exact test**. It yields an exact p-value when testing independence in 2×2 tables, and is especially useful when the sample size is small enough that the asymptotic χ^2 tests are not very accurate. See the next subsection for another example.

17.2.1 Tasting tea

Joan Fisher Box (1978) relates a story about Sir Ronald Fisher, her father, while he was at the Rothamstead Experimental Station in England in the 1920s. When the first woman researcher joined the staff, "No one in those days knew what to do with a woman worker in a laboratory; it was felt, however, that she must have tea, and so from the day of her arrival a tray of tea and a tin of Bath Oliver biscuits appeared each afternoon at four o'clock precisely." One afternoon, as Fisher and colleagues assembled for tea, he drew a cup of tea for one of the scientists, Dr. Muriel Bristol. "She declined it, saying she preferred a cup into which the milk had been poured first." This pronouncement created quite a stir. What difference should it make whether you put the milk in before or after the tea? They came up with an experiment (described in Fisher (1935)) to test whether she could tell the difference. They prepared eight cups of tea with milk. A random four had the milk put in the cup first, and the other four had the tea put in first. Dr. Bristol wasn't watching the randomization.

Once the cups were prepared, Dr. Bristol sampled each one, and for each tried to guess whether milk or tea had been put in first. She knew there were four cups of each type, which could have helped her in guessing. In any case, she got them all correct. Could she have just had lucky guesses? Let x_i indicate whether milk ($x_i = 0$) or tea ($x_i = 1$) was put into cup i first, and let z_i indicate her guess for cup i. Thus each of $\mathbf{x}$ and $\mathbf{z}$ consists of four ones and four zeroes. The null hypothesis is that she is just guessing, that is, she would have made the same guesses no matter which cups had the milk first. We will take the test statistic to be the number of correct guesses:

$$T(\mathbf{x}, \mathbf{z}) = \#\{i \mid x_i = z_i\}. \qquad (17.19)$$

The randomization permutes the $\mathbf{x}$, $\mathbf{Px}$, where $\mathbf{P} \sim \text{Uniform}(\mathcal{S}_8)$. We could try to model Dr. Bristol's thought process, but it will be enough to condition on her responses $\mathbf{z}$. Since she guessed all correctly ($T(\mathbf{x}, \mathbf{z}) = 8$), there is only one value $\mathbf{px}$ could be in order to do as well or better (and no one could do better):

$$\text{p-value}(\mathbf{z}) = P[T(\mathbf{Px}, \mathbf{z}) \geq 8] = P[\mathbf{Px} = \mathbf{z}] = \frac{1}{\binom{8}{4}} = \frac{1}{70} \approx 0.0143. \qquad (17.20)$$

(Note that here, $T(\mathbf{Px}, \mathbf{z})/2$ has a Hypergeometric(4,4,4) distribution.) This p-value is fairly small, so we conclude that it is unlikely she was guessing — She could detect which went into the cup first.

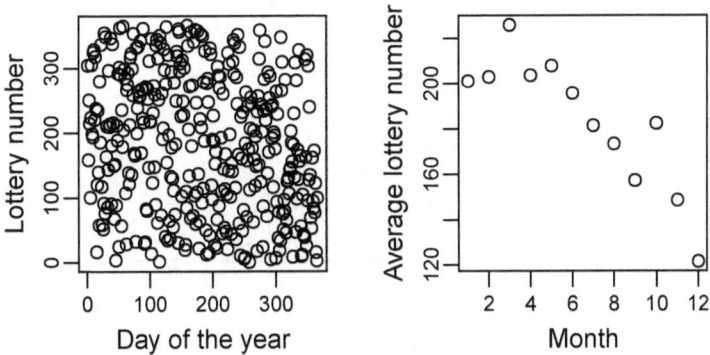

Figure 17.2: The 1969 draft lottery results. The first plot has day of the year on the X-axis and lottery number on the Y-axis. The second plots the average lottery number for each month.

17.3 Testing randomness

In 1969, a lottery was held in the United States that assigned a random draft number from 1 to 366 to each day of the year (including February 29). Young men who were at least 19 years old, but not yet 26 years old, were to be drafted into the military in the order of their draft numbers. The randomization used a box with 366 capsules, each capsule containing a slip of paper with one of the days of the year written on it. So there was one January 1, one January 2, etc. There were also slots numbered from 1 to 366 representing the draft numbers. Capsules were randomly chosen from the box, without replacement. The first one chosen (September 14) was assigned to draft slot 1, the second (April 24) was assigned to draft slot 2, ..., and the last one (June 8) was assigned draft slot 366. Some of the results are next:

Draft #	Day of the year
1	Sep. 14 (258)
2	Apr. 24 (115)
3	Dec. 30 (365)
⋮	⋮
364	May 5 (126)
365	Feb. 26 (57)
366	Jun. 8 (16)

(17.21)

The numbers in the parentheses are the day numbers, e.g., September 14 is the 258th day of the year.

There were questions about whether this method produced a completely random assignment. That is, when drawing capsules from the box, did each capsule have the same chance of being chosen as the others still left? The left-hand plot in Figure 17.2 shows the day of the year (from 1 to 366) on the X-axis, and the draft number on the

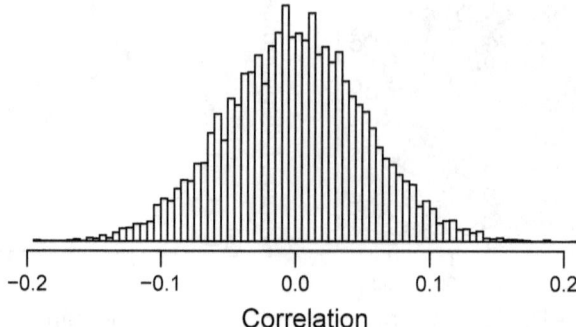

Figure 17.3: The histogram of correlations arising from 10,000 random permutations of the days.

Y-axis. It looks pretty random. But the correlation (12.81) is r = −0.226. If things are completely random, this correlation should be about 0. Is this −0.226 too far from 0? If we look at the average draft number for each month as in the right-hand plot of Figure 17.2, we see a pattern. There is a strong negative correlation, −0.807. Draft numbers earlier in the year tend to be higher than those later in the year. It looks like it might not be completely random.

In the walking exercise experiment, the null hypothesis together with the randomization design implied that every allocation of the data values to the two groups was equally likely. Here, the null hypothesis is that the randomization in the lottery was totally random, specifically, that each possible assignment of days to lottery numbers was equally likely. Thus the two null hypothesis have very similar implications.

To test the randomness of the lottery, we will use the absolute value of the correlation coefficient between the days of the year and the lottery numbers. Let $\mathbf{e}_N = (1, 2, \ldots, N)'$, where here $N = 366$. The lottery numbers are then represented by $\mathbf{e}_N$, and the days of the year assigned to the lottery numbers are a permutation of the elements of $\mathbf{e}_N$, $\mathbf{pe}_N$ for $\mathbf{p} \in \mathcal{S}_N$. Let $\mathbf{p}_0$ be the observed permutation of the days, so that $\mathbf{p}_0 \mathbf{e}_N$ is the vector of day numbers as in the second column of (17.21).

Letting r denote the sample correlation coefficient, our test statistic is the absolute correlation coefficient between the lottery numbers and the day numbers:

$$T(\mathbf{p}) = |r(\mathbf{e}_N, \mathbf{pe}_N)|. \tag{17.22}$$

The observed value is $T(\mathbf{p}_0) = 0.226$. The p-value is the proportion of permutation matrices that yield absolute correlations that size or larger:

$$\text{p-value}(\mathbf{p}_0) = P[T(\mathbf{P}) \geq T(\mathbf{p}_0)], \quad \mathbf{P} \sim \text{Uniform}(\mathcal{S}_N). \tag{17.23}$$

Since there are $366! \approx \infty$ such permutations, it is impossible to calculate the p-value exactly. Instead, we generated 10,000 random permutations of the days of the year, each time calculating the correlation with the lottery numbers. Figure 17.3 is the histogram of those correlations. The maximum correlation is 0.210 and the minimum is -0.193, hence none have absolute value even very close to the observed 0.226. Thus we estimate the p-value to be 0, leading to conclude that the lottery was *not* totally random.

What went wrong? Fienberg (1971) describes the actual randomization process. Briefly, the capsules with the January dates were first prepared, placed in a large box, mixed, and shoved to one end of the box. Then the February capsules were prepared, put in the box and shoved to the end with the January capsules, mixing them all together. This process continued until all the capsules were mixed into the box. The box was shut, further shaken, and carried up three flights of stairs, to await the day of the drawing. Before the televised drawing, the box was brought down the three flights, and the capsules were poured from one end of the box into a bowl and further mixed. The actual drawing consisted of drawing capsules one-by-one from the bowl, assigning them sequentially to the lottery numbers. The draws tended to be from near the top of the bowl. The p-value shows that the capsules were not mixed as thoroughly as possible. The fact that there was a significant negative correlation suggests that the box was emptied into the bowl from which end?

17.4 Randomization tests for sampling models

The procedures developed for randomization models can also be used to find exact tests in sampling models. For an example, consider the two-sample problem where $X_1, \ldots, X_n$ are iid $N(\mu_X, \sigma^2)$, $Y_1, \ldots, Y_m$ are iid $N(\mu_Y, \sigma^2)$, and the X_i's are independent of the Y_i's, and we test

$$H_0 : \mu_X = \mu_Y, \sigma^2 > 0 \text{ versus } H_A : \mu_X \neq \mu_Y, \sigma^2 > 0. \qquad (17.24)$$

Under the null, the entire set of $N = n + m$ observations constitutes an iid sample. We know that the vector of order statistics for an iid sample is sufficient. Conditioning on the order statistics, we have a mathematically identical situation as that for the randomization model in Section 17.1. That is, under the null, every arrangement of the $n + m$ observations with n observations being x_i's and the rest being y_i's has the same probability. Thus we can find a conditional p-value using a calculation similar to that in (17.9). If we consider rejecting the null if the conditional p-value is less or equal to a given α, then the conditional probability of rejecting the null is less than or equal to α, hence so is the unconditional probability.

Next we formalize that idea. In the two-sample model above, under the null, the distribution of the observations is invariant under permutations. That is, for iid observations, any ordering of them has the same distribution. Generalizing, we will look at groups of matrices under whose multiplication the observations' distributions do not change under the null. Suppose the data is an $N \times 1$ vector $\mathbf{Z}$, and $\mathcal{G}$ is an algebraic group of $N \times N$ matrices. (A set of matrices $\mathcal{G}$ is a group if $\mathbf{g} \in \mathcal{G}$ then $\mathbf{g}^{-1} \in \mathcal{G}$, and if $\mathbf{g}_1, \mathbf{g}_2 \in \mathcal{G}$ then $\mathbf{g}_1\mathbf{g}_2 \in \mathcal{G}$. The symmetric group $\mathcal{S}_N$ of $N \times N$ permutation matrices is indeed such a group. See (22.61) for a general definition of groups.) Then the distribution of $\mathbf{Z}$ is said to be **invariant** under $\mathcal{G}$ if

$$\mathbf{gZ} =^{\mathcal{D}} \mathbf{Z} \text{ for all } \mathbf{g} \in \mathcal{G}. \qquad (17.25)$$

Consider testing $H_0 : \boldsymbol{\theta}_0 \in \mathcal{T}_0$ based on $\mathbf{Z}$. Suppose we have a finite group $\mathcal{G}$ such that for any $\boldsymbol{\theta}_0 \in \mathcal{T}_0$, the distribution of $\mathbf{Z}$ is invariant under $\mathcal{G}$. Then for a given test statistic $T(\mathbf{z})$, we obtain the **randomization p-value** in a manner analogous to (17.9) and (17.23):

$$\text{p-value}(\mathbf{z}) = \frac{1}{\#\mathcal{G}} \#\{\mathbf{g} \in \mathcal{G} \mid T(\mathbf{gz}) \geq T(\mathbf{z})\}$$
$$= P[T(\mathbf{Gz}) \geq T(\mathbf{z})], \quad \mathbf{G} \sim \text{Uniform}(\mathcal{G}). \tag{17.26}$$

That is, $\mathbf{G}$ is a random matrix distributed uniformly over $\mathcal{G}$, which is independent of $\mathbf{Z}$. To see that this p-value acts like it should, we first use Lemma 15.2 on page 256, emphasizing that $\mathbf{G}$ is the random element:

$$P[\text{p-value}(\mathbf{Gz}) \leq \alpha] \leq \alpha \tag{17.27}$$

for given α. Next, we magically transfer the randomness in $\mathbf{G}$ over to $\mathbf{Z}$. Write the probability conditionally,

$$P[\text{p-value}(\mathbf{Gz}) \leq \alpha] = P[\text{p-value}(\mathbf{GZ}) \leq \alpha \mid \mathbf{Z} = \mathbf{z}], \tag{17.28}$$

then use (17.27) to show that, unconditionally,

$$P[\text{p-value}(\mathbf{GZ}) \leq \alpha \mid \boldsymbol{\theta}_0] \leq \alpha. \tag{17.29}$$

By (17.25), $[\mathbf{GZ} \mid \mathbf{G} = \mathbf{g}] =^{\mathcal{D}} \mathbf{Z}$ for any $\mathbf{g}$, hence $\mathbf{GZ} =^{\mathcal{D}} \mathbf{Z}$, giving us

$$P[\text{p-value}(\mathbf{Z}) \leq \alpha \mid \boldsymbol{\theta}_0] \leq \alpha. \tag{17.30}$$

Finally, we can take the supremum over $\boldsymbol{\theta}_0 \in \mathcal{T}_0$ to show that the randomization p-value does yield a level α test as in (15.49).

Returning to the two-sample testing problem (17.24), we let $\mathbf{Z}$ be the $N \times 1$ vector of all the observations, with the first sample listed first: $\mathbf{Z} = (X_1, \ldots, X_n, Y_1, \ldots, Y_m)'$. Then under the null, the Z_i's are iid, hence $\mathbf{Z}$ is invariant (17.25) under the group of $N \times N$ permutation matrices $\mathcal{S}_N$. If our statistic is the difference in means, we have $T(\mathbf{z}) = \mathbf{a}'\mathbf{z}$ for $\mathbf{a}$ as in (17.7), and the randomization p-value is as in (17.9) but for a two-sided test, i.e.,

$$\text{p-value}(\mathbf{a}'\mathbf{z}) = P[|\mathbf{a}'\mathbf{Pz}| \geq |\mathbf{a}'\mathbf{z}|], \quad \mathbf{P} \sim \text{Uniform}(\mathcal{S}_N). \tag{17.31}$$

This p-value depends only on the permutation invariance of the combined sample, so works for any distribution, not just normal. See Section 18.2.2 for a more general statement of the problem. Also, it is easily extended to any two-sample statistic. It will not work, though, if the variances of the two samples are not equal.

17.4.1 Paired comparisons

Stichler, Richey, and Mandel (1953) describes a study wherein each of $n = 16$ tires was subject to measurement of tread wear by two methods, one based on weight loss and one on groove wear. Thus the data are $(X_1, Y_1), \ldots, (X_n, Y_n)$, with the X_i and Y_i's representing the measurements based on weight loss and groove wear, respectively. We assume the tires (i.e., (X_i, Y_i)'s) are iid under both hypotheses. We wish to test whether the two measurement methods are equivalent in some sense. We could take the null hypothesis to be that X_i and Y_i have the same distribution, but for our purposes need to take the stronger null hypothesis that the two measurement methods are exchangeable, which here means that (X_i, Y_i) and (Y_i, X_i) have the same distribution:

$$H_0 : (X_i, Y_i) =^{\mathcal{D}} (Y_i, X_i), \quad i = 1, \ldots, n. \tag{17.32}$$

Exercise 17.6.7 illustrates the difference between having the same distribution and exchangeability. See Gibbons and Chakraborti (2011) for another example. Note that we are not assuming X_i is independent of Y_i. In fact, the design is taking advantage of the likely high correlation between X_i and Y_i.

The test statistic we will use is the absolute value of median of the differences:

$$T(\mathbf{z}) = |\operatorname{Median}(z_1, \ldots, z_n)|, \tag{17.33}$$

where $z_i = x_i - y_i$. Here are the data:

i	x_i	y_i	z_i	i	x_i	y_i	z_i	
1	45.9	35.7	10.2	9	30.4	23.1	7.3	
2	41.9	39.2	2.7	10	27.3	23.7	3.6	
3	37.5	31.1	6.4	11	20.4	20.9	−0.5	
4	33.4	28.1	5.3	12	24.5	16.1	8.4	(17.34)
5	31.0	24.0	7.0	13	20.9	19.9	1.0	
6	30.5	28.7	1.8	14	18.9	15.2	3.7	
7	30.9	25.9	5.0	15	13.7	11.5	2.2	
8	31.9	23.3	8.6	16	11.4	11.2	0.2	

Just scanning the differences, we see that in only one case is y_i larger than x_i, hence the evidence is strong that the null is not true. But we will illustrate with our statistic, which is observed to be $T(\mathbf{z}) = 4.35$.

To find the randomization p-value, we need a group. Exchangeability in the null implies that $X_i - Y_i$ has the same distribution as $Y_i - X_i$, that is, $Z_i =^{\mathcal{D}} -Z_i$. Since the Z_i's are iid, we can change the signs of any subset of them without changing the null distribution of the vector $\mathbf{Z}$. Thus the invariance group $\mathcal{G}_\pm$ consists of all $N \times N$ diagonal matrices with ± 1's on the diagonal:

$$\mathbf{g} = \begin{pmatrix} \pm 1 & 0 & \cdots & 0 \\ 0 & \pm 1 & \cdots & 0 \\ \vdots & \vdots & \ddots & \vdots \\ 0 & 0 & \cdots & \pm 1 \end{pmatrix}. \tag{17.35}$$

There are 2^{16} matrices in $\mathcal{G}_\pm$, though due to the symmetry in the null and the statistic, we can hold one of the diagonals at $+1$. The exact randomization p-value as in (17.26) is 0.0039 ($= 128/2^{15}$). It is less than 0.05 by quite a bit, in fact, we can easily reject the null hypothesis for $\alpha = 0.01$.

17.4.2 Regression

In Section 12.6 we looked at the regression with X = damage and Y = deaths for the hurricane data. There was a notable outlier in Katrina. Here we test independence of X and Y using randomization. The model is that $(X_1, Y_1), \ldots, (X_N, Y_N)$, $N = 94$, are iid. The null hypothesis is that the X_i's are independent of the Y_i's, which means the distribution of the data set is invariant under permutation of the X_i's, or of the Y_i's, or both. For test statistics, we will try the four slopes we used in Section 12.6, which are given in (12.74). If $\widehat{\beta}(\mathbf{x}, \mathbf{y})$ is the estimator of the slope, then the one-sided randomization p-value is given by

$$\text{p-value}(\mathbf{x}, \mathbf{y}) = P[\widehat{\beta}(\mathbf{x}, \mathbf{Py}) \geq \widehat{\beta}(\mathbf{x}, \mathbf{y})], \quad \mathbf{P} \sim \text{Uniform}(\mathcal{S}_N). \tag{17.36}$$

Note the similarity to (17.22) and (17.23) for the draft lottery data, for which the statistic was absolute correlation. We use 10,000 random permutations of the x_i's to estimate the randomization p-values. Table (17.37) contains the results. The second column contains one-sided p-values estimated using Student's t.

	Estimate	p-value Student's t	p-value Randomization
Least squares	7.7435	0.0000	0.0002
Least squares w/o outlier	1.5920	0.0002	0.0039
Least absolute deviations	2.0930	0.0213	0.0000
Least absolute deviations w/o outlier	0.8032	0.1593	0.0005

$$\tag{17.37}$$

We see that the randomization p-values are consistently very small whether using least squares or least absolute deviations, with or without the outlier. The p-values using the Student's t estimate are larger for least absolute deviations, especially with no outlier.

17.5   Large sample approximations

When the number of randomizations is too large to perform all, we generated a number of random randomizations. Another approach is to use a normal approximation that uses an extension of the central limit theorem. We will treat statistics that are linear functions of $\mathbf{z} = (z_1, \ldots, z_N)'$, and look at the distribution under the group of permutation matrices. See Section 17.5.2 for the sign change group.

Let $\mathbf{a}_N = (a_1, \ldots, a_N)'$ be the vector of constants defining the linear test statistic:

$$T_N(\mathbf{z}_N) = \mathbf{a}_N' \mathbf{z}_N, \tag{17.38}$$

so that the randomization distribution of T is given by

$$T(\mathbf{P}_N \mathbf{z}_N) = \mathbf{a}_N' \mathbf{P}_N \mathbf{z}_N \quad \text{where} \quad \mathbf{P}_N \sim \text{Uniform}(\mathcal{S}_N). \tag{17.39}$$

The $\mathbf{a}$ in (17.8) illustrates the $\mathbf{a}_N$ when comparing two treatments. For the draft lottery example in Section 17.3, $\mathbf{a}_N = \mathbf{e}_N = (1, 2, \ldots, N)'$. This idea can also be used in the least squares case as in Section 17.4.2, where $\mathbf{z}_N = \mathbf{x}$ and $\mathbf{a}_N = \mathbf{y}$, or vice versa.

The first step is to find the mean and variance of T in (17.39). Consider the random vector $\mathbf{U}_N = \mathbf{P}_N \mathbf{z}_N$, which is just a random permutation of the elements of $\mathbf{z}_N$. Since each permutation has the same probability, each U_i is equally likely to be any one of the z_k's. Thus

$$E[U_i] = \frac{1}{N} \sum z_k = \bar{z}_N \quad \text{and} \quad Var[U_i] = \frac{1}{N} \sum (z_k - \bar{z}_N)^2 = s_{z_N}^2. \tag{17.40}$$

Also, each pair (U_i, U_j) for $i \neq j$ is equally likely to be equal to any pair (z_k, z_l), $k \neq l$. Thus $Cov[U_i, U_j] = c_N$ is the same for any $i \neq j$. To figure out c_N, note that

$\sum U_i = \sum z_k$ no matter what $\mathbf{P}_N$ is. Since the z_k's are constant, $Var[\sum U_i] = 0$. Thus

$$0 = Var[\sum_{i=1}^{N} U_i]$$

$$= \sum_{i=1}^{N} Var[U_i] + \sum\sum_{i \neq j} Cov[U_i, U_j]$$

$$= N s_{z_N}^2 + N(N-1)c_N$$

$$\Longrightarrow c_N = -\frac{1}{N-1} s_{z_N}^2. \tag{17.41}$$

Exercise 17.6.1 shows the following:

Lemma 17.1. *Let* $\mathbf{U}_N = \mathbf{P}_N \mathbf{z}_N$, *where* $\mathbf{P}_N \sim \text{Uniform}(\mathcal{S}_N)$. *Then*

$$E[\mathbf{U}_N] = \bar{z}_N \mathbf{1}_N \quad \text{and} \quad Cov[\mathbf{U}_N] = \frac{N}{N-1} s_{z_N}^2 \mathbf{H}_N, \tag{17.42}$$

where $\mathbf{H}_N = \mathbf{I}_N - (1/N)\mathbf{1}_N\mathbf{1}_N'$ *is the centering matrix from (7.38).*

Since T in (17.39) equals $\mathbf{a}_N' \mathbf{U}_N$, the lemma shows that

$$E[T(\mathbf{P}_N\mathbf{z}_N)] = \mathbf{a}_N'\bar{z}_N\mathbf{1}_N = N\bar{a}_N\bar{z}_N \tag{17.43}$$

and

$$Var[T(\mathbf{P}_N\mathbf{z}_N)] = \frac{N}{N-1} s_N^2 \mathbf{a}_N'\mathbf{H}_N\mathbf{a}_N = \frac{N^2}{N-1} s_{a_N}^2 s_{z_N}^2, \tag{17.44}$$

where $s_{a_N}^2$ is the variance of the elements of $\mathbf{a}_N$. We will standardize the statistic to have mean 0 and variance 1 (see Exercise 17.6.2):

$$V_N = \frac{\mathbf{a}_N'\mathbf{P}_N\mathbf{z}_N - N\bar{a}_N\bar{z}_N}{\frac{N}{\sqrt{N-1}} s_{a_N} s_{z_N}} = \sqrt{N-1}\, r(\mathbf{a}_N, \mathbf{P}_N\mathbf{z}_N), \tag{17.45}$$

where $r(\mathbf{x}, \mathbf{y})$ is the usual Pearson correlation coefficient between $\mathbf{x}$ and $\mathbf{y}$ as in (12.81). Under certain conditions discussed in Section 17.5.1, this $V_N \to N(0,1)$ as $N \to \infty$, so that we can estimate the randomization p-value using the normal approximation.

For example, consider the two-treatment situation in Section 17.1, where $\mathbf{a}_N = (1/n, \ldots, 1/n, -1/m, \ldots, -1/m)'$ as in (17.8). Since there are n of $1/n$ and m of $-1/m$,

$$\bar{a}_N = 0 \quad \text{and} \quad s_{a_N}^2 = \frac{1}{nm}. \tag{17.46}$$

A little manipulation (see Exercise 17.6.3) shows that the observed V_n is

$$v_n = \frac{\bar{x} - \bar{y}}{\frac{N}{\sqrt{N-1}} \frac{1}{\sqrt{nm}} s_{z_N}} = \frac{\bar{x} - \bar{y}}{s_* \sqrt{\frac{1}{n} + \frac{1}{m}}}, \tag{17.47}$$

where $s_* = \sqrt{\sum(z_i - \bar{z}_N)^2 / (N-1)}$ from (7.34). The second expression is interesting because it is very close to the t-statistic in (15.16) used for the normal case, where the only difference is that there a pooled standard deviation was used instead of the s_*.

Fisher considered that this similarity helps justify the use of the statistic in sampling situations (for large N) even when the data are not normal.

For the the data in (17.1) on walking exercises for infants, we have $n = m = 6$, $\bar{x} - \bar{y} = -1.25$, and $s_{z_N}^2 = 2.7604$. Thus $v_N = -1.2476$, which yields an approximate one-sided p-value of $P[N(0,1) \leq v_N] = 0.1061$. The exact p-value was 0.133, so the approximation is fairly good even for this small sample.

In the draft lottery example of Section 17.3, we based the test on the Pearson correlation coefficient between the days and lottery numbers. We could have also used $r(\mathbf{x}, \mathbf{y})$ in the regression model in Section 17.4.2. In either case, (17.45) immediately gives the normalized statistic as $v_N = \sqrt{N-1}\,r(\mathbf{x}, \mathbf{y})$. For the draft lottery, $N = 366$ and $r = -0.226$, so that $v_N = \sqrt{365}(-0.226) = -4.318$. This statistic yields a two-sided p-value of 0.000016. The p-value we found earlier by sampling 10,000 random permutations was 0, hence the results do not conflict: Reject the null hypothesis that the lottery was totally random.

17.5.1   Technical conditions

Here we consider the conditions for the asymptotic normality of V_N in (17.45), where $\mathbf{P}_N \sim \text{Uniform}(\mathcal{S}_N)$. We assume that we have sequences $\mathbf{a}_N$ and $\mathbf{z}_N$, both with $N = 1, 2, \ldots$, where $\mathbf{a}_N = (a_{N1}, \ldots, a_{NN})'$ and $\mathbf{z}_N$ are both $N \times 1$. Fraser (1957), Chapter 6, summarizes various conditions that imply asymptotic normality. We will look at one specific condition introduced by Hoeffding (1952):

$$\frac{1}{N} h(\mathbf{z}_N) h(\mathbf{a}_N) \to 0 \ \text{ as } \ N \to \infty, \tag{17.48}$$

where for any $N \times 1$ vector $\mathbf{c}_N$,

$$h(\mathbf{c}_N) = \frac{\max_{i=1,\ldots,N}(c_{Ni} - \bar{c}_N)^2}{s_{c_N}^2}. \tag{17.49}$$

Fraser's Theorem 6.5 implies the following.

Theorem 17.2. *If (17.48) holds for the sequences $\mathbf{a}_N$ and $\mathbf{z}_N$, then $V_N \to N(0,1)$ as $N \to \infty$, where V_N is given in (17.45).*

We look at some special cases. In the two-treatment case where $\mathbf{a}_N$ is as in (17.8), assume that the proportions of observations in each treatment is roughly constant as $N \to \infty$, that is, $n/N \to p \in (0,1)$ as $N \to \infty$. Then Exercise 17.6.5 shows that

$$h(\mathbf{a}_N) = \frac{\max\{1/n^2, 1/m^2\}}{1/nm} \to \max\left\{\frac{1-p}{p}, \frac{p}{1-p}\right\} \in (0, \infty). \tag{17.50}$$

Thus the condition (17.48) holds if

$$\frac{1}{N} h(\mathbf{z}_N) \to 0. \tag{17.51}$$

It may be a bit problematic to decide whether it is reasonable to assume this condition. According to Theorem 6.7 of Fraser (1957) it will hold if the z_i's are the observations from an iid sample with positive variance and finite $E|Z_i|^3$. The data (17.1) in the walking exercise example looks consistent with this assumption.

In the draft lottery example, both $\mathbf{a}_N$ and $\mathbf{z}_N$ are $\mathbf{e}_N = (1, 2, \ldots, N)'$, or some permutation thereof. Thus we can find the means and variances from those of the Discrete Uniform$(1, N)$ in Table 1.2 (page 9):

$$\bar{e}_N = \frac{N+1}{2} \text{ and } s_{\bar{e}_N}^2 = \frac{N^2 - 1}{12}. \tag{17.52}$$

The $\max_{i=1,\ldots,N} (i - (N+1)/2)^2 = (N-1)^2 / 4$, hence

$$h(\mathbf{e}_N) = \frac{3(N-1)^2}{N^2 - 1} \to 3, \tag{17.53}$$

and it is easy to see that (17.48) holds.

If we are in the general correlation situation, $V_N = \sqrt{N-1}\, r(\mathbf{x}_N, \mathbf{P}_N \mathbf{y}_N)$, a sufficient condition for asymptotic normality of V_N is that

$$\frac{1}{\sqrt{N}} h(\mathbf{x}_N) \to 0 \text{ and } \frac{1}{\sqrt{N}} h(\mathbf{y}_N) \to 0. \tag{17.54}$$

These conditions will hold if the x_i's and y_i's are from iid sequences with tails that are "exponential" or less. The right tail of a distribution is exponential if as $x \to \infty$, $1 - F(x) \le a \exp(-bx)$ for some a and b. Examples include the normal, exponential, and logistic. The raw hurricane data analyzed in Section 17.4.2 may not conform to these assumptions due to the presence of extreme outliers. See Figure 12.2 on page 192. The assumptions do look reasonable if we take logs of the y-variable as we did in Section 12.5.2.

17.5.2 Sign changes

Similar asymptotics hold if the group used is $\mathcal{G}_\pm$, the group of sign-change matrices (17.35). Take the basic statistic to be $T(\mathbf{z}_N) = \mathbf{a}_N' \mathbf{z}_N$ as in (17.38) for some set of constants $\mathbf{a}_N$. The randomization distribution is then of $\mathbf{a}_N' \mathbf{G} \mathbf{z}_N$, where $\mathbf{G} \sim$ Uniform$(\mathcal{G}_\pm)$. Since the diagonals of $\mathbf{G}$ all have distribution $P[G_{ii} = -1] = P[G_{ii} = +1] = 1/2$, $E[G_{ii}] = 0$ and $Var[G_{ii}] = 1$. Thus the normalized statistic here is

$$V_N = \frac{\mathbf{a}_N' \mathbf{G} \mathbf{z}_N}{\sqrt{\sum a_i^2 z_i^2}}. \tag{17.55}$$

The **Lyapunov condition** for asymptotic normality is useful for sums of independent but not necessarily identically distributed random variables. See Serfling (1980), or any text book on probability.

Theorem 17.3. *Suppose $X_1, X_2, \ldots$ are independent with $E[X_i] = \mu_i$ and $Var[X_i] = \sigma_i^2 < \infty$ for each i. Then*

$$\frac{\sum_{i=1}^N (X_i - \mu_i)}{\sqrt{\sum_{i=1}^N \sigma_i^2}} \to^{\mathcal{D}} N(0, 1) \text{ if } \frac{\sum_{i=1}^N E[|X_i - \mu_i|^\nu]}{\sqrt{\sum_{i=1}^N \sigma_i^2}^\nu} \to 0 \text{ for some } \nu > 2. \tag{17.56}$$

For (17.55), Lyapunov says that

$$V_N \to^{\mathcal{D}} N(0, 1) \text{ if } \frac{\sum_{i=1}^N E[|a_i z_i|^\nu]}{\sqrt{\sum a_i^2 z_i^2}^\nu} \to 0 \text{ for some } \nu > 2. \tag{17.57}$$

17.6   Exercises

Exercise 17.6.1. (a) Show that $(N/(N-1))\mathbf{H}_N$ has 1's on the diagonal and $-1/(N-1)$'s on the off-diagonals. (b) Prove Lemma 17.1.

Exercise 17.6.2. (a) For $N \times 1$ vectors $\mathbf{x}$ and $\mathbf{y}$, show that the Pearson correlation coefficient can be written $r(\mathbf{x},\mathbf{y}) = (\mathbf{x}'\mathbf{y} - N\bar{x}\bar{y})/(Ns_xs_y)$, where s_x and s_y are the standard deviations of the elements of $\mathbf{x}$ and $\mathbf{y}$, respectively. (b) Verify (17.45).

Exercise 17.6.3. Let $\mathbf{a}_N = (1/n,\ldots,1/n,-1/m,\ldots,-1/m)'$ as in (17.8), where $N = n + m$. (a) Show that the mean of the elments is $\bar{a}_N = 0$ and the variance is

$$s_{a_N}^2 = \frac{1}{N}\left(\frac{1}{n} + \frac{1}{m}\right) = \frac{1}{nm}. \tag{17.58}$$

(b) Verify (17.47).

Exercise 17.6.4. The Affordable Care Act (ACA) is informally called "Obamacare." Even though they are the same thing, do some people feel more positive toward the Affordable Care Act than Obamacare? Each student in a group of 841 students was given a survey with fifteen questions. All the questions were the same, except for one — Students were randomly asked one of the two questions:

- What are your feelings toward The Affordable Care Act?
- What are your feelings toward Obamacare?

Whichever question was asked, the response is a number from 1 to 5, where 1 means one's feelings are very negative, and a 5 means very positive. Consider the randomization model in Section 17.1. Here, x_i would be person i's response to the question referring to the ACA, and y_i the response to the question referring to Obamacare. Take the exact null (17.3) that $x_i = y_i$ for all i, and use the difference in means of the two groups as the test statistic. There were $n = 416$ people assigned to the ACA group, with $\sum x_i = 1349$ and $\sum x_i^2 = 4797$ for those people. The Obamacare group had $m = 425$, $\sum y_i = 1285$, and $\sum y_i^2 = 4443$. (a) Find the difference in means for the observed groups, $\bar{x} - \bar{y}$, and the normalized version v_n as in (17.47). (b) Argue that the condition (17.51) is reasonable here (if it is). Find the p-value based on the normal approximation for the statistic in part (a). What do you conclude?

Exercise 17.6.5. Let $\mathbf{a}_N' = ((1/n)\mathbf{1}_n', -(1/m)\mathbf{1}_m')$. (a) Show that $\max\{(a_{Ni} - \bar{a}_N)^2\} = \max\{1/n^2, 1/m^2\}$. (b) Suppose $n/N \to p \in (0,1)$ as $N \to \infty$. Show that

$$h(\mathbf{a}_N) \equiv \frac{\max\{(a_{Ni} - \bar{a}_N)^2\}}{s_{a_N}^2} \to \max\left\{\frac{1-p}{p}, \frac{p}{1-p}\right\} \in (0,\infty) \tag{17.59}$$

as in (17.50). [Recall $s_{a_N}^2$ in (17.46).]

Exercise 17.6.6. The next table has the data from a study (Mendenhall, Million, Sharkey, and Cassisi, 1984) comparing surgery and radiation therapy for treating cancer of the larynx.

	Cancer controlled	Cancer not controlled	Total
Surgery	$X_{11} = 21$	$X_{12} = 2$	$X_{1+} = 23$
Radiation therapy	$X_{21} = 15$	$X_{22} = 3$	$X_{2+} = 18$
Total	$X_{+1} = 36$	$X_{+2} = 5$	$X_{++} = 41$

$$\tag{17.60}$$

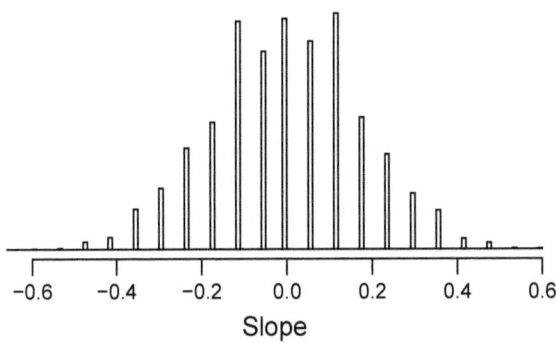

Figure 17.4: Histogram of 10,000 randomizations of the slope of Tukey's resistant line for the grades data.

The question is whether surgery is better than radiation therapy for controlling cancer. Use X_{11}, the upper left variable in the table, as the test statistic. (a) Conditional on the marginals $X_{1+} = 23$ and $X_{+1} = 36$, what is the range of X_{11}? (b) Find the one-sided p-value using Fisher's exact test. What do you conclude?

Exercise 17.6.7. In the section on paired comparisons, Section 17.4.1, we noted that we needed exchangeability rather than just equal distributions. To see why, consider (X, Y), with joint pmf and space

$$f(x,y) = \tfrac{1}{5}, (x,y) \in W = \{(1,2), (1,3), (2,1), (3,4), (4,1)\}. \qquad (17.61)$$

(a) Show that marginally, X and Y have the same distribution. (b) Show that X and Y are not exchangeable. (c) Find the pmf of $Z = X - Y$. Show that Z is not symmetric about zero, i.e., Z and $-Z$ have different distributions. (d) What is the median of Z? Is it zero?

Exercise 17.6.8. For the grades in a statistics class of 107 students, let X = score on hourly exams, Y = score on final exam. We wish to test whether these two variables are independent. (We would not expected them to be.) (a) Use the test statistic $\sum (X_i - \bar{X})(Y_i - \bar{Y})$. The data yield the following: $\sum (x_i - \bar{x})(y_i - \bar{y}) = 6016.373$, $\sum (x_i - \bar{x})^2 = 9051.411$, $\sum (y_i - \bar{y})^2 = 11283.514$. Find the normalized version of the test statistic, normalized according to the randomization distribution. Do you reject the null hypothesis? (b) Tukey (1977) proposed a **resistant-line** estimate of the fit in a simple linear regression. The data are rearranged so that the x_i's are in increasing order, then the data are split into three approximately equal-sized groups based on the values of x_i: the lower third, middle third, and upper third. With 107 observations, the group sizes are 36, 35, and 36. Then for each group, the median of the x_i's and median of the y_i's are calculated. The resistant slope is the slope between the two extreme points: $\hat{\beta}(\mathbf{x}, \mathbf{y}) = (y_3^* - y_1^*)/(x_3^* - x_1^*)$, where x_j^* is the median of the x_i's in the jth group, and similarly for the y_j^*'s. The R routine line calculates this

Chapter 17. Randomization Testing

slope, as well as an intercept. For the data here, $\widehat{\beta}(\mathbf{x}, \mathbf{y}) = 0.8391$. In order to use this slope as a test statistic, we simulated 10,000 randomizations of $\widehat{\beta}(\mathbf{x}, \mathbf{Py})$. Figure 17.4 contains the histogram of these values. What do you estimate the randomization p-value is?

Exercise 17.6.9. Suppose $T_N \sim$ Hypergeometric(k, l, n) as in (17.16), where $N = k + l$. Set $m = N - n$. As we saw in Section 17.2, we can represent this distribution with a randomization distribution of 0-1 vectors. Let $\mathbf{a}'_N = (\mathbf{1}'_n, \mathbf{0}'_m)$ and $\mathbf{z}'_N = (\mathbf{1}'_k, \mathbf{0}'_l)$, so that $T_N =^{\mathcal{D}} \mathbf{a}'_N \mathbf{P}_N \mathbf{z}_N$, where $\mathbf{P}_N \sim \text{Uniform}(\mathcal{S}_N)$. (a) Show that

$$E[T_N] = \frac{kn}{N} \quad \text{and} \quad Var[T_N] = \frac{klmn}{N^2(N-1)}. \tag{17.62}$$

[Hint: Use (17.44). What are $s^2_{a_N}$ and $s^2_{z_N}$?] (b) Suppose $k/N \to \kappa \in (0, 1)$ and $n/N \to p \in (0, 1)$. Show that Theorem 17.2 can be used to prove that

$$\sqrt{N-1}\, \frac{NT_N - nm}{\sqrt{klmn}} \longrightarrow^{\mathcal{D}} N(0, 1). \tag{17.63}$$

[Hint: Show that a result similar to that in (17.50) holds for the $\mathbf{a}_N$ and $\mathbf{z}_N$ here, which helps verify (17.48).]

Chapter 18

Nonparametric Tests Based on Signs and Ranks

18.1 Sign test

There are a number of useful and robust testing procedures based on signs and ranks of the data that traditionally go under the umbrella of **nonparametric tests**. They can be used as analogs to the usual normal-based testing situations when one does not wish to depend too highly on the normal assumption. These test statistics have the property that their randomization distribution is the same as their sampling distribution, so the work we have done so far will immediately apply here.

One nonparametric analog to hypotheses on the mean are hypotheses on the median. Assume that $Z_1, \ldots, Z_N$ are iid, $Z_i \sim F \in \mathcal{F}$, where $\mathcal{F}$ is the set of continuous distribution functions that are symmetric about their median η. We test

$$H_0 : F \in \mathcal{F} \text{ with } \eta = 0 \text{ versus } H_A : F \in \mathcal{F} \text{ with } \eta \neq 0. \qquad (18.1)$$

(What follows is easy to extend to tests of $\eta = \eta_0$; just subtract η_0 from all the Z_i's.)

If the median is zero, then one would expect about half of the observations to be positive and half negative. The sign test uses the signs of the data, $\text{Sign}(Z_1)$, ..., $\text{Sign}(Z_N)$, where for any $z \in \mathbb{R}$,

$$\text{Sign}(z) = \begin{cases} +1 & \text{if} \quad z > 0 \\ 0 & \text{if} \quad z = 0 \\ -1 & \text{if} \quad z < 0 \end{cases} . \qquad (18.2)$$

The basic sign statistic is

$$S(\mathbf{z}) = \sum_{i=1}^{N} \text{Sign}(z_i). \qquad (18.3)$$

The corresponding two-sided test statistic for (18.1) is $|S(\mathbf{z})|$. If F is continuous at 0, i.e., $P[Z_i = 0] = 0$, then the exact null distribution is the same as $|2W - N|$ where $W \sim \text{Binomial}(N, 1/2)$. Thus it is easy to perform the test, either exactly or asymptotically.

The test is often used in paired comparisons, as for the tread wear data in Section 17.4.1. There the data are iid (X_i, Y_i)'s, and the null hypothesis (17.32) is that X_i and Y_i are exchangeable $((X_i, Y_i) =^D (Y_i, X_i))$, hence $Z_i = X_i - Y_i$ is symmetric about the median of zero. From (17.34) we see that of the $N = 16$ z_i's, one is negative and 15 are positive, so that $w = 15$ and $S(\mathbf{z}) = 14$. The p-value is

$$P[|S(\mathbf{Z})| \geq 14] = P[W \in \{0, 1, 15, 16\}] = 0.0005, \tag{18.4}$$

leading to rejection of the null as before.

18.2   Rank transform tests

A popular approach to nonparametric tests is to take a familiar test statistic, and recreate it by replacing the observed values with their ranks. Such tests, called **rank transform tests** are often robust with null distributions that are easy to apply, either exactly or asymptotically. First, we need to define rank.

We saw rankings in Section 3.2.1, where people ranked from 1 to 3 where they like to live. For an $N \times 1$ vector $\mathbf{z}$ with no ties (i.e., no two z_i's are equal), the corresponding rank vector is $\mathbf{r}$, where $r_i = 1$ if z_i is the smallest observation, $r_i = 2$ if z_i is the second smallest, ..., $r_i = N$ if z_i is the largest. When there are ties, we will define ranks using **midranks**, where the midrank of z_i is the regular rank if no other observation equals z_i, and equals the average of what would be the ranks if there are multiple observations equalling z_i. An illustration is easier to understand. Suppose $\mathbf{z} = (12, 15, 14, 13, 16, 12, 14, 14)$. We order the z_i's, assign "ranks" from 1 to N to the order statistics, then average the ranks over tied observations:

$z_{(i)}$'s	12	12	13	14	14	14	15	16
"ranks"	1	2	3	4	5	6	7	8
midranks	1.5	1.5	3	5	5	5	7	8

$$\tag{18.5}$$

That is, the two 12's would be ranked 1 and 2, so their midranks are both $(1+2)/2$. Likewise the three 14's share the average of the ranks 4, 5, and 6. The rank vector then arranges the ranks in the order of the original $\mathbf{z}$:

$\mathbf{z}$	12	15	14	13	16	12	14	14
Rank($\mathbf{z}$)	1.5	7	5	3	8	1.5	5	5

$$\tag{18.6}$$

To be precise,

$$\text{Rank}(\mathbf{z})_i = 1 + \sum_{j \neq i} \left(I[z_i > z_j] + \tfrac{1}{2} I[z_i = z_j] \right)$$

$$= \tfrac{1}{2} \left(\sum_{j \neq i} \text{Sign}(z_i - z_j) + n + 1 \right). \tag{18.7}$$

Below we show how to use ranks in the testing situations previously covered in Chapter 17.

18.2.1   Signed-rank test

The sign test for testing the median is zero as in (18.1) is fairly crude in that it looks only at which side of zero the observations are, not how far from zero they are. If under the null, the distribution of the Z_i's is symmetric about zero, we can generalize the test. The usual mean difference can be written as a sign statistic weighted by the magnitudes of the observations:

$$\bar{z} = \frac{1}{N} \sum_{i=1}^{N} |z_i| \, \text{Sign}(z_i). \tag{18.8}$$

A modification of the sign test introduced by Wilcoxon (1945) is the **signed-rank test**, which uses ranks of the magnitudes instead of the magnitudes themselves. Letting

$$\mathbf{R} = \text{Rank}(|Z_1|, \ldots, |Z_N|), \quad \text{and} \quad \mathbf{S} = (\text{Sign}(Z_1), \ldots, \text{Sign}(Z_N))', \tag{18.9}$$

the statistic is

$$T(\mathbf{Z}) = \sum_{i=1}^{N} R_i S_i = \mathbf{R}'\mathbf{S}. \tag{18.10}$$

We will look at the randomization distribution of T under the null. As in Section 17.4.1, the distribution of $\mathbf{Z}$ is invariant under the group $\mathcal{G}_{\pm}$ of sign-change matrices as in (17.35). For fixed $z_i \neq 0$, if $P[G_i = -1] = P[G_i = +1] = 1/2$, then $P[\text{Sign}(G_i z_i) = -1] = P[\text{Sign}(G_i z_i) = +1] = 1/2$. If $z_i = 0$, then $\text{Sign}(G_i z_i) = 0$. Thus the randomization distribution of T is given by

$$T(\mathbf{Gz}) =^{\mathcal{D}} \sum_{i \,|\, z_i \neq 0} r_i G_i, \quad \mathbf{G} \sim \text{Uniform}(\mathcal{G}_{\pm}). \tag{18.11}$$

(In practice one can just ignore the zeroes, and proceed with a smaller sample size.) Thus we are in the same situation as Section 17.4.1, and can find the exact distribution if N is small, or simulate if not. Exercise 18.5.1 finds the mean and variance:

$$E[T(\mathbf{Gz})] = 0 \quad \text{and} \quad Var[T(\mathbf{Gz})] = \sum_{i \,|\, z_i \neq 0} r_i^2. \tag{18.12}$$

When there are no zero z_i's and no ties among the z_i's, there are efficient algorithms (e.g., wilcox.test in R) to calculate the exact distribution for larger N, up to 50. To use the asymptotic normal approximation, note that under these conditions, $\mathbf{r}$ is some permutation of $1, \ldots, N$. Exercise 18.5.1 shows that

$$V_N = \frac{T(\mathbf{Gz})}{\sqrt{N(N+1)(2N+1)/6}} \longrightarrow N(0,1). \tag{18.13}$$

If the distribution of the Z_i's is continuous (so there are no zeroes and no ties, with probability one), then the randomization distribution and sampling distribution of T are the same.

18.2.2 Mann-Whitney/Wilcoxon two-sample test

The normal-based two-sample testing situation (17.24) tests the null that the two means are equal versus either a one-sided or two-sided alternative. There are various nonparametric analogs of this problem. The one we will deal with has

$$X_1, \ldots, X_n \sim \text{iid } F_X \quad \text{and} \quad Y_1, \ldots, Y_m \sim \text{iid } F_Y, \tag{18.14}$$

where the X_i's and Y_i's are independent. The null hypothesis is that the distribution functions are equal, and the alternative is that F_X is **stochastically larger** than F_Y:

Definition 18.1. *The distribution function F is **stochastically larger** than the distribution function G, written*

$$F >_{st} G, \tag{18.15}$$

if

$$F(c) \leq G(c) \ \text{ for all } c \in \mathbb{R}, \ \text{ and}$$
$$F(c) < G(c) \ \text{ for some } c \in \mathbb{R}. \tag{18.16}$$

It looks like the inequality is going the wrong way, but the idea is that if $X \sim F$ and $Y \sim G$, then F being stochastically larger than G means that the X tends to be larger than the Y, or

$$P[X > c] > P[Y > c], \ \text{ which implies that } \ 1 - F(c) > 1 - G(c) \implies F(c) < G(c). \tag{18.17}$$

On can also say that "X is stochastically larger than Y." For example, if X and Y are both from the same location family, but X's parameter is larger than Y's, then X is stochastically larger than Y.

Back to the testing problem. With the data in (18.14), the hypotheses are

$$H_0 : F_X = F_Y, \ \text{ versus } \ H_A : F_X >_{st} F_Y. \tag{18.18}$$

We reject the null if the x_i's are too much larger than the y_i's in some sense. Wilcoxon (1945) proposed, and Mann and Whitney (1947) further studied, the rank transform statistic that replaces the difference in averages of the two groups, $\bar{x} - \bar{y}$, with the difference in averages of their ranks. That is, let

$$\mathbf{r} = \text{Rank}(x_1, \ldots, x_n, y_1, \ldots, y_m), \tag{18.19}$$

the ranks of the data combining the two samples. Then the statistic is

$$W_N = \frac{1}{n} \sum_{i=1}^{n} r_i - \frac{1}{m} \sum_{i=n+1}^{n+m} r_i = \mathbf{a}'\mathbf{r}, \ \ \mathbf{a}' = \left(\frac{1}{n}, \ldots, \frac{1}{n}, -\frac{1}{m}, \ldots, -\frac{1}{m} \right), \tag{18.20}$$

the difference in the average of the ranks assigned to the x_i's and those assigned to the y_i's. For example, return to the walking exercises study from Section 17.1. If we take ranks of the data in (17.1), then find the difference in average ranks, we obtain $W_N = -4$. Calculating this statistic for all possible allocations as in (17.8), we find 26 of the 924 values are less than or equal to -4. (The alternative here is $F_Y >_{st} F_X$.) Thus the randomization p-value for the one-sided test is 0.028. This may be marginally statistically significant, though the two-sided p-value is a bit over 5%.

The statistic in (18.20) can be equivalently represented, at least if there are no ties, by the number of x_i's larger than y_j's:

$$W_N^* = \sum_{i=1}^{n} \sum_{j=1}^{m} I[x_i > y_j]. \tag{18.21}$$

Exercise 18.5.4 shows that $W_N = N(W_N^*/(nm) - 1/2)$.

We are back in the two-treatment situation of Section 17.1, but using the ranks in place of the z_i's. The randomization distribution involves permuting the combined data vector, which similarly permutes the ranks. Thus the normalized statistic is as in (17.47):

$$V_N = \frac{W_N}{\frac{N}{\sqrt{N-1}} \frac{1}{\sqrt{nm}} s_{r_N}}, \tag{18.22}$$

where $s_{r_N}^2$ is the variance of the ranks. Exercise 18.5.5 shows that if there are no ties among the observations,

$$V_N = \frac{W_N}{N\sqrt{\frac{N+1}{12nm}}}, \tag{18.23}$$

which approaches $N(0,1)$ under the randomization distribution.

As for the signed-rank test, if there are no ties, the R routine wilcox.test can calculate the exact distribution of the statistic for $N \leq 50$. Also, if F_X and F_Y are continuous, the sampling distribution of W_N under the null is the same as the randomization distribution.

18.2.3 Spearman's ρ independence test

One nonparametric analog of testing for zero correlation between two variables is testing independence versus positive (or negative) association. The data are the iid pairs $(X_1, Y_1), \ldots, (X_N, Y_N)$. The null hypothesis is that the X_i's are independent of the Y_i's. There are a number of ways to define positive association between the variables. A regression-oriented approach looks at the conditional distribution of Y_i given $X_i = x$, say $F_x(y)$. Independence implies that $F_x(y)$ does not depend on x. A positive association could be defined by having $F_x(y)$ being stochstically larger than $F_{x^*}(y)$ if $x > x^*$.

Here we look at **Spearman's** ρ, which is the rank transform of the usual Pearson correlation coefficient. That is, letting $\mathbf{r}_x = \text{Rank}(\mathbf{x})$ and $\mathbf{r}_y = \text{Rank}(\mathbf{y})$,

$$\hat{\rho}(\mathbf{x}, \mathbf{y}) = r(\mathbf{r}_x, \mathbf{r}_y), \tag{18.24}$$

where r is the usual Pearson correlation coefficient from (12.81). The $\hat{\rho}$ measures a general monotone relationship between X and Y, rather than the linear relation the Pearson coefficient measures.

The randomization distribution here is the same as for the draft lottery data in Section 17.3. Thus again we can use the large-sample approximation to the randomization distribution for the statistic in (17.45): $\sqrt{N-1}\, r(\mathbf{r}_x, \mathbf{P}_N \mathbf{r}_y) \approx N(0,1)$, $\mathbf{P}_N \sim \text{Uniform}(\mathcal{S}_N)$. If the distributions of X_i and Y_i are continuous, then again the randomization distribution and sampling distribution under the null coincide. Also, in the no-tie case, the R routine cor.test will find an exact p-value for $N \leq 10$.

Continuing from Section 17.4.2 with the hurricane data, we look at the correlation between damage and deaths with and without Katrina:

	Pearson (raw)	Spearman	Pearson (transformed)	
All data	0.6081	0.7605	0.7379	(18.25)
Without Katrina	0.3563	0.7527	0.6962	

Comparing the Pearson coefficient on the raw data and the Spearman coefficient on the ranks, we can see how much more robust Spearman is. The one outlier adds 0.25 to the Pearson coefficient. Spearman is hardly affected at all by the outlier. We

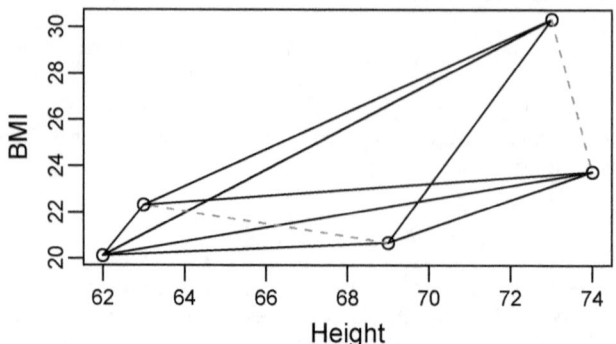

Figure 18.1: Connecting the points with line segments in a scatter plot. Kendall's τ equals the number of positive slope minus the number of negative slopes, divided by the total number of segments. See (18.27).

also include the Pearson coefficient where we take square root of the damage and log of the deaths. These coefficients are similar to the Spearman coefficients, though slightly less robust. These numbers suggest that Spearman's ρ gives a good simple and robust measure of association.

18.3 Kendall's τ independence test

Consider the same setup as for Spearman's ρ in Section 18.2.3, that is, $(X_1, Y_1), \ldots, (X_N, Y_N)$ are iid, and we test the null hypothesis that the X_i's are independent of the Y_i's. The alternative here is based on concordance. Given two of the pairs (X_i, Y_i) and (X_j, Y_j) $(i \neq j)$, they are **concordant** if the line connecting the points in $\mathbb{R}^2$ has positive slope, and **discordant** if the line has negative slope. For example, Figure 18.1 plots data for five students, with x_i being height in inches and y_i being body mass index (BMI). Each pair of point is connected by a line segment. Eight of these segments have positive slope, and two have negative slope.

A measure of concordance is

$$\tau = P[(X_i - X_j)(Y_i - Y_j) > 0] - P[(X_i - X_j)(Y_i - Y_j) < 0]$$
$$= E[\text{Sign}(X_i - X_j)\,\text{Sign}(Y_i - Y_j)]. \tag{18.26}$$

If $\tau > 0$, we tend to see larger x_i's going with larger y_i's, and smaller x_i's going with smaller y_i's. If $\tau < 0$, the x_i's and y_i's are more likely to go in different directions. If the X_i's are independent of the Y_i's, then $\tau = 0$. Kendall's τ, which we saw briefly in Exercise 4.4.2, is a statistic tailored to testing $\tau = 0$ versus $\tau > 0$ or < 0.

Kendall's τ test statistic is an unbiased estimator of τ:

$$\widehat{\tau}(\mathbf{x}, \mathbf{y}) = \frac{\sum\sum_{1 \leq i < j \leq N} \text{Sign}(x_i - x_j)\,\text{Sign}(y_i - y_j)}{\binom{N}{2}}. \tag{18.27}$$

This numerator is the number of positive slopes minus the number of negative slopes, so for the data in Figure 18.1, we have $\hat{\tau} = (8-2)/10 = 0.6$. As for Spearman's ρ, this statistic measures any kind of positive association, rather than just linear. To find the p-value based on the randomization distribution of $\hat{\tau}(\mathbf{x}, \mathbf{Py})$, $\mathbf{P} \sim \text{Uniform}(\mathcal{S}_N)$, we can enumerate the values if N is small, or simulate if N is larger, or use asymptotic considerations. The R function cor.test also handles Kendall's τ, exactly for $N \leq 50$.

To use the asymptotic normality approximation, we need the mean and variance under the randomization distribution. We will start by assuming that there are no ties among the x_i's, and no ties among the y_i's. Recall Kendall's distance in Exercises 4.4.1 and 4.4.2, defined by

$$d(\mathbf{x}, \mathbf{y}) = \sum\sum_{1 \leq i < j \leq N} I[(x_i - x_j)(y_i - y_j) < 0]. \tag{18.28}$$

With no ties,

$$\hat{\tau}(\mathbf{x}, \mathbf{y}) = 1 - \frac{4d(\mathbf{x}, \mathbf{y})}{N(N-1)}. \tag{18.29}$$

Arrange the observations so that the x_i's are in increasing order, $x_1 < x_2 < \cdots < x_N$. Then we can write

$$d(\mathbf{x}, \mathbf{y}) = \sum_{i=1}^{N-1} U_i, \text{ where } U_i = \sum_{j=i+1}^{N} I[y_i > y_j]. \tag{18.30}$$

Extending the result in Exercise 4.4.3 for $N = 3$, it can be shown that under the randomization distribution, the U_i's are independent with $U_i \sim \text{Discrete Uniform}(0, N-i)$. Exercise 18.5.11 shows that

$$E[d(\mathbf{x}, \mathbf{Py})] = \frac{N(N-1)}{4} \text{ and } Var[d(\mathbf{x}, \mathbf{Py})] = \frac{N(N-1)(2N+5)}{72}, \tag{18.31}$$

hence

$$E[\hat{\tau}(\mathbf{x}, \mathbf{Py}))] = 0 \text{ and } Var[\hat{\tau}(\mathbf{x}, \mathbf{Py}))] = \frac{2}{9}\frac{2N+5}{N(N-1)}, \tag{18.32}$$

then uses Lyapunov's condition to show that

$$V_N = \frac{\hat{\tau}(\mathbf{x}, \mathbf{Py})}{\sqrt{\frac{2}{9}\frac{2N+5}{N(N-1)}}} \longrightarrow^{\mathcal{D}} N(0, 1). \tag{18.33}$$

18.3.1 Ties

When X and Y are continuous, then the measure of concordance τ in (18.27) acts like a correlation coefficient, going from -1 if X and Y are perfectly negatively related, $Y = g(X)$ for a monotone decreasing function g, to $+1$ if they are perfectly positively related. The same goes for the data-based Kendall's τ in (18.28) if there are no ties. If the random variables are not continuous, or the observations of either or both variables contain ties, then the two measures are unable to achieve either -1 or $+1$. In practice, if ties are fairly scarce, then there is no need to make any modifications. For example, in data on 165 people's height and weight, heights are given to the nearest inch and weights to the nearest pound. There are 19 different heights and

48 different weights, hence quite a few ties. Yet the randomization values of $\hat{\tau}$ range from $-.971$ and $+.973$, which is close enough to the ± 1 range.

If there are extensive ties, which would occur if either variable is categorical, then modifications would be in order. The traditional approach is to note that τ is the covariance of $\text{Sign}(X_i - X_j)$ and $\text{Sign}(Y_i - Y_j)$, of which the correlation is a natural normalization. That is, in general we define

$$
\begin{aligned}
\tau &= Corr[\text{Sign}(X_i - X_j), \text{Sign}(Y_i - Y_j)] \\
&= \frac{E[\text{Sign}(X_i - X_j)\,\text{Sign}(Y_i - Y_j)]}{\sqrt{E[\text{Sign}(X_i - X_j)^2]\,E[\text{Sign}(Y_i - Y_j)^2]}},
\end{aligned} \tag{18.34}
$$

noting that $E[\text{Sign}(X_i - X_j)] = E[\text{Sign}(Y_i - Y_j)] = 0$. If the distribution of X is continuous, $P[\text{Sign}(X_i - X_j)^2 = 1] = 1$, and similarly for Y. Thus if both distributions are continuous, the denominator is 1, so that τ is as in (18.26). Suppose X is not continuous. Let $a_1, \dots, a_K$ be the points at which X has positive probability, and $p_k = P[X = a_k]$. If Y is not continuous, let $b_1, \dots, b_L$ be the corresponding points, and $q_l = P[Y = b_l]$. Either or both of K and L could be ∞. Exercise 18.5.8 shows that

$$
E[\text{Sign}(X_i - X_j)^2] = 1 - \sum p_k^2 \quad \text{and} \quad E[\text{Sign}(Y_i - Y_j)^2] = 1 - \sum q_l^2. \tag{18.35}
$$

The $\hat{\tau}$ in (18.27) is similarly modified. Let $\mathbf{u}_x$ be the $N(N-1) \times 1$ vector with all the $\text{Sign}(x_i - x_j)$ for $i \neq j$,

$$
\begin{aligned}
\mathbf{u}'_x = (&\text{Sign}(x_1 - x_2), \dots, \text{Sign}(x_1 - x_N), \\
&\text{Sign}(x_2 - x_1), \text{Sign}(x_2 - x_3), \dots, \text{Sign}(x_2 - x_N), \\
&\dots, \text{Sign}(x_N - x_1), \dots, \text{Sign}(x_N - x_{N-1})), \tag{18.36}
\end{aligned}
$$

and $\mathbf{u}_y$ be the corresponding vector of the $\text{Sign}(y_i - y_j)$'s. Then $\mathbf{u}'_x \mathbf{u}_y$ equals twice the numerator of $\hat{\tau}$ in (18.27) since it counts $i < j$ and $i > j$. Kendall's τ modified for ties is then Pearson's correlation coefficient of these signed difference vectors:

$$
\hat{\tau} = r(\mathbf{u}_x, \mathbf{u}_y) = \frac{\mathbf{u}'_x \mathbf{u}_y}{\|\mathbf{u}_x\|\,\|\mathbf{u}_y\|}, \tag{18.37}
$$

since the means of the elements of $\mathbf{u}_x$ and $\mathbf{u}_y$ are 0.

Now Exercise 18.5.8 shows that

$$
\|\mathbf{u}_x\|^2 = \sum\sum_{i \neq j} \text{Sign}(x_i - x_j)^2 = N(N-1) - \sum_{k=1}^{K} c_k(c_k - 1) \quad \text{and}
$$

$$
\|\mathbf{u}_y\|^2 = \sum\sum_{i \neq j} \text{Sign}(y_i - y_j)^2 = N(N-1) - \sum_{l=1}^{L} d_l(d_l - 1), \tag{18.38}
$$

where $(c_1, \dots, c_K)$ is the pattern of ties for the x_i's and $(d_1, \dots, d_L)$ is the pattern of ties for $\mathbf{y}$. That is, letting $a_1, \dots, a_K$ be the values that appear at least twice in the vector $\mathbf{x}$, set $c_k = \#\{i \mid x_i = a_k\}$. Similarly for $\mathbf{y}$. Finally,

$$
\hat{\tau}(\mathbf{x}, \mathbf{y}) = \frac{2\sum\sum_{1 \le i < j \le N} \text{Sign}(x_i - x_j)\,\text{Sign}(y_i - y_j)}{\sqrt{N(N-1) - \sum_{k=1}^{K} c_k(c_k - 1)}\,\sqrt{N(N-1) - \sum_{l=1}^{L} d_l(d_l - 1)}}. \tag{18.39}
$$

If there are no ties in one of the vectors, there is nothing to subtract from its $N(N-1)$, hence if neither vector has ties, we are back to the original $\hat{\tau}$ in (18.27). The statistic in (18.39) is often called **Kendall's τ_B**, the original one in (18.27) being then referred to as **Kendall's τ_A**. This modified statistic will generally not range all the way from -1 to $+1$, though it can get closer to those limits than without the modification. See Exercise 18.5.9.

For testing independence, the randomization distribution of $\hat{\tau}(\mathbf{x}, \mathbf{Py})$ still has mean 0, but the variance is a bit trickier if ties are present. In Section 18.3.2, we deal with ties in just one of the vectors. Here we give the answer for the general case without proof. Let

$$S(\mathbf{x}, \mathbf{y}) = \sum\sum_{1 \leq i < j \leq N} \text{Sign}(x_i - x_j)\,\text{Sign}(y_i - y_j). \tag{18.40}$$

The expectation under the randomization distribution is $E[S(\mathbf{x}, \mathbf{Py})] = 0$ whatever the ties situation is. Since S is a sum of signed differences, the variance is a sum of the variances plus the covariances of the signed differences, each of which can be calculated, though the process is tedious. Rather than go through the details, we will present the answer, and refer the interested reader to Chapter 5 of Kendall and Gibbons (1990). The variance in general is given by

$$
\begin{aligned}
Var[S(\mathbf{x}, \mathbf{Py})] ={}& \frac{N(N-1)(2N+5) - \sum c_k(c_k-1)(c_k+5) - \sum d_l(d_l-1)(2d_l+5)}{18} \\
&+ \frac{[\sum c_k(c_k-1)(c_k-2)][\sum d_l(d_l-1)(d_l-2)]}{9N(N-1)(N-2)} \\
&+ \frac{[\sum c_k(c_k-1)][\sum d_l(d_l-1)]}{2N(N-1)}.
\end{aligned} \tag{18.41}
$$

Notice that if there are ties in only one of the variables, the variance simplifies substantially, as we will see in (18.48). Also, if the ties are relative sparse, the last two terms are negligible relative to the first term for large N. The variance of $\hat{\tau}$ in (18.39) can then be obtained from (18.41). Dropping those last two terms and rearranging to make easy comparison to (18.32), we have

$$Var[\hat{\tau}(\mathbf{x}, \mathbf{Py})] \approx \frac{2}{9}\frac{2N+5}{N(N-1)}\frac{1 - c^{**} - d^{**}}{(1-c^*)(1-d^*)}, \tag{18.42}$$

where

$$c^* = \sum \frac{c_k(c_k-1)}{N(N-1)}, \quad c^{**} = \sum \frac{c_k(c_k-1)(2c_k+5)}{N(N-1)(2N+5)}, \tag{18.43}$$

and similarly for d^* and d^{**}.

18.3.2 Jonckheere-Terpstra test for trend among groups

Consider the two-sample situation in Section 18.2.2. We will switch to the notation in that section, where we have $x_1, \ldots, x_n$ for group 1 and $y_1, \ldots, y_m$ for group 2, so that $\mathbf{z} = (\mathbf{x}', \mathbf{y}')'$ and $N = n + m$. Let $\mathbf{a} = (1, \ldots, 1, 2, \ldots, 2)'$, where there are n 1's and m 2's. Then since $a_i = a_j$ if i and j are both between 1 and n, or both between $n+1$ and

$n + m$,

$$d(\mathbf{a}, \mathbf{z}) = \sum_{1 \le i < j \le n} \sum I[(a_i - a_j)(x_i - x_j) < 0] + \sum_{i=1}^{n} \sum_{j=1}^{m} I[(a_i - a_{n+j})(x_i - y_j) < 0]$$
$$+ \sum_{1 \le i < j \le m} \sum I[(a_{n+i} - a_{n+j})(y_i - y_j) < 0]$$
$$= \sum_{i=1}^{n} \sum_{j=1}^{m} I[(a_i - a_{n+j})(x_i - y_j) < 0]$$
$$= \sum_{i=1}^{n} \sum_{j=1}^{m} I[x_i > y_j], \tag{18.44}$$

which equals W_N^*, the representation in (18.21) of the Mann-Whitney/Wilcoxon statistic.

If there are several groups, and ordered in such a way that we are looking for a trend across groups, then we can again use $d(\mathbf{a}, \mathbf{z})$, where $\mathbf{a}$ indicates the group number. For example, if there are K groups, and n_k observations in group k, we would have

$$\begin{array}{c|ccccccccc} \mathbf{a} & 1 & \cdots & 1 & 2 & \cdots & 2 & \cdots & K & \cdots & K \\ \hline \mathbf{z} & z_{11} & \cdots & z_{1n_1} & z_{22} & \cdots & z_{2n_2} & \cdots & z_{K1} & \cdots & z_{Kn_K} \end{array}. \tag{18.45}$$

The $d(\mathbf{a}, \mathbf{z})$ is called the **Jonckheere-Terpstra** statistic (Terpstra (1952), Jonckheere (1954)), and is an extension of the two-sample statistic, summing the W_N^* for the pairs of groups:

$$d(\mathbf{a}, \mathbf{z}) = \sum_{1 \le k < l \le K} \sum \sum_{i=1}^{n_k} \sum_{j=1}^{n_l} I[z_{ki} > z_{lj}]. \tag{18.46}$$

To find the mean and variance of d, it is convenient to first imagine the sum of all pairwise comparisons $d(\mathbf{e}_N, \mathbf{z})$, where $\mathbf{e}_N = (1, 2, \ldots, N)'$. We can then decompose this sum into the parts comparing observations between groups and those comparing observations within groups:

$$d(\mathbf{e}_N, \mathbf{z}) = d(\mathbf{a}, \mathbf{z}) + \sum_{k=1}^{K} d(\mathbf{e}_{n_k}, \mathbf{z}_k), \quad \text{where } \mathbf{z}_k = (z_{k1}, \ldots, z_{kn_k})'. \tag{18.47}$$

Still assuming no ties in $\mathbf{z}$, it can be shown that using the randomization distribution on $\mathbf{z}$, the $K + 1$ random variables $d(\mathbf{a}, \mathbf{Pz})$, $d(\mathbf{e}_{n_1}, (\mathbf{Pz})_1)$, ..., $d(\mathbf{e}_{n_K}, (\mathbf{Pz})_K)$ are mutually independent. See Terpstra (1952), Lemma I and Theorem I. The idea is that the relative rankings of z_{ki}'s within one group are independent of those in any other group, and that the rankings within groups are independent of relative sizes of elements in one group to another.

The independence of the d's on the right-hand side of (18.47) implies that their variances sum, hence we can find the mean and variance of $d(\mathbf{a}, \mathbf{Pz})$ by subtraction:

$$E[d(\mathbf{a}, \mathbf{Pz})] = \frac{N(N-1)}{4} - \sum_{k=1}^{K} \frac{n_k(n_k - 1)}{4} \equiv \mu(\mathbf{n}), \quad \text{and}$$

$$Var[d(\mathbf{a}, \mathbf{Pz})] = \frac{N(N-1)(2N+5)}{72} - \sum_{k=1}^{K} \frac{n_k(n_k - 1)(2n_k + 5)}{72} \equiv \sigma^2(\mathbf{n}), \tag{18.48}$$

$\mathbf{n} = (n_1, \ldots, n_K)$, using (18.31) on $d(\mathbf{e}_N, \mathbf{Pz})$ and the $d(\mathbf{e}_{n_k}, (\mathbf{Pz})_k)'$s. For asymptotics, as long as $n_k/N \to \lambda_k \in (0,1)$ for each k, we have that

$$JT_N \equiv \frac{d(\mathbf{a}, \mathbf{Pz}) - \mu(\mathbf{n})}{\sigma(\mathbf{n})} \longrightarrow^{\mathcal{D}} N(0,1). \tag{18.49}$$

See Exercise 18.5.12. Since we based this statistic on d rather than $\hat{\tau}$, testing for positive association means rejecting for small values of JT_N, and testing for negative association means rejecting for large values of JT_N. We could also use $\hat{\tau}$ (A or B), noting that

$$\sum\sum_{1 \le i < j \le N} \text{Sign}(a_i - a_j)\,\text{Sign}(z_i - z_j) = \binom{N}{2} - \sum_{k=1}^{K} \binom{n_k}{2} - 2d(\mathbf{a}, \mathbf{z}). \tag{18.50}$$

As long as there are no ties in $\mathbf{z}$, the test here works for any $\mathbf{a}$, where in (18.48) the $(n_1, \ldots, n_K)$ is replaced by the pattern of ties for $\mathbf{a}$.

18.4 Confidence intervals

As in Section 15.6, we can invert these nonparametric tests to obtain nonparametric confidence intervals for certain parameters. To illustrate, consider the sign test based on $Z_1, \ldots, Z_N$ iid. We assume the distribution is continuous, and find a confidence interval for the median η. It is based on the order statistics $z_{(1)}, \ldots, z_{(N)}$. The idea stems from looking at $(z_{(1)}, z_{(N)})$ as a confidence interval for η. Note that the median is between the minimum and maximum unless all observations are greater than the median, or all are less than the median. By continuity, $P[Z_i > \eta] = P[Z_i < \eta] = 1/2$, hence the chance η is not in the interval is $2/2^N$. That is, $(z_{(1)}, z_{(N)})$ is a $100(1 - 2^{N-1})\%$ confidence interval for the median. By using other order statistics as limits, other percentages are obtained.

For the general interval, we test the null hypothesis $H_0 : \eta = \eta_0$ using the sign statistic in (18.3): $S(\mathbf{z}; \eta_0) = \sum_{i=1}^{N} \text{Sign}(z_i - \eta_0)$. An equivalent statistic is the number of z_i's larger than η_0:

$$S(\mathbf{z}; \eta_0) = 2\sum_{i=1}^{N} I[z_i > \eta_0] - N. \tag{18.51}$$

Under the null, $\sum_{i=1}^{N} I[Z_i > \eta_0] \sim \text{Binomial}(N, 1/2)$, so that for level α, the exact test rejects when

$$\sum_{i=1}^{N} I[z_i > \eta_0] \le A \quad \text{or} \quad \sum_{i=1}^{N} I[z_i > \eta_0] \ge B, \tag{18.52}$$

where

$$A = \max\{\text{integer } k \mid P[\text{Binomial}(N, \tfrac{1}{2}) \le k] \le \tfrac{1}{2}\alpha\} \text{ and}$$
$$B = \min\{\text{integer } l \mid P[\text{Binomial}(N, \tfrac{1}{2}) \ge l] \le \tfrac{1}{2}\alpha\}. \tag{18.53}$$

Exercise 18.5.17 shows that the confidence interval then consists of the η_0's for which (18.52) fails:

$$C(\mathbf{z}) = \{\eta_0 \mid A + 1 \le \sum_{i=1}^{N} I[z_i > \eta_0] \le B - 1\}$$

$$= [z_{(N-B+1)}, z_{(N-A)}). \tag{18.54}$$

Since the Z_i's are assumed continuous, it does not matter whether the endpoints are closed or open, so typically one would use $(z_{(N-B+1)}, z_{(N-A)})$.

For large N, we can use the normal approximation to estimate A and B. These are virtually never exact integers, so a good idea is to choose the closest integers that give the widest interval. That is, use

$$A = \text{floor}\left(\frac{N}{2} - z_{\alpha/2}\sqrt{\frac{N}{4}}\right) \quad \text{and} \quad B = \text{ceiling}\left(\frac{N}{2} + z_{\alpha/2}\sqrt{\frac{N}{4}}\right), \tag{18.55}$$

where floor(x) is the largest integer less than or equal to x, and ceiling(x) is the smallest integer greater than or equal to x.

18.4.1   Kendall's τ and the slope

A similar approach yields a nonparametric confidence interval for the slope in a regression. We will assume fixed x and random Y, so that the model is

$$Y_i = \alpha + \beta x_i + E_i, \quad i = 1, \dots, N, \quad E_i \text{ are iid with continuous distribution } F. \tag{18.56}$$

We allow ties in $\mathbf{x}$, but with continuous F are assuming there are no ties in the $\mathbf{Y}$. Kendall's τ and Spearman's ρ can be used to test that $\beta = 0$, and can also be repurposed to test $H_0 : \beta = \beta_0$ for any β_0. Writing $Y_i - \beta_0 x_i = \alpha + (\beta - \beta_0)x_i + E_i$, we can test that null by replacing Y_i with $Y_i - \beta_0 x_i$ in the statistic. For Kendall, it is easiest to use the distance d, where

$$d(\mathbf{x}, \mathbf{y}; \beta_0) = \sum_{1 \le i < j \le N} \sum I[(x_i - x_j)((y_i - \beta_0 x_i) - (y_j - \beta_0 x_j)) > 0]. \tag{18.57}$$

Exercise 18.5.18 shows that we can write

$$d(\mathbf{x}, \mathbf{y}; \beta_0) = \sum_{\substack{1 \le i < j \le N \\ x_i \ne x_j}} \sum I[b_{ij} > \beta_0], \tag{18.58}$$

where b_{ij} is the slope of the line segment connecting points i and j, $b_{ij} = (y_i - y_j)/(x_i - x_j)$. Figure 18.1 illustrates these segments. Under the null, this d has the distribution of the Jonckheere-Terpstra statistic in Section 18.3.2. The analog to the confidence interval defined in (18.54) is $(b_{(N-B+1)}, b_{(N-A)})$, where we use the null distribution of $d(\mathbf{x}, \mathbf{Y}; \beta_0)$ in place of the Binomial($N, 1/2$)'s in (18.53). Here, $b_{(k)}$ is the kth order statistic of the b_{ij}'s for i, j in the summation in (18.58).

Using the asymptotic approximation in (18.49), we have

$$A = \text{floor}(\mu(\mathbf{c}) - z_{\alpha/2}\sigma(\mathbf{c})) \quad \text{and} \quad B = \text{ceiling}(\mu(\mathbf{c}) + z_{\alpha/2}\sigma(\mathbf{c})). \tag{18.59}$$

The mean $\mu(\mathbf{c})$ and variance $\sigma^2(\mathbf{c})$ are given in (18.48), where $\mathbf{c}$ is the pattern of ties for $\mathbf{x}$.

An estimate of β can be obtained by shrinking the confidence interval down to a point. Equivalently, it is the β_0 for which the normalized test statistic JT_N of (18.49) is 0, or as close to 0 as possible. This value can be seen to be the median of the b_{ij}'s, and is known as the **Sen-Theil estimator** of the slope (Theil, 1950; Sen, 1968). See Exercise 18.5.18.

Ties in the $\mathbf{y}$ are technically not addressed in the above work, since the distributions are assumed continuous. The problem of ties arises only when there are ties in the $(y_i - \beta_0 x_i)$'s. Except for a finite set of β_0's, such ties occur only for i and j with $(x_i, y_i) = (x_j, y_j)$. Thus if such tied pairs are nonexistent or rare, we can proceed as if there are no ties.

To illustrate, consider the hurricane data with x = damage and y = deaths as in Figure 12.2 (page 192) and the table in (12.74). The pattern of ties for $\mathbf{x}$ is $\mathbf{c}$ = $(2,2,2,2)$, and for $\mathbf{y}$ is $(2, 2, 2, 2, 2, 2, 2, 2, 3, 4, 4, 5, 8, 10, 12, 13)$, indicating quite a few ties. But there is only one tie among the (x_i, y_i) pairs, so we use the calculations for no ties in $\mathbf{y}$. Here, $N = 94$, so that $\mu(\mathbf{c}) = 2183.5$ and $\sigma^2(\mathbf{c}) = 23432.42$, and from (18.59), $A = 1883$ and $B = 2484$. There are 4367 b_{ij}'s, so as in (18.54) but with N replaced by 4367, we have confidence interval and estimate

$$C(\mathbf{x}, \mathbf{y}) = (b_{(1884)}, b_{(2484)}) = (0.931, 2.706) \text{ and } \hat{\beta} = \text{Median}\{b_{ij}\} = 1.826. \quad (18.60)$$

In (12.74) we have estimates and standard errors of the slope using least squares or median regression, with and without the outlier, Katrina. Below we add the current Sen-Theil estimator as above, and that without the outlier. The confidence intervals for the original estimates use $\pm 1.96\, se$ for consistency.

	Estimate	Confidence interval	
Least squares	7.743	$(4.725, 10.761)$	
Least squares w/o outlier	1.592	$(0.734, 2.450)$	
Least absolute deviations	2.093	$(-0.230, 4.416)$	(18.61)
Least absolute deviations w/o outlier	0.803	$(-0.867, 2.473)$	
Sen-Theil	1.826	$(0.931, 2.706)$	
Sen-Theil w/o outlier	1.690	$(0.884, 2.605)$	

Of the three estimating techniques, Sen-Theil is least affected by the outlier. It is also quite close to least squares when the outlier is removed.

18.5   Exercises

Exercise 18.5.1. Consider the randomization distribution of the signed-rank statistic given in (18.11), where there are no ties in the z_i's. That is, $\mathbf{r}$ is a permutation of the integers from 1 to N, and

$$T(\mathbf{Gz}) =^{\mathcal{D}} \sum_{i=1}^{N} iG_i, \quad (18.62)$$

where the G_i's are independent with $P[G_i = -1] = P[G_i = +1] = 1/2$. (a) Show that $E[T(\mathbf{Gz})] = 0$ and $Var[T(\mathbf{Gz})] = N(N+1)(2N+1)/6$. [Hint: Use the formula for $\sum_{i=1}^{K} i^2$.] (b) Apply Theorem 17.3 on page 299 with $\nu = 3$ to show that $V_N \longrightarrow^{\mathcal{D}} N(0,1)$ in (18.13), where $V_N = T(\mathbf{Gz})/\sqrt{Var[T(\mathbf{Gz})]}$. [Hint: In this case, $X_i = iG_i$.

Use the bound $E[|X_i|^3] \leq i^3$ and the formula for $\sum_{i=1}^{K} i^3$ to show (17.56).] (c) Find $T(\mathbf{z})$ and V_N for the tread-wear data in (17.34). What is the two-sided p-value based on the normal approximation? What do you conclude?

Exercise 18.5.2. In the paper Student (1908), W.S. Gosset presented the Student's t distribution, using the alias Student because his employer, Guinness, did not normally allow employees to publish (fearing trade secrets would be revealed). One example compared the yields of barley depending on whether the seeds were kiln-dried or not. Eleven varieties of barley were used, each having one batch sown with regular seeds, and one with kiln-dried seeds. Here are the results, in pounds per acre:

Regular	Kiln-dried	
1903	2009	
1935	1915	
1910	2011	
2496	2463	
2108	2180	
1961	1925	(18.63)
2060	2122	
1444	1482	
1612	1542	
1316	1443	
1511	1535	

Consider the differences, $X_i = \text{Regular}_i - \text{Kiln-dried}_i$. Under the null hypothesis that the two methods are exchangeable as in (17.32), these X_i's are distributed symmetrically about the median 0. Calculate the test statistics for the sign test, signed-rank test, and regular t-test. (For the first two, normalize by subtracting the mean and dividing by the standard deviation of the statistic, both calculated under the null.) Find p-values of these statistics for the two-sided alternative, where for the first two statistics you can approximate the null distribution with the standard normal. What do you see?

Exercise 18.5.3. Suppose $\mathbf{z}$, $N \times 1$, has no ties and no zeroes. This exercise shows that the signed-rank statistic in (18.10) can be equivalently represented by

$$T^*(\mathbf{z}) = \sum_{1 \leq i \leq j \leq N} I[z_i + z_j > 0]. \tag{18.64}$$

Note that the summation includes $i = j$. (a) Letting $\mathbf{r} = \text{Rank}(|z_1|, \ldots, |z_N|)$, show that

$$T(\mathbf{z}) \equiv \sum_{i=1}^{N} \text{Sign}(z_i) r_i = 2 \sum_{i=1}^{N} I[z_i > 0] r_i - \frac{N(N+1)}{2}. \tag{18.65}$$

[Hint: Note that with no zeroes, $\text{Sign}(z_i) = 2I[z_i > 0] - 1$.] (b) Use the definition of rank given in (18.7) to show that

$$\sum_{i=1}^{N} I[z_i > 0] r_i = \#\{z_i > 0\} + \sum \sum_{j \neq i} I[z_i > 0] I[|z_i| > |z_j|]. \tag{18.66}$$

(c) Show that $I[z_i > 0] I[|z_i| > |z_j|] = I[z_i > |z_j|]$, and

$$\sum \sum_{j \neq i} I[z_i > 0] I[|z_i| > |z_j|] = \sum \sum_{i < j} (I[z_i > |z_j|] + I[z_j > |z_i|]). \tag{18.67}$$

(d) Show that $I[z_i > |z_j|] + I[z_j > |z_i|] = I[z_i + z_j > 0]$. [Hint: Write out all the possibilities, depending on the signs of z_i and z_j and their relative absolute values.]
(e) Verify that $T(\mathbf{z}) = 2T^*(\mathbf{z}) - N(N+1)/2$. (f) Use (18.12) and Exercise 18.5.1 to show that under the null, $E[T^*(\mathbf{Gz})] = N(N+1)/2$ and $Var[T^*(\mathbf{Gz})] = N(N+1)(2N+1)/24]$.

Exercise 18.5.4. This exercise shows the equivalence of the two Mann-Whitney/ Wilcoxon statistics in (18.20) and (18.21). We have $\mathbf{r} = \text{Rank}(x_1,\dots,x_n,y_1,\dots,y_m)$, where $N = n + m$. We assume there are no ties among the N observations. (a) Show that

$$W_N = \frac{1}{n}\sum_{i=1}^{n} r_i - \frac{1}{m}\sum_{i=n+1}^{N} r_i = \left(\frac{1}{n} + \frac{1}{m}\right)\sum_{i=1}^{n} r_i - \frac{N(N+1)}{2m}. \tag{18.68}$$

[Hint: Since there are no ties, $\sum_{i=n+1}^{N} r_i = c_N - \sum_{i=1}^{n} r_i$ for a known constant c_N.] (b) Using the definition of rank in (18.7), show that

$$\sum_{i=1}^{n} r_i = \sum_{i=1}^{n}\left(1 + \sum_{1 \le j \le n, j \ne i} I[x_i > x_j]\right) + W_N^*, \quad \text{where } W_N^* = \sum_{i=1}^{n}\sum_{j=1}^{m} I[x_i > y_j]. \tag{18.69}$$

Then note that the summation on the right-hand side of the first equation in (18.69) equals $n(n+1)/2$. (c) Conclude that $W_N = N(W_N^*/(nm) - 1/2)$.

Exercise 18.5.5. Continue with consideration of the Mann-Whitney/Wilcoxon statistic as in (18.20), and again assume there are no ties. (a) Show that $Var[W_N] = N^2(N+1)/(12nm)$ as used in (18.23). (b) Assume that $n/N \to p \in (0,1)$ as $N \to \infty$. Apply Theorem 17.2 (page 298) to show that $V_N \xrightarrow{D} N(0,1)$ for V_N in (18.23). [Hint: The condition (17.50) holds here, hence only (17.51) needs to be verified for $\mathbf{z}$ being $\mathbf{r}$.]

Exercise 18.5.6. The BMI (Body Mass Index) was collected on 165 people, $n = 62$ men (the x_i's) and $m = 103$ women (the y_i's). The W_N from (18.20) is 43.38. (You can assume there were no ties.) (a) Under the null hypothesis in (18.18), what are the mean and variance of W_N? (b) Use the statistic V_N from (18.23) to test the null hypothesis in part (a), using a two-sided alternative. What do you conclude?

Exercise 18.5.7. Let $d(\mathbf{x},\mathbf{y}) = \sum\sum_{1 \le i < j \le N} I[(x_i - x_j)(y_i - y_j) < 0]$ as in (18.28). Assume that there are no ties among the x_i's nor among the y_i's. (a) Show that

$$\sum\sum_{1 \le i < j \le N} I[(x_i - x_j)(y_i - y_j) > 0] = \binom{N}{2} - d(\mathbf{x},\mathbf{y}). \tag{18.70}$$

(b) Show that

$$\sum\sum_{1 \le i < j \le N} \text{Sign}(x_i - x_j)\text{Sign}(y_i - y_j) = \binom{N}{2} - 2d(\mathbf{x},\mathbf{y}). \tag{18.71}$$

(c) Conclude that $\hat{\tau}(\mathbf{x},\mathbf{y})$ in (18.27) equals $1 - 4d(\mathbf{x},\mathbf{y})/(N(N-1))$ as in (18.29).

Exercise 18.5.8. (a) Suppose X is a random variable, and $a_1,\dots,a_K$ (K may be ∞) are the points at which X has positive probability. Let X_1 and X_2 be iid with the same distribution as X. Show that $E[\text{Sign}(X_1 - X_2)^2] = 1 - \sum_{k=1}^{K} P[X = a_k]^2$, proving (18.35). (b) Let $\mathbf{x}$ be $N \times 1$ with pattern of ties $(c_1, \ldots, c_K)$ as below (18.38). Show that $\sum\sum_{i \neq j} \text{Sign}(x_i - x_j)^2 = N(N-1) - \sum_{k=1}^{K} c_k(c_k - 1)$. [Hint: There are $N(N-1)$ terms in the summation, and they are 1 unless $x_i = x_j$. So you need to subtract the number of such tied pairs.]

Exercise 18.5.9. Let $\mathbf{x} = (1,2,2,3)'$ and $\mathbf{y} = (1,1,2,3)'$, and $\widehat{\tau}(\mathbf{x}, \mathbf{y})$ be Kendall's τ_B from (18.39). Note $\mathbf{x}$ and $\mathbf{y}$ have the same pattern of ties. (a) What is the value of $\widehat{\tau}(\mathbf{x}, \mathbf{y})$? (b) What are the minimum and maximum values of $\widehat{\tau}(\mathbf{x}, \mathbf{py})$ as $\mathbf{p}$ ranges over the 4×4 permutation matrices? Are they ± 1? (c) Let $\mathbf{w} = (3,3,4,7)'$. What is $\widehat{\tau}(\mathbf{w}, \mathbf{y})$? What are the minimum and maximum values of $\widehat{\tau}(\mathbf{w}, \mathbf{py})$? Are they ± 1?

Exercise 18.5.10. Show that (18.42) holds for $\widehat{\tau}$ in (18.39) when we drop the final two summands in the expression for $Var[S(\mathbf{x}, \mathbf{Py})]$ in (18.41).

Exercise 18.5.11. Let $U_1, \ldots, U_{N-1}$ be independent with $U_i \sim$ Discrete Uniform$(0, N-i)$, as in (18.30), and set $U = \sum_{i=1}^{N} U_i$. (a) Show that $E[U] = N(N-1)/4$. (b) Show that $Var[U] = N(N-1)(2N+5)/72$. [Hint: Table 1.2 on page 9 has the mean and variance of the discrete uniform. Then use the formulas for $\sum_{i=1}^{K} i$ for the mean and $\sum_{i=1}^{K} i^2$ for the variance.] (c) Show that $\sum_{i=1}^{N} E[|U_i - E[U_i]|^3] / \sqrt{Var[U]}^3 \to 0$ as $N \to \infty$. [Hint: First show that $E[|U_i - E[U_i]|^3] \leq |N-i|^3/8$. Then use the formula for $\sum_{i=1}^{K} i^3$ to show that $\sum_{i=1}^{N} E[|U_i - E[U_i]|^3] \leq N^4/32$. Finally, use part (b) to show the desired limit.] (d) Let $K_N = (U - E[U])/\sqrt{Var[U]}$. Use Theorem 17.3 on page 299 and part (c) to show that $K_N \longrightarrow^{\mathcal{D}} N(0,1)$ as $N \to \infty$. Argue that therefore the asymptotic normality of $d(\mathbf{x}, \mathbf{Py})$ as in (18.33) holds.

Exercise 18.5.12. This exercise proves the asymptotic normality of the Jonckheere-Terpstra statistic. Assume there are no ties in the z_i's, and $n_k/N \to \lambda_k \in (0,1)$ for each k. (a) Start with the representation given in (18.47). Find the constants $c_N, d_{1N}, \ldots, d_{KN}$ such that

$$W_N = c_N JT_N + W_N^* \quad \text{where} \quad W_N^* = \sum_{k=1}^{K} d_{kN} W_{kN}, \qquad (18.72)$$

JT_N is the normalized Jonckheere-Terpstra statistic in (18.49), and the W's are the normalized Kendall distances,

$$W_N = \frac{d(\mathbf{e}_N, \mathbf{Pz}_N) - E[d(\mathbf{e}_N, \mathbf{Pz}_N)]}{\sqrt{Var[d(\mathbf{e}_N, \mathbf{Pz}_N)]}} \quad \text{and} \quad W_{kN} = \frac{d(\mathbf{e}_{n_k}, (\mathbf{Pz})_k) - E[d(\mathbf{e}_{n_k}, (\mathbf{Pz})_k)]}{\sqrt{Var[d(\mathbf{e}_{n_k}, (\mathbf{Pz})_k)]}},$$

$$\qquad (18.73)$$

$k = 1, \ldots, K$. (b) Show that $d_{kN}^2 \to \lambda_k^3$ and $c_N^2 \to 1 - \sum_{k=1}^{K} \lambda_k^3$. (c) We know that the W's are asymptotically $N(0,1)$. Why is $W_N^* \longrightarrow^{\mathcal{D}} N(0, \sum_{k=1}^{K} \lambda_k^3)$? (d) Show that $c_N JT_N \longrightarrow^{\mathcal{D}} N(0, 1 - \sum_{k=1}^{K} \lambda_k^3)$. [Hint: Use moment generating functions on both sides of the expression for W_N in (18.72), noting that JT_N and W_N^* are independent. Then we know the mgfs of the asymptotic limits of W_N and W_N^*, hence can find that of $c_N JT_N$.] (e) Finally, use parts (b) and (d) to show that $JT_N \longrightarrow^{\mathcal{D}} N(0,1)$.

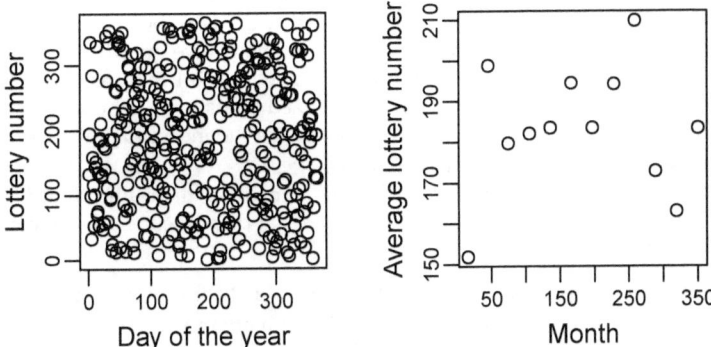

Figure 18.2: The 1970 draft lottery results. The first plot has day of the year on the X-axis and lottery number on the Y-axis. The second plots the average lottery number for each month.

Exercise 18.5.13. Consider the draft lottery data of Section 17.3. Letting **x** be the days of the year and **y** be the corresponding lottery numbers, we calculate $d(\mathbf{x}, \mathbf{y}) = 38369$ for d being Kendall's distance (18.28). (a) Find Kendall's τ. (Here, $N = 366$.) (b) Find the value of the standard deviation of Kendall's τ under the null hypothesis that the lottery was totally random, and the normalized statistic $\tau / \sqrt{Var[\tau]}$. What is the approximate p-value based on Kendall's τ? Is the conclusion different than that found below (17.23)? (c) In 1970, they tried to do a better randomization. There were two bowls of capsules, one for the days of the year, and one for the lottery numbers. They were both thoroughly mixed, and lottery numbers were assigned to dates by choosing one capsule from each bowl. See Figure 18.2. This time Kendall's distance between the days and the lottery numbers was 32883. Now $N = 365$, since the lottery was only for people born in 1951. What is Kendall's τ and its standard deviation for this year? What is the approximate two-sided p-value for testing complete randomness? What do you conclude? (d) Which year had a better randomization? Or were they about equally good?

Exercise 18.5.14. Continue with the draft lottery example from Exercise 18.5.13. The description of the randomization process at the end of Section 17.3 suggests that there may be a trend over months, but leaves open the possibility that there is no trend within months. To assess this idea, decompose the overall distance $d(\mathbf{x}, \mathbf{y}) = 38369$ from above as in (18.47), where **a** as in (18.45) indicates month, so that $K = 12$ and n_k is the number of days in the kth month. The between-month Kendall distance is $d(\mathbf{a}, \mathbf{z}) = 35787$, and the within-month distances are given in the table:

Month	1	2	3	4	5	6	7	8	9	10	11	12
n_k	31	29	31	30	31	30	31	31	30	31	30	31
$d(\mathbf{e}_{n_k}, \mathbf{z}_k)$	260	172	202	195	212	215	237	195	186	278	173	257

$$(18.74)$$

(a) Normalize the Jonckheere-Terpstra statistic $d(\mathbf{a}, \mathbf{z})$ to obtain JT_N as in (18.49). Find the two-sided p-value based on the normal approximation. What do you conclude?

(b) Test for trend within each month. Do there appear to be any months where there is a significant trend? (c) It may be that the monthly data can be combined for a more powerful test. There are various ways to combine the twelve months into one overall test for trend. If looking for the same direction within all months, then summing the $d(\mathbf{e}_{n_k}, \mathbf{z}_k)$'s is reasonable. Find the normalized statistic based on this sum. Is it statistically significant? (d) Another way to combine the individual months is to find the sum of squares of the individual statistics, i.e., $\sum_{k=1}^{12} W_{kN}^2$, for W_{kN}'s as in (18.73). What is the asymptotic distribution of this statistic under the null? What is its value for these data? Is it statistically significant? (e) What do you conclude concerning the issue of within-month and between-month trends.

Exercise 18.5.15. Suppose $Z_1, \ldots, Z_N$ are iid with distribution symmetric about the median η, which means $Z_i - \eta =^{\mathcal{D}} \eta - Z_i$. (If the median is not unique, let η be the middle of the interval of medians.) We will use the signed-rank statistic as expressed in (18.64) to find a confidence interval and estimate for η. (a) Show that for testing $H_0 : \eta = \eta_0$, we can use the statistic

$$T^*(\mathbf{z}; \eta_0) = \sum\sum_{1 \leq i \leq j \leq N} I[(z_i + z_j)/2 > \eta_0]. \qquad (18.75)$$

(b) Using the normal approximation, find a and b so that $C(\mathbf{z}) = \{\eta_0 \,|\, a < T^*(\mathbf{z}; \eta_0) < b\}$ is an approximate $100(1 - \alpha)\%$ confidence interval for η. [Hint: See Exercise 18.5.3(f).] (c) With $A = \text{floor}(a)$ and $B = \text{ceiling}(b)$, the confidence interval becomes $(w_{(N-B+1)}, w_{(N-A)})$, where the $w_{(i)}$'s are the order statistics of which quantities? (d) What is the corresponding estimate of η?

Exercise 18.5.16. Here we look at a special case of the two-sample situation as in (18.14). Let F be a continuous distribution function, and suppose that for "shift" parameter δ, $X_1, \ldots, X_n$ are iid with distribution function $F(x_i - \delta)$, and $Y_1, \ldots, Y_m$ are iid with distribution function $F(y_i)$. Also, the X_i's are independent of the Y_i's. The goal is to find a nonparametric confidence interval and estimate for δ. (a) Show that X_i is stochastically larger than Y_j if $\delta > 0$, and Y_j is stochastically larger than X_i if $\delta < 0$. (b) Show that $\text{Median}(X_i) = \text{Median}(Y_j) + \delta$. (c) Consider testing the hypotheses $H_0 : \delta = \delta_0$ versus $H_A : \delta \neq \delta_0$. Show that we can base the test on the Wilcoxon/Mann-Whitney statistic as in (18.21) given by

$$W_N^*(\mathbf{z}; \delta_0) = \sum_{i=1}^{n}\sum_{j=1}^{m} I[x_i - y_j > \delta_0], \qquad (18.76)$$

where $\mathbf{z} = (x_1, \ldots, x_n, y_1, \ldots, y_m)'$. (d) Find $E[W_N^*(\mathbf{Pz}; \delta_0)]$ and $Var[W_N^*(\mathbf{Pz}; \delta_0)]$ under the null. [Hint: See (18.21) and therebelow.] (e) Using the normal approximation, find A and B so that $(w_{(N-B+1)}, w_{(N-A)})$ is an approximate $100(1 - \alpha)\%$ confidence interval for δ. What are the $w_{(i)}$'s? (f) What is the corresponding estimate of δ?

Exercise 18.5.17. Let $z_{(1)}, \ldots, z_{(N)}$ be the order statistics from the sample $z_1, \ldots, z_N$, and A and B integers between 0 and N inclusive. (a) Show that for integer K,

$$\eta_0 < z_{(K)} \quad \text{if and only if} \quad \#\{i \,|\, \eta_0 < z_i\} \geq N - K + 1, \qquad (18.77)$$

hence $\sum_{i=1}^{N} I[\eta_0 < z_i] \geq A + 1$ if and only if $\eta_0 < z_{(N-A)}$. (b) Show that (18.77) is equivalent to

$$\eta_0 \geq z_{(K)} \quad \text{if and only if} \quad \#\{i \,|\, \eta_0 < z_i\} \leq N - K. \qquad (18.78)$$

Conclude that $\sum_{i=1}^{N} I[\eta_0 < z_i] \leq B - 1$ if and only if $\eta_0 \geq z_{(N-B+1)}$. (c) Argue that parts (a) and (b) prove the confidence interval formula in (18.54).

Exercise 18.5.18. Consider the linear model $Y_i = \alpha + \beta x_i + E_i$, where the E_i are iid F, and F is continuous. Suppose that the x_i's are distinct (no ties). The least squares estimates are those for which the estimated residuals have zero correlation with the x_i's. An alternative (and more robust) estimator of the slope β finds $\widehat{\beta}$ so that the estimated residuals have zero Kendall's τ with the x_i's. It is called the Sen-Theil estimator, as mentioned in Section 18.4.1. The residuals are the $Y_i - \alpha - \beta x_i$, so the numerator of Kendall's τ between the residuals and the x_i's (which depends on β but not α) is

$$U(\beta) = \sum_{i=1}^{n-1} \sum_{j=i+1}^{n} \text{Sign}((y_i - \beta x_i) - (y_j - \beta x_j)) \, \text{Sign}(x_i - x_j). \qquad (18.79)$$

Then $\widehat{\beta}$ satisfies $U(\widehat{\beta}) = 0$. (Or at least as close to 0 as possible.) (a) Show that

$$\text{Sign}((y_i - \beta x_i) - (y_j - \beta x_j)) \, \text{Sign}(x_i - x_j) = \text{Sign}(b_{ij} - \beta), \qquad (18.80)$$

where $b_{ij} = (y_i - y_j)/(x_i - x_j)$ is the slope of the line segment connecting points i and j. (Note that $\text{Sign}(a) \, \text{Sign}(b) = \text{Sign}(ab)$.) (b) Now $\widehat{\beta}$ is what familiar statistic of the b_{ij}'s?

Part III

Optimality

Chapter 19

Optimal Estimators

So far, our objectives have been mostly methodological, that is, we have a particular model and wish to find and implement procedures to infer something about the parameters. A more meta goal of mathematical statistics is to consider the procedures as the data, and try to find the best procedure(s), or at least evaluate how effective procedures are relative to each other or absolutely. Statistical decision theory is an offshoot of game theory that attempts to do such comparison of procedures. We have seen a little of this idea when comparing the asymptotic efficiencies of the median and the mean in (9.32), (14.74), and (14.75).

Chapter 20 presents the general decision-theoretic approach. Before then we will look at some special optimality results for estimation: best unbiased estimators and best shift-equivariant estimators.

Our first goal is to find the best unbiased estimator, by which we mean the unbiased estimator that has the lowest possible variance for all values of the parameter among all unbiased estimators. Such an estimator is called a **uniformly minimum variance unbiased estimator (UMVUE)**. In Section 19.2 we introduce the concept of *completeness* of a model, which together with sufficiency is key to finding UMVUEs. Section 19.5 gives a lower bound on the variance of an unbiased estimator, which can be used as a benchmark even if there is no UMVUE.

Throughout this chapter we assume we have the basic statistical model:

$$\text{Random vector } \mathbf{X}, \text{ space } \mathcal{X}, \text{ and set of distributions } \mathcal{P} = \{P_{\boldsymbol{\theta}} \mid \boldsymbol{\theta} \in \mathcal{T}\}, \qquad (19.1)$$

where $\mathcal{T}$ is the parameter space. We wish to estimate a function of the parameter, $g(\boldsymbol{\theta})$. Repeating (11.3), an estimator is a function δ of $\mathbf{x}$:

$$\delta : \mathcal{X} \longrightarrow \mathcal{A}, \qquad (19.2)$$

where here $\mathcal{A}$ is generally the space of $g(\boldsymbol{\theta})$. It may be that $\mathcal{A}$ is somewhat larger than that space, e.g., in the binomial case where $\theta \in (0,1)$, one may wish to allow estimators in the range [0,1]. As in (10.10), an estimator is unbiased if

$$E_{\boldsymbol{\theta}}[\delta(\mathbf{X})] = g(\boldsymbol{\theta}) \text{ for all } \boldsymbol{\theta} \in \mathcal{T}. \qquad (19.3)$$

Next is the formal definition of UMVUE.

Definition 19.1. *The procedure δ in (19.2) is a **uniformly minimum variance unbiased estimator (UMVUE)** of the function $g(\theta)$ if it is unbiased, has finite variance, and for any other unbiased estimator δ',*

$$Var_{\theta}[\delta(\mathbf{X})] \leq Var_{\theta}[\delta'(\mathbf{X})] \quad for\ all\ \ \theta \in \mathcal{T}. \tag{19.4}$$

19.1 Unbiased estimators

The first step in finding UMVUEs is to find the unbiased estimators. There is no general automatic method for finding unbiased estimators, in contrast to maximum likelihood estimators or Bayes estimators. The latter two types may be difficult to calculate, but it is a mathematical or computational problem, not a statistical one.

One method that occasionally works for finding unbiased estimators is to find a power series based on the definition of unbiasedness. For example, suppose $X \sim$ Binomial(n, θ), $\theta \in (0, 1)$. We know X/n is an unbiased estimator for θ, but what about estimating the variance of X, $g(\theta) = \theta(1 - \theta)/n$? An unbiased δ will satisfy

$$\sum_{x=0}^{n} \delta(x) \binom{n}{x} \theta^{x}(1-\theta)^{n-x} = \frac{1}{n}\theta(1-\theta) \quad for\ all\ \ \theta \in (0,1). \tag{19.5}$$

Both sides of equation (19.5) are polynomials in θ, hence for equality to hold for every θ, the coefficients of each power θ^{k} must match.

First, suppose $n = 1$, so that we need

$$\delta(0)(1-\theta) + \delta(1)\theta = \theta(1-\theta) = \theta - \theta^{2}. \tag{19.6}$$

There is no δ that will work, because the coefficient of θ^{2} on the right-hand side is -1, and on the left-hand side is 0. That is, with $n = 1$, there is no unbiased estimator of $\theta(1 - \theta)$. With $n = 2$, we have

$$
\begin{array}{ccc}
\delta(0)(1-\theta)^{2} + 2\delta(1)\theta(1-\theta) + \delta(2)\theta^{2} & = & \theta(1-\theta)/2 \\
\| & & \| \\
\delta(0) + (-2\delta(0) + 2\delta(1))\theta + (\delta(0) - 2\delta(1) + \delta(2))\theta^{2} & \leftrightarrow & (\theta - \theta^{2})/2.
\end{array}
\tag{19.7}
$$

Matching coefficents of θ^{k}:

k	Left-hand side	Right-hand side
0	$\delta(0)$	0
1	$-2\delta(0) + 2\delta(1)$	1/2
2	$\delta(0) - 2\delta(1) + \delta(2)$	-1/2

(19.8)

we see that the only solution is $\delta_{0}(0) = \delta_{0}(2) = 0$ and $\delta_{0}(1) = 1/4$, which actually is easy to see directly from the first line of (19.7). Thus the δ_{0} must be the best unbiased estimator, being the only one. In fact, because for any function $\delta(x)$, $E_{\theta}[\delta(X)]$ is an nth-degree polynomial in θ, the only functions $g(\theta)$ that have unbiased estimators are those that are themselves polynomials of degree n or less (see Exercise 19.8.2), and each one is a UMVUE by the results in Section 19.2. For example, $1/\theta$ and e^{θ} do not have unbiased estimators.

Another approach to finding an unbiased estimator is to take a biased one and see if it can be modified. For example, suppose $X_{1}, \dots, X_{n}$ are iid Poisson(θ), and consider estimating $g(\theta) = P_\theta[X_1 = 0] = \exp(-\theta)$. Thus if X_i is the number of phone calls in a hour, $g(\theta)$ is the chance that there are no calls in the next hour. (We saw something similar in Exercise 11.7.6.) The MLE of $g(\theta)$ is $\exp(-\overline{X})$, which is biased. But one can find a c such that $\exp(-c\overline{X})$ is unbiased. See Exercise 19.8.3.

We have seen that there may or may not be an unbiased estimator. Often, there are many. For example, suppose $X_1, \ldots, X_n$ are iid Poisson(θ). Then $\overline{X}$ and S_*^2 (the sample variance with $n - 1$ in the denominator) are both unbiased, being unbiased estimators of the mean and variance, respectively. Weighted averages of the X_i's are also unbiased, e.g., X_1, $(X_1 + X_2)/2$, and $X_1/2 + X_2/6 + X_3/3$ are all unbiased. The rest of this chapter uses sufficiency and the concept of completeness to find UMVUEs in many situations.

19.2 Completeness and sufficiency

We have already answered the question of finding the best unbiased estimator in certain situations without realizing it. We know from the Rao-Blackwell theorem (Theorem 13.8 on page 210) that any estimator that is not a function of just the sufficient statistic can be improved. That is, if $\delta(\mathbf{X})$ is an unbiased estimator of $g(\boldsymbol{\theta})$, and $\mathbf{S} = s(\mathbf{X})$ is a sufficient statistic, then

$$\delta^*(\mathbf{s}) = E[\delta(\mathbf{X}) \mid s(\mathbf{X}) = \mathbf{s}] \tag{19.9}$$

is also unbiased, and has no larger variance than δ.

Also, if there is only one unbiased estimator that is a function of the sufficient statistic, then it must be the best one that depends on only the sufficient statistic. Furthermore, it must be the best overall, because it is better than any estimator that is not a function of just the sufficient statistic.

The concept of "there being only one unbiased estimator" is called **completeness**. It is a property attached to a model. Consider the model with random vector $\mathbf{Y}$ and parameter space $\mathcal{T}$. Suppose there are at least two unbiased estimators of some function $g(\boldsymbol{\theta})$ in this model, say $\delta_g(\mathbf{y})$ and $\delta_g^*(\mathbf{y})$. That is,

$$P_{\boldsymbol{\theta}}[\delta_g(\mathbf{Y}) \neq \delta_g^*(\mathbf{Y})] > 0 \text{ for some } \boldsymbol{\theta} \in \mathcal{T}, \tag{19.10}$$

and

$$E_{\boldsymbol{\theta}}[\delta_g(\mathbf{Y})] = g(\boldsymbol{\theta}) = E_{\boldsymbol{\theta}}[\delta_g^*(\mathbf{Y})] \text{ for all } \boldsymbol{\theta} \in \mathcal{T}. \tag{19.11}$$

Then

$$E_{\boldsymbol{\theta}}[\delta_g(\mathbf{Y}) - \delta_g^*(\mathbf{Y})] = 0 \text{ for all } \boldsymbol{\theta} \in \mathcal{T}. \tag{19.12}$$

This $\delta_g(\mathbf{y}) - \delta_g^*(\mathbf{y})$ is an unbiased estimator of 0. Now suppose $\delta_h(\mathbf{y})$ is an unbiased estimator of the function $h(\boldsymbol{\theta})$. Then so is

$$\delta_h^*(\mathbf{y}) = \delta_h(\mathbf{y}) + \delta_g(\mathbf{y}) - \delta_g^*(\mathbf{y}), \tag{19.13}$$

because

$$E_{\boldsymbol{\theta}}[\delta_h^*(\mathbf{Y})] = E_{\boldsymbol{\theta}}[\delta_h(\mathbf{Y})] + E_{\boldsymbol{\theta}}[\delta_g(\mathbf{Y}) - \delta_g^*(\mathbf{Y})] = h(\boldsymbol{\theta}) + 0. \tag{19.14}$$

That is, if there is more than one unbiased estimator of one function, then there is more than one unbiased estimator of any other function (that has at least one unbiased estimator). Logically, it follows that if there is only one unbiased estimator of some function, then there is only one (or zero) unbiased estimator of any function. That one function may as well be the zero function.

Definition 19.2. *Suppose for the model on* **Y** *with parameter space* $\mathcal{T}$, *the only unbiased estimator of 0 is 0 itself. That is, suppose*

$$E_{\boldsymbol\theta}[\delta(\mathbf{Y})] = 0 \ \text{ for all } \ \boldsymbol\theta \in \mathcal{T} \tag{19.15}$$

implies that

$$P_{\boldsymbol\theta}[\delta(\mathbf{Y}) = 0] = 1 \ \text{ for all } \ \boldsymbol\theta \in \mathcal{T}. \tag{19.16}$$

Then the model is **complete**.

The important implication follows.

Lemma 19.3. *Suppose the model is complete. Then there exists at most one unbiased estimator of any function* $g(\boldsymbol\theta)$.

Illustrating with the binomial again, suppose $X \sim \text{Binomial}(n, \theta)$ with $\theta \in (0, 1)$, and $\delta(x)$ is an unbiased estimator of 0:

$$E_\theta[\delta(X)] = 0 \ \text{ for all } \ \theta \in (0, 1). \tag{19.17}$$

We know the left-hand side is a polynomial in θ, as is the right-hand side. All the coefficients of θ^i are zero on the right, hence on the left. Write

$$E_\theta[\delta(X)] = \delta(0)\binom{n}{0}(1-\theta)^n + \delta(1)\binom{n}{1}\theta(1-\theta)^{n-1}$$
$$+ \cdots + \delta(n-1)\binom{n}{n-1}\theta^{n-1}(1-\theta) + \delta(n)\binom{n}{n}\theta^n. \tag{19.18}$$

The coefficient of θ^0, i.e, the constant, arises from just the first term, so is $\delta(0)\binom{n}{0}$. For that to be 0, we have $\delta(0) = 0$. Erasing that first term, we see that the coefficient of θ is $\delta(1)\binom{n}{1}$, hence $\delta(1) = 0$. Continuing, we see that $\delta(2) = \cdots = \delta(n) = 0$, which means that $\delta(x) = 0$, which means that the only unbiased estimator of 0 is 0 itself. Hence this model is complete, verifying (with Exercise 19.8.2) the fact mentioned below (19.8) that $g(\theta)$ has a UMVUE if and only if it is a polynomial in θ of degree less than or equal to n.

19.2.1   Poisson distribution

Suppose $X_1, \ldots, X_n$ are iid Poisson(θ), $\theta \in (0, \infty)$, with $n > 1$. Is this model complete? No. Consider $\delta(\mathbf{X}) = x_1 - x_2$:

$$E_\theta[\delta(\mathbf{X})] = E_\theta[X_1] - E_\theta[X_2] = \theta - \theta = 0, \tag{19.19}$$

but

$$P_\theta[\delta(\mathbf{X}) = 0] = P_\theta[X_1 = X_2] > 0. \tag{19.20}$$

Thus "0" is not the only unbiased estimator of 0; $X_1 - X_2$ is another. You can come up with an infinite number, in fact. Note that no iid model is complete when $n > 1$, unless the distribution is just a constant.

Now let $S = X_1 + \cdots + X_n$, which is a sufficient statistic. Then $S \sim \text{Poisson}(n\theta)$ for $\theta \in (0, \infty)$. Is *the model for S* complete? Suppose $\delta^*(s)$ is an unbiased estimator of 0. Then

$$E_\theta[\delta^*(S)] = 0 \text{ for all } \theta \in (0, \infty) \Rightarrow \sum_{s=0}^\infty \delta^*(s)e^{-n\theta}\frac{(n\theta)^s}{s!} = 0 \text{ for all } \theta \in (0, \infty)$$

$$\Rightarrow \sum_{s=0}^\infty \delta^*(s)\frac{n^s}{s!}\theta^s = 0 \text{ for all } \theta \in (0, \infty)$$

$$\Rightarrow \delta^*(s)\frac{n^s}{s!} = 0 \text{ for all } s = 0, 1, 2, \ldots$$

$$\Rightarrow \delta^*(s) = 0 \text{ for all } s = 0, 1, 2, \ldots. \tag{19.21}$$

Thus the only unbiased estimator of 0 that is a function of S is 0, meaning the model for S is complete.

19.2.2 Uniform distribution

Suppose $X_1, \ldots, X_n$ are iid $\text{Uniform}(0, \theta)$, $\theta \in (0, \infty)$, with $n > 1$. This model again is not complete. Consider the sufficient statistic $S = \max\{X_1, \ldots, X_n\}$. The model for S has space $(0, \infty)$ and pdf

$$f_\theta(s) = \begin{cases} ns^{n-1}/\theta^n & \text{if } 0 < s < \theta \\ 0 & \text{if } \text{not} \end{cases}. \tag{19.22}$$

To see if the model for S is complete, suppose that δ^* is an unbiased estimator of 0. Then

$$E_\theta[\delta^*(S)] = 0 \text{ for all } \theta \in (0, \infty) \Rightarrow \int_0^\theta \delta^*(s)s^{n-1}ds/\theta^n = 0 \text{ for all } \theta \in (0, \infty)$$

$$\Rightarrow \int_0^\theta \delta^*(s)s^{n-1}ds = 0 \text{ for all } \theta \in (0, \infty)$$

$$(\text{taking } d/d\theta) \Rightarrow \delta^*(\theta)\theta^{n-1} = 0 \text{ for (almost) all } \theta \in (0, \infty)$$

$$\Rightarrow \delta^*(\theta) = 0 \text{ for (almost) all } \theta \in (0, \infty). \tag{19.23}$$

That is, δ^* must be 0, so that the model for S is complete. [The "(almost)" means that one can deviate from zero for a few values (with total Lebesgue measure 0) without changing the fact that $P_\theta[\delta^*(S) = 0] = 1$ for all θ.]

19.3 Uniformly minimum variance estimators

This section contains the key result of this chapter:

Theorem 19.4. *Suppose* $\mathbf{S} = s(\mathbf{X})$ *is a sufficient statistic for the model (19.1) on* $\mathbf{X}$, *and the model for* $\mathbf{S}$ *is complete. If* $\delta^*(\mathbf{s})$ *is an unbiased estimator (depending on* $\mathbf{s}$*) of the function* $g(\theta)$, *then* $\delta_0(\mathbf{X}) = \delta^*(s(\mathbf{X}))$ *is the UMVUE of* $g(\theta)$.

Proof. Let δ be any unbiased estimator of $g(\theta)$, and consider

$$e_\delta(\mathbf{s}) = E[\delta(\mathbf{X}) \,|\, \mathbf{S} = \mathbf{s}]. \tag{19.24}$$

Because **S** is sufficient, e_δ does not depend on θ, so it is an estimator. Furthermore, since

$$E_{\boldsymbol{\theta}}[e_\delta(\mathbf{S})] = E_{\boldsymbol{\theta}}[\delta(\mathbf{X})] = g(\boldsymbol{\theta}) \quad \text{for all } \theta \in \mathcal{T}, \tag{19.25}$$

it is unbiased, and because it is a conditional expectation, as in (13.61),

$$Var_{\boldsymbol{\theta}}[e_\delta(\mathbf{S})] \leq Var_{\boldsymbol{\theta}}[\delta(\mathbf{X})] \quad \text{for all } \theta \in \mathcal{T}. \tag{19.26}$$

But by completeness of the model for **S**, there is only one unbiased estimator that is a function of just **S**, δ^*, hence $\delta^*(\mathbf{s}) = e_\delta(\mathbf{s})$ (with probability 1), and

$$Var_{\boldsymbol{\theta}}[\delta_0(\mathbf{X})] = Var_{\boldsymbol{\theta}}[\delta^*(\mathbf{S})] \leq Var_{\boldsymbol{\theta}}[\delta(\mathbf{X})] \quad \text{for all } \theta \in \mathcal{T}. \tag{19.27}$$

This equation holds for any unbiased δ, hence δ_0 is best.   □

This proof is actually constructive in a sense. If you do not know δ^*, but have any unbiased δ, then you can find the UMVUE by using the Rao-Blackwell theorem (Theorem 13.8 on page 210), conditioning on the sufficient statistic. Or, if you can by any means find an unbiased estimator that is a function of **S**, then it is UMVUE.

19.3.1 Poisson distribution

Consider again the Poisson case in Section 19.2.1. We have that $S = X_1 + \cdots + X_n$ is sufficient, and the model for S is complete. Because

$$E_\theta[S/n] = \theta, \tag{19.28}$$

we know that S/n is the UMVUE of θ.

Now let $g(\theta) = e^{-\theta} = P_\theta[X_1 = 0]$. We have from Exercise 19.8.3 an unbiased estimator of $g(\theta)$ that is a function of S, hence it is UMVUE. But finding the estimator took a bit of work, and luck. Instead, we could start with a very simple estimator,

$$\delta(\mathbf{X}) = I[X_1 = 0], \tag{19.29}$$

which indicates whether there were 0 calls in the first minute. That estimator is unbiased, but obviously not using all the data. From (13.36) we have that **X** given $S = s$ is multinomial, hence X_1 is binomial:

$$X_1 \mid S = s \sim \text{Binomial}(s, \tfrac{1}{n}). \tag{19.30}$$

Then the UMVUE is

$$E[\delta(\mathbf{X}) \mid s(\mathbf{X}) = s] = P[X_1 = 0 \mid S = s] = P[\text{Binomial}(s, \tfrac{1}{n}) = 0] = (1 - \tfrac{1}{n})^s. \tag{19.31}$$

As must be, it is the same as the estimator in Exercise 19.8.3.

19.4 Completeness for exponential families

Showing completeness of a model in general is not an easy task. Fortunately, exponential families as in (13.25) usually do provide completeness for the natural sufficient statistic. To present the result, we start by assuming the model for the $p \times 1$ vector **S** has exponential family density with $\boldsymbol{\theta} \in \mathcal{T} \subset \mathbb{R}^p$ as the natural parameter, and **S** itself as the natural sufficient statistic:

$$f_{\boldsymbol{\theta}}(\mathbf{s}) = a(\mathbf{s}) \, e^{\theta_1 s_1 + \cdots + \theta_p s_p - \psi(\boldsymbol{\theta})}. \tag{19.32}$$

Lemma 19.5. *In the above model for* **S***, suppose that the parameter space contains a nonempty open p-dimensional rectangle. Then the model is complete.*

If $\mathcal{S}$ is a lattice, e.g., the set of vectors containing all nonnegative integers, then the lemma can be proven by looking at the power series in $\exp(\theta_i)$'s. See Theorem 4.3.1 of Lehmann and Romano (2005) for the general case.

The requirement on the parameter space guards against exact constraints among the parameters. For example, suppose $X_1, \ldots, X_n$ are iid $N(\mu, \sigma^2)$, where μ is known to be positive and the coefficient of variation, σ/μ, is known to be 10%. The exponent in the pdf, with $\mu = 10\sigma$, is

$$\frac{\mu}{\sigma^2} \sum x_i - \frac{1}{2\sigma^2} \sum x_i^2 = \frac{10}{\sigma} \sum x_i - \frac{1}{2\sigma^2} \sum x_i^2. \tag{19.33}$$

The exponential family terms are then

$$\boldsymbol{\theta} = \left(\frac{10}{\sigma}, -\frac{1}{2\sigma^2}\right) \quad \text{and} \quad \mathbf{S} = \left(\sum X_i, \sum X_i^2\right). \tag{19.34}$$

The parameter space, with $p = 2$, is

$$\mathcal{T} = \left\{\left(\frac{10}{\sigma}, -\frac{1}{2\sigma^2}\right) \mid \sigma > 0\right\}, \tag{19.35}$$

which does *not* contain a two-dimensional open rectangle, because $\theta_2 = -\theta_1^2/100$. Thus we cannot use the lemma to show completeness.

Note. It is important to read the lemma carefully. It does *not* say that if the requirement is violated, then the model is not complete. (Although realistic counterexamples are hard to come by.) To prove a model is not complete, you must produce a nontrivial unbiased estimator of 0. For example, in (19.34),

$$E_{\boldsymbol{\theta}}\left[\left(\frac{S_1}{n}\right)^2\right] = E_{\boldsymbol{\theta}}[\overline{X}^2] = \mu^2 + \frac{\sigma^2}{n} = (10\sigma)^2 + \frac{\sigma^2}{n} = \left(100 + \frac{1}{n}\right)\sigma^2 \tag{19.36}$$

and

$$E_{\boldsymbol{\theta}}\left[\frac{S_2}{n}\right] = E_{\boldsymbol{\theta}}[X_i^2] = \mu^2 + \sigma^2 = (10\sigma)^2 + \sigma^2 = 101\sigma^2. \tag{19.37}$$

Then

$$\delta(s_1, s_2) = \frac{s_1}{100 + 1/n} - \frac{s_2}{101} \tag{19.38}$$

has expected value 0, but is not zero itself. Thus the model is not complete.

19.4.1 Examples

Suppose $p = 1$, so that the natural sufficient statistic and parameter are both scalars. Then all that is needed is that $\mathcal{T}$ contains an interval (a, b), $a < b$. The table below has some examples, where $X_1, \ldots, X_n$ are iid with the given distribution (the parameter space is assumed to be the most general):

Distribution	Sufficient statistic T	Natural parameter θ	$\mathcal{T}$	
$N(\mu, 1)$	$\sum X_i$	μ	$\mathbb{R}$	
$N(0, \sigma^2)$	$\sum X_i^2$	$1/(2\sigma^2)$	$(0, \infty)$	(19.39)
Poisson(λ)	$\sum X_i$	$\log(\lambda)$	$\mathbb{R}$	
Exponential(λ)	$\sum X_i$	$-\lambda$	$(-\infty, 0)$	

The parameter spaces all contain open intervals; in fact they are open intervals. Thus the models for the T's are all complete.

For $X_1, \ldots, X_n$ iid $N(\mu, \sigma^2)$, with $(\mu, \sigma^2) \in \mathbb{R} \times (0, \infty)$, the exponential family terms are

$$\mathcal{T} = \left(\frac{\mu}{\sigma}, -\frac{1}{2\sigma^2} \right) \quad \text{and} \quad \mathbf{T} = \left(\sum X_i, \sum X_i^2 \right). \tag{19.40}$$

Here, without any extra constraints on μ and σ^2, we have

$$\mathcal{T} = \mathbb{R} \times (-\infty, 0), \tag{19.41}$$

because for any (a, b) with $a \in \mathbb{R}$, $b < 0$, $\mu = a/\sqrt{-2b} \in \mathbb{R}$ and $\sigma^2 = -1/(2b) \in (0, \infty)$ are valid parameter values. Thus the model is complete for (T_1, T_2). From Theorem 19.4, we then have that any function of (T_1, T_2) is the UMVUE for its expected value. For example, $\overline{X}$ is the UMVUE for μ, $S_*^2 = \sum (X_i - \overline{X})^2 / (n - 1)$ is the UMVUE for σ^2. Also, since $E[\overline{X}^2] = \mu^2 + \sigma^2/n$, the UMVUE for μ^2 is $\overline{X}^2 - S_*^2/n$.

19.5 Cramér-Rao lower bound

If you find a UMVUE, then you know you have the best unbiased estimator. Oftentimes, there is no UMVUE, or there is one but it is too difficult to calculate. Then what? At least it would be informative to have an idea of whether a given estimator is very poor, or just not quite optimal. One approach is to find a lower bound for the variance. The closer an estimator's variance is to the lower bound, the better.

The model for this section has random vector $\mathbf{X}$, space $\mathcal{X}$, densities $f_\theta(\mathbf{x})$, and parameter space $\mathcal{T} = (a, b) \subset \mathbb{R}$. We need the likelihood assumptions from Section 14.4 to hold. In particular, the pdf should always be positive (so that Uniform$(0, \theta)$'s are not allowed) and have several derivatives with respect to θ, and certain expected values must be finite. Suppose δ is an unbiased estimator of $g(\theta)$, so that

$$E_\theta[\delta(\mathbf{x})] = \int_{\mathcal{X}} \delta(\mathbf{x}) f_\theta(\mathbf{x}) d\mathbf{x} = g(\theta) \quad \text{for all} \quad \theta \in \mathcal{T}. \tag{19.42}$$

(Use a summation if the density is a pmf.) Now take the derivative with respect to θ of both sides. Assuming interchanging the derivative and integral below is valid, we have

$$g'(\theta) = \frac{\partial}{\partial \theta} \int_{\mathcal{X}} \delta(\mathbf{x}) f_\theta(\mathbf{x}) d\mathbf{x}$$

$$= \int_{\mathcal{X}} \delta(\mathbf{x}) \frac{\partial}{\partial \theta} f_\theta(\mathbf{x}) d\mathbf{x}$$

$$= \int_{\mathcal{X}} \delta(\mathbf{x}) \frac{\frac{\partial}{\partial \theta} f_\theta(\mathbf{x})}{f_\theta(\mathbf{x})} f_\theta(\mathbf{x}) d\mathbf{x}$$

$$= \int_{\mathcal{X}} \delta(\mathbf{x}) \frac{\partial}{\partial \theta} \log(f_\theta(\mathbf{x})) f_\theta(\mathbf{x}) d\mathbf{x}$$

$$= E_\theta \left[\delta(\mathbf{X}) \frac{\partial}{\partial \theta} \log(f_\theta(\mathbf{X})) \right]$$

$$= \text{Cov}_\theta \left[\delta(\mathbf{X}), \frac{\partial}{\partial \theta} \log(f_\theta(\mathbf{X})) \right]. \tag{19.43}$$

The last step follows from Lemma 14.1 on page 226, which shows that

$$E_\theta \left[\frac{\partial}{\partial \theta} \log(f_\theta(\mathbf{X})) \right] = 0, \tag{19.44}$$

recalling that the derivative of the log of the density is the score function.

Now we can use the correlation inequality (2.31), $Cov[U, V]^2 \le Var[U]Var[V]$:

$$g'(\theta)^2 \le Var_\theta[\delta(\mathbf{X})] \, Var_\theta \left[\frac{\partial}{\partial \theta} \log(f_\theta(\mathbf{X})) \right] = Var_\theta[\delta(\mathbf{X})]\mathcal{I}(\theta), \tag{19.45}$$

where $\mathcal{I}(\theta)$ is the Fisher information, which is the variance of the score function (see Lemma 14.1 again). We leave off the "1" in the subscript. Thus

$$Var_\theta[\delta(\mathbf{X})] \ge \frac{g'(\theta)^2}{\mathcal{I}(\theta)} \equiv CRLB_g(\theta), \tag{19.46}$$

the **Cramér-Rao lower bound (CRLB)** for unbiased estimators of $g(\theta)$. Note that Theorem 14.7 on page 232 uses the same bound for the asymptotic variance for consistent and asymptotically normal estimators.

If an unbiased estimator achieves the CRLB, that is,

$$Var_\theta[\delta(\mathbf{X})] = \frac{g'(\theta)^2}{\mathcal{I}(\theta)} \quad \text{for all } \theta \in \mathcal{T}, \tag{19.47}$$

then it is the UMVUE. The converse is not necessarily true. The UMVUE may not achieve the CRLB, which of course means that no unbiased estimator does. There are other more accurate bounds, called Bhattacharya bounds, that the UMVUE may achieve in such cases. See Lehmann and Casella (2003), page 128.

In fact, basically the only time an estimator achieves the CRLB is when we can write the density of $\mathbf{X}$ as an exponential family with natural sufficient statistic $\delta(\mathbf{x})$, in which case we already know it is UMVUE by completeness. Wijsman (1973) has a precise statement and proof. Here we give a sketch of the idea under the current assumptions. The correlation inequality shows that equality in (19.43) for each θ implies that there is a linear relationship between δ and the score function as functions of $\mathbf{x}$:

$$\frac{\partial}{\partial \theta} \log(f_\theta(\mathbf{x})) = a(\theta) + b(\theta)\delta(\mathbf{x}) \quad \text{for almost all } \mathbf{x} \in \mathcal{X}. \tag{19.48}$$

Now looking at (19.48) as a function of θ for fixed $\mathbf{x}$, we solve the simple differential equation, remembering the constant (which may depend on $\mathbf{x}$):

$$\log(f_\theta(\mathbf{x})) = A(\theta) + B(\theta)\delta(\mathbf{x}) + C(\mathbf{x}), \tag{19.49}$$

so that $f_\theta(\mathbf{x})$ has a one-dimensional exponential family form (19.32) with natural parameter $B(\theta)$ and natural sufficient statistic $\delta(\mathbf{x})$. (Some regard should be given to the space of $B(\theta)$ to make sure it is an interval.)

19.5.1 Laplace distribution

Suppose $X_1, \dots, X_n$ are iid Laplace(θ), $\theta \in \mathbb{R}$, so that

$$f_\theta(\mathbf{x}) = \frac{1}{2^n} e^{-\Sigma |x_i - \theta|}. \tag{19.50}$$

This density is not an exponential family one, and if $n > 1$, the sufficient statistic is the vector of order statistics, which does not have a complete model. Thus the UMVUE approach does not bear fruit. Because $E_\theta[X_i] = \theta$, $\overline{X}$ is an unbiased estimator of θ, with

$$Var_\theta[\overline{X}] = \frac{Var_\theta[X_i]}{n} = \frac{2}{n}. \tag{19.51}$$

Is this variance reasonable? We will compare it to the CRLB:

$$\frac{\partial}{\partial \theta} \log(f_\theta(x)) = \frac{\partial}{\partial \theta} - \sum |x_i - \theta| = \sum \text{Sign}(x_i - \theta), \tag{19.52}$$

because the derivative of $|x|$ is $+1$ if $x > 0$ and -1 if $x < 0$. (We are ignoring the possibility that $x_i = \theta$, where the derivative does not exist.) By symmetry of the density around θ, each $\text{Sign}(X_i - \theta)$ has a probabilty of $\frac{1}{2}$ to be either -1 or $+1$, so it has mean 0 and variance 1. Thus

$$\mathcal{I}(\theta) = Var_\theta[\sum \text{Sign}(X_i - \theta)] = n. \tag{19.53}$$

Here, $g(\theta) = \theta$, hence

$$CRLB_g(\theta) = \frac{g'(\theta)^2}{\mathcal{I}(\theta)} = \frac{1}{n}. \tag{19.54}$$

Compare this bound to the variance in (19.51). The variance of $\overline{X}$ is twice the CRLB, which is not very good. It appears that there should be a better estimator. In fact, there is, which we will see later in Section 19.7, although even that estimator does not achieve the CRLB. We also know that the median is better asymptotically.

19.5.2 Normal μ^2

Suppose $X_1, \ldots, X_n$ are iid $N(\mu, 1)$, $\mu \in \mathbb{R}$. We know that $\overline{X}$ is sufficient, and the model for it is complete, hence any unbiased estimator based on the sample mean is UMVUE. In this case, $g(\mu) = \mu^2$, and $E_\mu[\overline{X}^2] = \mu^2 + 1/n$, hence

$$\delta(\mathbf{x}) = \overline{x}^2 - \frac{1}{n} \tag{19.55}$$

is the UMVUE. To find the variance of the estimator, start by noting that $\sqrt{n}\overline{X} \sim N(\sqrt{n}\mu, 1)$, so that its square is noncentral χ^2 (Definition 7.7 on page 113),

$$n\overline{X}^2 \sim \chi_1^2(n\mu^2). \tag{19.56}$$

Thus from (7.75),

$$Var_\mu[n\overline{X}^2] = 2 + 4n\mu^2 (= 2\nu + 4\Delta). \tag{19.57}$$

Finally,

$$Var_\mu[\delta(\mathbf{X})] = Var_\mu[\overline{X}^2] = \frac{2}{n^2} + \frac{4\mu^2}{n}. \tag{19.58}$$

For the CRLB, we first need Fisher's information. Start with

$$\frac{\partial}{\partial \mu} \log(f_\mu(\mathbf{X})) = \frac{\partial}{\partial \mu} - \frac{1}{2} \sum (x_i - \mu)^2 = \sum (x_i - \mu). \tag{19.59}$$

The X_i's are independent with variance 1, so that $\mathcal{I}_n(\mu) = n$. Thus with $g(\mu) = \mu^2$,

$$CRLB_g(\mu) = \frac{(2\mu)^2}{n} = \frac{4\mu^2}{n}. \tag{19.60}$$

Comparing (19.58) to (19.60), we see that the UMVUE does not achieve the CRLB, which implies that no unbiased estimator will. But note that the variance of the UMVUE is only off by $2/n^2$, so that for large n, the ratio of the variance to the CRLB is close to 1.

19.6 Shift-equivariant estimators

As we have seen, pursuing UMVUEs is generally most successful when the model is an exponential family. Location family models (see Section 4.2.4) provide another opportunity for optimal estimation, but instead of unbiasedness we look at **shift-equivariance**.

Recall that a location family of distributions is one for which the only parameter is the center; the shape of the density stays the same. Start with the random variable Z. We will assume it has density f and take the space to be $\mathbb{R}$, where it may be that $f(z) = 0$ for some values of z. (E.g., Uniform$(0,1)$ is allowed.) The model on X has parameter $\theta \in \mathcal{T} \equiv \mathbb{R}$, where $X = \theta + Z$, hence has density $f(x - \theta)$.

The data will be n iid copies of X, $\mathbf{X} = (X_1, \ldots, X_n)$, so that the density of $\mathbf{X}$ is

$$f_\theta(\mathbf{x}) = \prod_{i=1}^{n} f(x_i - \theta). \tag{19.61}$$

This θ may or may not be the mean or median. Examples include the X_i's being $N(\theta, 1)$, or Uniform$(\theta, \theta + 1)$, or Laplace(θ) as in (19.50). By contrast, Uniform$(0, \theta)$ and Exponential(θ) are not location families since the spread as well as the center is affected by the θ.

A location family model is **shift-invariant** in that if we add the same constant a to all the X_i's, the resulting random vector has the same model as $\mathbf{X}$. That is, suppose $a \in \mathbb{R}$ is fixed, and we look at the transformation

$$\mathbf{X}^* = (X_1^*, \ldots, X_n^*) = (X_1 + a, \ldots, X_n + a). \tag{19.62}$$

The Jacobian of the transformation is 1, so the density of $\mathbf{X}^*$ is

$$f_\theta^*(\mathbf{x}^*) = \prod_{i=1}^{n} f(x_i^* - a - \theta) = \prod_{i=1}^{n} f(x_i^* - \theta^*), \quad \text{where } \theta^* = \theta + a. \tag{19.63}$$

Thus adding a to everything just shifts everything by a. Note that the space of $\mathbf{X}^*$ is the same as the space of $\mathbf{X}$, $\mathbb{R}^n$, and the space for the parameter θ^* is the same as that for θ, $\mathbb{R}$:

	Model	Model*
Data	$\mathbf{X}$	$\mathbf{X}^*$
Sample space	$\mathbb{R}^n$	$\mathbb{R}^n$
Parameter space	$\mathbb{R}$	$\mathbb{R}$
Density	$\prod_{i=1}^{n} f(x_i - \theta)$	$\prod_{i=1}^{n} f(x_i^* - \theta^*)$

$$\tag{19.64}$$

Thus the two models are the same. Not that $\mathbf{X}$ and $\mathbf{X}^*$ are equal, but that the sets of distributions considered for them are the same. This is the sense in which the **model**

is shift-invariant. (If the model includes a prior on θ, then the two models are not the same, because the prior distribution on θ^* would be different than that on θ.)

If $\delta(x_1, \ldots, x_n)$ is an estimate of θ in the original model, then $\delta(x_1^*, \ldots, x_n^*) = \delta(x_1 + c, \ldots, x_n + c)$ is an estimate of $\theta^* = \theta + c$ in Model*. Thus it may seem reasonable that $\delta(\mathbf{x}^*) = \delta(\mathbf{x}) + c$. This idea leads to **shift-equivariant estimators**, where $\delta(\mathbf{x})$ is shift-equivariant if for any $\mathbf{x}$ and c,

$$\delta(x_1 + c, x_2 + c, \ldots, x_n + c) = \delta(x_1, x_2, \ldots, x_n) + c. \qquad (19.65)$$

The mean and the median are both shift-equivariant. For example, suppose we have iid measurements in degrees Kelvin and wish to estimate the population mean. If someone else decides to redo the data in degrees Celsius, which entails adding 273.15 to each observation, then you would expect the new estimate of the mean would be the old one plus 273.15.

Our goal is to find the best shift-equivariant estimator, where we evaluate estimators based on the mean square error:

$$\text{MSE}_\theta[\delta] = E_\theta[(\delta(\mathbf{X}) - \theta)^2]. \qquad (19.66)$$

If δ is shift-equivariant, and has a finite variance, then Exercise 19.8.19 shows that

$$E_\theta[\delta(\mathbf{X})] = \theta + E_0[\delta(\mathbf{X})] \quad \text{and} \quad Var_\theta[\delta(\mathbf{X})] = Var_0[\delta(\mathbf{X})], \qquad (19.67)$$

hence (using Exercise 11.7.1),

$$\text{Bias}_\theta[\delta] = E_0[\delta(\mathbf{X})] \quad \text{and} \quad \text{MSE}_\theta[\delta] = Var_0[\delta(\mathbf{X})] + E_0[\delta(\mathbf{X})]^2 = E_0[\delta(\mathbf{X})^2], \qquad (19.68)$$

which do not depend on θ.

19.7   The Pitman estimator

For any biased equivariant estimator, it is fairly easy to find one that is unbiased but has the same variance by just shifting it a bit. Suppose δ is biased, and $b = E_0[\delta(\mathbf{X})] \neq 0$ is the bias. Then $\delta^* = \delta - b$ is also shift-equivariant, with

$$Var_\theta[\delta^*(\mathbf{X})] = Var_0[\delta(\mathbf{X})] \quad \text{and} \quad \text{Bias}_\theta[\delta^*(\mathbf{X})] = 0, \qquad (19.69)$$

hence

$$\text{MSE}_\theta[\delta^*] = Var_0[\delta(\mathbf{X})] < Var_0[\delta(\mathbf{X})] + E_0[\delta(\mathbf{X})]^2 = \text{MSE}_\theta[\delta]. \qquad (19.70)$$

Thus if a shift-equivariant estimator is biased, it can be improved, ergo ...

Lemma 19.6. *If δ is the best shift-equivariant estimator, and it has a finite expected value, then it is unbiased.*

In the normal case, $\overline{X}$ is the best shift-equivariant estimator, because it is the best unbiased estimator, and it is shift-equivariant. But, of course, in the normal case, $\overline{X}$ is always the best at everything. Now consider the general location family case. We will assume that $Var_0[X_i] < \infty$. To find the best estimator, we first characterize the shift-equivariant estimators. We can use a similar trick as we did when finding the UMVUE when we took a simple unbiased estimator, then found its expected value given the sufficient statistic.

Start with an arbitrary shift-equivariant estimator δ. Let "X_n" be another shift-equivariant estimator, and look at the difference:

$$\delta(x_1,\ldots,x_n) - x_n = \delta(x_1 - x_n, x_2 - x_n, \ldots, x_{n-1} - x_n, 0). \tag{19.71}$$

Define the function v on the differences $y_i = x_i - x_n$,

$$v : \mathbb{R}^{n-1} \longrightarrow \mathbb{R},$$
$$\mathbf{y} \longrightarrow v(\mathbf{y}) = \delta(y_1, \ldots, y_{n-1}, 0). \tag{19.72}$$

Then what we have is that for any equivariant δ, there is a v such that

$$\delta(\mathbf{x}) = x_n + v(x_1 - x_n, \ldots, x_{n-1} - x_n). \tag{19.73}$$

Now instead of trying to find the best δ, we will look for the best v, then use (19.73) to get the δ. From (19.68), the best v is found by minimizing

$$E[\delta(\mathbf{Z})^2] = E_0[(Z_n + v(Z_1 - Z_n, \ldots, Z_{n-1} - Z_n))^2]. \tag{19.74}$$

(The Z_i's are iid with pdf f, i.e., $\theta = 0$.) The trick is to condition on the differences $Z_1 - Z_n, \ldots, Z_{n-1} - Z_n$:

$$E[(Z_n + v(Z_1 - Z_n, \ldots, Z_{n-1} - Z_n))^2] = E[e(Z_1 - Z_n, \ldots, Z_{n-1} - Z_n)], \tag{19.75}$$

where

$$\begin{aligned} e(y_1, \ldots, y_{n-1}) &= E[(Z_n + v(Z_1 - Z_n, \ldots, Z_{n-1} - Z_n))^2 \,| \\ &\qquad Z_1 - Z_n = y_1, \ldots, Z_{n-1} - Z_n = y_{n-1}] \\ &= E[(Z_n + v(y_1, \ldots, y_{n-1}))^2 \,| \\ &\qquad Z_1 - Z_n = y_1, \ldots, Z_{n-1} - Z_n = y_{n-1}]. \end{aligned} \tag{19.76}$$

It is now possible to minimize e for each fixed set of y_i's, e.g., by differentiating with respect to $v(y_1, \ldots, y_{n-1})$. But we know the minimum is minus the (conditional) mean of Z_n, that is, the best v is

$$v(y_1, \ldots, y_{n-1}) = -E[Z_n \,|\, Z_1 - Z_n = y_1, \ldots, Z_{n-1} - Z_n = y_{n-1}]. \tag{19.77}$$

To find that conditional expectation, we need the joint pdf divided by the marginal to get the conditional pdf. For the joint, let

$$Y_1 = Z_1 - Z_n, \ldots, \; Y_{n-1} = Z_{n-1} - Z_n, \tag{19.78}$$

and find the pdf of $(Y_1, \ldots, Y_{n-1}, Z_n)$. We use the Jacobian approach, so need the inverse function:

$$z_1 = y_1 + z_n, \ldots, \; z_{n-1} = y_{n-1} + z_n, \; z_n = z_n. \tag{19.79}$$

Exercise 19.8.20 verifies that the Jacobian of this transformation is 1, so that the pdf of $(Y_1, \ldots, Y_{n-1}, Z_n)$ is

$$f^*(y_1, \ldots, y_{n-1}, z_n) = f(z_n) \prod_{i=1}^{n-1} f(y_i + z_n), \tag{19.80}$$

and the marginal of the Y_i's is

$$f_Y^*(y_1, \ldots, y_{n-1}) = \int_{-\infty}^{\infty} f(z_n) \prod_{i=1}^{n-1} f(y_i + z_n) dz_n. \tag{19.81}$$

The conditional pdf is then

$$f^*(z_n | y_1, \ldots, y_{n-1}) = \frac{f(z_n) \prod_{i=1}^{n-1} f(y_i + z_n)}{\int_{-\infty}^{\infty} f(u) \prod_{i=1}^{n-1} f(y_i + u) du}, \tag{19.82}$$

hence conditional mean is

$$E[Z_n | Y_1 = y_1, \ldots, Y_{n-1} = y_{n-1}] = \frac{\int_{-\infty}^{\infty} z_n f(z_n) \prod_{i=1}^{n-1} f(y_i + z_n) dz_n}{\int_{-\infty}^{\infty} f(z_n) \prod_{i=1}^{n-1} f(y_i + z_n) dz_n}$$

$$= -v(y_1, \ldots, y_{n-1}). \tag{19.83}$$

The best δ uses that v, so that from (19.73), with $y_i = x_i - x_n$,

$$\delta(\mathbf{x}) = x_n - \frac{\int_{-\infty}^{\infty} z_n f(z_n) \prod_{i=1}^{n-1} f(x_i - x_n + z_n) dz_n}{\int_{-\infty}^{\infty} f(z_n) \prod_{i=1}^{n-1} f(x_i - x_n + z_n) dz_n}$$

$$= \frac{\int_{-\infty}^{\infty} (x_n - z_n) f(z_n) \prod_{i=1}^{n-1} f(x_i - x_n + z_n) dz_n}{\int_{-\infty}^{\infty} f(z_n) \prod_{i=1}^{n-1} f(x_i - x_n + z_n) dz_n}. \tag{19.84}$$

Exercise 19.8.20 derives a more pleasing expression:

$$\delta(\mathbf{x}) = \frac{\int_{-\infty}^{\infty} \theta \prod_{i=1}^{n} f(x_i - \theta) d\theta}{\int_{-\infty}^{\infty} \prod_{i=1}^{n} f(x_i - \theta) d\theta}. \tag{19.85}$$

This best estimator is called the **Pitman estimator** (Pitman, 1939). Although we assumed finite variance for X_i, the following theorem holds more generally. See Theorem 3.1.20 of Lehmann and Casella (2003).

Theorem 19.7. *In the location-family model (19.61), if there exists any equivariant estimator with finite MSE, then the equivariant estimator with lowest MSE is given by the Pitman estimator (19.85).*

The final expression in (19.85) is very close to a Bayes posterior mean. With prior π, the posterior mean is

$$E[\theta \mid \mathbf{X} = \mathbf{x}] = \frac{\int_{-\infty}^{\infty} \theta f_\theta(\mathbf{x}) \pi(\theta) d\theta}{\int_{-\infty}^{\infty} f_\theta(\mathbf{x}) \pi(\theta) d\theta}. \tag{19.86}$$

Thus the Pitman estimator can be thought of as the posterior mean for improper prior $\pi(\theta) = 1$. Note that we do not need iid observations, but only that the pdf is of the form $f_\theta(\mathbf{x}) = f(x_1 - \theta, \ldots, x_n - \theta)$.

One thing special about this theorem is that it is constructive, that is, there is a formula telling exactly how to find the best. By contrast, when finding the UMVUE, you have to do some guessing. If you find an unbiased estimator that is a function of a complete sufficient statistic, then you have it, but it may not be easy to find. You might be able to use the power-series approach, or find an easy one then use Rao-Blackwell, but maybe not. A drawback to the formula for the Pitman estimator is that the integrals may not be easy to perform analytically, although in practice it would be straightforward to use numerical integration. The exercises have a couple of examples, the normal and uniform, in which the integrals are doable. Additional examples follow.

19.7.1 Shifted exponential distribution

Here, $f(x) = e^{-x}I[x > 0]$, i.e., the Exponential(1) pdf. The location family is *not* Exponential(θ), but a shifted Exponential(1) that starts at θ rather than at 0:

$$f(x - \theta) = e^{-(x-\theta)}I[x - \theta > 0] = \begin{cases} e^{-(x-\theta)} & \text{if } x > \theta \\ 0 & \text{if } x \leq \theta \end{cases}. \tag{19.87}$$

Then the pdf of **X** is

$$f(\mathbf{x} \mid \theta) = \prod_{i=1}^{n} e^{-(x_i-\theta)}I[x_i > \theta] = e^{n\theta}e^{-\Sigma x_i}I[x_{(1)} > \theta], \tag{19.88}$$

because all x_i's are greater than θ if and only if the minimum $x_{(1)}$ is. Then the Pitman estimator is, from (19.85),

$$\delta(\mathbf{x}) = \frac{\int_{-\infty}^{\infty} \theta e^{n\theta} e^{-\Sigma x_i}I[x_{(1)} > \theta]d\theta}{\int_{-\infty}^{\infty} e^{n\theta} e^{-\Sigma x_i}I[x_{(1)} > \theta]d\theta}$$

$$= \frac{\int_{-\infty}^{x_{(1)}} \theta e^{n\theta} d\theta}{\int_{-\infty}^{x_{(1)}} e^{n\theta} d\theta}. \tag{19.89}$$

Use integration by parts in the numerator, so that

$$\int_{-\infty}^{x_{(1)}} \theta e^{n\theta} d\theta = \frac{\theta}{n} e^{n\theta} \Big|_{-\infty}^{x_{(1)}} - \frac{1}{n} \int_{-\infty}^{x_{(1)}} e^{n\theta} d\theta = \frac{x_{(1)}}{n} e^{nx_{(1)}} - \frac{1}{n^2} e^{nx_{(1)}}, \tag{19.90}$$

and the denominator is $(1/n)e^{nx_{(1)}}$, hence

$$\delta(\mathbf{x}) = \frac{\frac{x_{(1)}}{n} e^{nx_{(1)}} - \frac{1}{n^2} e^{nx_{(1)}}}{\frac{1}{n} e^{nx_{(1)}}} = x_{(1)} - \frac{1}{n}. \tag{19.91}$$

Is that estimator unbiased? Yes, because it is the best equivariant estimator. Actually, it is the UMVUE, because $x_{(1)}$ is a complete sufficient statistic.

19.7.2 Laplace distribution

Now

$$f_\theta(\mathbf{x}) = \prod_{i=1}^{n} \frac{1}{2} e^{-|x_i-\theta|} = \frac{1}{2^n} e^{-\Sigma|x_i-\theta|}, \tag{19.92}$$

and the Pitman estimator is

$$\delta(\mathbf{x}) = \frac{\int_{-\infty}^{\infty} \theta e^{-\Sigma|x_i - \theta|} d\theta}{\int_{-\infty}^{\infty} e^{-\Sigma|x_i - \theta|} d\theta} = \frac{\int_{-\infty}^{\infty} \theta e^{-\Sigma|x_{(i)} - \theta|} d\theta}{\int_{-\infty}^{\infty} e^{-\Sigma|x_{(i)} - \theta|} d\theta}, \tag{19.93}$$

where in the last expression we just substituted the order statistics. These integrals need to be broken up, depending on which $(x_{(i)} - \theta)$'s are positive and which negative. We'll do the $n = 2$ case in detail, where we have three regions of integration:

$$\begin{aligned} \theta < x_{(1)} &\Rightarrow \Sigma|x_{(i)} - \theta| = x_{(1)} + x_{(2)} - 2\theta; \\ x_{(1)} < \theta < x_{(2)} &\Rightarrow \Sigma|x_{(i)} - \theta| = -x_{(1)} + x_{(2)}; \\ x_{(2)} < \theta &\Rightarrow \Sigma|x_{(i)} - \theta| = -x_{(1)} - x_{(2)} + 2\theta. \end{aligned} \tag{19.94}$$

The numerator in (19.93) is

$$e^{-x_{(1)} - x_{(2)}} \int_{-\infty}^{x_{(1)}} \theta e^{2\theta} d\theta + e^{x_{(1)} - x_{(2)}} \int_{x_{(1)}}^{x_{(2)}} \theta d\theta + e^{x_{(1)} + x_{(2)}} \int_{x_{(2)}}^{\infty} \theta e^{-2\theta} d\theta$$

$$= e^{-x_{(1)} - x_{(2)}} \left(\frac{x_{(1)}}{2} - \frac{1}{4} \right) e^{2x_{(1)}} + e^{x_{(1)} - x_{(2)}} \frac{x_{(2)}^2 - x_{(1)}^2}{2} + e^{x_{(1)} + x_{(2)}} \left(\frac{x_{(2)}}{2} + \frac{1}{4} \right) e^{-2x_{(2)}}$$

$$= \frac{1}{2} e^{x_{(1)} - x_{(2)}} \left(x_{(1)} + x_{(2)}^2 - x_{(1)}^2 + x_{(2)} \right). \tag{19.95}$$

The denominator is

$$e^{-x_{(1)} - x_{(2)}} \int_{-\infty}^{x_{(1)}} e^{2\theta} d\theta + e^{x_{(1)} - x_{(2)}} \int_{x_{(1)}}^{x_{(2)}} d\theta + e^{x_{(1)} + x_{(2)}} \int_{x_{(2)}}^{\infty} e^{-2\theta} d\theta$$

$$= e^{-x_{(1)} - x_{(2)}} \tfrac{1}{2} e^{2x_{(1)}} + e^{x_{(1)} - x_{(2)}} \left(x_{(2)} - x_{(1)} \right) + e^{x_{(1)} + x_{(2)}} \tfrac{1}{2} e^{-2x_{(2)}}$$

$$= e^{x_{(1)} - x_{(2)}} \left(1 + x_{(2)} - x_{(1)} \right). \tag{19.96}$$

Finally,

$$\delta(\mathbf{x}) = \frac{1}{2} \frac{x_{(1)} + x_{(2)}^2 - x_{(1)}^2 + x_{(2)}}{1 + x_{(2)} - x_{(1)}} = \frac{x_{(1)} + x_{(2)}}{2}, \tag{19.97}$$

which is a rather long way to calculate the mean! The answer for $n = 3$ is not the mean. Is it the median?

19.8 Exercises

Exercise 19.8.1. Let $X \sim$ Geometric(θ), $\theta \in (0, 1)$, so that the space of X is $\{0, 1, 2, \ldots\}$, and the pmf is $f_\theta(x) = (1 - \theta)\theta^x$. For each $g(\theta)$ below, find an unbiased estimator of $g(\theta)$ based on X, if it exists. (Notice there is only one X.) (a) $g(\theta) = \theta$. (b) $g(\theta) = 0$. (c) $g(\theta) = 1/\theta$. (d) $g(\theta) = \theta^2$. (e) $g(\theta) = \theta^3$. (f) $g(\theta) = 1/(1 - \theta)$.

Exercise 19.8.2. At the start of Section 19.1, we noted that if $X \sim$ Binomial(n, θ) with $\theta \in (0, 1)$, the only functions $g(\theta)$ for which there is an unbiased estimator are the polynomials in θ of degree less than or equal to n. This exercise finds these estimators. To that end, suppose $Z_1, \ldots, Z_n$ are iid Bernoulli(θ) and $X = Z_1 + \cdots + Z_n$. (a) For integer $1 \le k \le n$, show that $\delta_k(\mathbf{z}) = z_1 \cdots z_k$ is an unbiased estimator of θ^k. (b) What is the distribution of $\mathbf{Z}$ given $X = x$? Why is this distribution independent of θ? (c) Show that

$$E[\delta_k(\mathbf{Z}) \mid X = x] = \frac{x}{n} \frac{x-1}{n-1} \cdots \frac{x-k+1}{n-k+1}. \tag{19.98}$$

[Hint: Note that $\delta_k(\mathbf{z})$ is either 0 or 1, and it equals 1 if and only if the first k z_i's are all 1. This probability is the same as that of drawing without replacement k 1's from a box with x 1's and $n - x$ 0's.] (d) Find the UMVUE of $g(\theta) = \sum_{i=0}^{n} a_i \theta^i$. Why is it UMVUE? (e) Specialize to $g(\theta) = Var_\theta[X] = n\theta(1 - \theta)$. Find the UMVUE of $g(\theta)$ if $n \geq 2$.

Exercise 19.8.3. Suppose $X_1, \ldots, X_n$ are iid Poisson(θ), $\theta \in (0, \infty)$. We wish to find an unbiased estimator of $g(\theta) = \exp(-\theta)$. Let $\delta_c(\mathbf{x}) = \exp(-c\bar{x})$. (a) Show that

$$E[\delta_c(\mathbf{X})] = e^{-n\theta} e^{n\theta e^{-c/n}}. \tag{19.99}$$

[Hint: We know that $S = X_1 + \cdots + X_n \sim$ Poisson$(n\theta)$, so the expectation of $\delta_c(\mathbf{X})$ can be found using an exponential summation.] (b) Show that the MLE, $\exp(-\bar{x})$, is biased. What is its bias as $n \to \infty$? (c) For $n \geq 2$, find c^* so that $\delta_{c^*}(\mathbf{x})$ is unbiased. Show that $\delta_{c^*}(\mathbf{x}) = (1 - 1/n)^{n\bar{x}}$. (d) What happens when you try to find such a c for $n = 1$? What is the unbiased estimator when $n = 1$?

Exercise 19.8.4. Suppose $X \sim$ Binomial(n, θ), $\theta \in (0, 1)$. Find the Cramér-Rao lower bound for estimating θ. Does the UMVUE achieve the CRLB?

Exercise 19.8.5. Suppose $X_1, \ldots, X_n$ are iid Poisson(λ), $\lambda \in (0, \infty)$. (a) Find the CRLB for estimating λ. Does the UMVUE achieve the CRLB? (b) Find the CRLB for estimating $e^{-\lambda}$. Does the UMVUE achieve the CRLB?

Exercise 19.8.6. Suppose $X_1, \ldots, X_n$ are iid Exponential(λ), $\lambda \in (0, \infty)$. (a) Find the CRLB for estimating λ. (b) Find the UMVUE of λ. (c) Find the variance of the UMVUE from part (b). Does it achieve the CRLB? (d) What is the limit of $CRLB/Var[UMVUE]$ as $n \to \infty$?

Exercise 19.8.7. Suppose the data consist of just one $X \sim$ Poisson(θ), and the goal is to estimate $g(\theta) = \exp(-2\theta)$. Thus you want to estimate the probability that no calls will come in the next two hours, based on just one hour's data. (a) Find the MLE of $g(\theta)$. Find its expected value. Is it unbiased? (b) Find the UMVUE of $g(\theta)$. Does it make sense? [Hint: Show that if the estimator $\delta(x)$ is unbiased, $\exp(-\theta) = \sum_{x=1}^{\infty} \delta(x)\theta^k/k!$ for all θ. Then write the left-hand side as a power series in θ, and match coefficients of θ^k on both sides.]

Exercise 19.8.8. For each situation, decide how many different unbiased estimators of the given $g(\theta)$ there are (zero, one, or lots). (a) $X \sim$ Binomial(n, θ), $\theta \in (0, 1)$, $g(\theta) = \theta^{n+1}$. (b) $X \sim$ Poisson(θ_1) and $Y \sim$ Poisson(θ_2), and X and Y are independent, $\theta = (\theta_1, \theta_2) \in (0, \infty) \times (0, \infty)$. (i) $g(\theta) = \theta_1$; (ii) $g(\theta) = \theta_1 + \theta_2$; (iii) $g(\theta) = 0$. (c) $(X_{(1)}, X_{(n)})$ derived from a sample of iid Uniform$(\theta, \theta + 1)$'s, $\theta \in \mathbb{R}$. (i) $g(\theta) = \theta$; (ii) $g(\theta) = 0$.

Exercise 19.8.9. Let $X_1, \ldots, X_n$ be iid Exponential(λ), $\lambda \in (0, \infty)$, and

$$g(\lambda) = P_\lambda[X_i > 1] = e^{-\lambda}. \tag{19.100}$$

Then $\delta(\mathbf{X}) = I[X_1 > 1]$ is an unbiased estimator of $g(\lambda)$. Also, $T = X_1 + \cdots + X_n$ is sufficient. (a) What is the distribution of $U = X_1/T$? (b) Is T independent of the U in part (a)? (c) Find $P[U > u]$ for $u \in (0, \infty)$. (d) $P[X_1 > 1 \mid T = t] = P[U > u \mid T = t] = P[U > u]$ for what u (which is a function of t)? (e) Find $E[\delta(\mathbf{X}) \mid T = t]$. It that the UMVUE of $g(\lambda)$?

Exercise 19.8.10. Suppose X and Y are independent, with $X \sim$ Poisson(2θ) and $Y \sim$ Poisson$(2(1 - \theta))$, $\theta \in (0, 1)$. (This is the same model as in Exercise 13.8.2.) (a) How many unbiased estimators are there of $g(\theta)$ when (i) $g(\theta) = \theta$? (ii) $g(\theta) = 0$? (b) Is (X, Y) sufficient? Is (X, Y) minimal sufficient? Is the model complete for (X, Y)? (c) Find Fisher's information. (d) Consider the unbiased estimators $\delta_1(x, y) = x/2$, $\delta_2(x, y) = 1 - y/2$, and $\delta_3(x, y) = (x - y)/4 + 1/2$. None of these achieve the CRLB. But for each estimator, consider $CRLB_\theta / Var_\theta[\delta_i]$. For each estimator, this ratio equals 1 for some $\theta \in [0, 1]$. Which θ for which estimator?

Exercise 19.8.11. For each of the models below, say whether it is an exponential family model. If it is, give the natural parameter and sufficient statistic, and say whether the model for the sufficient statistic is complete or not, justifying your assertion. (a) $X_1, \ldots, X_n$ are iid $N(\mu, \mu^2)$, where $\mu \in \mathbb{R}$. (b) $X_1, \ldots, X_n$ are iid $N(\theta, \theta)$, where $\theta \in (0, \infty)$. (c) $X_1, \ldots, X_n$ are iid Laplace(μ), where $\mu \in \mathbb{R}$. (d) $X_1, \ldots, X_n$ are iid Beta(α, β), where $(\alpha, \beta) \in (0, \infty)^2$.

Exercise 19.8.12. Suppose $\mathbf{X} \sim$ Multinomial$(n, \mathbf{p})$, where n is fixed and

$$\mathbf{p} \in \{\mathbf{p} \in \mathbb{R}^K \mid 0 < p_i \text{ for all } i = 1, \ldots, K, \text{ and } p_1 + \cdots + p_K = 1\}. \qquad (19.101)$$

(a) Find a sufficient statistic for which the model is complete, and show that it is complete. (b) Is the model complete for $\mathbf{X}$? Why or why not? (c) Find the UMVUE, if it exists, for p_1. (d) Find the UMVUE, if it exists, for $p_1 p_2$.

Exercise 19.8.13. Go back to the fruit fly example, where in Exercise 14.9.2 we have that $(N_{00}, N_{01}, N_{10}, N_{11}) \sim$ Multinomial$(n, \mathbf{p}(\theta))$, with

$$\mathbf{p}(\theta) = (\tfrac{1}{2}(1 - \theta)(2 - \theta), \tfrac{1}{2}\theta(1 - \theta), \tfrac{1}{2}\theta(1 - \theta), \tfrac{1}{2}\theta(1 + \theta)). \qquad (19.102)$$

(a) Show that for $n > 2$, the model is a two-dimensional exponential family with sufficient statistic (N_{00}, N_{11}). Give the natural parameter $(\eta_1(\theta), \eta_2(\theta))$. (b) Sketch the parameter space of the natural parameter. Does it contain a two-dimensional open rectangle? (c) The Dobzhansky estimator for θ is given in (6.65). It is unbiased with variance $3\theta(1 - \theta)/4n$. The Fisher information is given in (14.110). Does the Dobzhansky estimator achieve the CRLB? What is the ratio of the CLRB to the variance? Does the ratio approach 1 as $n \to \infty$? [Hint: The ratio is the same as the asymptotic efficiency found in Exercise 14.9.2(c).]

Exercise 19.8.14. Suppose

$$\begin{pmatrix} X_1 \\ Y_1 \end{pmatrix}, \ldots, \begin{pmatrix} X_n \\ Y_n \end{pmatrix} \text{ are iid } N_2 \left(\begin{pmatrix} 0 \\ 0 \end{pmatrix}, \sigma^2 \begin{pmatrix} 1 & \rho \\ \rho & 1 \end{pmatrix} \right), \qquad (19.103)$$

where $(\sigma^2, \rho) \in (0, \infty) \times (-1, 1)$, and $n > 2$. (a) Write the model as a two-dimensional exponential family. Give the natural parameter and sufficient statistic. (b) Sketch the parameter space for the natural parameter. (First, show that $\theta_2 = -2\theta_1 \rho$.) Does it contain an open nonempty two-dimensional rectangle? (c) Is the model for the sufficient statistic complete? Why or why not? (d) Find the expected value of the sufficient statistic (there are two components). (e) Is there a UMVUE for σ^2? If so, what is it?

Exercise 19.8.15. Take the same model as in Exercise 19.8.14, but with $\sigma^2 = 1$, so that the parameter is just $\rho \in (-1, 1)$. (a) Write the model as a two-dimensional exponential family. Give the natural parameter and sufficient statistic. (b) Sketch the parameter space for the natural parameter. Does it contain an open nonempty two-dimensional rectangle? (c) Find the expected value of the sufficient statistic. (d) Is the model for the sufficient statistic complete? Why or why not? (e) Find an unbiased estimator of ρ that is a function of the sufficient statistic. Either show this estimator is UMVUE, or find another unbiased estimator of ρ that is a function of the sufficient statistic.

Exercise 19.8.16. Suppose $\mathbf{Y} \sim N(\mathbf{x}\boldsymbol{\beta}, \sigma^2 \mathbf{I}_n)$ as in (12.10), where $\mathbf{x}'\mathbf{x}$ is invertible, so that we have the normal linear model. Exercise 13.8.19 shows that $(\hat{\boldsymbol{\beta}}, SS_e)$ is the sufficient statistic, where $\hat{\boldsymbol{\beta}}$ is the least squares estimate of $\boldsymbol{\beta}$, and $SS_e = \|\mathbf{y} - \hat{\boldsymbol{\beta}}\mathbf{x}\|^2$, the sum of squared residuals. (a) Show that the pdf of $\mathbf{Y}$ is an exponential family density with natural statistic $(\hat{\boldsymbol{\beta}}, SS_e)$. What is the natural parameter? [Hint: See the likelihood in (13.89).] (b) Show that the model is complete for this sufficient statistic. (c) Argue that $\hat{\beta}_i$ is the UMVUE of β_i for each i. (d) What is the UMVUE of σ^2? Why?

Exercise 19.8.17. Suppose X_1 and X_2 are iid Uniform$(0, \theta)$, with $\theta \in (0, \infty)$. (a) Let $(X_{(1)}, X_{(2)})$ be the order statistics. Find the space and pdf of the order statistics. Are the order statistics sufficient? (b) Find the space and pdf of $W = X_{(1)}/X_{(2)}$. [Find $P[W \leq w \,|\, \theta]$ in terms of $(X_{(1)}, X_{(2)})$, then differentiate. The pdf should not depend on θ.] What is $E[W]$? (c) Is the model for $(X_{(1)}, X_{(2)})$ complete? Why or why not? (d) We know that $T = X_{(2)}$ is a sufficient statistic. Find the space and pdf of T. (e) Find $E_\theta[T]$, and an unbiased estimator of θ. (f) Show that the model for T is complete. (Note that if $\int_0^\theta h(t)dt = 0$ for all θ, then $\frac{\partial}{\partial \theta} \int_0^\theta h(t)dt = 0$ for almost all θ. See Section 19.2.2 for general n.) (g) Is the estimator in part (e) the UMVUE?

Exercise 19.8.18. Suppose X_{ijk}'s are independent, $i = 1, 2; j = 1, 2; k = 1, 2$. Let $X_{ijk} \sim N(\mu + \alpha_i + \beta_j, 1)$, where $(\mu, \alpha_1, \alpha_2, \beta_1, \beta_2) \in \mathbb{R}^5$. These data can be thought of as observations from a two-way analysis of variance with two observations per cell:

	Column 1	Column 2	
Row 1	X_{111}, X_{112}	X_{121}, X_{122}	(19.104)
Row 2	X_{211}, X_{212}	X_{221}, X_{222}	

Let

$$X_{i++} = X_{i11} + X_{i12} + X_{i21} + X_{i22}, i = 1, 2 \text{ (row sums)},$$
$$X_{+j+} = X_{1j1} + X_{1j2} + X_{2j1} + X_{2j2}, j = 1, 2 \text{ (column sums)}, \qquad (19.105)$$

and X_{+++} be the sum of all eight observations. (a) Write the pdf of the data as a $K = 5$ dimensional exponential family, where the natural parameter is $(\mu, \alpha_1, \alpha_2, \beta_1, \beta_2)$. What is the natural sufficient statistic? (b) Rewrite the model as a $K = 3$ dimensional

344

Chapter 19. Optimal Estimators

exponential family. (Note that $X_{2++} = X_{+++} - X_{1++}$, and similarly for the columns.) What are the natural parameter and natural sufficient statistic? (c) Is the model in part (b) complete? (What is the space of the natural parameter?) (d) Find the expected values of the natural sufficient statistics from part (b) as functions of $(\mu, \alpha_1, \alpha_2, \beta_1, \beta_2)$. (e) For each of the following, find an unbiased estimator if possible, and if you can, say whether it is UMVUE or not. (Two are possible, two not.) (i) μ; (ii) $\mu + \alpha_1 + \beta_1$; (iii) β_1; (iv) $\beta_1 - \beta_2$.

Exercise 19.8.19. Suppose $X_1, \ldots, X_n$ are iid with the location family model given by density f, and $\delta(\mathbf{x})$ is a shift-equivariant estimator of the location parameter θ. Let $Z_1, \ldots, Z_n$ be the iid random variables with pdf $f(z_i)$, so that $X_i =^{\mathcal{D}} Z_i + \theta$. (a) Show that $\delta(\mathbf{X}) =^{\mathcal{D}} \delta(\mathbf{Z}) + \theta$. (b) Assuming the mean and variance exists, show that $E_\theta[\delta(\mathbf{X})] = E[\delta(\mathbf{Z})] + \theta$ and $Var_\theta[\delta(\mathbf{X})] = Var[\delta(\mathbf{Z})]$. (c) Show that therefore $Bias_\theta[\delta(\mathbf{X})] = E[\delta(\mathbf{Z})]$ and $MSE_\theta[\delta(\mathbf{X})] = E[\delta(\mathbf{Z})^2]$, which imply (19.68).

Exercise 19.8.20. Suppose $Z_1, \ldots, Z_n$ are iid with density $f(z_i)$, and let $Y_1 = Z_1 - Z_n, \ldots, Y_{n-1} = Z_{n-1} - Z_n$ as in (19.78). (a) Consider the one-to-one transformation $(z_1, \ldots, z_n) \leftrightarrow (y_1, \ldots, y_{n-1}, z_n)$. Show that the inverse transformation is given as in (19.79), its Jacobian is 1, and its pdf is $f(z_n) \prod_{i=1}^{n-1} f(y_i + z_n)$ as in (19.80). (b) Show that

$$\frac{\int_{-\infty}^{\infty} (x_n - z_n) f(z_n) \prod_{i=1}^{n-1} f(x_i - x_n + z_n) dz_n}{\int_{-\infty}^{\infty} f(z_n) \prod_{i=1}^{n-1} f(x_i - x_n + z_n) dz_n} = \frac{\int_{-\infty}^{\infty} \theta \prod_{i=1}^{n} f(x_i - \theta) d\theta}{\int_{-\infty}^{\infty} \prod_{i=1}^{n} f(x_i - \theta) d\theta}, \quad (19.106)$$

verifying the expression of the Pitman estimator in (19.85). [Hint: Use the substitution $\theta = x_n - z_n$ in the integrals.]

Exercise 19.8.21. Consider the location family model with just one X. Show that the Pitman estimator is $X - E_0[X]$ if the expectation exists.

Exercise 19.8.22. Show that $\overline{X}$ is the Pitman estimator in the $N(\theta, 1)$ location family model. [Hint: In the pdf of the normal, write $\sum(x_i - \theta)^2 = n(\overline{x} - \theta)^2 + \sum(x_i - \overline{x})^2$.]

Exercise 19.8.23. Find the Pitman estimator in the Uniform$(\theta, \theta + 1)$ location family model, so that $f(x) = I[0 < x < 1]$. [Hint: Note that the density $\prod I[\theta < x_i < \theta + 1] = 1$ if $x_{(n)} - 1 < \theta < x_{(1)}$, and 0 otherwise.]

Chapter 20

The Decision-Theoretic Approach

20.1 Binomial estimators

In the previous chapter, we found the best estimators when looking at a restricted set of estimators (unbiased or shift-equivariant) in some fairly simple models. More generally, one may not wish to restrict choices to just unbiased estimators, say, or there may be no obvious equivariance or other structural limitations to impose. Decision theory is one approach to this larger problem, where the key feature is that some procedures do better for some values of the parameter, and other procedures do better for other values of the parameter.

For example, consider the simple situation where Z_1 and Z_2 are iid Bernoulli(θ), $\theta \in (0,1)$, and we wish to estimate θ using the mean square error as the criterion. Here are five possible estimators:

$$\begin{array}{lll}
\delta_1(\mathbf{z}) = (z_1 + z_2)/2 & \text{(The MLE, UMVUE);} & \\
\delta_2(\mathbf{z}) = z_1/3 + 2z_2/3 & \text{(Unbiased);} & \\
\delta_3(\mathbf{z}) = (z_1 + z_2 + 1)/6 & \text{(Bayes wrt Beta(1,3) prior);} & (20.1) \\
\delta_4(\mathbf{z}) = (z_1 + z_2 + 1/\sqrt{2})/(2 + \sqrt{2}) & \text{(Bayes wrt Beta($1/\sqrt{2}, 1/\sqrt{2}$) prior);} & \\
\delta_5(\mathbf{z}) = 1/2 & \text{(Constant at $1/2$).} &
\end{array}$$

Figure 20.1 graphs the MSEs, here called the "risks." Notice the risk functions cross, that is, for the most part when comparing two estimators, sometimes one is better, sometimes the other is. The one exception is that δ_1 is always better than δ_2. We know this because both are unbiased, and the first one is the UMVUE. Decision-theoretically, we say that δ_2 is **inadmissible** among these five estimators. The other four are **admissible** among these five, since none of them is dominated by any of the others. Even $\delta_5(\mathbf{z}) = 1/2$ is admissible, since its MSE at $\theta = 1/2$ is zero, and no other estimator can claim such.

Rather than evaluate the entire curve, one may wish to know what is the worst risk each estimator has. An estimator with the lowest worst risk is called **minimax**. The table contains the maximum risk for each of the estimators:

	δ_1	δ_2	δ_3	δ_4	δ_5	
Maximum risk	0.1250	0.1389	0.1600	0.0429	0.2500	(20.2)

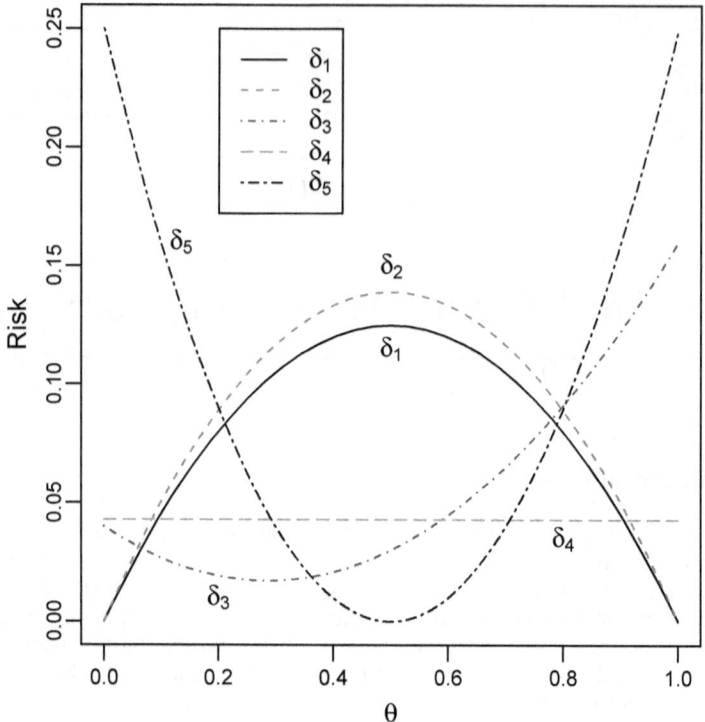

Figure 20.1: MSEs for the estimators given in (20.1).

As can be seen from either the table or the graph, the minimax procedure is δ_4, with a maximum risk of $0.0429(= 1/(12 + 8\sqrt{2}))$. In fact, the risk for this estimator is constant, which the $\sqrt{2}$'s in the estimator were chosen to achieve.

This example exhibits the three main concepts of statistical decision theory: Bayes, admissibility, and minimaxity. The next section presents the formal setup.

20.2   Basic setup

We assume the usual statistical model: random vector $\mathbf{X}$, space $\mathcal{X}$, and set of distributions $\mathcal{P} = \{P_{\boldsymbol{\theta}} \,|\, \boldsymbol{\theta} \in \mathcal{T}\}$, where $\mathcal{T}$ is the parameter space. The decision-theoretic approach supposes an **action space** $\mathcal{A}$ that specifies the possible "actions" we might take, which represent the possible outputs of the inference. For example, in estimation the action is the estimate, in testing the action is accept or reject the null, and in model selection the action is the model selected.

A **decision procedure** specifies which action to take for each possible value of the data. Formally, a decision procedure is a function $\delta(\mathbf{x})$,

$$\delta : \mathcal{X} \longrightarrow \mathcal{A}. \tag{20.3}$$

The above is a *nonrandomized* decision procedure. A *randomized* procedure would depend on not just the data **x**, but also some outside randomization element. See Section 20.7.

A good procedure is one that takes good actions. To measure how good, we need a **loss function** that specifies a penalty for taking a particular action when a particular distribution obtains. Formally, the loss function L is a function

$$L : \mathcal{A} \times \mathcal{T} \longrightarrow [0, \infty). \tag{20.4}$$

When estimating a function $g(\boldsymbol{\theta})$, common loss functions are squared-error loss,

$$L(a, \boldsymbol{\theta}) = (a - g(\boldsymbol{\theta}))^2, \tag{20.5}$$

and absolute-error loss,

$$L(a, \boldsymbol{\theta}) = |a - g(\boldsymbol{\theta})|. \tag{20.6}$$

In hypothesis testing or model selection, a "$0 - 1$" loss is common, where you lose 0 by making the correct decision, and lose 1 if you make a mistake.

A frequentist evaluates procedures by their behavior over experiments. In this decision-theoretic framework, the **risk function** for a particular decision procedure δ is key. The risk is the expected loss, where δ takes place of the a. It is a function of $\boldsymbol{\theta}$:

$$R(\boldsymbol{\theta}; \delta) = E_{\boldsymbol{\theta}}[L(\delta(\mathbf{X}), \boldsymbol{\theta})] = E[L(\delta(\mathbf{X}), \boldsymbol{\Theta}) \mid \boldsymbol{\Theta} = \boldsymbol{\theta}]. \tag{20.7}$$

The two expectations are the same, but the first is written for frequentists, and the second for Bayesians. In estimation problems with L being squared-error loss (20.5), the risk is the mean square error:

$$R(\boldsymbol{\theta}; \delta) = E_{\boldsymbol{\theta}}[(\delta(\mathbf{X}) - g(\boldsymbol{\theta}))^2] = MSE_{\boldsymbol{\theta}}[\delta]. \tag{20.8}$$

The idea is to choose a δ with small risk. The challenge is that usually there is no one procedure that is best for every $\boldsymbol{\theta}$, as we saw in Section 20.1. One way to choose is to restrict consideration to a subset of procedures, e.g., unbiased estimators as in Section 19.1, or shift-equivariant ones as in Section 19.6, or in hypothesis testing, tests with set level α. Often, one either does not wish to use such restrictions, or cannot. In the absence of a uniquely defined best procedure, frequentists have several possible strategies, among them determining admissible procedures, minimax procedures, or Bayes procedures.

20.3 Bayes procedures

One method for selecting among various δ's is to find one that minimizes the average of the risk, where the average is taken over $\boldsymbol{\theta}$. This averaging needs a probability measure on $\mathcal{T}$. From a Bayesian perspective, this distribution is the prior. From a frequentist perspective, it may or may not reflect prior belief, but it should be "reasonable." The procedure that minimizes this average is the **Bayes procedure** corresponding to the distribution. We first extend the definition of risk to a function of distributions π on $\mathcal{T}$, where the **Bayes risk** at π is the expectation of the risk over $\boldsymbol{\theta}$:

$$R(\pi; \delta) = E_{\pi}[R(\boldsymbol{\theta}; \delta)]. \tag{20.9}$$

Definition 20.1. *For given risk function* $R(\theta; \delta)$, *set of procedures* $\mathcal{D}$, *and distribution* π *on* $\mathcal{T}$, *a* **Bayes procedure** *with respect to (wrt)* $\mathcal{D}$ *and* π *is a procedure* $\delta_\pi \in \mathcal{D}$ *with* $R(\pi, \delta_\pi) < \infty$ *that minimizes the Bayes risk over* δ, *i.e.,*

$$R(\pi; \delta_\pi) \le R(\pi; \delta) \ \text{for any} \ \delta \in \mathcal{D}. \tag{20.10}$$

It might look daunting to minimize over an entire function δ, but we can reduce the problem to minimizing over a single value by using an iterative expected value. With both $\mathbf{X}$ and Θ random, the Bayes risk is the expected value of the loss over the joint distribution of $(\mathbf{X}, \Theta)$, hence can be written as the expected value of the conditional expected value of L given $\mathbf{X}$:

$$\begin{aligned} R(\pi; \delta) &= E[L(\delta(\mathbf{X}), \Theta)] \\ &= E[e_L(\mathbf{X})], \end{aligned} \tag{20.11}$$

where

$$e_L(\mathbf{x}) = E[L(\delta(\mathbf{x}), \Theta) \mid \mathbf{X} = \mathbf{x}]. \tag{20.12}$$

In that final expectation, Θ is random, having the posterior distribution given $\mathbf{X} = \mathbf{x}$. In (20.12), because $\mathbf{x}$ is fixed, $\delta(\mathbf{x})$ is just a constant, hence it may not be too difficult to minimize $e_L(\mathbf{x})$ over $\delta(\mathbf{x})$. If we find the $\delta(\mathbf{x})$ to minimize $e_L(\mathbf{x})$ for each $\mathbf{x}$, then we have also minimized the overall Bayes risk (20.11). Thus a Bayes procedure is δ_π such that

$$\delta_\pi(\mathbf{x}) \ \text{minimizes} \ E[L(\delta(\mathbf{x}), \Theta) \mid \mathbf{X} = \mathbf{x}] \ \text{over} \ \delta(\mathbf{x}) \ \text{for each} \ \mathbf{x} \in \mathcal{X}. \tag{20.13}$$

In estimation with squared-error loss, (20.12) becomes

$$e_L(\mathbf{x}) = E[(\delta(\mathbf{x}) - g(\Theta))^2 \mid \mathbf{X} = \mathbf{x}]. \tag{20.14}$$

That expression is minimized with $\delta(\mathbf{x})$ being the mean, in this case the conditional (posterior) mean of g:

$$\delta_\pi(\mathbf{x}) = E[g(\Theta) \mid \mathbf{X} = \mathbf{x}]. \tag{20.15}$$

If the loss is absolute error (20.6), the Bayes procedure is the posterior median.

A Bayesian does not care about $\mathbf{x}$'s not observed, hence would immediately go to the conditional equation (20.13), and use the resulting $\delta_\pi(\mathbf{x})$. It is interesting that the decision-theoretic approach appears to bring Bayesians and frequentists together. They do end up with the same procedure, but from different perspectives. The Bayesian is trying to limit expected losses given the data, while the frequentist is trying to limit average expected losses, taking expected values as the experiment is repeated.

20.4 Admissibility

Recall Figure 20.1 in Section 20.1. The important message in the graph is that none of the five estimators is obviously "best" in terms of MSE. In addition, only one of them is discardable in the sense that another estimator is better. A fairly weak criterion is this lack of discardability. Formally, the estimator δ' is said to **dominate** the estimator δ if

$$\begin{aligned} R(\theta; \delta') &\le R(\theta; \delta) \ \text{for all} \ \theta \in \mathcal{T}, \ \text{and} \\ R(\theta; \delta') &< R(\theta; \delta) \ \text{for at least one} \ \theta \in \mathcal{T}. \end{aligned} \tag{20.16}$$

In Figure 20.1, δ_1 dominates δ_2, but there are no other such dominations. The concept of admissibility is based on lack of domination.

Definition 20.2. *Let $\mathcal{D}$ be a set of decision procedures. A $\delta \in \mathcal{D}$ is **inadmissible** among procedures in $\mathcal{D}$ if there is another $\delta' \in \mathcal{D}$ that dominates δ. If there is no such δ', then δ is **admissible** among procedures in $\mathcal{D}$.*

A handy corollary of the definition is that δ is admissible if

$$R(\boldsymbol{\theta}; \delta') \leq R(\boldsymbol{\theta}; \delta) \text{ for all } \theta \in \mathcal{T} \Rightarrow R(\boldsymbol{\theta}; \delta') = R(\boldsymbol{\theta}; \delta) \text{ for all } \theta \in \mathcal{T}. \quad (20.17)$$

If $\mathcal{D}$ is the set of unbiased estimators, then the UMVUE is admissible in $\mathcal{D}$, and any other unbiased estimator is inadmissible. Similarly, if $\mathcal{D}$ is the set of shift-equivariant procedures (assuming that restriction makes sense for the model), then the best shift-invariant estimator is the only admissible estimator in $\mathcal{D}$. In the example of Section 20.1 above, if $\mathcal{D}$ consists of the five given estimators, then all but δ_2 are admissible in $\mathcal{D}$.

The presumption is that one would not want to use an inadmissible procedure, at least if risk is the only consideration. Other considerations, such as intuitivity or computational ease, may lead one to use an inadmissible procedure, provided it cannot be dominated by much. Conversely, any admissible procedure is presumed to be at least plausible, although there are some strange ones.

It is generally not easy to decide whether a procedure is admissible or not. For the most part, Bayes procedures are admissible. A Bayes procedure with respect to a prior π has good behavior averaging over the $\boldsymbol{\theta}$. Any procedure that dominated that procedure would also have to have at least as good Bayes risk, hence would also be Bayes. The next lemma collects some sufficient conditions for a Bayes estimator to be admissible. The lemma also holds if everything is stated relative to a restricted set of procedures $\mathcal{D}$.

Lemma 20.3. *Suppose δ_π is Bayes wrt to the prior π. Then it is admissible if any of the following hold: (a) It is admissible among the set of estimators that are Bayes wrt π. (b) It is the unique Bayes procedure, up to equivalence, wrt π. That is, if δ'_π is also Bayes wrt π, then $R(\boldsymbol{\theta}; \delta_\pi) = R(\boldsymbol{\theta}; \delta'_\pi)$ for all $\boldsymbol{\theta} \in \mathcal{T}$. (c) The parameter space is finite or countable, $\mathcal{T} = \{\boldsymbol{\theta}_1, \ldots, \boldsymbol{\theta}_K\}$, and the pmf π is positive, $\pi(\boldsymbol{\theta}_k) > 0$ for each $k = 1, \ldots, K$. (K may be $+\infty$.) (d) The parameter space is $\mathcal{T} = (a, b)$ $(-\infty \leq a < b \leq \infty)$, the risk function for any procedure δ is continuous in $\boldsymbol{\theta}$, and for any nonempty interval $(c, d) \subset \mathcal{T}, \pi(c, d) > 0$. (e) The parameter space $\mathcal{T}$ is an open subset of $\mathbb{R}^p$, the risk function for any procedure δ is continuous in $\boldsymbol{\theta}$, and for any nonempty open set $\mathcal{B} \subset \mathcal{T}, \pi(\mathcal{B}) > 0$.*

The proofs of parts (b) and (c) are found in Exercises 20.9.1 and 20.9.2. Note that the condition on π in part (d) holds if π has a pdf that is positive for all θ. Part (e) is a multivariate analog of part (d), left to the reader.

Proof. (a) Suppose the procedure δ' satisfies $R(\boldsymbol{\theta}; \delta') \leq R(\boldsymbol{\theta}; \delta_\pi)$ for all $\boldsymbol{\theta} \in \mathcal{T}$. Then by taking expected values over θ wrt π,

$$R(\pi; \delta') = E_\pi[R(\boldsymbol{\Theta}; \delta')] \leq E_\pi[R(\boldsymbol{\Theta}; \delta_\pi)] = R(\pi; \delta_\pi). \quad (20.18)$$

Thus δ' is also Bayes wrt π. But by assumption, δ_π is admissible among Bayes procedures, hence δ' must have the same risk as δ_π for all θ, which by (20.17) proves δ_π is admissible.

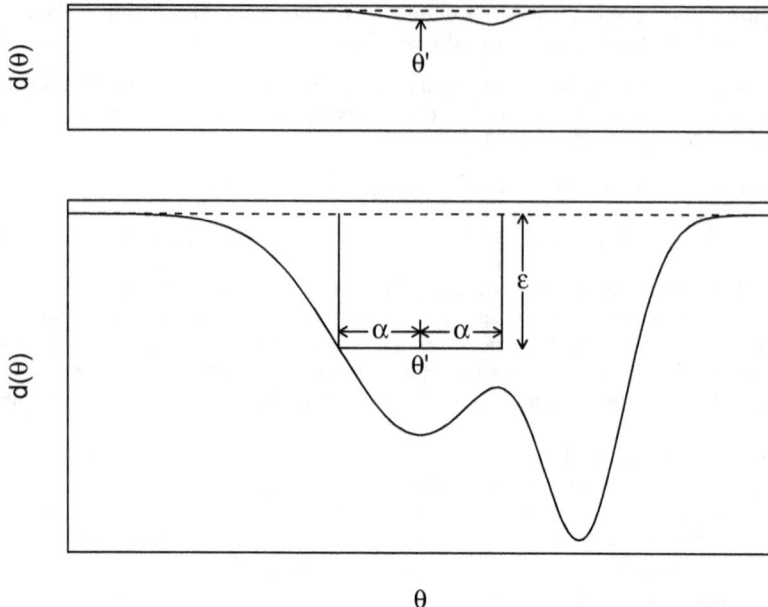

Figure 20.2: Illustration of domination with a continuous risk function. Here, δ' dominates δ. The function graphs the difference in risks, $d(\theta) = R(\theta; \delta') - R(\theta; \delta)$. The top graph shows the big view, where the dashed line represents zero. The θ' is a point at which δ' is strictly better than δ. The bottom graph zooms in on the area near θ', showing $d(\theta) \leq -\epsilon$ for $\theta' - \alpha < \theta < \theta' + \alpha$.

(d) Again suppose δ' has risk no larger than δ_π's, and for some $\theta' \in \mathcal{T}$, $R(\theta'; \delta') < R(\theta'; \delta_\pi)$. By the continuity of the risk functions, the inequality must hold for an interval around θ'. That is, there exist $\alpha > 0$ and $\epsilon > 0$ such that

$$R(\theta; \delta') - R(\theta; \delta_\pi) \leq -\epsilon \text{ for } \theta' - \alpha < \theta < \theta' + \alpha. \tag{20.19}$$

See Figure 20.2. (Since $\mathcal{T}$ is open, α can be taken small enough so that $\theta' \pm \alpha$ are in $\mathcal{T}$.) Then integrating over θ,

$$R(\pi; \delta') - R(\pi; \delta_\pi) \leq -\epsilon \pi(\theta' - \alpha < \theta < \theta' + \alpha) < 0, \tag{20.20}$$

meaning δ' has better Bayes risk than δ_π. This is a contradiction, hence there is no such θ', i.e., δ_π is admissible.   □

In Section 20.8, we will see that when the parameter space is finite (and other conditions hold), all admissible tests are Bayes. In general, not all admissible procedures are Bayes, but they are at least limits of Bayes procedures in some sense. Exactly which limits are admissible is a bit delicate, though. In any case, at least approximately, one can think of a procedure being admissible if there is some Bayesian somewhere, or a sequence of Bayesians, who would use it. See Section 22.3 for the hypothesis testing case. Ferguson (1967) is an accessible introduction to the relationship between Bayes procedures and admissibility; Berger (1993) and Lehmann and Casella (2003) contain additional results and pointers to more recent work.

20.5 Estimating a normal mean

Consider estimating a normal mean based on an iid sample with known variance using squared-error loss, so that the risk is the mean square error. The sample mean is the obvious estimator, which is the UMVUE, MLE, best shift-equivariant estimator, etc. It is not a Bayes estimator because it is unbiased, as seen in Exercise 11.7.21. It is the posterior mean for the improper prior Uniform$(-\infty, \infty)$ on μ, as in Exercise 11.7.16. A posterior mean using an improper prior is in this context called a **generalized Bayes procedure**. It may or may not be admissible, but any admissible procedure here has to be Bayes or generalized Bayes wrt some prior, proper or improper as the case may be. See Sacks (1963) or Brown (1971).

We first simplify the model somewhat by noting that the sample mean is a sufficient statistic. From the Rao-Blackwell theorem (Theorem 13.8 on page 210), any estimator not a function of just the sufficient statistic can be improved upon by a function of the sufficient statistic that has the same bias but smaller variance, hence smaller mean square error. Thus we may as well limit ourselves to function of the mean, which can be represented by the model

$$X \sim N(\mu, 1), \quad \mu \in \mathbb{R}. \tag{20.21}$$

We wish to show that $\delta_0(x) = x$ is admissible. Since it is unbiased and has variance 1, its risk is $R(\mu ; \delta_0) = 1$.

We will use the method of Blyth (1951) to show admissibility. The first step is to find a sequence of proper priors that approximates the uniform prior. We will take π_n to be the $N(0, n)$ prior on μ. Exercise 7.8.14 shows that the Bayes estimator wrt π_n is

$$\delta_n(x) = E_{\pi_n}[M \mid X = x] = \frac{n}{n+1} x, \tag{20.22}$$

which is admissible. Exercise 20.9.5 shows that the Bayes risks are

$$R(\pi_n ; \delta_0) = 1 \text{ and } R(\pi_n ; \delta_n) = \frac{n}{n+1}. \tag{20.23}$$

Thus the Bayes risk of δ_n is very close to that of δ_0 for large n, suggesting that δ_0 must be close to admissible.

As in the proof of Lemma 20.3(d), assume that the estimator δ' dominates δ_0. Since $R(\mu ; \delta_0) = 1$, it must be that

$$R(\mu ; \delta') \leq 1 \text{ for all } \mu, \text{ and } R(\mu' ; \delta') < 1 \text{ for some } \mu'. \tag{20.24}$$

We look at the difference in Bayes risks:

$$R(\pi_n ; \delta_0) - R(\pi_n ; \delta') = (R(\pi_n ; \delta_0) - R(\pi_n ; \delta_n)) + (R(\pi_n ; \delta_n) - R(\pi_n ; \delta'))$$
$$\leq R(\pi_n ; \delta_0) - R(\pi_n ; \delta_n)$$
$$= \frac{1}{n+1}, \tag{20.25}$$

the inequality holding since δ_n is Bayes wrt π_n. Using π_n explicitly, we have

$$\int_{-\infty}^{\infty} (1 - R(\mu\,;\delta')) \frac{1}{\sqrt{2\pi n}} e^{-\frac{\mu^2}{2n}}\, d\mu \le \frac{1}{n+1}. \tag{20.26}$$

Both sides look like they would go to zero, but multiplying by $\sqrt{n}$ may produce a nonzero on the left:

$$\int_{-\infty}^{\infty} (1 - R(\mu\,;\delta')) \frac{1}{\sqrt{2\pi}} e^{-\frac{\mu^2}{2n}}\, d\mu \le \frac{\sqrt{n}}{n+1}. \tag{20.27}$$

The risk function $R(\mu\,;\delta')$ is continuous in μ (see Ferguson (1967), Section 3.7), which together with (20.23) shows that there exist $\alpha > 0$ and $\epsilon > 0$ such that

$$1 - R(\mu\,;\delta') \ge \epsilon \text{ for } \mu' - \alpha < \mu < \mu' + \alpha. \tag{20.28}$$

See Figure 20.2. Hence from (20.27) we have that

$$\frac{\epsilon}{\sqrt{2\pi}} \int_{\mu'-\alpha}^{\mu'+\alpha} e^{-\frac{\mu^2}{2n}}\, d\mu \le \frac{\sqrt{n}}{n+1}. \tag{20.29}$$

Letting $n \to \infty$, we obtain

$$\frac{2\alpha\epsilon}{\sqrt{2\pi}} \le 0, \tag{20.30}$$

which is a contradiction. Thus there is no δ' that dominates δ_0, hence δ_0 is admissible.

To recap, the basic idea in using Blyth's method to show δ_0 is admissible is to find a sequence of priors π_n and constants c_n (which are $\sqrt{n}$ in our example) so that if δ' dominates δ_0,

$$c_n(R(\pi_n\,;\delta_0) - R(\pi_n\,;\delta')) \to C > 0 \text{ and } c_n(R(\pi_n\,;\delta_0) - R(\pi_n\,;\delta_n)) \to 0. \tag{20.31}$$

If the risk function is continuous and $\mathcal{T} = (a, b)$, then the first condition in (20.31) holds if

$$c_n \pi_n(c, d) \to C' > 0 \tag{20.32}$$

for any $c < d$ such that $(c, d) \subset \mathcal{T}$.

It may not always be possible to use Blyth's method, as we will see in the next section.

20.5.1 Stein's surprising result

Admissibility is generally considered a fairly weak criterion. An admissible procedure does not have to be very good everywhere, but just have something going for it. Thus the statistical community was rocked when Charles Stein (Stein, 1956b) showed that in the multivariate normal case, the usual estimator of the mean could be inadmissible.

The model has random vector

$$\mathbf{X} \sim N(\boldsymbol{\mu}, \mathbf{I}_p), \tag{20.33}$$

with $\mu \in \mathbb{R}^p$. The objective is to estimate μ with squared-error loss, which in this case is multivariate squared error:

$$L(\mathbf{a}, \mu) = \|\mathbf{a} - \mu\|^2 = \sum_{i=1}^{p} (a_i - \mu_i)^2. \tag{20.34}$$

The risk is then the sum of the mean square errors for the individual estimators of the μ_i's. Again the obvious estimator is $\delta_0(\mathbf{x}) = \mathbf{x}$, which has risk

$$R(\mu; \delta) = E_\mu[\|\mathbf{X} - \mu\|^2] = p \tag{20.35}$$

because the X_i's all have variance 1. As we saw above, when $p = 1$ δ_0 is admissible.

When $p = 2$, we could try to use Blyth's method with prior on the bivariate μ being $N(0, n\mathbf{I}_2)$, but in the step analogous to (20.27), we would multiply by n instead of $\sqrt{n}$, hence the limit on the right-hand side of (20.30) would be 1 instead of 0, thus not necessarily be a contradiction. Brown and Hwang (1982) present a more complicated prior for which Blyth's method does prove admissibility of $\delta_0(\mathbf{x}) = \mathbf{x}$.

The surprise is that when $p \geq 3$, δ is inadmissible. The most famous estimator that dominates it is the **James-Stein estimator** (James and Stein, 1961),

$$\delta_{JS}(\mathbf{x}) = \left(1 - \frac{p-2}{\|\mathbf{x}\|^2}\right) \mathbf{x}. \tag{20.36}$$

It is a *shinkage* estimator, because it takes the usual estimator, and shrinks it (towards 0 in this case), at least when $(p-2)/\|\mathbf{x}\|^2 < 1$. Throughout the 1960s and 1970s, there was a frenzy of work on various shrinkage estimators. They are still quite popular. The domination result is not restricted to normality. It is quite broad. The general notion of shrinkage is very important in machine learning, where better predictions are found by restraining estimators from becoming too large using regularization (Section 12.5).

To find the risk function for the James-Stein estimator when $p \geq 3$, start by writing

$$R(\mu; \delta_{JS}) = E_\mu[\|\delta_{JS}(\mathbf{X}) - \mu\|^2]$$

$$= E_\mu\left[\left\|\mathbf{X} - \mu - \frac{p-2}{\|\mathbf{X}\|^2}\mathbf{X}\right\|^2\right]$$

$$= E_\mu[\|\mathbf{X} - \mu\|^2] + E_\mu\left[\frac{(p-2)^2}{\|\mathbf{X}\|^2}\right] - 2(p-2)E_\mu\left[\frac{\mathbf{X}'(\mathbf{X} - \mu)}{\|\mathbf{X}\|^2}\right]. \tag{20.37}$$

The first term we recognize from (20.35) to be p.

Consider the third term, where

$$E_\mu\left[\frac{\mathbf{X}'(\mathbf{X} - \mu)}{\|\mathbf{X}\|^2}\right] = \sum_{i=1}^{p} E_\mu\left[\frac{X_i(X_i - \mu_i)}{\|\mathbf{X}\|^2}\right]. \tag{20.38}$$

We take each term in the summation separately. The first one can be written

$$E_\mu\left[\frac{X_1(X_1 - \mu_1)}{\|\mathbf{X}\|^2}\right] = \int_{-\infty}^{\infty} \cdots \int_{-\infty}^{\infty} \frac{x_1(x_1 - \mu_1)}{\|\mathbf{x}\|^2} \phi_{\mu_1}(x_1)dx_1 \phi_{\mu_2}(x_2)dx_2 \cdots \phi_{\mu_p}(x_p)dx_p, \tag{20.39}$$

where ϕ_{μ_i} is the $N(\mu_i, 1)$ pdf,

$$\phi_{\mu_i}(x_i) = \frac{1}{\sqrt{2\pi}} e^{-\frac{1}{2}(x_i - \mu_i)^2}. \tag{20.40}$$

Exercise 20.9.9 looks at the innermost integral, showing that

$$\int_{-\infty}^{\infty} \frac{x_1(x_1 - \mu_1)}{\|\mathbf{x}\|^2} \phi_{\mu_1}(x_1)dx_1 = \int_{-\infty}^{\infty} \left(\frac{1}{\|\mathbf{x}\|^2} - \frac{2x_1^2}{\|\mathbf{x}\|^4} \right) \phi_{\mu_1}(x_1)dx_1. \tag{20.41}$$

Replacing the innermost integral in (20.39) with (20.41) yields

$$E_{\boldsymbol{\mu}}\left[\frac{X_1(X_1 - \mu_1)}{\|\mathbf{X}\|^2} \right] = E_{\boldsymbol{\mu}}\left[\frac{1}{\|\mathbf{X}\|^2} - \frac{2X_1^2}{\|\mathbf{X}\|^4} \right]. \tag{20.42}$$

The same calculation works for $i = 2, \ldots, p$, so that from (20.38),

$$E_{\boldsymbol{\mu}}\left[\frac{\mathbf{X}'(\mathbf{X} - \boldsymbol{\mu})}{\|\mathbf{X}\|^2} \right] = \sum_{i=1}^{p} E_{\boldsymbol{\mu}}\left[\frac{1}{\|\mathbf{X}\|^2} - \frac{2X_i^2}{\|\mathbf{X}\|^4} \right]$$

$$= E_{\boldsymbol{\mu}}\left[\frac{p}{\|\mathbf{X}\|^2} \right] - E_{\boldsymbol{\mu}}\left[\frac{2\sum X_i^2}{\|\mathbf{X}\|^4} \right]$$

$$= E_{\boldsymbol{\mu}}\left[\frac{p-2}{\|\mathbf{X}\|^2} \right]. \tag{20.43}$$

Exercise 20.9.10 verifies that from (20.37),

$$R(\boldsymbol{\mu}; \delta_{JS}) = p - E_{\boldsymbol{\mu}}\left[\frac{(p-2)^2}{\|\mathbf{X}\|^2} \right]. \tag{20.44}$$

That's it! The expected value at the end is positive, so that the risk is less than p. That is

$$R(\boldsymbol{\mu}; \delta_{JS}) < p = R(\boldsymbol{\mu}; \delta) \quad \text{for all } \boldsymbol{\mu} \in \mathbb{R}^p, \tag{20.45}$$

meaning $\delta(\mathbf{x}) = \mathbf{x}$ is inadmissible.

How much does the James-Stein estimator dominate δ? It shrinks towards zero, so if the true mean is zero, one would expect the James-Stein estimator to be quite good. In fact, Exercise 20.9.10 shows that $R(\mathbf{0}; \delta_{JS}) = 2$. Especially when p is large, this risk is much less than that of δ, which is always p. Even for $p = 3$, the James-Stein risk is $2/3$ of δ's. The farther from $\mathbf{0}$ the $\boldsymbol{\mu}$ is, the less advantage the James-Stein estimator has. As $\|\boldsymbol{\mu}\| \to \infty$, with $\|\mathbf{X}\| \sim \chi_p^2(\|\boldsymbol{\mu}\|^2)$, the $E_{\boldsymbol{\mu}}[1/\|\mathbf{X}\|^2] \to 0$, so

$$\lim_{\|\boldsymbol{\mu}\| \to \infty} R(\boldsymbol{\mu}; \delta_{JS}) \longrightarrow p = R(\boldsymbol{\mu}; \delta). \tag{20.46}$$

If rather than having good risk at zero, one has a "prior" idea that the mean is near some fixed $\boldsymbol{\mu}_0$, one can instead shrink towards that vector:

$$\delta_{JS}^*(\mathbf{x}) = \left(1 - \frac{p-2}{\|\mathbf{x} - \boldsymbol{\mu}_0\|^2} \right) (\mathbf{x} - \boldsymbol{\mu}_0) + \boldsymbol{\mu}_0. \tag{20.47}$$

This estimator has the same risk as the regular James-Stein estimator, but with shifted parameter:

$$R(\mu\,; \delta_{JS}^*) = p - E_\mu \left[\frac{(p-2)^2}{\|\mathbf{X} - \mu_0\|^2} \right] = p - E_{\mu - \mu_0} \left[\frac{(p-2)^2}{\|\mathbf{X}\|^2} \right], \tag{20.48}$$

and has risk of 2 when $\mu = \mu_0$.

The James-Stein estimator itself is not admissible. There are many other similar estimators in the literature, some that dominate δ_{JS} but are not admissible (such as the "positive part" estimator that does not allow the shrinking factor to be negative), and many admissible estimators that dominate δ. See, e.g., Strawderman and Cohen (1971) and Brown (1971) for overviews.

20.6 Minimax procedures

Using a Bayes procedure involves choosing a prior π. When using an admissible estimator, one is implicitly choosing a Bayes, or close to a Bayes, procedure. One attempt to objectifying the choice of a procedure is for each procedure, see what its worst risk is. Then you choose the procedure that has the best worst, i.e., the minimax procedure. Next is the formal definition.

Definition 20.4. *Let $\mathcal{D}$ be a set of decision procedures. A $\delta \in \mathcal{D}$ is **minimax** among procedures in $\mathcal{D}$ if for any other $\delta' \in \mathcal{D}$,*

$$\sup_{\theta \in \mathcal{T}} R(\theta\,; \delta) \leq \sup_{\theta \in \mathcal{T}} R(\theta\,; \delta'). \tag{20.49}$$

For the binomial example with $n = 2$ in Section 20.1, Figure 20.1 graphs the risk functions of five estimators. Their maximum risks are given in (20.2), repeated here:

	δ_1	δ_2	δ_3	δ_4	δ_5	
Maximum risk	0.1250	0.1389	0.1600	0.0429	0.2500	(20.50)

Of these, δ_4 (the Bayes procedure wrt Beta$(1/\sqrt{2}, 1/\sqrt{2})$) has the lowest maximum, hence is minimax among these five procedures.

Again looking at Figure 20.1, note that the minimax procedure is the flattest. In fact, it is also **maximin** in that it has the worst best risk. It looks as if when trying to limit bad risk everywhere, you give up very good risk somewhere. This idea leads to one method for finding a minimax procedure: A Bayes procedure with flat risk is minimax. The next lemma records this result and some related ones.

Lemma 20.5. *Suppose δ_0 has a finite and constant risk,*

$$R(\theta\,; \delta_0) = c < \infty \text{ for all } \theta \in \mathcal{T}. \tag{20.51}$$

Then δ_0 is minimax if any of the following conditions hold: (a) δ_0 is Bayes wrt a proper prior π. (b) δ_0 is admissible. (c) There exists a sequence of Bayes procedures δ_n wrt priors π_n such that their Bayes risks approach c, i.e.,

$$R(\pi_n\,; \delta_n) \longrightarrow c. \tag{20.52}$$

Proof. (a) Suppose δ_0 is not minimax, so that there is a δ' such that

$$\sup_{\theta \in \mathcal{T}} R(\theta; \delta') < \sup_{\theta \in \mathcal{T}} R(\theta; \delta_0) = c. \tag{20.53}$$

But then

$$E_\pi[R(\Theta; \delta')] < c = E_\pi[R(\Theta; \delta_0)], \tag{20.54}$$

meaning that δ_0 is **not** Bayes wrt π. Hence we have a contradiction, so that δ_0 is minimax. $\square$

Exercise 20.9.14 verifies parts (b) and (c).

Continuing with the binomial, let $X \sim \text{Binomial}(n, \theta)$, $\theta \in (0, 1)$, so that the Bayes estimator using the Beta(α, β) prior is $\delta_{\alpha,\beta} = (\alpha + x)/(\alpha + \beta + n)$. See (11.43) and (11.44). Exercise 20.9.15 shows that the mean square error is

$$R(\theta; \delta_{\alpha,\beta}) = \frac{n\theta(1-\theta)}{(n+\alpha+\beta)^2} + \left(\frac{(\alpha+\beta)\theta - \alpha}{n+\alpha+\beta}\right)^2. \tag{20.55}$$

If we can find an (α, β) so that this risk is constant, then the corresponding estimator is minimax. As in the exercise, the risk is constant if $\alpha = \beta = \sqrt{n}/2$, hence $(x + \sqrt{n}/2)/(n + \sqrt{n})$ is minimax.

Note that based on the results from Section 20.5, the usual estimator $\delta(\mathbf{x}) = \mathbf{x}$ for estimating μ based on $\mathbf{X} \sim N(\mu, \mathbf{I}_p)$ is minimax for $p = 1$ or 2 since it is admissible in those cases. It is also minimax for $p \geq 3$. See Exercise 20.9.19.

20.7   Game theory and randomized procedures

We take a brief look at simple two-person zero-sum games. The two players we will call "the house" and "you." Each has a set of possible actions to take: You can choose from the set $\mathcal{A}$, and the house can choose from the set $\mathcal{T}$. Each player chooses an action without knowledge of the other's choice. There is a loss function, $L(a, \theta)$ (as in (20.4), but negative losses are allowed), where if you choose a and the house chooses θ, you lose $L(a, \theta)$ and the house wins the $L(a, \theta)$. ("Zero-sum" refers to the fact that whatever you lose, the house gains, and vice versa.) Your aim is to minimize L, while the house wants to maximize L.

Consider the game with $\mathcal{A} = \{a_1, a_2\}$ and $\mathcal{T} = \{\theta_1, \theta_2\}$, and loss function

House	You	
$\downarrow$	a_1	a_2
θ_1	2	0
θ_2	0	1

(20.56)

If you play this game once, deciding which action to take involves trying to psych out your opponent. E.g., you might think that a_2 is your best choice, since at worst you lose only 1. But then you realize the house may be thinking that's what you are thinking, so you figure the house will pick θ_2 so you will lose. Which leads you to choose a_1. But then you wonder if the house is thinking two steps ahead as well. And so on.

To avoid such circuitous thinking, the mathematical analysis of such games presumes the game is played repeatedly, and each player can see what the other's overall strategy is. Thus if you always play a_2, the house will catch on and always play θ_2, and you would always lose 1. Similarly, if you always play a_1, the house would always play θ_1, and you'd lose 2. An alternative is to not take the same action each time, nor to have any regular repeated pattern, but to randomly choose an action each time. The house does the same.

Let $p_i = P[\text{You choose } a_i]$ and $\pi_i = P[\text{House chooses } \theta_i]$. Then if both players use these probabilities each time, independently, your long-run average loss would be

$$R(\pi ; \mathbf{p}) = \sum_{i=1}^{2} \sum_{j=1}^{2} p_i \pi_j L(a_i, \theta_j) = 2\pi_1 p_1 + \pi_2 p_2. \tag{20.57}$$

If the house knows your $\mathbf{p}$, which it would after playing the game enough, it can adjust its π, to $\pi_1 = 1$ if $2p_1 > p_2$ and $\pi_2 = 1$ if $2p_1 < p_2$, yielding the average loss of $\max\{2p_1, p_2\}$. You realize that the house will take that strategy, so choose $\mathbf{p}$ to minimize that maximum, i.e., take $2p_1 = p_2$, so $\mathbf{p} = (1/3, 2/3)$. Then no matter what the house does, $R(\pi ; \mathbf{p}) = 2/3$, which is better than 1 or 2. Similarly, if you know the house's π, you can choose the $\mathbf{p}$ to minimize the expected loss, hence the house will choose π to maximize that minimum. We end up with $\pi = \mathbf{p} = (1/3, 2/3)$.

The fundamental theorem analyzing such games (two-person, zero-sum, finite $\mathcal{A}$ and $\mathcal{T}$) by John von Neumann (von Neumann and Morgenstern, 1944) states that there always exists a minimax strategy $\mathbf{p}_0$ for you, and maximin strategy π_0 for the house, such that

$$V \equiv R(\pi_0 ; \mathbf{p}_0) = \min_{\mathbf{p}} \max_{\pi} R(\pi ; \mathbf{p}) = \max_{\pi} \min_{\mathbf{p}} R(\pi ; \mathbf{p}). \tag{20.58}$$

This V is called the **value** of the game, and the distribution π_0 is called the **least favorable distribution**. If either player deviates from their optimum strategy, the other player will benefit, hence the theorem guarantees that the game with rational players will always have you losing V on average.

The statistical decision theory we have seen so far in this chapter is based on game theory, but has notable differences. The first is that in statistics, we have data that gives us some information about what θ the "house" has chosen. Also, the action spaces are often infinite. Either of these modifications easily fit into the vast amount of research done in game theory since 1944.

The most important difference is the lack of a house trying to subvert you, the statistician. You may be cautious or pessimistic, and wish to minimize your maximum expected loss, but it is perfectly rational to use non-minimax procedures. Another difference is that, for us so far, actions have not been randomized. Once we have the data, $\delta(\mathbf{x})$ gives us the estimate of θ, say. We don't randomize to decide between several possible estimates. In fact, a client would be quite upset if after setting up a carefully designed experiment, the statistician flipped a coin to decide whether to accept or reject the null hypothesis. But theoretically, randomized procedures have some utility in statistics, which we will see in Chapter 21 on hypothesis testing. It is possible for a non-randomized test to be dominated by a randomized test, especially in discrete models where the actual size of a nonrandomized test is lower than the desired level.

A more general formulation of statistical decision theory does allow randomization. A decision procedure is defined to be a function from the sample space $\mathcal{X}$ to the space of probability measures on the action space $\mathcal{A}$. For our current purposes, we can instead explicitly incorporate randomization into the function. The idea is that in addition to the data $\mathbf{x}$, we can make a decision based on spinning a spinner as often as we want. Formally, suppose $\mathbf{U} = \{U_k \,|\, i = 1, 2, \ldots\}$ is an infinite sequence of independent Uniform$(0, 1)$'s, all independent of $\mathbf{X}$. Then a randomized procedure is a function of $\mathbf{X}$ and possibly a finite number of the U_k's:

$$\delta(\mathbf{x}, u_1, \ldots, u_K) \in \mathcal{A}, \tag{20.59}$$

for some $K \geq 0$.

In estimation with squared-error loss, the Rao-Blackwell theorem (Theorem 13.8 on page 210) shows that any such estimator that non-trivially depends on $\mathbf{u}$ is inadmissible. Suppose $\delta(\mathbf{x}, u_1, \ldots, u_K)$ has finite mean square error, and let

$$\delta^*(\mathbf{x}) = E[\delta(\mathbf{X}, U_1, \ldots, U_K) \,|\, \mathbf{X} = \mathbf{x}] = E[\delta(\mathbf{x}, U_1, \ldots, U_K))], \tag{20.60}$$

where the expectation is over $\mathbf{U}$. Then δ^* has the same bias and lower variance than δ, hence lower MSE. This result holds if $L(\mathbf{a}, \boldsymbol{\theta})$ is strictly convex in $\mathbf{a}$ for each $\boldsymbol{\theta}$. If it is convex but not strictly so, then at least δ^* is no worse than δ.

20.8   Minimaxity and admissibility when $\mathcal{T}$ is finite

Here we present a statistical analog of the minimax theorem for game theory in (20.58), which assumes a finite parameter space, and from that show that all admissible procedures are Bayes. Let $\mathcal{D}$ be the set of procedures under consideration.

Theorem 20.6. *Suppose the parameter space is finite*

$$\mathcal{T} = \{\boldsymbol{\theta}_1, \ldots, \boldsymbol{\theta}_K\}. \tag{20.61}$$

*Define the **risk set** to be the set of all achievable vectors of risks for the procedures:*

$$\mathcal{R} = \{(R(\boldsymbol{\theta}_1 \,;\, \delta), \ldots, R(\boldsymbol{\theta}_K \,;\, \delta)) \,|\, \delta \in \mathcal{D}\}. \tag{20.62}$$

Suppose the risk set $\mathcal{R}$ is closed, convex, and bounded from below. Then there exists a minimax procedure δ_0 and a least favorable distribution (prior) π_0 on $\mathcal{T}$ such that δ_0 is Bayes wrt π_0.

The assumption of a finite parameter space is too restrictive to have much practical use in typical statistical models, but the theorem does serve as a basis for more general situations, as for hypothesis testing in Section 22.3. The convexity of the risk set often needs the use of randomized procedures. For example, if $\mathcal{D}$ is closed under randomization (see Exercise 20.9.25), then the risk set is convex. Most loss functions we use in statistics are nonnegative, so that the risks are automatically bounded from below. The closedness of the risk set depends on how limits of procedures behave. Again, for testing see Section 22.3.

We will sketch the proof, which is a brief summary of the thorough but accessible proof is found in Ferguson (1967) for his Theorem 2.9.1. A set $\mathcal{C}$ is convex if every line segment connecting two points in $\mathcal{C}$ is in $\mathcal{C}$. That is,

$$\mathbf{b}, \mathbf{c} \in \mathcal{C} \implies \alpha \mathbf{b} + (1 - \alpha)\mathbf{c} \in \mathcal{C} \text{ for all } 0 \leq \alpha \leq 1. \tag{20.63}$$

It is closed if any limit of points in $\mathcal{C}$ is also in $\mathcal{C}$:

$$\mathbf{c}_n \in \mathcal{C} \text{ for } n = 1, 2, \ldots \text{ and } \mathbf{c}_n \longrightarrow \mathbf{c} \implies \mathbf{c} \in \mathcal{C}. \tag{20.64}$$

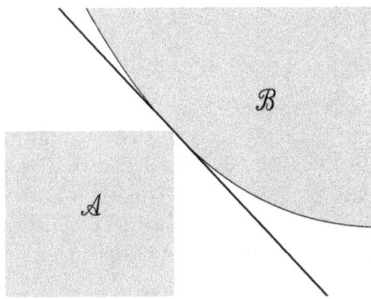

Figure 20.3: This plot illustrates the separating hyperplane theorem, Theorem 20.7. Two convex sets $\mathcal{A}$ and $\mathcal{B}$ have empty intersection, so there is a hyperplane, in this case a line, that separates them.

Bounded from below means there is a finite κ such that $\mathbf{c} = (c_1, \ldots, c_K) \in \mathcal{C}$ implies that $c_i \geq \kappa$ for all i.

For real number s, consider the points in the risk set whose maximum risk is no larger than s. We can express this set as an intersection of $\mathcal{R}$ with the set $\mathcal{L}_s$ defined below:

$$\{\mathbf{r} \in \mathcal{R} \mid r_i \leq s, i = 1, \ldots, K\} = \mathcal{L}_s \cap \mathcal{R}, \quad \mathcal{L}_s = \{\mathbf{x} \in \mathbb{R}^K \mid x_i \leq s, i = 1, \ldots, K\}. \quad (20.65)$$

We want to find the minimax s, i.e., the smallest s obtainable:

$$s_0 = \inf\{s \mid \mathcal{L}_s \cap \mathcal{R} \neq \emptyset\}. \quad (20.66)$$

It does exist, since $\mathcal{R}$ being bounded from below implies that the set $\{s \mid \mathcal{L}_s \cap \mathcal{R} \neq \emptyset\}$ is bounded from below. Also, there exists an $\mathbf{r}_0 \in \mathcal{R}$ with $s_0 = \max\{r_{0i}, i = 1, \ldots, K\}$ because we have assumed $\mathcal{R}$ is closed. Let δ_0 be a procedure that achieves this risk, so that it is minimax:

$$\max_{i=1,\ldots,K} R(\boldsymbol{\theta}_i ; \delta_0) = s_0 \leq \max_{i=1,\ldots,K} R(\boldsymbol{\theta}_i ; \delta), \quad \delta \in \mathcal{D}. \quad (20.67)$$

Next we argue that this procedure is Bayes. Let $\text{int}(\mathcal{L}_{s_0})$ be the interior of $\mathcal{L}_{s_0}$: $\{\mathbf{x} \in \mathbb{R}^K \mid x_i < s_0, i - 1, \ldots, K\}$. It can be shown that $\text{int}(\mathcal{L}_{s_0})$ is convex, and

$$\text{int}(\mathcal{L}_{s_0}) \cap \mathcal{R} = \emptyset \quad (20.68)$$

by definition of s_0. Now we need to bring out a famous result, the **separating hyperplane theorem**:

Theorem 20.7. *Suppose $\mathcal{A}$ and $\mathcal{B}$ are two nonempty convex sets in $\mathbb{R}^K$ such that $\mathcal{A} \cap \mathcal{B} = \emptyset$. Then there exists a nonzero vector $\boldsymbol{\gamma} \in \mathbb{R}^K$ such that*

$$\boldsymbol{\gamma} \cdot \mathbf{x} \leq \boldsymbol{\gamma} \cdot \mathbf{y} \text{ for all } \mathbf{x} \in \mathcal{A} \text{ and } \mathbf{y} \in \mathcal{B}. \quad (20.69)$$

See Figure 20.3 for an illustration, and Exercises 20.9.28 through 20.9.30 for a proof. The idea is that if two convex sets do not intersect, then there is a hyperplane separating them. In the theorem, such a hyperplane is the set $\{\mathbf{x} \mid \boldsymbol{\gamma} \cdot \mathbf{x} = a\}$, where a is any constant satisfying

$$a_L \equiv \sup\{\gamma \cdot \mathbf{x} \mid \mathbf{x} \in \mathcal{A}\} \leq a \leq \inf\{\gamma \cdot \mathbf{y} \mid \mathbf{y} \in \mathcal{B}\} \equiv a_U. \qquad (20.70)$$

Neither γ nor a is necessarily unique.

Apply the theorem with $\mathcal{A} = \mathrm{int}(\mathcal{L}_{s_0})$ and $\mathcal{B} = \mathcal{R}$. In this case, the elements of γ must be nonnegative: Suppose $\gamma_j < 0$, and take $\mathbf{x} \in \mathrm{int}(\mathcal{L}_{s_0})$. Note that we can let $x_j \to -\infty$, and $\mathbf{x}$ will still be in $\mathrm{int}(\mathcal{L}_{s_0})$. Thus $\gamma \cdot \mathbf{x} \to +\infty$, which contradicts the bound a_L in (20.70). So $\gamma_j \geq 0$ for all j. Since $\gamma \neq \mathbf{0}$, we can define $\pi_0 = \gamma / \sum \gamma_i$ and have (20.69) hold for π_0 in place of γ. Note that π_0 is a legitimate pmf on $\boldsymbol{\theta}$ with $P[\boldsymbol{\theta} = \boldsymbol{\theta}_i] = \pi_{0i}$.

By the definition in (20.65), the points $(s, s, \ldots, s) \in \mathrm{int}(\mathcal{L}_{s_0})$ for all $s < s_0$. Now $\sum \pi_{0i} s = s$, hence that sum can get arbitrarily close to s_0, meaning $a_L = s_0$ in (20.70). Translating back,

$$s_0 \leq \pi_0 \cdot \mathbf{r} \text{ for all } \mathbf{r} \in \mathcal{R} \implies s_0 \leq \sum \pi_{0i} R(\boldsymbol{\theta}_i \,;\, \delta)(= R(\pi_0 \,;\, \delta)) \text{ for all } \delta \in \mathcal{D}, \qquad (20.71)$$

hence from (20.67),

$$\max_{i=1,\ldots,K} R(\boldsymbol{\theta}_i \,;\, \delta_0) \leq R(\pi_0 \,;\, \delta) \text{ for all } \delta \in \mathcal{D}, \qquad (20.72)$$

which implies that

$$R(\pi_0 \,;\, \delta_0) \leq R(\pi_0 \,;\, \delta) \text{ for all } \delta \in \mathcal{D}. \qquad (20.73)$$

That is, δ_0 is Bayes wrt π_0.

To complete the proof of Theorem 20.6, we need only that π_0 is the least favorable distribution, which is shown in Exercise 20.9.26.

The next result shows that under the same conditions as Theorem 20.6, any admissible procedure is Bayes.

Theorem 20.8. *Suppose the parameter space is finite*

$$\mathcal{T} = \{\boldsymbol{\theta}_1, \ldots, \boldsymbol{\theta}_K\}, \qquad (20.74)$$

and define the risk set $\mathcal{R}$ as in (20.62). Suppose the risk set $\mathcal{R}$ is closed, convex, and bounded from below. If δ_0 is admissible, then it is Bayes wrt some prior π_0.

Proof. Assume δ_0 is admissible. Consider the same setup, but with risk function

$$R^*(\boldsymbol{\theta} \,;\, \delta) = R(\boldsymbol{\theta} \,;\, \delta) - R(\boldsymbol{\theta} \,;\, \delta_0). \qquad (20.75)$$

Then the new risk set, $\mathcal{R}^* = \{R^*(\boldsymbol{\theta} \,;\, \delta) \mid \delta \in \mathcal{D}\}$ is also closed, convex, and bounded from below. (See Exercise 20.9.27.) Since $R^*(\boldsymbol{\theta} \,;\, \delta_0) = 0$ for all $\boldsymbol{\theta} \in \mathcal{T}$, the maximum risk of δ^* is zero. Suppose δ is another procedure with smaller maximum risk:

$$\max_{\boldsymbol{\theta} \in \mathcal{T}} R^*(\boldsymbol{\theta} \,;\, \delta) < 0. \qquad (20.76)$$

But then we would have

$$R(\boldsymbol{\theta} \,;\, \delta) < R(\boldsymbol{\theta} \,;\, \delta_0) \text{ for all } \boldsymbol{\theta} \in \mathcal{T}, \qquad (20.77)$$

which contradicts the assumption that δ_0 is admissible. Thus (20.76) cannot hold, which means that δ_0 is minimax for R^*. Then by Theorem 20.6, there exists a π_0 wrt which δ_0 is Bayes under risk R^*, so that $R^*(\pi_0 \,;\, \delta_0) \leq R^*(\pi_0 \,;\, \delta)$ for any $\delta \in \mathcal{D}$. But since $R^*(\pi_0 \,;\, \delta_0) = 0$, we have $R(\pi_0 \,;\, \delta_0) \leq R(\pi_0 \,;\, \delta)$ for all $\delta \in \mathcal{D}$, hence δ_0 is Bayes wrt π_0 under the original risk R. $\qquad \square$

20.9 Exercises

Exercise 20.9.1. (Lemma 20.3(b).) Suppose that δ_π is a Bayes procedure wrt π, and if δ'_π is also Bayes wrt π, then $R(\theta; \delta_\pi) = R(\theta; \delta'_\pi)$ for all $\theta \in \mathcal{T}$. Argue that δ_π is admissible.

Exercise 20.9.2. (Lemma 20.3(c).) Suppose that the parameter space is finite or countable, $\mathcal{T} = \{\theta_1, \ldots, \theta_K\}$ (K possibly infinite), and π is a prior on $\mathcal{T}$ such that $\pi_k = P[\Theta = \theta_k] > 0$ for $k = 1, \ldots, K$. Show that δ_π, the Bayes procedure wrt π, is admissible.

Exercise 20.9.3. Consider estimating θ with squared-error loss based on $X \sim$ Discrete Uniform$(0, \theta)$, where $\mathcal{T} = \{0, 1, 2\}$. Let π be the prior with $\pi(0) = \pi(1) = 1/2$ (so that it places no probability on $\theta = 2$). (a) Show that any Bayes estimator wrt π satisfies $\delta_c(0) = 1/3$, $\delta_c(1) = 1$, and $\delta_c(2) = c$ for any c. (b) Find the risk function for δ_c, and show that its Bayes risk $R(\pi, \delta_c) = 1/6$. (c) Let $\mathcal{D}$ be the set of estimators δ_c for $0 \le c \le 2$. For which c is δ_c the only estimator admissible among those in $\mathcal{D}$? Is it admissible among all estimators?

Exercise 20.9.4. Here we have $\mathbf{X}$ with density $f(\mathbf{x} \mid \theta)$, $\theta \in (b, c)$, and wish to estimate θ with a weighted squared-error loss:

$$L(a, \theta) = g(\theta)(a - \theta)^2, \tag{20.78}$$

where $g(\theta) \ge 0$ is the weight function. (a) Show that if the prior π has pdf $\pi(\theta)$ and the integrals below exists, then the Bayes estimator δ_π is given by

$$\delta_\pi(\mathbf{x}) = \frac{\int_b^c \theta f(\mathbf{x} \mid \theta) g(\theta) \pi(\theta) d\theta}{\int_b^c f(\mathbf{x} \mid \theta) g(\theta) \pi(\theta) d\theta}. \tag{20.79}$$

(b) Suppose $g(\theta) > 0$ for all $\theta \in \mathcal{T}$. Show that δ is admissible for squared-error loss if and only if it is admissible for the weighted loss in (20.78).

Exercise 20.9.5. Suppose $X \sim N(\mu, 1)$, $\mu \in \mathbb{R}$, and we wish to estimate μ using squared-error loss. Let the prior π_n on μ be $N(0, n)$. (a) Show that the risk, hence Bayes risk, of $\delta_0(x) = x$ is constant at 1. (b) The Bayes estimator wrt π_n is given in (20.22) to be $\delta_n(x) = nx/(n+1)$. Show that $R(\mu; \delta_n) = (\mu^2 + n)/(n+1)^2$, and $R(\pi_n; \delta_n) = n/(n+1)$.

Exercise 20.9.6. This exercise shows that an admissible estimator can sometimes be surprising. Let X_1 and X_2 be independent, $X_i \sim N(\mu_i, \sigma_i^2)$, $i = 1, 2$, with unrestricted parameter space $\boldsymbol{\theta} = (\mu_1, \mu_2, \sigma_1^2, \sigma_2^2) \in \mathbb{R}^2 \times (0, \infty)^2$. We are interested in estimating just μ_1 under squared-error loss, so that $L(a, \boldsymbol{\theta}) = (a - \mu_1)^2$. Let $\mathcal{D}$ be the set of linear estimators $\delta_{a,b,c}(\mathbf{x}) = ax_1 + bx_2 + c$ for constants $a, b, c \in \mathbb{R}$. Let $\delta_1(\mathbf{x}) = x_1$ and $\delta_2(\mathbf{x}) = x_2$. (a) Show that the risk of $\delta_{a,b,c}$ is $((a-1)\mu_1 + b\mu_2 + c)^2 + a^2\sigma_1^2 + b^2\sigma_2^2$. (b) Find the risks of δ_1 and δ_2. Show that neither one of these two estimators dominates the other. (c) Show that δ_1 is admissible among the estimators in $\mathcal{D}$. Is this result surprising? [Hint: Suppose $R(\boldsymbol{\theta}; \delta_{a,b,c}) \le R(\boldsymbol{\theta}; \delta_1)$ for all $\boldsymbol{\theta} \in \mathcal{T}$. Let $\mu_2 \to \infty$ to show b must be 0; let $\mu_1 \to \infty$ to show that a must be 1, then argue further that $c = 0$. Thus $\delta_{a,b,c}$ must be δ_1.] (d) Show that δ_2 is admissible among the estimators in $\mathcal{D}$, even though the distribution of X_2 does not depend on μ_1. [Hint: Proceed as in the hint to part (b), but let $\sigma_1^2 \to \infty$, then $\mu_1 = \mu_2 = \mu \to \infty$.]

Exercise 20.9.7. Continue with the setup in Exercise 20.9.6, but without the restriction to linear estimators. Again let $\delta_2(\mathbf{x}) = x_2$. The goal here is to show δ_2 is a limit of Bayes estimators. (a) For fixed $\sigma_0^2 > 0$, let $\pi_{\sigma_0^2}$ be the prior on $\boldsymbol{\theta}$ where $\mu_1 = \mu_2 = \mu$, $\sigma_1^2 = \sigma_0^2$, $\sigma_2^2 = 1$, and $\mu \sim N(0, \sigma_0^2)$. Show that the Bayes estimator wrt $\pi_{\sigma_0^2}$ is

$$\delta_{\sigma_0^2}(x_1, x_2) = \frac{x_1/\sigma_0^2 + x_2}{2/\sigma_0^2 + 1}. \tag{20.80}$$

(b) Find the risk of $\delta_{\sigma_0^2}$ as a function of $\boldsymbol{\theta}$. (c) Find the Bayes risk of $\delta_{\sigma_0^2}$ wrt $\pi_{\sigma_0^2}$. (d) What is the limit of $\delta_{\sigma_0^2}$ as $\sigma_0^2 \to \infty$?

Exercise 20.9.8. Let $X_1, \ldots, X_n$ be iid $N(\mu, 1)$, with $\mu \in \mathbb{R}$. The analog to the regularized least squares in (12.43) for this simple situation defines the estimator $\delta_\kappa(\mathbf{x})$ to be the value of m that minimizes

$$obj_\kappa(m; x_1, \ldots, x_n) = \sum_{i=1}^{n}(x_i - m)^2 + \kappa m^2, \tag{20.81}$$

where $\kappa \geq 0$ is some fixed constant. (a) What is $\delta_\kappa(\mathbf{x})$? (b) For which value of κ is δ_κ the MLE? (c) For $\kappa > 0$, δ_κ is the Bayes posterior mean using the $N(\mu_0, \sigma_0^2)$ for which μ_0 and σ_0^2? (d) For which $\kappa \geq 0$ is δ_κ admissible among all estimators?

Exercise 20.9.9. Let $\mathbf{x}$ be $p \times 1$, and $\phi_{\mu_1}(x_1)$ be the $N(\mu_i, 1)$ pdf. Fixing $x_2, \ldots, x_p$, show that (20.41) holds, i.e.,

$$\int_{-\infty}^{\infty} \frac{x_1(x_1 - \mu_1)}{\|\mathbf{x}\|^2} \phi_{\mu_1}(x_1)dx_1 = \int_{-\infty}^{\infty}\left(\frac{1}{\|\mathbf{x}\|^2} - \frac{2x_1^2}{\|\mathbf{x}\|^4}\right)\phi_{\mu_1}(x_1)dx_1. \tag{20.82}$$

[Hint: Use integration by parts, where $u = x_1/\|\mathbf{x}\|^2$ and $dv = (x_1 - \mu_1)\phi_{\mu_1}(x_1)$.]

Exercise 20.9.10. (a) Use (20.37) and (20.43) to show that the risk of the James-Stein estimator is

$$R(\boldsymbol{\mu}; \delta_{JS}) = p - E_{\boldsymbol{\mu}}\left[\frac{(p-2)^2}{\|\mathbf{X}\|^2}\right]. \tag{20.83}$$

as in (20.44). (b) Show that $R(\mathbf{0}; \delta_{JS}) = 2$. [Hint: When $\boldsymbol{\mu} = \mathbf{0}$, $\|\mathbf{X}\|^2 \sim \chi_p^2$. What is $E[1/\chi_p^2]$?]

Exercise 20.9.11. In Exercise 11.7.17 we found that the usual estimator of the binomial parameter θ is a Bayes estimator wrt the improper prior $1/(\theta(1-\theta))$, at least when $x \neq 0$ or n. Here we look at a truncated version of the binomial, where the usual estimator is proper Bayes. The truncated binomial is given by the usual binomial conditioned to be between 1 and $n-1$. That is, take the pmf of X to be

$$f^*(x \mid \theta) = \frac{f(x \mid \theta)}{\alpha(\theta)}, \quad x = 1, \ldots, n-1 \tag{20.84}$$

for some $\alpha(\theta)$, where $f(x \mid \theta)$ is the usual Binomial(n, θ) pmf. (Assume $n \geq 2$.) The goal is to estimate $\theta \in (0, 1)$ using squared-error loss. For estimator δ, the risk is denoted

$$R^*(\theta; \delta) = \sum_{x=1}^{n-1}(\delta(x) - \theta)^2 f^*(x \mid \theta). \tag{20.85}$$

(a) Find $\alpha(\theta)$. (b) Let $\pi^*(\theta) = c\alpha(\theta)/(\theta(1-\theta))$. Find the constant c so that π^* is a proper pdf on $\theta \in (0,1)$. [Hint: Note that $\alpha(\theta)$ is $\theta(1-\theta)$ times a polynomial in θ.] (c) Show that the Bayes estimator wrt π^* for the risk R^* is $\delta_0(x) = x/n$. Argue that therefore δ_0 is admissible for the truncated binomial, so that for estimator δ',

$$R^*(\theta;\delta') \le R^*(\theta;\delta_0) \text{ for all } \theta \in (0,1) \implies R^*(\theta;\delta') = R^*(\theta;\delta_0) \text{ for all } \theta \in (0,1).$$
$$(20.86)$$

Exercise 20.9.12. This exercise proves the admissibility of $\delta_0(x) = x/n$ for the usual binomial using two stages. Here we have $X \sim \text{Binomial}(n,\theta)$, $\theta \in (0,1)$, and estimate θ with squared-error loss. Suppose δ' satisfies

$$R(\theta;\delta') \le R(\theta;\delta_0) \text{ for all } \theta \in (0,1). \tag{20.87}$$

(a) Show that for any estimator,

$$\lim_{\theta \to 0} R(\theta;\delta) = \delta(0)^2 \text{ and } \lim_{\theta \to 1} R(\theta;\delta) = (1-\delta(n))^2, \tag{20.88}$$

hence (20.87) implies that $\delta'(0) = 0$ and $\delta'(n) = 1$. Thus δ_0 and δ' agree at $x = 0$ and n. [Hint: What are the limits in (20.88) for δ_0?] (b) Show that for any estimator δ,

$$R(\theta;\delta) = (\delta(0) - \theta)^2(1-\theta)^2 + \alpha(\theta)R^*(\theta;\delta) + (\delta(n) - \theta)^2\theta^n, \tag{20.89}$$

where R^* and α are given in Exercise 20.9.11. (c) Use the conclusion in part (a) to show that (20.87) implies

$$R^*(\theta;\delta') \le R^*(\theta;\delta_0) \text{ for all } \theta \in (0,1). \tag{20.90}$$

(d) Use (20.86) to show that (20.90) implies $R(\theta;\delta') = R(\theta;\delta_0)$ for all $\theta \in (0,1)$, hence δ_0 is admissible in the regular binomial case. (See Johnson (1971) for this two-stage idea in the binomial, and Brown (1981) for a generalization to problems with finite sample space.)

Exercise 20.9.13. Suppose $X \sim \text{Poisson}(2\theta)$, $Y \sim \text{Poisson}(2(1-\theta))$, where X and Y are independent, and $\theta \in (0,1)$. The MLE of θ is $x/(x+y)$ if $x+y > 0$, but not unique if $x+y = 0$. For fixed c, define the estimator δ_c by

$$\delta_c(x,y) = \begin{cases} c & \text{if } x+y = 0 \\ \frac{x}{x+y} & \text{if } x+y > 0 \end{cases}. \tag{20.91}$$

This question looks at the decision-theoretic properties of these estimators under squared-error loss. (a) Let $T = X + Y$. What is the distribution of T? Note that it does not depend on θ. (b) Find the conditional distribution of $X \mid T = t$. (c) Find $E_\theta[\delta_c(X,Y)]$ and $Var_\theta[\delta_c(X,Y)]$, and show that

$$\text{MSE}_\theta[\delta_c(X,Y)] = \theta(1-\theta)r + (c-\theta)^2 p \text{ where } r = \sum_{t=1}^{\infty} \frac{1}{t} f_T(t) \text{ and } p = f_t(0), \tag{20.92}$$

where f_T is the pmf of T. [Hint: First find the conditional mean and variance of δ_c given $T = t$.] (d) For which value(s) of c is δ_c unbiased, if any? (e) Sketch the MSEs for δ_c when $c = 0, .5, .75, 1$, and 2. Among these four estimators, which are admissible and which inadmissible? Which is minimax? (f) Now consider the set of estimators δ_c for $c \in \mathbb{R}$. Show that δ_c is admissible among these if and only if $0 \leq c \leq 1$. (g) Which δ_c, if any, is minimax among the set of all δ_c's? What is its maximum risk? [Hint: You can restrict attention to $0 \leq c \leq 1$. First show that the maximum risk of δ_c over $\theta \in (0,1)$ is $(r - 2cp)^2/(4(r - p)) + c^2p$, then find the c for which the maximum is minimized.] (h) Show that $\delta_0(x,y) = (x - y)/4 + 1/2$, is unbiased, and find its MSE. Is it admissible among all estimators? [Hint: Compare it to those in part (e).] (i) The Bayes estimator with respect to the prior $\Theta \sim \text{Beta}(\alpha, \beta)$ is $\delta_{\alpha,\beta}(x, y) = (x + \alpha)/(x + y + \alpha + \beta)$. (See Exercise 13.8.2.) None of the δ_c's equals a $\delta_{\alpha,\beta}$. However, some δ_c's are limits of $\delta_{\alpha,\beta}$'s for some sequences of (α, β)'s. For which c's can one find such a sequence? (Be sure that the α's and β's are positive.)

Exercise 20.9.14. (Lemma 20.5(b) and (c).) Suppose δ has constant risk, $R(\theta; \delta) = c$. (a) Show that if δ admissible, it is minimax. (b) Suppose there exists a sequence of Bayes procedures δ_n wrt π_n such that $R(\pi_n; \delta_n) \to c$ and $n \to \infty$. Show that δ is minimax. [Hint: Suppose δ' has better maximum risk than δ, $R(\theta; \delta') < c$ for all $\theta \in \mathcal{T}$. Show that for large enough n, $R(\theta; \delta') < R(\pi_n; \delta_n)$, which can be used to show that δ' has better Bayes risk than δ_n wrt π_n.]

Exercise 20.9.15. Let $X \sim \text{Binomial}(n, \theta)$, $\theta \in (0, 1)$. The Bayes estimator using the $\text{Beta}(\alpha, \beta)$ prior is $\delta_{\alpha,\beta} = (\alpha + x)/(\alpha + \beta + n)$. (a) Show that the risk $R(\theta; \delta_{\alpha,\beta})$ is as in (20.55). (b) Show that if $\alpha = \beta = \sqrt{n}/2$, the risk has constant value $1/(4(\sqrt{n} + 1)^2)$.

Exercise 20.9.16. Consider a location-family model, where the object is to estimate the location parameter θ with squared-error loss. Suppose the Pitman estimator has finite variance. (a) Is the Pitman estimator admissible among shift-equivariant estimators? (b) Is the Pitman estimator minimax among shift-equivariant estimators? (c) Is the Pitman estimator Bayes among shift-equivariant estimators?

Exercise 20.9.17. Consider the normal linear model as in (12.9), where $\mathbf{Y} \sim N(\mathbf{x}\boldsymbol{\beta}, \sigma^2 \mathbf{I}_n)$, $\mathbf{Y}$ is $n \times 1$, $\mathbf{x}$ is a fixed known $n \times p$ matrix, $\boldsymbol{\beta}$ is the $p \times 1$ vector of coefficients, and $\sigma^2 > 0$. Assume that $\mathbf{x}'\mathbf{x}$ is invertible. The objective is to estimate $\boldsymbol{\beta}$ using squared-error loss,

$$L(\mathbf{a}, (\boldsymbol{\beta}, \sigma^2)) = \sum_{j=1}^{p}(a_j - \beta_j)^2 = \|\mathbf{a} - \boldsymbol{\beta}\|^2. \tag{20.93}$$

Argue that the ridge regression estimator of $\boldsymbol{\beta}$ in (12.45), $\widehat{\boldsymbol{\beta}}_\kappa = (\mathbf{x}'\mathbf{x} + \kappa\mathbf{I}_p)^{-1}\mathbf{x}'\mathbf{Y}$, is admissible for $\kappa > 0$. (You can assume the risk function is continuous in $\boldsymbol{\beta}$.) [Hint: Choose the appropriate $\boldsymbol{\beta}_0$ and $\mathbf{K}_0$ in (12.37).]

Exercise 20.9.18. Let $X \sim \text{Poisson}(\lambda)$, $\lambda > 0$. We wish to estimate $g(\lambda) = \exp(-2\lambda)$ with squared-error loss. Recall from Exercise 19.8.7 that the UMVUE is $\delta_U(x) = (-1)^x$. (a) Find the variance and risk of δ_U. (They are the same.) (b) For prior density $\pi(\lambda)$ on λ, write down the expression for the Bayes estimator of $g(\lambda)$. Is it possible to find a prior so that the Bayes estimate equals δ_U? Why or why not? (c) Find an estimator δ^* that dominates δ_U. [Hint: Which value of δ_U is way outside of the range of $g(\lambda)$? What other value are you sure must be closer to $g(\lambda)$ than that one?] (d) Is the estimator δ^* in part (c) unbiased? Is δ_U admissible?

Exercise 20.9.19. Let $\mathbf{X} \sim N(\boldsymbol{\mu}, \mathbf{I}_p)$ with parameter space $\boldsymbol{\mu} \in \mathbb{R}^p$, and consider estimating $\boldsymbol{\mu}$ using squared-error loss. Then $\delta_0(\mathbf{x})$ has the constant risk of p. (a) Find the

Bayes estimator for prior π_n being $N(0_p, n\mathbf{I}_p)$. (b) Show that the Bayes risk of π_n is $pn/(n+1)$. (c) Show that δ_0 is minimax. [Hint: What is the limit of the Bayes risk in part (b) as $n \to \infty$?]

Exercise 20.9.20. Suppose $X_1, \ldots, X_p$ are independent, $X_i \mid \mu_i \sim N(\mu_i, 1)$, so that $\mathbf{X} \mid \boldsymbol{\mu} \sim N_p(\boldsymbol{\mu}, \mathbf{I}_p)$. The parameter space for $\boldsymbol{\mu}$ is $\mathbb{R}^p$. Consider the prior on $\boldsymbol{\mu}$ where the μ_i are iid $N(0, V)$, so that $\boldsymbol{\mu} \sim N_p(0_p, V\mathbf{I}_p)$. The goal is to estimate $\boldsymbol{\mu}$ using squared-error loss as in (20.34), $L(\mathbf{a}, \boldsymbol{\mu}) = \|\mathbf{a} - \boldsymbol{\mu}\|^2$. (a) For known prior variance $V \in (0, \infty)$, show that the Bayes estimator is

$$\delta_V(\mathbf{x}) = (1 - c_V)\mathbf{x}, \text{ where } c_V = \frac{1}{V+1}. \tag{20.94}$$

(b) Now suppose that V is not known, and you wish to estimate c_V based on the marginal of distribution of $\mathbf{X}$. The marginal distribution (i.e., not conditional on the $\boldsymbol{\mu}$) is $\mathbf{X} \sim N(0_p, d_V \mathbf{I}_p)$ for what d_V? (c) Using the marginal distribution in $\mathbf{X}$, find the a_p and b_V so that

$$E_V\left[\frac{1}{\|\mathbf{X}\|^2}\right] = \frac{1}{a_p} \frac{1}{b_V}. \tag{20.95}$$

(d) From part (c), find an unbiased estimator of c_V: $\hat{c}_V = f_p/\|\mathbf{X}\|^2$ for what f_p? (e) Now put that estimator in for c_V in δ_V. Is the result an estimator for $\boldsymbol{\mu}$? It is called an **empirical Bayes estimator**, because it is similar to a Bayes estimator, but uses the data to estimate the parameter V in the prior. What other name is there for this estimator?

Exercise 20.9.21. Let $\mathbf{X} \sim N(\boldsymbol{\mu}, \mathbf{I}_p)$, $\boldsymbol{\mu} \in \mathbb{R}^p$. This problem will use a different risk function than in Exercise 20.9.20, one based on *prediction*. The data are $\mathbf{X}$, but imagine predicting a new vector $\mathbf{X}^{New}$ that is independent of $\mathbf{X}$ but has the same distribution as $\mathbf{X}$. This $\mathbf{X}^{New}$ is not observed, so it cannot be used in the estimator. An estimator $\delta(\mathbf{x})$ of $\boldsymbol{\mu}$ can be thought of as a predictor of the new vector $\mathbf{X}^{New}$. The loss is how far off the prediction is,

$$PredSS \equiv \|\mathbf{X}^{New} - \delta(\mathbf{X})\|^2, \tag{20.96}$$

which is unobservable, and the risk is the expected value over both the data and the new vector,

$$R(\boldsymbol{\mu}; \delta) = E[PredSS \mid \boldsymbol{\mu}] = E[\|\mathbf{X}^{New} - \delta(\mathbf{X})\|^2 \mid \boldsymbol{\mu}]. \tag{20.97}$$

(a) Suppose $\delta(\mathbf{x}) = \mathbf{x}$ itself. What is $R(\boldsymbol{\mu}; \delta)$? (b) Suppose $\delta(\mathbf{x}) = 0_p$. What is $R(\boldsymbol{\mu}; \delta)$? (c) For a subset $\mathcal{A} \subset \{1, 2, \ldots, p\}$, define the estimator $\delta_{\mathcal{A}}(\mathbf{x})$ by setting $\delta_i(\mathbf{x}) = x_i$ for $i \in \mathcal{A}$ and $\delta_i(\mathbf{x}) = 0$ for $i \notin \mathcal{A}$. That is, the estimator starts with $\mathbf{x}$, then sets the components with indices not in $\mathcal{A}$ to zero. For example, if $p = 4$, then

$$\delta_{\{1,4\}}(\mathbf{x}) = \begin{pmatrix} x_1 \\ 0 \\ 0 \\ x_4 \end{pmatrix} \text{ and } \delta_{\{2\}}(\mathbf{x}) = \begin{pmatrix} 0 \\ x_2 \\ 0 \\ 0 \end{pmatrix}. \tag{20.98}$$

In particular, $\delta_\varnothing(\mathbf{x}) = 0_p$ and $\delta_{\{1,2,\ldots,p\}}(\mathbf{x}) = \mathbf{x}$. Let $q = \#\mathcal{A}$, that is, q is the number of μ_i's being estimated rather than being set to 0. For general p, find $R(\boldsymbol{\mu}; \delta_{\mathcal{A}})$ as a function of p, q, and the μ_i's. (d) Let $\mathcal{D}$ be the set of estimators $\delta_{\mathcal{A}}$ as in part (c). Which (if any) are admissible among those in $\mathcal{D}$? Which (if any) are minimax among those in $\mathcal{D}$?

Exercise 20.9.22. Continue with the setup in Exercise 20.9.21. One approach to deciding which estimator to use is to try to estimate the risk for each δ_A, then choose the estimator with the smallest estimated risk. A naive estimator of the *PredSS* just uses the observed $\mathbf{x}$ in place of the $\mathbf{X}^{New}$, which gives the observed error:

$$ObsSS = \|\mathbf{x} - \delta(\mathbf{x})\|^2. \tag{20.99}$$

(a) What is *ObsSS* for δ_A as in Exercise 20.9.21(c)? For which such estimator is *ObsSS* minimized? (b) Because we want to use *ObsSS* as an estimator of $E[PredSS\,|\,\mu]$, it would be helpful to know whether it is a good estimator. What is $E[ObsSS\,|\,\mu]$ for a given δ_A? Is *ObsSS* an unbiased estimator of $E[PredSS\,|\,\mu]$? What is $E[PredSS\,|\,\mu] - E[ObsSS\,|\,\mu]$? (c) Find a constant C_A (depending on the subset A) so that $ObsSS + C_A$ is an unbiased estimator of $E[PredSS\,|\,\mu]$. (The quantity $ObsSS + C_A$ is a special case of Mallows' C_p statistic from Section 12.5.3.) (d) Let $\delta^*(\mathbf{x})$ be $\delta_{A(\mathbf{x})}(\mathbf{x})$, where $A(\mathbf{x})$ is the subset that minimizes $ObsSS + C_A$ for given $\mathbf{x}$. Give $\delta^*(\mathbf{x})$ explicitly as a function of $\mathbf{x}$.

Exercise 20.9.23. The generic two-person zero-sum game has loss function given by the following table:

House	You	
$\downarrow$	a_1	a_2
θ_1	a	c
θ_2	b	d

(20.100)

(a) If $a \geq c$ and $b > d$, then what should your strategy be? (b) Suppose $a > c$ and $d > b$, so neither of your actions is always better than the other. Find your minimax strategy $\mathbf{p}_0$, the least favorable distribution π_0, and show that the value of the game is $V = (ad - bc)/(a - b - c + d)$.

Exercise 20.9.24. In the two-person game rock-paper-scissors, each player chooses one of the three options (rock, paper, or scissors). If they both choose the same option, then the game is a tie. Otherwise, rock beats scissors (by crushing them); scissors beats paper (by cutting it); and paper beats rock (by wrapping it). If you are playing the house, your loss is 1 if you lose, 0 if you tie, and -1 if you win. (a) Write out the loss function as a 3×3 table. (b) Show that your minimax strategy (and the least favorable distribution) is to choose each option with probability $1/3$. (c) Find the value of the game.

Exercise 20.9.25. The set of randomized procedures $\mathcal{D}$ is closed under randomization if given any two procedures in $\mathcal{D}$, the procedure that randomly chooses between the two is also in $\mathcal{D}$. For this exercise, suppose the randomized procedures can be represented as in (20.59) by $\delta(\mathbf{X}, U_1, \ldots, U_K)$, where $U_1, U_2, \ldots$ are iid Uniform$(0, 1)$ and independent of $\mathbf{X}$. Suppose that if $\delta_1(\mathbf{x}, u_1, \ldots, u_K)$ and $\delta_2(\mathbf{x}, u_1, \ldots, u_L)$ are both in $\mathcal{D}$, then for any $\alpha \in [0, 1]$, so is δ defined by

$$\delta(\mathbf{x}, u_1, \ldots, u_{M+1}) = \begin{cases} \delta_1(\mathbf{x}, u_1, \ldots, u_K) & \text{if } u_{M+1} < \alpha \\ \delta_2(\mathbf{x}, u_1, \ldots, u_L) & \text{if } u_{M+1} \geq \alpha \end{cases}, \tag{20.101}$$

where $M = \max\{K, L\}$. (a) Show that if δ_1 and δ_2 both have finite risk at θ, then

$$R(\theta; \delta) = \alpha R(\theta; \delta_1) + (1 - \alpha) R(\theta; \delta_2). \tag{20.102}$$

(b) Show that the risk set is convex.

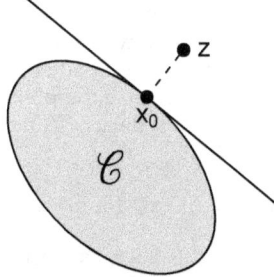

Figure 20.4: This plot illustrates the result in Exercise 20.9.28. The x_0 is the closest point in C to $\mathbf{z}$. The solid line is the set $\{\mathbf{x} \mid \boldsymbol{\gamma} \cdot \mathbf{x} = \boldsymbol{\gamma} \cdot \mathbf{x}_0\}$, where $\boldsymbol{\gamma} = (\mathbf{z} - \mathbf{x}_0)/\|\mathbf{z} - \mathbf{x}_0\|$.

Exercise 20.9.26. Consider the setup in Theorem 20.6. Let δ_0 and π_0 be as in (20.71) to (20.73), so that δ_0 is minimax and Bayes wrt π_0, with $R(\pi_0 ; \delta_0) = s_0 = \max_i R(\theta_i ; \delta_0)$. (a) Show that $R(\pi ; \delta_0) \leq s_0$ for any prior π. (b) Argue that

$$\inf_{\delta \in \mathcal{D}} R(\pi ; \delta) \leq \inf_{\delta \in \mathcal{D}} R(\pi_0 ; \delta), \qquad (20.103)$$

so that π_0 is a least favorable prior.

Exercise 20.9.27. Suppose the set $\mathcal{R} \subset \mathbb{R}^K$ is closed, convex, and bounded from below. For constant vector $\mathbf{a} \in \mathbb{R}^K$, set $\mathcal{R}^* = \{\mathbf{r} - \mathbf{a} \mid \mathbf{r} \in \mathcal{R}\}$. Show that $\mathcal{R}^*$ is also closed, convex, and bounded from below.

The next three exercises prove the separating hyperplane theorem, Theorem 20.7. Exercise 20.9.28 proves the theorem when one of the sets contains a single point separated from the other set. Exercise 20.9.29 extends the proof to the case that the single point is on the border of the other set. Exercise 20.9.30 then completes the proof.

Exercise 20.9.28. Suppose $C \in \mathbb{R}^K$ is convex, and $\mathbf{z} \notin$ closure(C). The goal is to show that there exists a vector $\boldsymbol{\gamma}$ with $\|\boldsymbol{\gamma}\| = 1$ such that

$$\boldsymbol{\gamma} \cdot \mathbf{x} < \boldsymbol{\gamma} \cdot \mathbf{z} \text{ for all } \mathbf{x} \in C. \qquad (20.104)$$

See Figure 20.4 for an illustration. Let $s_0 = \inf\{\|\mathbf{x} - \mathbf{z}\| \mid \mathbf{x} \in C\}$, the shortest distance from $\mathbf{z}$ to C. Then there exists a sequence $\mathbf{x}_n \in C$ and point $\mathbf{x}_0$ such that $\mathbf{x}_n \to \mathbf{x}_0$ and $\|\mathbf{x}_0 - \mathbf{z}\| = s_0$. [Extra credit: Prove that fact.] (Note: This $\mathbf{x}_0$ is unique, and called the projection of $\mathbf{0}$ onto C, analogous to the project $\hat{\mathbf{y}}$ in (12.14) for linear regression.) (a) Show that $\mathbf{x}_0 \neq \mathbf{z}$, hence $s_0 > 0$. [Hint: Note that $\mathbf{x}_0 \in$ closure(C).] (b) Take any $\mathbf{x} \in C$. Argue that for any $\alpha \in [0,1]$, $\alpha \mathbf{x} + (1 - \alpha)\mathbf{x}_n \in C$, hence $\|\alpha \mathbf{x} + (1 - \alpha)\mathbf{x}_n - \mathbf{z}\|^2 \geq s_0$. Then by letting $n \to \infty$, we have that $\|\alpha \mathbf{x} + (1 - \alpha)\mathbf{x}_0 - \mathbf{z}\|^2 \geq \|\mathbf{x}_0 - \mathbf{z}\|^2$. (c) Show that the last inequality in part (b) can be written $\alpha^2 \|\mathbf{x} - \mathbf{x}_0\|^2 - 2\alpha(\mathbf{z} - \mathbf{x}_0) \cdot (\mathbf{x} - \mathbf{x}_0) \geq 0$. For $\alpha \in (0,1)$, divide by α and let $\alpha \to 0$ to show that $(\mathbf{z} - \mathbf{x}_0) \cdot (\mathbf{x} - \mathbf{x}_0) \leq 0$. (d) Take $\boldsymbol{\gamma} = (\mathbf{z} - \mathbf{x}_0)/\|\mathbf{z} - \mathbf{x}_0\|$, so that $\|\boldsymbol{\gamma}\| = 1$. Part (c) shows that $\boldsymbol{\gamma} \cdot (\mathbf{x} - \mathbf{x}_0) \leq 0$ for $\mathbf{x} \in C$. Show that $\boldsymbol{\gamma} \cdot (\mathbf{z} - \mathbf{x}_0) > 0$. (e) Argue that therefore (20.104) holds.

Exercise 20.9.29. Now suppose $\mathcal{C}$ is convex and $\mathbf{z} \notin \mathcal{C}$, but $\mathbf{z} \in$ closure($\mathcal{C}$). It is a nontrivial fact that the interior of a convex set is the same as the interior of its closure. (You don't need to prove this. See Lemma 2.7.2 of Ferguson (1967).) Thus $\mathbf{z} \notin$ interior(closure($\mathcal{C}$)), which means that there exists a sequence $\mathbf{z}_n \to \mathbf{z}$ with $\mathbf{z}_n \notin$ closure($\mathcal{C}$). (Thus $\mathbf{z}$ is on the boundary of $\mathcal{C}$.) (a) Show that for each n there exists a vector γ_n, $\|\gamma_n\| = 1$, such that

$$\gamma_n \cdot \mathbf{x} < \gamma_n \cdot \mathbf{z}_n \text{ for all } \mathbf{x} \in \mathcal{C}. \tag{20.105}$$

[Hint: Use Exercise 20.9.28.] (b) Since the γ_n's exist in a compact space, there is a subsequence of them and a vector γ such that $\gamma_{n_i} \to \gamma$. Show that by taking the limit along this subsequence in (20.105), we have that

$$\gamma \cdot \mathbf{x} \le \gamma \cdot \mathbf{z} \text{ for all } \mathbf{x} \in \mathcal{C}. \tag{20.106}$$

The set $\{\mathbf{x} \mid \gamma \cdot \mathbf{x} = c\}$ where $c = \gamma \cdot \mathbf{z}$ is called a **supporting hyperplane** of $\mathcal{C}$ through $\mathbf{z}$.

Exercise 20.9.30. Let $\mathcal{A}$ and $\mathcal{B}$ be convex sets with $\mathcal{A} \cap \mathcal{B} = \emptyset$. Define $\mathcal{C}$ to be their difference:

$$\mathcal{C} = \{\mathbf{x} - \mathbf{y} \mid \mathbf{x} \in \mathcal{A} \text{ and } \mathbf{y} \in \mathcal{B}\}. \tag{20.107}$$

(a) Show that $\mathcal{C}$ is convex and $\mathbf{0} \notin \mathcal{C}$. (b) Use (20.106) to show that there exists a γ, $\|\gamma\| = 1$, such that

$$\gamma \cdot \mathbf{x} \le \gamma \cdot \mathbf{y} \text{ for all } x \in \mathcal{A} \text{ and } \mathbf{y} \in \mathcal{B}. \tag{20.108}$$

Chapter 21

Optimal Hypothesis Tests

21.1   Randomized tests

Chapter 15 discusses hypothesis testing, where we choose between the null and alternative hypotheses,

$$H_0 : \theta \in \mathcal{T}_0 \;\; versus \;\; H_A : \theta \in \mathcal{T}_A, \tag{21.1}$$

$\mathcal{T}_0$ and $\mathcal{T}_A$ being disjoint subsets of the overall parameter space $\mathcal{T}$. The goal is to make a good choice, so that we desire a procedure that has small probability of rejecting the null when it is true (the size), and large probability of rejecting the null when it is false (the power). This approach to hypothesis testing usually fixes a level α, and considers tests whose size is less than or equal to α. Among the level α tests, the ones with good power are preferable. In certain cases a best level α test exists in that it has the best power for any parameter value θ in the alternative space $\mathcal{T}_A$. More commonly, different tests are better for different values of the parameter, hence decision-theoretic concepts such as admissibility and minimaxity are relevant, which will be covered in Chapter 22.

In Chapter 15, hypothesis tests are defined using a test statistic and cutoff point, rejecting the null when the statistic is larger (or smaller) than the cutoff point. These are non-randomized tests, since once we have the data we know the outcome. Randomized tests, as in the randomized strategies for game theory presented in Section 20.7, are useful in the decision-theoretical analysis of testing. As we noted earlier, actual decisions in practice should not be randomized.

To understand the utility of randomized tests, let $X \sim \text{Binomial}(4, \theta)$, and test $H_0 : \theta = 1/2$ versus $H_A : \theta = 3/5$ at level $\alpha = 0.35$. The table below gives the pmfs under the null and alternative:

x	$f_{1/2}(x)$	$f_{3/5}(x)$
0	0.0625	0.0256
1	0.2500	0.1536
2	0.3750	0.3456
3	0.2500	0.3456
4	0.0625	0.1296

$$\tag{21.2}$$

If we reject the null when $X \geq 3$, the size is 0.3125, and if we reject when $X \geq 2$, the size is 0.6875. Thus to keep the size from exceeding α, we use $X \geq 3$. The power at

$\theta = 3/5$ is 0.4752. Consider a randomized test that rejects the null when $X \geq 3$, and if $X = 2$, it rejects the null with probability 0.1. Then the size is

$$P[X \geq 3 \mid \theta = \tfrac{1}{2}] + 0.1 \cdot P[X = 2 \mid \theta = \tfrac{1}{2}] = 0.3125 + 0.1 \cdot 0.3750 = 0.35, \qquad (21.3)$$

hence its level is α. But it has a larger power than the original test since it rejects more often:

$$P[X \geq 3 \mid \theta = \tfrac{3}{5}] + 0.1 \cdot P[X = 2 \mid \theta = \tfrac{3}{5}] = 0.50976 > 0.4752 = P[X \geq 3 \mid \theta = \tfrac{3}{5}].$$
$$(21.4)$$

In order to accommodate randomized tests, rather than using test statistics and cutoff points, we define a testing procedure as a function

$$\phi : \mathcal{X} \longrightarrow [0,1], \qquad (21.5)$$

where $\phi(\mathbf{x})$ is the probability of rejecting the null given $\mathbf{X} = \mathbf{x}$:

$$\phi(\mathbf{x}) = P[\text{Reject} \mid \mathbf{X} = \mathbf{x}]. \qquad (21.6)$$

For a nonrandomized test, $\phi(x) = I[T(\mathbf{x}) > c]$ as in (15.6). The test in (21.3) and (21.4) is given by

$$\phi'(x) = \begin{cases} 1 & \text{if} \quad x \geq 3 \\ 0.1 & \text{if} \quad x = 2 \\ 0 & \text{if} \quad x \leq 1 \end{cases}. \qquad (21.7)$$

Now the level and power are easy to represent, since $E_\theta[\phi(\mathbf{X})] = P_\theta[\phi \text{ rejects}]$.

21.2   Simple versus simple

We will start simple, where each hypothesis has exactly one distribution as in (15.25). We observe $\mathbf{X}$ with density $f(\mathbf{x})$, and test

$$H_0 : f = f_0 \quad versus \quad H_A : f = f_A. \qquad (21.8)$$

(The parameter space consists of just two points, $\mathcal{T} = \{0, A\}$.) In Section 15.3 we mentioned that the test based on the likelihood ratio,

$$LR(\mathbf{x}) = \frac{f_A(\mathbf{x})}{f_0(\mathbf{x})}, \qquad (21.9)$$

is optimal. Here we formalize this result.

Fix $\alpha \in [0,1]$. We wish to find a level α test that maximizes the power among level α tests. For example, suppose $X \sim \text{Binomial}(4, \theta)$, and the hypotheses are $H_0 : \theta = 1/2$ versus $H_A : \theta = 3/5$, so that the pmfs are given in (21.2). With $\alpha = 0.35$, the objective is to find a test ϕ that

$$\text{Maximizes } E_{3/5}[\phi(X)] \text{ subject to } E_{1/2}[\phi(X)] \leq \alpha = 0.35, \qquad (21.10)$$

that is, maximizes the power subject to being of level α. What should we look for? First, an analogy.

Bang for the buck. Imagine you have some bookshelves you wish to fill up as cheaply as possible, e.g., to use as props in a play. You do not care about the quality of the books, just their widths (in inches) and prices (in dollars). You have $3.50, and five books to choose from:

Book	Cost	Width	Inches/Dollar
0	0.625	0.256	0.4096
1	2.50	1.536	0.6144
2	3.75	3.456	0.9216
3	2.50	3.456	1.3824
4	0.625	1.296	2.0736

(21.11)

You are allowed to split the books lengthwise, and pay proportionately. You want to maximize the total number of inches for your $3.50. Then, for example, book 4 is a better deal than book 0, because they cost the same but book 4 is wider. Also, book 3 is more attractive than book 2, because they are the same width but book 3 is cheaper. Which is better between books 3 and 4? Book 4 is cheaper by the inch: It costs about 48¢ per inch, while book 3 is about 73¢ per inch. This suggests the strategy should be to buy the books that give you the most inches per dollar.

Let us definitely buy book 4, and book 3. That costs us $3.125, and gives us 1.296+3.456 = 4.752 inches. We still have 37.5¢ left, with which we can buy a tenth of book 2, giving us another 0.3456 inches, totaling 5.0976 inches. □

Returning to the hypothesis testing problem, we can think of having α to spend, and we wish to spend where we get the most bang for the buck. Here, bang is power. The key is to look at the likelihood ratio of the densities:

$$LR(x) = \frac{f_{3/5}(x)}{f_{1/2}(x)}, \qquad (21.12)$$

which turn out to be the same as the inches per dollar in table (21.11). (The cost is ten times the null pmf, and the width is ten times the alternative pmf.) If $LR(x)$ is large, then the alternative is much more likely than the null is. If $LR(x)$ is small, the null is more likely. One uses the likelihood ratio as the statistic, and finds the right cutoff point, randomizing at the cutoff point if necessary. The **likelihood ratio test** is then

$$\phi_{LR}(x) = \begin{cases} 1 & \text{if} \quad LR(x) > c \\ \gamma & \text{if} \quad LR(x) = c \\ 0 & \text{if} \quad LR(x) < c \end{cases}. \qquad (21.13)$$

Looking at the table (21.11), we see that taking $c = LR(2)$ works, because we reject if $x = 3$ or 4, and use up only .3125 of our α. Then the rest we put on $x = 2$, the $\gamma = 0.1$ since we have $.35 - .3125 = 0.0375$ left. Thus the test is

$$\phi_{LR}(x) = \begin{cases} 1 & \text{if} \quad LR(x) > 0.9216 \\ 0.1 & \text{if} \quad LR(x) = 0.9216 \\ 0 & \text{if} \quad LR(x) < 0.9216 \end{cases} = \begin{cases} 1 & \text{if} \quad x \geq 3 \\ 0.1 & \text{if} \quad x = 2 \\ 0 & \text{if} \quad x \leq 1 \end{cases}, \qquad (21.14)$$

which is ϕ' in (21.7). The last expression is easier to deal with, and valid since $LR(x)$ is a strictly increasing function of x. Then the power and level are 0.35 and 0.50976, as in (21.3) and (21.4). This is the same as for the books: The power is identified with the number of inches. Is this the best test? Yes, as we will see from the Neyman-Pearson lemma in the next section.

21.3   Neyman-Pearson lemma

Let $\mathbf{X}$ be the random variable or vector with density f, and f_0 and f_A be two possible densities for $\mathbf{X}$. We are interested in testing f_0 versus f_A as in (21.8). For given $\alpha \in [0,1]$, we wish to find a test function ϕ that

$$\text{Maximizes } E_A[\phi(\mathbf{X})] \text{ subject to } E_0[\phi(\mathbf{X})] \le \alpha. \tag{21.15}$$

A test function ψ has **Neyman-Pearson form** if for some constant $c \in [0,\infty]$ and function $\gamma(\mathbf{x}) \in [0,1]$,

$$\psi(\mathbf{x}) = \left\{ \begin{array}{ll} 1 & \text{if } f_A(\mathbf{x}) > cf_0(\mathbf{x}) \\ \gamma(\mathbf{x}) & \text{if } f_A(\mathbf{x}) = cf_0(\mathbf{x}) \\ 0 & \text{if } f_A(\mathbf{x}) < cf_0(\mathbf{x}) \end{array} \right\} = \left\{ \begin{array}{ll} 1 & \text{if } LR(\mathbf{x}) > c \\ \gamma(\mathbf{x}) & \text{if } LR(\mathbf{x}) = c \\ 0 & \text{if } LR(\mathbf{x}) < c \end{array} \right. , \tag{21.16}$$

with the caveat that

$$\text{if } c = \infty \text{ then } \gamma(\mathbf{x}) = 1 \text{ for all } \mathbf{x}. \tag{21.17}$$

Note that this form is the same as ϕ_{LR} in (21.13), but allows γ to depend on $\mathbf{x}$. Here,

$$LR(\mathbf{x}) = \frac{f_A(\mathbf{x})}{f_0(\mathbf{x})} \in [0,\infty] \tag{21.18}$$

is defined unless $f_A(\mathbf{x}) = f_0(\mathbf{x}) = 0$, in which case $\psi(\mathbf{x}) = \gamma(\mathbf{x})$. Notice that LR and c are allowed to take on the value ∞.

Lemma 21.1. Neyman-Pearson. *Any test ψ of Neyman-Pearson form (21.16,21.17) for which $E_0[\psi(\mathbf{X})] = \alpha$ satisfies (21.15).*

So basically, the likelihood ratio test is best. One can take the $\gamma(\mathbf{x})$ to be a constant, but sometimes it is convenient to have it depend on $\mathbf{x}$. Before getting to the proof, consider some special cases, of mainly theoretical interest.

- **α=0.** If there is no chance of rejecting when the null is true, then one must always accept if $f_0(\mathbf{x}) > 0$, and it always makes sense to reject when $f_0(\mathbf{x}) = 0$. Such actions invoke the caveat (21.17), that is, when $f_A(\mathbf{x}) > 0$,

$$\psi(\mathbf{x}) = \left\{ \begin{array}{ll} 1 & \text{if } LR(\mathbf{x}) = \infty \\ 0 & \text{if } LR(\mathbf{x}) < \infty \end{array} \right\} = \left\{ \begin{array}{ll} 1 & \text{if } f_0(\mathbf{x}) = 0 \\ 0 & \text{if } f_0(\mathbf{x}) > 0 \end{array} \right. . \tag{21.19}$$

- **α=1.** This one is silly from a practical point of view, but if you do not care about rejecting when the null is true, then you should always reject, i.e., take $\phi(\mathbf{x}) = 1$.

- **Power = 1.** If you want to be sure to reject if the alternative is true, then $\phi(\mathbf{x}) = 1$ when $f_A(\mathbf{x}) > 0$, so take the test (21.16) with $c = 0$. Of course, you may not be able to achieve your desired α.

Proof. (Lemma 21.1) If $\alpha = 0$, then the above discussion shows that taking $c = \infty, \gamma(\mathbf{x}) = 1$ as in (21.17) is best. For $\alpha \in (0,1]$, suppose ψ satisfies (21.16) for some

c and $\gamma(\mathbf{x})$ with $E_0[\psi(\mathbf{X})] = \alpha$, and ϕ is any other test function with $E_0[\phi(\mathbf{X})] \leq \alpha$. Look at

$$
\begin{aligned}
E_A[\psi(\mathbf{X}) - \phi(\mathbf{X})] - c\, E_0[\psi(\mathbf{X}) - \phi(\mathbf{X})] &= \int_{\mathcal{X}} (\psi(\mathbf{x}) - \phi(\mathbf{x})) f_A(\mathbf{x}) d\mathbf{x} \\
&\quad - c \int_{\mathcal{X}} (\psi(\mathbf{x}) - \phi(\mathbf{x})) f_0(\mathbf{x}) d\mathbf{x} \\
&= \int_{\mathcal{X}} (\psi(\mathbf{x}) - \phi(\mathbf{x}))(f_A(\mathbf{x}) - c f_0(\mathbf{x})) d\mathbf{x} \\
&\geq 0.
\end{aligned} \tag{21.20}
$$

The final inequality holds because $\psi = 1$ if $f_A(\mathbf{x}) - c f_0(\mathbf{x}) > 0$, and $\psi = 0$ if $f_A(\mathbf{x}) - c f_0(\mathbf{x}) < 0$, so that the final integrand is always nonnegative. Thus

$$
\begin{aligned}
E_A[\psi(\mathbf{X}) - \phi(\mathbf{X})] &\geq c\, E_0[\psi(\mathbf{X}) - \phi(\mathbf{X})] \\
&\geq 0,
\end{aligned} \tag{21.21}
$$

because $E_0[\psi(\mathbf{X})] = \alpha \geq E_0[\phi(\mathbf{X})]$. Hence $E_A[\psi(\mathbf{X})] \geq E_A[\phi(\mathbf{X})]$, i.e., any other level α test has lower or equal power. □

There are a couple of addenda to the lemma that we will not prove here, but Lehmann and Romano (2005) does in their Theorem 3.2.1. First, for any α, there is a test of Neyman-Pearson form. Second, if the ϕ in the proof is not essentially of Neyman-Pearson form, then the power of ψ is strictly better than that of ϕ. That is,

$$
P_0[\phi(\mathbf{X}) \neq \psi(\mathbf{X}) \,\&\, LR(\mathbf{X}) \neq c] > 0 \implies E_A[\psi(\mathbf{X})] > E_A[\phi(\mathbf{X})]. \tag{21.22}
$$

21.3.1 Examples

If $f_0(\mathbf{x}) > 0$ and $f_A(\mathbf{x}) > 0$ for all $\mathbf{x} \in \mathcal{X}$, then it is straightforward (though maybe not easy) to find the Neyman-Pearson test. It can get tricky if one or the other density is 0 at times.

Normal means

Suppose μ_0 and μ_A are fixed, $\mu_A > \mu_0$, and $X \sim N(\mu, 1)$. We wish to test

$$
H_0 : \mu = \mu_0 \quad versus \quad H_A : \mu = \mu_A \tag{21.23}
$$

with $\alpha = 0.05$. Here,

$$
LR(x) = \frac{e^{-\frac{1}{2}(x - \mu_A)^2}}{e^{-\frac{1}{2}(x - \mu_0)^2}} = e^{x(\mu_A - \mu_0) - \frac{1}{2}(\mu_A^2 - \mu_0^2)}. \tag{21.24}
$$

Because $\mu_A > \mu_0$, $LR(x)$ is strictly increasing in x, so $LR(x) > c$ is equivalent to $x > c^*$ for some c^*. For level 0.05, we know that $c^* = 1.645 + \mu_0$, so the test must reject when $LR(x) > LR(c^*)$, i.e.,

$$
\psi(\mathbf{x}) = \begin{cases} 1 & \text{if } e^{x(\mu_A - \mu_0) - \frac{1}{2}(\mu_A^2 - \mu_0^2)} > c \\ 0 & \text{if } e^{x(\mu_A - \mu_0) - \frac{1}{2}(\mu_A^2 - \mu_0^2)} \leq c \end{cases}, \quad c = e^{(1.645 + \mu_0)(\mu_A - \mu_0) - \frac{1}{2}(\mu_A^2 - \mu_0^2)}.
$$

$$
\tag{21.25}
$$

We have taken the $\gamma = 0$; the probability $LR(X) = c$ is 0, so it doesn't matter what happens then. Expression (21.25) is unnecessarily complicated. In fact, to find c we already simplified the test, that is

$$LR(x) > c \iff x - \mu_0 > 1.645, \tag{21.26}$$

hence

$$\psi(x) = \begin{cases} 1 & \text{if} \quad x - \mu_0 > 1.645 \\ 0 & \text{if} \quad x - \mu_0 \le 1.645 \end{cases}. \tag{21.27}$$

That is, we really do not care about c, as long as we have the ψ.

Laplace versus normal

Suppose f_0 is the Laplace pdf and f_A is the $N(0,1)$ pdf, and $\alpha = 0.1$. Then

$$LR(x) = \frac{\frac{1}{\sqrt{2\pi}} e^{-\frac{1}{2} x^2}}{\frac{1}{2} e^{-|x|}} = \sqrt{\frac{2}{\pi}} e^{|x| - \frac{1}{2} x^2}. \tag{21.28}$$

Now $LR(x) > c$ if and only if

$$|x| - \frac{1}{2} x^2 > c^* = \log(c) + \frac{1}{2} \log(\pi/2) \tag{21.29}$$

if and only if (completing the square)

$$(|x| - 1)^2 < c^{**} = -2c^* - 1 \iff ||x| - 1| < c^{***} = \sqrt{c^{**}}. \tag{21.30}$$

We need to find the constant c^{***} so that

$$P_0[||X| - 1| < c^{***}] = 0.10, \quad X \sim \text{Laplace}. \tag{21.31}$$

For a smallish c^{***}, using the Laplace pdf,

$$\begin{aligned} P_0[||X| - 1| < c^{***}] &= P_0[-1 - c^{***} < X < -1 + c^{***} \ \text{or} \ 1 - c^{***} < X < 1 + c^{***}] \\ &= 2\, P_0[-1 - c^{***} < X < -1 + c^{***}] \\ &= e^{-(1 - c^{***})} - e^{-(1 + c^{***})}. \end{aligned} \tag{21.32}$$

Setting that probability equal to 0.10, we find $c^{***} = 0.1355$. Figure 21.1 shows a horizontal line at 0.1355. The rejection region consists of the x's for which the graph of $||x| - 1|$ is below the line.

The power substitutes the normal for the Laplace in (21.32):

$$P_A[||N(0,1)| - 1| < 0.1355] = 2(\Phi(1.1355) - \Phi(0.8645)) = 0.1311.$$

Not very powerful, but at least it is larger than α! Of course, it is not surprising that it is hard to distinguish the normal from the Laplace with just one observation.

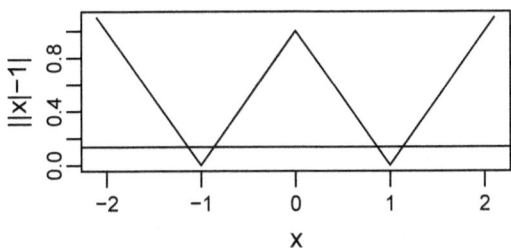

Figure 21.1: The rejection region for testing Laplace versus normal is $\{x \mid ||x| - 1| < 0.1355\}$. The horizontal line is at 0.1355.

Uniform versus uniform I

Suppose $X \sim \text{Uniform}(0, \theta)$, and the question is whether $\theta = 1$ or 2. We could then test

$$H_0 : \theta = 1 \ \textit{versus} \ H_A : \theta = 2. \tag{21.33}$$

The likelihood ratio is

$$LR(x) = \frac{f_A(x)}{f_0(x)} = \frac{\frac{1}{2} I[0 < x < 2]}{I[0 < x < 1]} = \begin{cases} \frac{1}{2} & \text{if} \ 0 < x < 1 \\ \infty & \text{if} \ 1 \le x < 2 \end{cases}. \tag{21.34}$$

No matter what, you would reject the null if $1 \le x < 2$, because it is impossible to observe an x in that region under the $\text{Uniform}(0, 1)$.

First try $\alpha = 0$. Usually, that would mean never reject, so power would be 0 as well, but here it is not that bad. We invoke (21.17), that is, take $c = \infty$ and $\gamma(x) = 1$:

$$\psi(x) = \begin{cases} 1 & \text{if} \ LR(x) = \infty \\ 0 & \text{if} \ LR(x) < \infty \end{cases} = \begin{cases} 1 & \text{if} \ 1 \le x < 2 \\ 0 & \text{if} \ 0 < x < 1 \end{cases}. \tag{21.35}$$

Then

$$\alpha = P[1 \le U(0, 1) < 2] = 0 \ \text{and} \ Power = P[1 \le U(0, 2) < 2] = \frac{1}{2}. \tag{21.36}$$

What if $\alpha = 0.1$? Then the Neyman-Pearson test would take $c = 1/2$:

$$\psi(x) = \begin{cases} 1 & \text{if} \ LR(x) > \frac{1}{2} \\ \gamma(x) & \text{if} \ LR(x) = \frac{1}{2} \\ 0 & \text{if} \ LR(x) < \frac{1}{2} \end{cases} = \begin{cases} 1 & \text{if} \ 1 \le x < 2 \\ \gamma(x) & \text{if} \ 0 < x < 1 \end{cases}, \tag{21.37}$$

because LR cannot be less than $1/2$. Notice that

$$E_0[\psi(X)] = E_0[\gamma(X)] = \int_0^1 \gamma(x)dx, \tag{21.38}$$

so that any γ that integrates to α works. Some examples:

$$\gamma(x) = 0.1, \ \gamma(x) = I[0 < x < 0.1], \ \gamma(x) = I[0.9 < x < 1], \ \gamma(x) = 0.2 \, x. \tag{21.39}$$

No matter which you choose, the power is the same:

$$Power = E_A[\psi(X)] = \frac{1}{2} \int_0^1 \gamma(x)dx + \frac{1}{2} \int_1^2 dx = \frac{1}{2} \alpha + \frac{1}{2} = 0.55. \tag{21.40}$$

Uniform versus uniform II

Now switch the null and alternative in (21.33), keeping $X \sim \text{Uniform}(0, \theta)$:

$$H_0 : \theta = 2 \ \ versus \ \ H_A : \theta = 1. \tag{21.41}$$

Then the likelihood ratio is

$$LR(x) = \frac{f_A(x)}{f_0(x)} = \frac{I[0 < x < 1]}{\frac{1}{2} I[0 < x < 2]} = \begin{cases} 2 & \text{if} \ \ 0 < x < 1 \\ 0 & \text{if} \ \ 1 \le x < 2 \end{cases}. \tag{21.42}$$

For level $\alpha = 0.1$, we have to take $c = 2$:

$$\psi(x) = \begin{cases} \gamma(x) & \text{if} \ \ LR(x) = 2 \\ 0 & \text{if} \ \ LR(x) < 2 \end{cases} = \begin{cases} \gamma(x) & \text{if} \ \ 0 < x < 1 \\ 0 & \text{if} \ \ 1 \le x < 2 \end{cases}. \tag{21.43}$$

Then any $\gamma(x)$ with

$$\frac{1}{2} \int_0^1 \gamma(x)dx = \alpha \implies \int_0^1 \gamma(x)dx = 0.2 \tag{21.44}$$

will work. And that is the power, 0.2.

21.4 Uniformly most powerful tests

Simple versus simple is too simple. Some testing problems are not so simple, and yet do have a best test. Here is the formal definition.

Definition 21.2. *The test function ψ is a **uniformly most power** (UMP) level α test for testing $H_0 : \theta \in \mathcal{T}_0$ versus $H_A : \theta \in \mathcal{T}_A$ if it is level α and*

$$E_\theta[\psi(\mathbf{X})] \ge E_\theta[\phi(\mathbf{X})] \ \text{for all} \ \theta \in \mathcal{T}_A \tag{21.45}$$

for any other level α test ϕ.

Often, one-sided tests do have a UMP test, while two-sided tests do not. For example, suppose $X \sim N(\mu, 1)$, and we test whether $\mu = 0$. In a one-sided testing problem, the alternative is one of $\mu > 0$ or $\mu < 0$, say

$$H_0 : \mu = 0 \ \ versus \ \ H_A^{(1)} : \mu > 0. \tag{21.46}$$

The corresponding two-sided testing problem is

$$H_0 : \mu = 0 \ \ versus \ \ H_A^{(2)} : \mu \ne 0. \tag{21.47}$$

The usual level $\alpha = 0.05$ tests for these are, respectively,

$$\phi^{(1)}(x) = \begin{cases} 1 & \text{if} \ \ x > 1.645 \\ 0 & \text{if} \ \ x \le 1.645 \end{cases} \ \text{and} \ \phi^{(2)}(x) = \begin{cases} 1 & \text{if} \ \ |x| > 1.96 \\ 0 & \text{if} \ \ |x| \le 1.96 \end{cases}. \tag{21.48}$$

Their powers are

$$E_\mu[\phi^{(1)}(X)] = P[N(\mu, 1) > 1.645] = \Phi(\mu - 1.645) \ \text{and}$$
$$E_\mu[\phi^{(2)}(X)] = P[|N(\mu, 1)| > 1.96] = \Phi(\mu - 1.96) + \Phi(-\mu - 1.96), \tag{21.49}$$

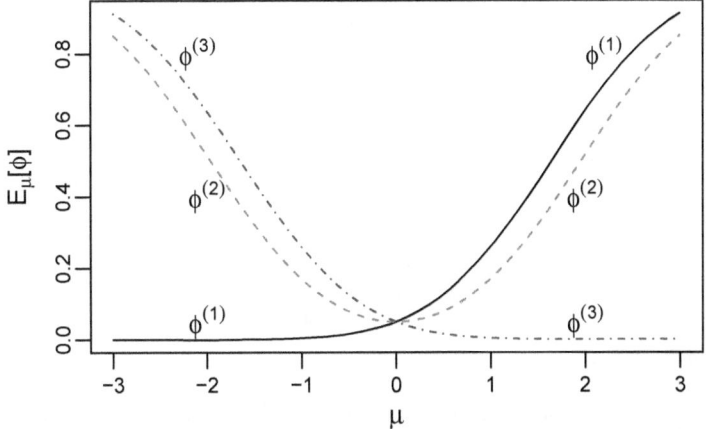

Figure 21.2: The probability of rejecting for testing a normal mean μ is zero. For alternative $\mu > 0$, $\phi^{(1)}$ is the best. For alternative $\mu < 0$, $\phi^{(3)}$ is the best. For alternative $\mu \neq 0$, the two-sided test is $\phi^{(2)}$.

where Φ is the N(0,1) distribution function. See Figure 21.2, or Figure 15.1.

For the one-sided problem, the power is good for $\mu > 0$, but bad (below α) for $\mu < 0$. But the alternative is just $\mu > 0$, so it does not matter what $\phi^{(1)}$ does when $\mu < 0$. For the two-sided test, the power is fairly good on both sided of $\mu = 0$, but it is not quite as good as the one-sided test when $\mu > 0$. The other line in the graph is the one-sided test $\phi^{(3)}$ for alternative $\mu < 0$, which mirrors $\phi^{(1)}$, rejecting when $x < -1.645$.

The following are true, and to be proved:

- For the one-sided problem (21.46), the test $\phi^{(1)}$ **is** the UMP level $\alpha = 0.05$ test.

- For the two-sided problem (21.47), there is **no** UMP level α test. Test $\phi^{(1)}$ is better on one side ($\mu > 0$), and test $\phi^{(3)}$ is better on the other side. None of the three tests is always best. In Section 21.6 we will see that $\phi^{(2)}$ is the UMP *unbiased* test.

We start with the null being simple and the alternative being composite (i. e., not simple). The way to prove a test is UMP level α is to show that it is level α, and that it is of Neyman-Pearson form for each simple versus simple subproblem derived from the big problem. That is, suppose we are testing

$$H_0 : \theta = \theta_0 \;\; versus \;\; H_A : \theta \in \mathcal{T}_A. \tag{21.50}$$

A simple versus simple subproblem takes a specific value from the alternative, so that for a given $\theta_A \in \mathcal{T}_A$, we consider

$$H_0 : \theta = \theta_0 \;\; versus \;\; H_A^{(\theta_A)} : \theta = \theta_A. \tag{21.51}$$

Theorem 21.3. *Suppose that for testing problem (21.50), ψ satisfies*

$$E_{\boldsymbol{\theta}_0}[\psi(\mathbf{X})] = \alpha, \tag{21.52}$$

and that for each $\boldsymbol{\theta}_A \in \mathcal{T}_A$,

$$\psi(\mathbf{x}) = \begin{cases} 1 & \text{if } \ LR(\mathbf{x};\boldsymbol{\theta}_A) > c(\boldsymbol{\theta}_A) \\ \gamma(\mathbf{x}) & \text{if } \ LR(\mathbf{x};\boldsymbol{\theta}_A) = c(\boldsymbol{\theta}_A) \\ 0 & \text{if } \ LR(\mathbf{x};\boldsymbol{\theta}_A) < c(\boldsymbol{\theta}_A) \end{cases} , \quad \text{for some constant } c(\boldsymbol{\theta}_A), \tag{21.53}$$

where

$$LR(x;\boldsymbol{\theta}_A) = \frac{f_{\boldsymbol{\theta}_A}(\mathbf{x})}{f_{\boldsymbol{\theta}_0}(\mathbf{x})}. \tag{21.54}$$

Then ψ is a UMP level α test for (21.50).

Proof. Suppose ϕ is another level α test. Then for the subproblem (21.51), ψ has at least as high power, i.e.,

$$E_{\boldsymbol{\theta}_A}[\psi(X)] \geq E_{\boldsymbol{\theta}_A}[\phi(X)]. \tag{21.55}$$

But that inequality is true for any $\boldsymbol{\theta}_A \in \mathcal{T}_A$, hence ψ is UMP level α. $\qquad\square$

The difficulty is to find a test ψ which is Neyman-Pearson for *all* $\boldsymbol{\theta}_A$. Consider the example with $X \sim N(\mu,1)$ and hypotheses

$$H_0 : \mu = 0 \ \textit{versus} \ H_A^{(1)} : \mu > 0, \tag{21.56}$$

and take $\alpha = 0.05$. For fixed $\mu_A > 0$, the Neyman-Pearson test is found as in (21.25) to (21.27) with $\mu_0 = 0$:

$$
\begin{aligned}
\phi^{(1)}(x) &= \begin{cases} 1 & \text{if } \ LR(\mathbf{x};\mu_A) > c(\mu_A) \\ 0 & \text{if } \ LR(\mathbf{x};\mu_A) \leq c(\mu_A) \end{cases} \\[4pt]
&= \begin{cases} 1 & \text{if } \ e^{-\frac{1}{2}((x-\mu_A)^2 - x^2)} > c(\mu_A) \\ 0 & \text{if } \ e^{-\frac{1}{2}((x-\mu_A)^2 - x^2)} \leq c(\mu_A) \end{cases} \\[4pt]
&= \begin{cases} 1 & \text{if } \ x > (\log(c(\mu_A)) + \frac{1}{2}\mu_A^2)/\mu_A \\ 0 & \text{if } \ x \leq (\log(c(\mu_A)) + \frac{1}{2}\mu_A^2)/\mu_A \end{cases} .
\end{aligned} \tag{21.57}
$$

The last step is valid because we know $\mu_A > 0$. That messy constant is chosen so that the level is 0.05, which we know must be

$$(\log(c(\mu_A)) + \tfrac{1}{2}\mu_A^2)/\mu_A = 1.645. \tag{21.58}$$

The key point is that (21.58) is true for *any* $\mu_A > 0$, that is,

$$\phi^{(1)}(x) = \begin{cases} 1 & \text{if } \ x > 1.645 \\ 0 & \text{if } \ x \leq 1.645 \end{cases} \tag{21.59}$$

is true for any μ_A. Thus $\phi^{(1)}$ is indeed UMP. Note that the constant $c(\mu_A)$ is different for each μ_A, but the test $\phi^{(1)}$ is the same. (See Figure 21.2 again for its power function.)

Why is there no UMP test for the two-sided problem,

$$H_0 : \mu = 0 \ versus \ H_A^{(2)} : \mu \neq 0? \tag{21.60}$$

The best test at alternative $\mu_A > 0$ is (21.59), but the best test at alternative $\mu_A < 0$ is found as in (21.57), except that the inequalities reverse in the fourth equality, yielding

$$\phi^{(3)}(x) = \begin{cases} 1 & \text{if} \quad x < -1.645 \\ 0 & \text{if} \quad x \geq -1.645 \end{cases}, \tag{21.61}$$

which is different than $\phi^{(1)}$ in (21.59). That is, there is no test that is best at both positive and negative values of μ_A, so there is no UMP test.

21.4.1 One-sided exponential family testing problems

The normal example above can be extended to general exponential families. The key to the existence of a UMP test is that the $LR(\mathbf{x}; \theta_A)$ is increasing in the same function of $\mathbf{x}$ no matter what the alternative. That is, suppose $X_1, \ldots, X_n$ are iid with a one-dimensional exponential family density

$$f(\mathbf{x} \mid \theta) = a(\mathbf{x}) \, e^{\theta \Sigma t(x_i) - n\rho(\theta)}. \tag{21.62}$$

A one-sided testing problem is

$$H_0 : \theta = \theta_0 \ versus \ H_A : \theta > \theta_0. \tag{21.63}$$

Then for fixed alternative $\theta_A > \theta_0$,

$$LR(\mathbf{x}; \theta_A) = \frac{f(\mathbf{x} \mid \theta_A)}{f(\mathbf{x} \mid \theta_0)} = e^{(\theta_A - \theta_0)\Sigma t(x_i) - n(\rho(\theta_A) - \rho(\theta_0))}. \tag{21.64}$$

Similar calculations as in (21.57) show that the best test at the alternative θ_A is

$$\psi(x) = \begin{cases} 1 & \text{if} \quad LR(\mathbf{x}; \theta_A) > c(\theta_A) \\ \gamma & \text{if} \quad LR(\mathbf{x}; \theta_A) = c(\theta_A) \\ 0 & \text{if} \quad LR(\mathbf{x}; \theta_A) < c(\theta_A) \end{cases}$$

$$= \begin{cases} 1 & \text{if} \quad \Sigma t(x_i) > c \\ \gamma & \text{if} \quad \Sigma t(x_i) = c \\ 0 & \text{if} \quad \Sigma t(x_i) < c \end{cases}. \tag{21.65}$$

Then c and γ are chosen to give the right level, but they are the same for any alternative $\theta_A > \theta_0$. Thus the test (21.65) is UMP level α.

If the alternative were $\theta < \theta_0$, then the same reasoning would work, but the inequalities would switch. For a two-sided alternative, there would not be a UMP test.

21.4.2 Monotone likelihood ratio

A generalization of exponential families, which guarantee UMP tests, are families with **monotone likelihood ratio**, which is a stronger condition than the stochastic increasing property we saw in Definition 18.1 on page 306. Non-exponential family examples include the noncentral χ^2 and F distributions.

Definition 21.4. *A family of densities* $f(\mathbf{x} \mid \theta), \theta \in \mathcal{T} \subset \mathbb{R}$ *has* **monotone likelihood ratio (MLR)** *with respect to* **parameter** θ *and statistic* $s(\mathbf{x})$ *if for any* $\theta' < \theta$,

$$\frac{f(\mathbf{x} \mid \theta)}{f(\mathbf{x} \mid \theta')} \tag{21.66}$$

is a function of just $s(\mathbf{x})$, *and is nondecreasing in* $s(\mathbf{x})$. *If the ratio is strictly increasing in* $s(\mathbf{x})$, *then the family has* **strict** *monotone likelihood ratio.*

Note in particular that this $s(\mathbf{x})$ is a sufficient statistic. It is fairly easy to see that one-dimensional exponential families have MLR. The general idea of MLR is that in some sense, as θ gets bigger, $s(\mathbf{X})$ gets bigger. The next lemma formalizes such a sense.

Lemma 21.5. *If the family* $f(\mathbf{x} \mid \theta)$ *has MLR with respect to* θ *and* $s(\mathbf{x})$, *then for any nondecreasing function* $g(w)$,

$$E_\theta[g(s(\mathbf{X}))] \ \text{is nondecreasing in } \theta. \tag{21.67}$$

If the family has strict MLR, and g *is strictly increasing, then the expected value in (21.67) is strictly increasing in* θ.

Proof. We present the proof using pdfs. Suppose $g(w)$ is nondecreasing, and $\theta' < \theta$. Then

$$E_\theta[g(s(\mathbf{X}))] - E_{\theta'}[g(s(\mathbf{X}))] = \int g(s(\mathbf{x}))(f_\theta(\mathbf{x}) - f_{\theta'}(\mathbf{x}))d\mathbf{x}$$

$$= \int g(s(\mathbf{x}))(r(s(\mathbf{x})) - 1)f_{\theta'}(\mathbf{x})d\mathbf{x}, \tag{21.68}$$

where $r(s(\mathbf{x})) = f_\theta(\mathbf{x})/f_{\theta'}(\mathbf{x})$, the ratio guaranteed to be a function of just $s(\mathbf{x})$ by the MLR definition. (It does depend on θ and θ'.) Since both f's are pdfs, neither one can always be larger than the other, hence the ratio $r(s)$ is either always 1, or sometimes less than 1 and sometimes greater. Thus there must be a constant s_0 such that

$$r(s) \le 1 \ \text{if} \ s \le s_0 \ \text{and} \ r(s) \ge 1 \ \text{if} \ s \ge s_0. \tag{21.69}$$

Note that if r is defined at s_0, then $r(s_0) = 1$. From (21.68),

$$E_\theta[g(s(\mathbf{X}))] - E_{\theta'}[g(s(\mathbf{X}))] = \int_{s(\mathbf{x})<s_0} g(s(\mathbf{x}))(r(s(\mathbf{x})) - 1)f_{\theta'}(\mathbf{x})d\mathbf{x}$$

$$+ \int_{s(\mathbf{x})>s_0} g(s(\mathbf{x}))(r(s(\mathbf{x})) - 1)f_{\theta'}(\mathbf{x})d\mathbf{x}$$

$$\ge \int_{s(\mathbf{x})<s_0} g(s_0)(r(s(\mathbf{x})) - 1)f_{\theta'}(\mathbf{x})d\mathbf{x}$$

$$+ \int_{s(\mathbf{x})>s_0} g(s_0)(r(s(\mathbf{x})) - 1)f_{\theta'}(\mathbf{x})d\mathbf{x}$$

$$= g(s_0) \int (r(s(\mathbf{x})) - 1)f_{\theta'}(\mathbf{x})d\mathbf{x} = 0. \tag{21.70}$$

The last equality holds because the integral is $\int (f_\theta(\mathbf{x}) - f_{\theta'}(\mathbf{x}))d\mathbf{x} = 0$. Thus $E_\theta[g(s(\mathbf{X}))]$ is nondecreasing in θ. The proof of the result for strict MLR and strictly increasing g is left to the reader, but basically replaces the "$\ge$" in (21.70) with a "$>$." $\qquad \square$

The key implication for hypothesis testing is the following, proved in Exercise 21.8.6.

Lemma 21.6. *Suppose the family $f(\mathbf{x} \mid \theta)$ has MLR with respect to θ and $s(\mathbf{x})$, and we are testing*

$$H_0 : \theta = \theta_0 \;\; versus \;\; H_A : \theta > \theta_0 \tag{21.71}$$

for some level α. Then the test

$$\psi(\mathbf{x}) = \begin{cases} 1 & if \;\; s(\mathbf{x}) > c \\ \gamma & if \;\; s(\mathbf{x}) = c \\ 0 & if \;\; s(\mathbf{x}) < c \end{cases}, \tag{21.72}$$

where c and γ are chosen to achieve level α, is UMP level α.

In the situation in Lemma 21.6, the power function $E_\theta[\psi(\mathbf{X})]$ is nondecreasing by Lemma 21.5, since ψ is nondecreasing in θ. In fact, MLR also can be used to show that the test (21.72) is UMP level α for testing

$$H_0 : \theta \le \theta_0 \;\; versus \;\; H_A : \theta > \theta_0. \tag{21.73}$$

21.5 Locally most powerful tests

We now look at tests that have the best power for alternatives very close to the null. Consider the one-sided testing problem

$$H_0 : \theta = \theta_0 \;\; versus \;\; H_A : \theta > \theta_0. \tag{21.74}$$

Suppose the test ψ has level α, and for any other level α test ϕ, there exists an $\epsilon_\phi > 0$ such that

$$E_\theta[\psi] \ge E_\theta[\phi] \;\; \text{for all} \;\; \theta \in (\theta_0, \theta_0 + \epsilon_\phi). \tag{21.75}$$

Then ψ is a **locally most powerful (LMP)** level α test. Note that the ϵ depends on ϕ, so there may not be an ϵ that works for all ϕ. A UMP test will be locally most powerful.

Suppose ϕ and ψ both have size α: $E_{\theta_0}[\phi] = E_{\theta_0}[\psi] = \alpha$. Then (21.75) implies than for any θ an arbitrarily small amount above θ_0,

$$\frac{E_\theta[\psi] - E_{\theta_0}[\psi]}{\theta} \ge \frac{E_\theta[\phi] - E_{\theta_0}[\phi]}{\theta}. \tag{21.76}$$

If the power function $E_\theta[\phi]$ is differentiable in θ for any ϕ, we can let $\theta \to \theta_0$ in (21.76), so that

$$\frac{\partial}{\partial \theta} E_\theta[\psi] \Big|_{\theta=\theta_0} \ge \frac{\partial}{\partial \theta} E_\theta[\phi] \Big|_{\theta=\theta_0}, \tag{21.77}$$

i.e., a LMP test will maximize the derivative of the power at $\theta = \theta_0$. Often the score tests of Section 16.3 are LMP.

To find the LMP test, we need to assume that the pdf $f_\theta(\mathbf{x})$ is positive and differentiable in θ for all $\mathbf{x}$, and that for any test ϕ, we can move the derivative under the integral:

$$\frac{\partial}{\partial \theta} E_\theta[\phi] \Big|_{\theta=\theta_0} = \int_{\mathcal{X}} \phi(\mathbf{x}) \left[\frac{\partial}{\partial \theta} f_\theta(\mathbf{x}) \right]_{\theta=\theta_0} d\mathbf{x}. \tag{21.78}$$

Consider the analog of (21.20), where f_A is replaced by f_θ, and the first summand on the left has a derivative. That is,

$$\frac{\partial}{\partial\theta} E_\theta[\psi(\mathbf{X}) - \phi(\mathbf{X})]\Big|_{\theta=\theta_0} - c\, E_{\theta_0}[\psi(\mathbf{X}) - \phi(\mathbf{X})]$$

$$= \int_{\mathcal{X}} (\psi(\mathbf{x}) - \phi(\mathbf{x})) \left(\left[\frac{\partial}{\partial\theta} f_\theta(\mathbf{x})\right]_{\theta=\theta_0} - c f_{\theta_0}(\mathbf{x}) \right) d\mathbf{x}$$

$$= \int_{\mathcal{X}} (\psi(\mathbf{x}) - \phi(\mathbf{x}))(l'(\theta_0 ; \mathbf{x}) - c) f_{\theta_0}(\mathbf{x}) d\mathbf{x}, \qquad (21.79)$$

where

$$l'(\theta ; \mathbf{x}) = \frac{\partial}{\partial\theta} \log(f_\theta(\mathbf{x})), \qquad (21.80)$$

the score function from Section 14.1. Now the final expression in (21.79) will be nonnegative if ψ is 1 or 0 depending on the sign of $l'(\theta_0 ; \mathbf{x}) - c$, which leads us to define the Neyman-Pearson-like test

$$\psi(\mathbf{x}) = \begin{cases} 1 & \text{if } l'(\theta_0 ; \mathbf{x}) > c \\ \gamma(\mathbf{x}) & \text{if } l'(\theta_0 ; \mathbf{x}) = c \\ 0 & \text{if } l'(\theta_0 ; \mathbf{x}) < c \end{cases}, \qquad (21.81)$$

where c and $\gamma(\mathbf{x})$ are chosen so that $E_{\theta_0}[\psi] = \alpha$, the desired level. Then using calculations as in the proof of the Neyman-Pearson lemma (Lemma 21.1), we have (21.77) for any other level α test ϕ.

Also, similar to (21.22),

$$P_{\theta_0}[\phi(\mathbf{X}) \neq \psi(\mathbf{X}) \,\&\, l'(\theta_0 ; \mathbf{X}) \neq c] > 0 \implies \frac{\partial}{\partial\theta} E_\theta[\psi]\Big|_{\theta=\theta_0} > \frac{\partial}{\partial\theta} E_\theta[\phi]\Big|_{\theta=\theta_0}. \qquad (21.82)$$

Satisfying (21.81) is necessary for ψ to be LMP level α, but it is not sufficient. For example, it could be that several tests have the same best first derivative, but not all have the highest second derivative. See Exercises 21.8.16 and 21.8.17. One sufficient condition is that if ϕ has the same derivative as ψ, then it has same risk for all θ. That is, ψ is LMP level α if for any other level α test ϕ^* of the form (21.81) but with γ^* in place of γ,

$$E_\theta[\gamma^*(\mathbf{X}) \,|\, l'(\theta_0 ; \mathbf{X}) = c] P[l'(\theta_0 ; \mathbf{X}) = c] = E_\theta[\gamma(\mathbf{X}) \,|\, l'(\theta_0 ; \mathbf{X}) = c] P[l'(\theta_0 ; \mathbf{X}) = c]$$
$$\text{for all } \theta > \theta_0. \qquad (21.83)$$

This condition holds immediately if the distribution of $l'(\theta_0 ; \mathbf{X})$ is continuous, or if there is one or zero $\mathbf{x}$'s with $l'(\theta_0 ; \mathbf{x}) = c$.

As an example, if $X_1, \ldots, X_n$ are iid Cauchy(θ), so that the pdf of X_i is $1/(1 + (x_i - \theta)^2)$, then the LMP level α test rejects when

$$l'_n(\theta_0 ; \mathbf{x}) \equiv \sum_{i=1}^{n} \frac{2x_i}{1 + x_i^2} > c, \qquad (21.84)$$

where c is chosen to achieve size α. See (16.54). As mentioned there, this test has poor power if θ is much larger than θ_0.

21.6 Unbiased tests

Moving on a bit from situations with UMP tests, we look at restricting consideration to tests that have power of at least α for all parameter values in the alternative, such as $\phi^{(2)}$ in Figure 21.2 for the alternative $\mu \neq 0$. That is, you are more likely to reject when you should than when you shouldn't. Such tests are **unbiased**, as in the next definition.

Definition 21.7. *Consider the general hypotheses*

$$H_0 : \boldsymbol{\theta} \in \mathcal{T}_0 \; \text{ versus } \; H_A : \boldsymbol{\theta} \in \mathcal{T}_A \tag{21.85}$$

*and fixed level α. The test ψ is **unbiased level α** if*

$$E_{\boldsymbol{\theta}_A}[\psi(\mathbf{X})] \geq \alpha \geq E_{\boldsymbol{\theta}_0}[\psi(\mathbf{X})] \; \text{ for any } \; \boldsymbol{\theta}_0 \in \mathcal{T}_0 \text{ and } \boldsymbol{\theta}_A \in \mathcal{T}_A. \tag{21.86}$$

In some two-sided testing problems, most prominently one-dimensional exponential families, there exists a uniformly most powerful unbiased level α test. Here we assume a one-dimensional parameter θ with parameter space $\mathcal{T}$ an open interval containing θ_0, and test

$$H_0 : \theta = \theta_0 \; \text{ versus } \; H_A : \theta \neq \theta_0. \tag{21.87}$$

We also assume that for any test ϕ, $E_\theta[\phi]$ is differentiable (and continuous) in θ. This last assumption holds in the exponential family case by Theorem 2.7.1 in Lehmann and Romano (2005). If ϕ is unbiased level α, then $E_\theta[\phi] \geq \alpha$ for $\theta \neq \theta_0$, hence by continuity $E_{\theta_0}[\phi] = \alpha$. Furthermore, the power must have relative minimum at $\theta = \theta_0$. Thus differentiability implies that the derivative is zero at $\theta = \theta_0$. That is, any unbiased level α test ϕ satisfies

$$E_{\theta_0}[\phi] = \alpha \; \text{ and } \; \frac{\partial}{\partial\theta} E_\theta[\phi]\Big|_{\theta=\theta_0} = 0. \tag{21.88}$$

Again, test $\phi^{(2)}$ in Figure 21.2 exemplifies these conditions.

Another assumption we need is that the derivative and integral in the latter equation can be switched (which holds in the exponential family case, or more generally under the Cramér conditions in Section 14.4):

$$\frac{\partial}{\partial\theta} E_\theta[\phi]\Big|_{\theta=\theta_0} = E_{\theta_0}[\phi(\mathbf{X})l(\mathbf{X}\,|\,\theta_0)], \; \text{ where } \; l(\mathbf{x}\,|\,\theta_0) - \frac{\partial}{\partial\theta} \frac{f(\mathbf{x}\,|\,\theta)}{f(\mathbf{x}\,|\,\theta_0)}\Big|_{\theta=\theta_0}. \tag{21.89}$$

A generalization of the Neyman-Pearson lemma (Lemma 21.1) gives conditions for the unbiased level α test with the highest power at specific alternative θ_A. Letting

$$LR(\mathbf{x}\,|\,\theta_A) = \frac{f(\mathbf{x}\,|\,\theta_A)}{f(\mathbf{x}\,|\,\theta_0)}, \tag{21.90}$$

the test has the form

$$\psi(\mathbf{x}) = \begin{cases} 1 & \text{if } LR(\mathbf{x}\,|\,\theta_A) > c_1 + c_2 l(\mathbf{x}\,|\,\theta_0) \\ \gamma(\mathbf{x}) & \text{if } LR(\mathbf{x}\,|\,\theta_A) = c_1 + c_2 l(\mathbf{x}\,|\,\theta_0) \\ 0 & \text{if } LR(\mathbf{x}\,|\,\theta_A) < c_1 + c_2 l(\mathbf{x}\,|\,\theta_0) \end{cases} \tag{21.91}$$

for some constants c_1 and c_2. Suppose we can choose c_1 and c_2 so that ψ is unbiased level α, i.e., it satisfies (21.88). If ϕ is also unbiased level α, then as in the proof of the Neyman-Pearson lemma,

$$
\begin{aligned}
E_{\theta_A}[\psi] - E_{\theta_A}[\phi] &= E_{\theta_A}[\psi] - E_{\theta_A}[\phi] - c_1(E_{\theta_0}[\psi] - E_{\theta_0}[\phi]) \\
&\quad - c_2(E_{\theta_0}[\psi(\mathbf{X})l(\mathbf{X}\,|\,\theta_0)] - E_{\theta_0}[\phi(\mathbf{X})l(\mathbf{X}\,|\,\theta_0)]) \\
&= \int_{\mathcal{X}} (\psi(\mathbf{x}) - \phi(\mathbf{x}))(LR(\mathbf{x}\,|\,\theta_A) - c_1 - c_2 l(\mathbf{x}\,|\,\theta_0))f(\mathbf{x}\,|\,\theta_0)d\mathbf{x} \\
&\geq 0.
\end{aligned}
\tag{21.92}
$$

Thus ψ has at least as good power at θ_A as ϕ. If we can show that the same ψ satisfies (21.88) for any $\theta_A \neq \theta_0$, then it must be a UMP unbiased level α test.

Now suppose $\mathbf{X}$ has a one-dimensional exponential family distribution with natural parameter θ and natural sufficient statistic $s(\mathbf{x})$:

$$
f(\mathbf{x}\,|\,\theta) = a(\mathbf{x})e^{\theta s(\mathbf{x}) - \rho(\theta)}.
\tag{21.93}
$$

Then since $\rho'(\theta) = \mu(\theta) = E_\theta[s(\mathbf{X})]$ (see Exercise 14.9.5), $l(\mathbf{x}\,|\,\theta_0) = s(\mathbf{x}) - \mu(\theta_0)$, hence

$$
LR(\mathbf{x}\,|\,\theta) - c_1 - c_2 l(\mathbf{x}\,|\,\theta) = e^{\rho(\theta_0) - \rho(\theta)}e^{(\theta - \theta_0)s(\mathbf{x})} - c_1 - c_2(s(\mathbf{x}) - \mu(\theta_0)).
\tag{21.94}
$$

If $\theta \neq \theta_0$, then the function in (21.94) is strictly convex in $s(\mathbf{x})$ (see Definition 14.2 on page 226). Thus the set of $\mathbf{x}$ for which it is less than 0 is either empty or an interval (possibly half-infinite or infinite) based on $s(\mathbf{x})$. In the latter case, ψ in (21.91) can be written

$$
\psi(\mathbf{x}) = \begin{cases} 1 & \text{if} \quad s(\mathbf{x}) < a \ \text{or} \ s(\mathbf{x}) > b \\ \gamma(\mathbf{x}) & \text{if} \quad s(\mathbf{x}) = a \ \text{or} \ s(\mathbf{x}) = b \\ 0 & \text{if} \quad a < s(\mathbf{x}) < b \end{cases}
\tag{21.95}
$$

for some $-\infty \leq a < b \leq \infty$. In fact, for any a and b, and any $\theta_A \neq \theta_0$, we can find c_1 and c_2 so that (21.91) equals (21.95). The implication is that if for some a, b, and $\gamma(\mathbf{x})$, the ψ in (21.95) satisfies the conditions in (21.88), then it is a UMP unbiased level α test.

To check the second condition in (21.88) for the exponential family case, (21.89) shows that for any level α test ϕ,

$$
\frac{\partial}{\partial\theta}E_\theta[\phi]\bigg|_{\theta=\theta_0} = E_{\theta_0}[\phi(\mathbf{X})l(\mathbf{X}\,|\,\theta_0)] = E_{\theta_0}[\phi(\mathbf{X})(s(\mathbf{X}) - \mu(\theta_0))]
$$

$$
= E_{\theta_0}[\phi(\mathbf{X})s(\mathbf{X})] - \alpha\mu(\theta_0).
\tag{21.96}
$$

For any $\alpha \in (0,1)$, we can find a test ψ of the form (21.95) such that (21.88) holds. See Exercise 21.8.18 for the continuous case.

21.6.1　Examples

In the normal mean case, where $X \sim N(\mu, 1)$ and we test $\mu = 0$ versus $H_A : \mu \neq 0$, the test that rejects when $|x| > z_{\alpha/2}$ is indeed UMP unbiased level α, since it is level α, unbiased, and of the form (21.95) with $s(x) = x$.

For testing a normal variance, suppose $U \sim \sigma^2\chi^2_\nu$. It may be that $U = \sum(X_i - \overline{X})^2$ for an iid normal sample. We test $H_0 : \sigma^2 = 1$ versus $H_A : \sigma^2 \neq 1$. A reasonable test is the equal-tailed test, where we reject the null when $U < a$ or $U > b$, a and b are chosen so that $P[\chi^2_\nu < a] = P[\chi^2_\nu > b] = \alpha/2$. Unfortunately, that test is not unbiased. The density is an exponential family type with natural statistic U and natural parameter $\theta = -1/(2\sigma^2)$, so that technically we are testing $\theta = -1/2$ versus $\theta \neq -1/2$. Because the distribution of U is continuous, we do not have to worry about the γ. Letting $f_\nu(u)$ be the χ^2_ν pdf, we wish to find a and b so that

$$\int_a^b f_\nu(u)du = 1 - \alpha \text{ and } \int_a^b u f_\nu(u)du = \nu(1 - \alpha). \tag{21.97}$$

These equations follow from (21.88) and (21.96). They cannot be solved in closed form. Exercise 21.8.23 suggests an iterative approach for finding the constants. Here are a few values:

ν	1	2	5	10	50	100	
a	0.0032	0.0847	0.9892	3.5162	32.8242	74.7436	
b	7.8168	9.5303	14.3686	21.7289	72.3230	130.3910	(21.98)
$P[\chi^2_\nu < a]$	0.0448	0.0415	0.0366	0.0335	0.0289	0.0277	
$P[\chi^2_\nu > b]$	0.0052	0.0085	0.0134	0.0165	0.0211	0.0223	

Note that as ν increases, the two tails become more equal.

Now let $X \sim \text{Poisson}(\lambda)$. We wish to find the UMP unbiased level $\alpha = 0.05$ test of $H_0 : \lambda = 1$ versus $H_A : \lambda \neq 1$. Here the natural sufficient statistic is X, and the natural parameter is $\theta = \log(\lambda)$, so we are testing $\theta = 0$ versus $\theta \neq 0$. We need to find the a and b, as well as the randomization values $\gamma(a)$ and $\gamma(b)$, in (21.95) so that (since $E_1[X] = 1$)

$$1 - \alpha = (1 - \gamma(a))p(a) + \sum_{i=a+1}^{b-1} p(i) + (1 - \gamma(b))p(b)$$

$$= a(1 - \gamma(a))p(a) + \sum_{i=a+1}^{b-1} i\,p(i) + b(1 - \gamma(b))p(b), \tag{21.99}$$

where $p(x)$ is the Poisson(1) pmf, $p(x) = e^{-1}/x!$. For given a and b, (21.99) is a linear system of two equations in $\gamma(a)$ and $\gamma(b)$, hence

$$\begin{pmatrix} \frac{\gamma(a)}{a!} \\ \frac{\gamma(b)}{b!} \end{pmatrix} = \begin{pmatrix} 1 & 1 \\ a & b \end{pmatrix}^{-1} \begin{pmatrix} \sum_{i=a}^b \frac{1}{i!} - e(1-\alpha) \\ \sum_{i=a}^b i\frac{1}{i!} - e(1-\alpha) \end{pmatrix}. \tag{21.100}$$

We can try pairs (a, b) until we find one for which the $\gamma(a)$ and $\gamma(b)$ in (21.100) are between 0 and 1. It turns out that $(0,4)$ works for $\alpha = 0.05$, yielding the UMP unbiased level 0.05 test

$$\phi(x) = \begin{cases} 1 & \text{if} & x \geq 5 \\ 0.5058 & \text{if} & x = 4 \\ 0 & \text{if} & 1 \leq x \leq 3 \\ 0.1049 & \text{if} & x = 0 \end{cases}. \tag{21.101}$$

21.7   Nuisance parameters

The optimal tests so far in this chapter applied to just one-parameter models. Usually, even if we are testing only one parameter, there are other parameters needed to describe the distribution. For example, testing problems on a normal mean usually need to deal with the unknown variance. Such extra parameters are called **nuisance parameters**. Often their presence prevents there from being UMP or UMP unbiased tests. Exceptions can be found in certain exponential family models in which there are UMP unbiased tests.

We will illustrate with Fisher's exact test from Section 17.2. We have X_1 and X_2 independent, with $X_i \sim \text{Binomial}(n_i, p_i)$, $i = 1, 2$, and test

$$H_0 : p_1 = p_2 \ \ versus \ \ H_A : p_1 > p_2, \tag{21.102}$$

where otherwise the only restriction on the p_i's is that they are in $(0,1)$. Fisher's exact test arises by conditioning on $T = X_1 + X_2$. First, we find the conditional distribution of X_1 given $T = t$. The joint pmf of (X_1, T) can be written as a two-dimensional exponential family, where the first parameter θ_1 is the log odds ratio (similar to that in Exercise 13.8.22),

$$\theta_1 = \log\left(\frac{p_1}{1 - p_1} \frac{1 - p_2}{p_2}\right). \tag{21.103}$$

The pmf is

$$
\begin{aligned}
f_{(\theta_1, \theta_2)}(x_1, t) &= \binom{n_1}{x_1}\binom{n_2}{t - x_1} p_1^{x_1}(1 - p_1)^{n_1 - x_1} p_2^{t - x_1}(1 - p_2)^{n_2 - t + x_1} \\
&= \binom{n_1}{x_1}\binom{n_2}{t - x_1} \left(\frac{p_1}{1 - p_1} \frac{1 - p_2}{p_2}\right)^{x_1} \left(\frac{p_2}{1 - p_2}\right)^{t} (1 - p_1)^{n_1}(1 - p_2)^{n_2} \\
&= \binom{n_1}{x_1}\binom{n_2}{t - x_1} e^{\theta_1 x_1 + \theta_2 t - \rho(\theta_1, \theta_2)}, \tag{21.104}
\end{aligned}
$$

where $\theta_2 = \log(p_2/(1 - p_2))$. Hence the conditional pmf is

$$
\begin{aligned}
f_{(\theta_1, \theta_2)}(x_1 \mid t) &= \frac{f_{(\theta_1, \theta_2)}(x_1, t)}{\sum_{y_1 \in \mathcal{X}_t} f_{(\theta_1, \theta_2)}(y_1, t)} \\
&= \frac{\binom{n_1}{x_1}\binom{n_2}{t - x_1} e^{\theta x_1}}{\sum_{y_1 \in \mathcal{X}_t} \binom{n_1}{y_1}\binom{n_2}{t - y_1} e^{\theta y_1}}, \quad \mathcal{X}_t = (\max\{0, t - n_2\}, \ldots, \min\{t, n_1\}).
\end{aligned}
\tag{21.105}
$$

Thus conditional on $T = t$, X_1 has a one-dimensional exponential family distribution with natural parameter θ_1, and the hypotheses in (21.102) become $H_0 : \theta_1 = 0$ versus $H_A : \theta_1 > 0$. The distribution for $X_1 \mid T = t$ in (21.105) is called the **noncentral hypergeometric distribution**. When $\theta_1 = 0$, it is the Hypergeometric(n_1, n_2, t) from (17.16). There are three main steps to showing the test is UMP unbiased level α.

Step 1: Show the test ψ is the UMP conditional test. The Neyman-Pearson test for the problem conditioning on $T = t$ is as in (21.65) for the exponential family case:

$$
\psi(x_1, t) = \begin{cases} 1 & \text{if} \quad x_1 > c(t) \\ \gamma(t) & \text{if} \quad x_1 = c(t) \ , \\ 0 & \text{if} \quad x_1 < c(t) \end{cases}
\tag{21.106}
$$

where the constants $c(t)$ and $\gamma(t)$ are chosen so that

$$E_{(0,\theta_2)}[\psi(X_1,t) \mid T = t] = \alpha. \qquad (21.107)$$

(Note that the conditional distribution does not depend on θ_2.) Thus by the Neyman-Pearson lemma (Lemma 21.1), for given t, $\psi(x_1,t)$ is the *conditional* UMP level α test given $T = t$. That is, if $\phi(x_1,t)$ is another test with conditional level α, it cannot have better conditional power:

$$E_{(0,\theta_2)}[\phi(X_1,t) \mid T = t] = \alpha \implies E_{(\theta_1,\theta_2)}[\psi(X_1,t) \mid T = t] \geq E_{(\theta_1,\theta_2)}[\phi(X_1,t) \mid T = t]$$
$$\text{for all } \theta_1 > 0, \theta_2 \in \mathbb{R}. \qquad (21.108)$$

Step 2: Show that any unbiased level α test has conditional level α for each t. Now let ϕ be any unbiased level α test for the unconditional problem. Since the power function is continuous in θ, ϕ must have size α:

$$E_{(0,\theta_2)}[\phi(X_1,T)] = \alpha \text{ for all } \theta_2 \in \mathbb{R}. \qquad (21.109)$$

Look at the conditional expected value of ϕ under the null, which is a function of just t:

$$e_\phi(t) = E_{(0,\theta_2)}[\phi(X_1,t) \mid T = t]. \qquad (21.110)$$

Thus from (21.109), if $\theta_1 = 0$,

$$E_{(0,\theta_2)}[e_\phi(T)] = \alpha \text{ for all } \theta_2 \in \mathbb{R}. \qquad (21.111)$$

The null $\theta_1 = 0$ is the same as $p_1 = p_2$, hence marginally, $T \sim \text{Binomial}(n_1 + n_2, p_2)$. Since this model is a one-dimensional exponential family model with parameter $\theta_2 \in \mathbb{R}$, we know from Lemma 19.5 on page 331 that the model is complete. That is, there is only one unbiased estimator of α, which is the constant α itself. Thus $e_\phi(t) = \alpha$ for all t, or by (21.110),

$$E_{(0,\theta_2)}[\phi(X_1,t) \mid T = t] = \alpha \text{ for all } t \in \{0,\ldots,n_1 + n_2\}. \qquad (21.112)$$

Step 3: Argue that conditionally best implies unconditionally best. Suppose ϕ is unbiased level α, so that (21.112) holds. Then by (21.108), for each t,

$$E_{(\theta_1,\theta_2)}[\psi(X_1,t) \mid T = t] \geq E_{(\theta_1,\theta_2)}[\phi(X_1,t) \mid T = t] \text{ for all } \theta_1 > 0, \theta_2 \in \mathbb{R}. \qquad (21.113)$$

Taking expectations over T yields

$$E_{(\theta_1,\theta_2)}[\psi(X_1,T)] \geq E_{(\theta_1,\theta_2)}[\phi(X_1,T)] \text{ for all } \theta_1 > 0, \theta_2 \in \mathbb{R} \qquad (21.114)$$

Thus ψ is indeed UMP unbiased level α. $\qquad \square$

If the alternative hypothesis in (21.102) is two sided, $p_1 \neq p_2$, the same idea will work, but where the ψ is conditionally the best unbiased level α test, so has form (21.95) for each t. This approach works for exponential families in general. We need to be able to write the exponential family so that with natural parameter $\theta = (\theta_1,\ldots,\theta_p)$ and natural statistic $(t_1(x),\ldots,t_p(x))$, the null hypothesis is $\theta_1 = 0$ and the alternative is either one-sided or two-sided. Then we condition on $(t_2(X),\ldots,t_p(X))$ to find the best conditional test. To prove that the test is UMP unbiased level α, the marginal model for $(t_2(X),\ldots,t_p(X))$ under the null needs to be complete, which will often be the case. Section 4.4 of Lehmann and Romano (2005) details and extends these ideas. Also, see Exercises 21.8.20 through 21.8.23.

21.8   Exercises

Exercise 21.8.1. Suppose $X \sim$ Exponential(λ), and consider testing $H_0: \lambda = 2$ versus $H_A: \lambda = 5$. Find the best level $\alpha = 0.05$ test and its power.

Exercise 21.8.2. Suppose X_1, X_2, X_3 are iid Poisson(λ), and consider testing $H_0: \lambda = 2$ versus $H_A: \lambda = 3$. Find the best level $\alpha = 0.05$ test and its power.

Exercise 21.8.3. Suppose $X \sim N(\theta, \theta)$ (just one observation). Find explicitly the best level $\alpha = 0.05$ test of $H_0: \theta = 1$ versus $H_A: \theta > 1$.

Exercise 21.8.4. Suppose $X \sim$ Cauchy(θ), i.e., has pdf $1/(\pi(1 + (x - \theta)^2))$. (a) Find the best level $\alpha = 0.05$ test of $H_0: \theta = 0$ versus $H_A: \theta = 1$. Find its power. (b) Consider using the test from part (a) for testing $H_0: \theta = 0$ versus $H_A: \theta > 0$. What is its power as $\theta \to \infty$? Is there a UMP level $\alpha = 0.05$ test for this situation?

Exercise 21.8.5. The table below describes the horses in a race. You have \$35 to bet, which you can distribute among the horses in any way you please as long as you do not bet more than the maximum bet for any horse. In the "ϕ" column, put down a number in the range [0,1] that indicates the proportion of the maximum bet you wish to bet on each horse. (Any money left over goes to me.) So if you want to bet the maximum bet on a particular horse, put "1," and if you want to bet nothing, put "0," or put something in between. If that horse wins, then you get \$100$\times\phi$. Your objective is to fill in the ϕ's to maximize your expected winnings,

$$\$100 \times \sum_{i=1}^{5} \phi_i P[\text{Horse } i \text{ wins}] \tag{21.115}$$

subject to the constraint that

$$\sum_{i=1}^{5} \phi_i \times (\text{Maximum bet})_i = \$35. \tag{21.116}$$

(a) Fill in the ϕ's and amount bet on the five horses to maximize the expected winnings subject to the constraints.

Horse	Maximum Bet	Probability of winning	ϕ	Amount Bet
Trigger	\$6.25	0.0256		
Man-o-War	\$25.00	0.1356		
Mr. Ed	\$37.50	0.3456		
Silver	\$25.00	0.3456		
Sea Biscuit	\$6.25	0.1296		

$$\tag{21.117}$$

(b) What are the expected winnings for the best strategy?

Exercise 21.8.6. Prove Lemma 21.6.

Exercise 21.8.7. Suppose $X_1, \ldots, X_n$ are iid Beta(β, β) for $\beta > 0$. (a) Show that this family has monotone likelihood ratio with respect to T and β, and give the statistic T. (b) Find the form of the UMP level α test of $H_0: \beta = 1$ versus $H_A: \beta < 1$. (c) For $n = 1$ and $\alpha = 0.05$, find the UMP level α test explicitly. Find and sketch the power function.

Exercise 21.8.8. Suppose $(X_i, Y_i), i = 1, \ldots, n$, are iid

$$N_2 \left(\begin{pmatrix} 0 \\ 0 \end{pmatrix}, \begin{pmatrix} 1 & \rho \\ \rho & 1 \end{pmatrix} \right), \rho \in (-1, 1). \tag{21.118}$$

(a) Show that a sufficient statistic is (T_1, T_2), where $T_1 = \sum(X_i^2 + Y_i^2)$ and $T_2 = \sum X_i Y_i$.
(b) Find the form of the best level α test for testing $H_0: \rho = 0$ versus $H_A: \rho = 0.5$.
(The test statistic is a linear combination of T_1 and T_2.) (c) Find the form of the best level α test for testing $H_0: \rho = 0$ versus $H_A: \rho = 0.7$. (d) Does there exist a UMP level α test of $H_0: \rho = 0$ versus $H_A: \rho > 0$? If so, find it. If not, why not? (e) Find the form of the LMP level α test for testing $H_0: \rho = 0$ versus $H_A: \rho > 0$.

Exercise 21.8.9. Consider the null hypothesis to be that X is Discrete Uniform$(0, 4)$, so that it has pmf

$$f_0(x) = 1/5, \ x = 0, 1, 2, 3, 4, \tag{21.119}$$

and 0 otherwise. The alternative is that $X \sim$ Geometric$(1/2)$, so that

$$f_A(x) = \frac{1}{2^{x+1}}, \ x = 0, 1, 2, \ldots. \tag{21.120}$$

(a) Give the best level $\alpha = 0$ test ϕ. What is the power of this test? (b) Give the best level $\alpha = 0.30$ test ϕ. What is the power of this test?

Exercise 21.8.10. Now reverse the hypotheses from Exercise 21.8.9, so that the null hypothesis is that $X \sim$ Geometric$(1/2)$, and the alternative is that $X \sim$ Discrete Uniform$(0, 4)$. (a) Give the best level $\alpha = 0$ test ϕ. What is the power of this test? (b) Give the best level $\alpha = 0.30$ test ϕ. What is the power of this test? (c) Among tests with power=1, find that with the smallest level. What is the size of this test?

Exercise 21.8.11. Suppose $X \sim N(\mu, \mu^2)$, so that the absolute value of the mean and the standard deviation are the same. (There is only one observation.) Consider testing $H_0: \mu = 1$ versus $H_A: \mu > 1$. (a) Find the level $\alpha = 0.10$ test with the highest power at $\mu = 2$. (b) Find the level $\alpha = 0.10$ test with the highest power at $\mu = 3$. (c) Find the powers of the two tests in parts (a) and (b) at $\mu = 2$ and 3. (d) Is there a UMP level 0.10 test for this hypothesis testing problem?

Exercise 21.8.12. For each testing problem, say whether there is a UMP level 0.05 test or not. (a) $X \sim$ Uniform$(0, \theta)$, $H_0: \theta = 1$ versus $H_A: \theta < 1$. (b) $X \sim$ Poisson(λ), $H_0: \lambda = 1$ versus $H_A: \lambda > 1$. (c) $X \sim$ Poisson(λ), $H_0: \lambda = 1$ versus $H_A: \lambda \neq 1$. (d) $X \sim N(\mu, \sigma^2)$, $H_0: \mu = 0, \sigma^2 = 1$ versus $H_A: \mu > 0, \sigma^2 > 0$. (e) $X \sim N(\mu, \sigma^2)$, $H_0: \mu = 0, \sigma^2 = 1$ versus $H_A: \mu > 0, \sigma^2 = 1$. (f) $X \sim N(\mu, \sigma^2)$, $H_0: \mu = 0, \sigma^2 = 1$ versus $H_A: \mu = 1, \sigma^2 = 10$.

Exercise 21.8.13. This exercise shows that the noncentral chisquare and noncentral F distributions have monotone likelihood ratio. Assume the degrees of freedom are fixed, so that the noncentrality parameter $\Delta \geq 0$ is the only parameter. From (7.130) and (7.134) we have that the pdfs, with $w > 0$ as the variable, can be written as

$$f(w \mid \Delta) = f(w \mid 0)e^{-\frac{1}{2}\Delta} \sum_{k=0}^{\infty} c_k \Delta^k w^k, \tag{21.121}$$

where $c_k > 0$ for each k. For given $\Delta > \Delta'$, write

$$\frac{f(w \mid \Delta)}{f(w \mid \Delta')} = e^{\frac{1}{2}(\Delta' - \Delta)} R(w, \Delta, \Delta'), \quad \text{where } R(w, \Delta, \Delta') = \frac{\sum_{k=0}^{\infty} c_k \Delta^k w^k}{\sum_{k=0}^{\infty} c_k \Delta'^k w^k}. \qquad (21.122)$$

For fixed Δ', consider the random variable K with space the nonnegative integers, parameter w, and pmf

$$g(k \mid w) = \frac{c_k \Delta'^k w^k}{\sum_{l=0}^{\infty} c_l \Delta'^l w^l}. \qquad (21.123)$$

(a) Is $g(k \mid w)$ a legitimate pmf? Show that it has strict monotone likelihood ratio with respect to k and w. (b) Show that $R(w, \Delta, \Delta') = E_w[(\Delta/\Delta')^K]$ where K has pdf g. (c) Use part (a) and Lemma 21.5 to show that for fixed $\Delta > \Delta'$, $E_w[(\Delta/\Delta')^K]$ is increasing in w. (d) Argue that $f(w \mid \Delta)/f(w \mid \Delta')$ is increasing in w, hence f has strict monotone likelihood ratio wrt w and Δ.

Exercise 21.8.14. Find the form of the LMP level α test testing $H_0 \colon \theta = 0$ versus $H_A \colon \theta > 0$ based on $X_1, \ldots, X_n$ iid Logistic(θ). (So the pdf of X_i is $\exp(x_i - \theta)/(1 + \exp(x_i - \theta))^2$.

Exercise 21.8.15. Recall the fruit fly example in Exercise 14.9.2 (and points earlier). Here, $(N_{00}, N_{01}, N_{10}, N_{11})$ is Multinomial($n, \mathbf{p}(\theta)$) with

$$\mathbf{p}(\theta) = (\tfrac{1}{2}(1 - \theta)(2 - \theta), \tfrac{1}{2}\theta(1 - \theta), \tfrac{1}{2}\theta(1 - \theta), \tfrac{1}{2}\theta(1 + \theta)). \qquad (21.124)$$

Test the hypotheses $H_0 \colon \theta = 1/2$ versus $H_A \colon \theta > 1/2$. (a) Show that there is no UMP level α test for $\alpha \in (0, 1)$. (b) Show that any level α test that maximizes the derivative of $E_\theta[\phi]$ at $\theta = 1/2$ can be written as

$$\phi(\mathbf{n}) = \begin{cases} 1 & \text{if} \quad n_{11} - n_{00} > c \\ \gamma(\mathbf{n}) & \text{if} \quad n_{11} - n_{00} = c \\ 0 & \text{if} \quad n_{11} - n_{00} < c \end{cases} \qquad (21.125)$$

for some constant c and function $\gamma(\mathbf{n})$. (c) Do you think ϕ in (21.125) is guaranteed to be the LMP level α test? Or does it depend on what γ is.

Exercise 21.8.16. Suppose $X \sim N(\theta^2, 1)$ and we wish to test $H_0 \colon \theta = 0$ versus $H_A \colon \theta > 0$ for level $\alpha = 0.05$. (a) Show that $[\partial E_\theta[\phi]/\partial \theta]|_{\theta=0} = 0$ for any test ϕ. (b) Argue that the test $\phi^*(x) = \alpha$ has level α and maximizes the derivative of $E_\theta[\phi]$ at $\theta = 0$. Is it LMP level α? (c) Find the UMP level α test ψ. Is it LMP level α? (d) Find $[\partial^2 E_\theta[\phi]/\partial \theta^2]|_{\theta=0}$ for $\phi = \phi^*$ and $\phi = \psi$. Which is larger?

Exercise 21.8.17. This exercise provides an example of finding an LMP test when the condition (21.83) fails. Suppose X_1 and X_2 are independent, with $X_1 \sim$ Binomial($4, \theta$) and $X_2 \sim$ Binomial($3, \theta^2$). We test $H_0 \colon \theta = 1/2$ versus $H_A \colon \theta > 1/2$. (a) Show that any level α test that maximizes $[\partial E_\theta[\phi]/\partial \theta]|_{\theta=1/2}$ has the form

$$\phi(x_1, x_2) = \begin{cases} 1 & \text{if} \quad 3x_1 + 4x_2 > c \\ \gamma(x_1, x_2) & \text{if} \quad 3x_1 + 4x_2 = c \\ 0 & \text{if} \quad 3x_1 + 4x_2 < c \end{cases}. \qquad (21.126)$$

(b) Show that for tests of the form (21.126),

$$E_\theta[\phi] = P_\theta[3X_1 + 4X_2 > c] + E_\theta[\gamma(X_1, X_2) \mid 3X_1 + 4X_2 = c]P_\theta[3X_1 + 4X_2 = c].$$
(21.127)

(c) For level $\alpha = 0.25$, the $c = 12$. Then $P_{1/2}[3X_1 + 4X_2 > 12] = 249/2^{10} \approx 0.2432$ and $P_{1/2}[3X_1 + 4X_2 = 12] = 28/2^{10} \approx 0.02734$. Show that in order for the test (21.126) to have level 0.25, we need

$$E_{1/2}[\gamma(X_1, X_2) \mid 3X_1 + 4X_2 = 12] = \frac{1}{4}.$$
(21.128)

(d) Show that $\{(x_1, x_2) \mid 3x_1 + 4x_2 = 12\}$ consists of just $(4, 0)$ and $(0, 3)$, and

$$P_\theta[(X_1, X_2) = (4, 0) \mid 3X_1 + 4X_2 = 12] = \frac{(1+\theta)^3}{(1+\theta)^3 + \theta^3},$$
(21.129)

hence

$$E_\theta[\gamma(X_1, X_2) \mid 3X_1 + 4X_2 = 12] = \frac{\gamma(4,0)(1+\theta)^3 + \gamma(0,3)\theta^3}{(1+\theta)^3 + \theta^3}.$$
(21.130)

(e) Using (21.128) and (21.130), to obtain size 0.25, we need $27\gamma(4,0) + \gamma(0,3) = 7$. Find the range of such possible $\gamma(0,3)$'s. [Don't forget that $0 \le \phi(x) \le 1$.] (f) Show that among the level 0.25 tests of the form (21.126), the one with $\gamma(0,3) = 1$ maximizes (21.130) for all $\theta \in (0.5, 1)$. Call this test ψ. (g) Argue that ψ from part (f) is the LMP level 0.25 test.

Exercise 21.8.18. Suppose X has an exponential family pdf, where X itself is the natural sufficient statistic, so that $f(x \mid \theta) = a(x)\exp(x\theta - \rho(\theta))$. We test $H_0: \theta = \theta_0$ versus $H_A: \theta \ne \theta_0$. Assume that $\mathcal{X} = (k, l)$, where k or l could be infinite, and $a(x) > 0$ for $x \in \mathcal{X}$. Consider tests ψ of the form (21.95) for some a, b, where by continuity we can set $\gamma(x) = 0$. Fix $\alpha \in (0, 1)$. (a) Show that

$$1 - E_{\theta_0}[\psi] = \int_a^b f(x \mid \theta_0)dx = F_{\theta_0}(b) - F_{\theta_0}(a) \text{ and}$$

$$\frac{\partial}{\partial \theta} E_\theta[\psi]\bigg|_{\theta=\theta_0} = -\int_a^b (x - \mu(\theta_0))f(x \mid \theta_0)dx.$$
(21.131)

(b) Let a^* be the lower α cutoff point, i.e., $F_{\theta_0}(a^*) = \alpha$. For $a \le a^*$, define $b(a) = F_{\theta_0}^{-1}(F_{\theta_0}(a) + 1 - \alpha)$. Show that $b(a)$ is well-defined and continuous in $a \in (k, a^*)$, and that $F_{\theta_0}(b(a)) - F_{\theta_0}(a) = 1 - \alpha$. (c) Show that $\lim_{a \to k} b(a) = b^*$, where $1 - F_{\theta_0}(b^*) = \alpha$, and $\lim_{a \to a^*} b(a) = l$. (d) Consider the function of a,

$$d(a) = \int_a^{b(a)} (x - \mu(\theta_0))f(x \mid \theta_0)dx.$$
(21.132)

Show that

$$\lim_{a \to k} d(a) = \int_k^{b^*} (x - \mu(\theta_0))f(x \mid \theta_0)dx < 0 \text{ and}$$

$$\lim_{a \to a^*} d(a) = \int_{a^*}^l (x - \mu(\theta_0))f(x \mid \theta_0)dx > 0.$$
(21.133)

[Hint: Note that the integral from k to l is 0.] Argue that by continuity of $d(a)$, there must be an a_0 such that $d(a_0) = 0$. (e) Using ψ with $a = a_0$ from part (d) and $b = b(a_0)$, show that (21.88) holds, proving that ψ is the UMP unbiased level α test.

Exercise 21.8.19. Continue the setup from Exercise 21.8.18, where now $\theta_0 = 0$, so that we test $H_0: \theta = 0$ versus $H_A: \theta \neq 0$. Also, suppose the distribution under the null is symmetric about 0, i.e., $f(x \mid 0) = f(-x \mid 0)$, so that $\mu(0) = 0$. Let a be the upper $\alpha/2$ cutoff point for the null distribution of X, so that $P_0[|X| > a] = \alpha$. Show that the UMP level α test rejects the null when $|X| > a$.

Exercise 21.8.20. Suppose $X_1, \ldots, X_n$ are iid $N(\mu, \sigma^2)$, and we test

$$H_0: \mu = 0, \sigma^2 > 0 \quad versus \quad H_A: \mu > 0, \sigma^2 > 0. \tag{21.134}$$

The goal is to find the UMP unbiased level α test. (a) Write the density of $\mathbf{X}$ as a two-parameter exponential family, where the natural parameter is (θ_1, θ_2) with $\theta_1 = n\mu/\sigma^2$ and $\theta_2 = -1/(2\sigma^2)$, and the natural sufficient statistic is $(\overline{x}, w)$ with $w = \sum x_i^2$. Thus we are testing $\theta_1 = 0$ versus $\theta_1 \neq 0$, with θ_2 as a nuisance parameter. (b) Show that the conditional distribution of $\overline{X}$ given $W = w$ has space $(-\sqrt{w/n}, \sqrt{w/n})$ and pdf

$$f_{\theta_1}(\overline{x} \mid w) = \frac{(w - n\overline{x}^2)^{(n-3)/2} e^{\theta_1 \overline{x}}}{\int_{-\sqrt{w/n}}^{\sqrt{w/n}} (w - nz^2)^{(n-3)/2} e^{\theta_1 z} dz}. \tag{21.135}$$

[Hint: First write down the joint pdf of $(\overline{X}, V)$ where $V = \sum (X_i - \overline{X})^2$, then use the transformation $w = v + n\overline{x}^2$.] (c) Argue that, conditioning on $W = w$, the conditional UMP level α test of the null is

$$\phi(\overline{x}, w) = \begin{cases} 1 & \text{if} \quad \overline{x} \geq c(w) \\ 0 & \text{if} \quad \overline{x} < c(w) \end{cases}, \tag{21.136}$$

where $c(w)$ is the constant such that $P_0[\overline{X} > c(w) \mid W = w] = \alpha$. (d) Show that under the null, W has a one-dimensional exponential family distribution, and the model is complete. Thus the test $\phi(\overline{x}, w)$ is the UMP unbiased level α test.

Exercise 21.8.21. Continue with the testing problem in Exercise 21.8.20. Here we show that the test ϕ in (21.136) is the t test. We take $\theta_1 = 0$ throughout this exercise. (a) First, let $u = \sqrt{n}\,\overline{x}/\sqrt{w}$, and show that the conditional distribution of U given $W = w$ is

$$g(u \mid w) = du^{-\frac{1}{2}}(1 - u^2)^{\frac{n-3}{2}}, \tag{21.137}$$

where d is a constant not depending on w. Note that this conditional distribution does not depend on W, hence U is independent of W. (b) Show that

$$T \equiv \sqrt{n-1}\,\frac{U}{\sqrt{1 - U^2}} = \frac{\sqrt{n}\,\overline{X}}{\sqrt{\sum (X_i - \overline{X})^2/(n-1)}}, \tag{21.138}$$

the usual t statistic. Why is T independent of W? (c) Show that T is a function of $(\overline{X}, W)$, and argue that

$$\phi(\overline{x}, w) = \begin{cases} 1 & \text{if} \quad t(\overline{x}, w) \geq t_{n-1, \alpha} \\ 0 & \text{if} \quad t(\overline{x}, w) < t_{n-1, \alpha} \end{cases}, \tag{21.139}$$

where $t_{n-1, \alpha}$ is the upper α cutoff point of a t_{n-1} distribution. Thus the one-sided t test is the UMP unbiased level α test.

Exercise 21.8.22. Suppose $X_1, \ldots, X_n$ are iid $N(\mu, \sigma^2)$ as in Exercise 21.8.20, but here we test the two-sided hypotheses

$$H_0: \mu = 0, \sigma^2 > 0 \;\; versus \;\; H_A: \mu \neq 0, \sigma^2 > 0. \tag{21.140}$$

Show that the two-sided t test, which rejects the null when $|T| > t_{n-1, \alpha/2}$, is the UMP unbiased level α test for (21.140). [Hint: Follow Exercises 21.8.20 and 21.8.21, but use Exercise 21.8.19 as well.]

Exercise 21.8.23. Let $U \sim \sigma^2 \chi_\nu^2$. We wish to find the UMP unbiased level α test for testing $H_0: \sigma^2 = 1$ versus $H_A: \sigma^2 \neq 1$. The test is to reject the null when $u < a_0$ or $u > b_0$, where a_0 and b_0 satisfy the conditions in (21.97). (a) With f_ν being the χ_ν^2 pdf, show that

$$\int_a^b u f_\nu(u) du = \nu \int_a^b f_{\nu+2}(u) du. \tag{21.141}$$

(b) Letting F_ν be the χ_ν^2 distribution function, show that the conditions in (21.97) can be written

$$F_\nu(b) - F_\nu(a) = 1 - \alpha = F_{\nu+2}(b) - F_{\nu+2}(a). \tag{21.142}$$

Thus with $b(a) = F_\nu^{-1}(F_\nu(a) + 1 - \alpha)$, we wish to find a_0 so that

$$g(a_0) = 0 \;\; where \;\; g(a) = F_{\nu+2}(b(a)) - F_{\nu+2}(a) - (1 - \alpha). \tag{21.143}$$

Based on an initial guess a_1 for a_0, the Newton-Raphson iteration for obtaining a new guess a_{i+1} from guess a_i is $a_{i+1} = a_i - g(a_i)/g'(a_i)$. (c) Show that $g'(a) = f_\nu(a)(b(a) - a)/\nu$. [Hint: Note that $dF_\nu^{-1}(x)/dx = f_\nu(F_\nu^{-1}(x))$.] Thus the iterations are

$$a_{i+1} = a_i - \nu \frac{F_{\nu+2}(b(a)) - F_{\nu+2}(a) - (1 - \alpha)}{f_\nu(a)(b(a) - a)}, \tag{21.144}$$

which can be implemented using just the χ^2 pdf and distribution function.

Exercise 21.8.24. In testing hypotheses of the form $H_0: \theta = 0$ versus $H_A: \theta > 0$, an **asymptotically most powerful** level α test is a level α test ψ such that for any other level α test ϕ, there exists an N_ϕ such that

$$E_\theta[\psi] \geq E_\theta[\phi] \;\; for \; all \;\; \theta > N_\phi. \tag{21.145}$$

Let $X_1, \ldots, X_n$ be iid Laplace(θ), so that the pdf of X_i is $(1/2) \exp(-|x_i - \theta|)$. Consider the test

$$\psi(\mathbf{x}) = \begin{cases} 1 & \text{if} \;\; \sum \max\{0, x_i\} > c \\ 0 & \text{if} \;\; \sum \max\{0, x_i\} \leq c \end{cases}, \tag{21.146}$$

where c is chosen to obtain size α. (a) Sketch the acceptance region for ψ when $n = 2$ and $c = 1$. (b) Show that for each $\mathbf{x}$,

$$\lim_{\theta \to \infty} e^{n\theta - 2c} \frac{f(\mathbf{x} \mid \theta)}{f(\mathbf{x} \mid 0)} = e^{2(\sum \max\{0, x_i\} - c)}. \tag{21.147}$$

(c) Let ϕ be level α. Show that

$$e^{n\theta - 2c} E_\theta[\psi(\mathbf{X}) - \phi(\mathbf{X})] - E_0[\psi(\mathbf{X}) - \phi(\mathbf{X})]$$
$$= \int_{\mathcal{X}} (\psi(\mathbf{x}) - \phi(\mathbf{x})) \left(e^{n\theta - 2c} \frac{f(\mathbf{x} \mid \theta)}{f(\mathbf{x} \mid 0)} - 1 \right) f(\mathbf{x} \mid 0) d\mathbf{x}, \tag{21.148}$$

and since ψ has size α and ϕ has level α,

$$e^{n\theta-2c}E_\theta[\psi(\mathbf{X}) - \phi(\mathbf{X})] \geq \int_{\mathcal{X}} (\psi(\mathbf{x}) - \phi(\mathbf{x})) \left(e^{n\theta-2c}\frac{f(\mathbf{x}\,|\,\theta)}{f(\mathbf{x}\,|\,0)} - 1\right) f(\mathbf{x}\,|\,0)d\mathbf{x}. \quad (21.149)$$

(d) Let $\theta \to \infty$ on the right-hand side of (21.149). Argue that the limit is nonnegative, and unless $P_0[\phi(\mathbf{X}) = \psi(\mathbf{X})] = 1$, the limit is positive. (e) Explain why part (d) shows that ψ is asymptotically most powerful level α.

Chapter 22

Decision Theory in Hypothesis Testing

22.1 A decision-theoretic framework

Again consider the general hypothesis testing problem

$$H_0 : \boldsymbol{\theta} \in \mathcal{T}_0 \quad versus \quad H_A : \boldsymbol{\theta} \in \mathcal{T}_A. \tag{22.1}$$

The previous chapter exhibits a number of best-test scenarios, all where the essential part of the null hypothesis was based on a single parameter. This chapter deals with multiparametric hypotheses, where there typically is no UMP or UMP unbiased test. Admissibility and minimaxity are then relevant concepts.

The typical decision-theoretic framework used for testing has action space $\mathcal{A} = \{\text{Accept}, \text{Reject}\}$, denoting accepting or rejecting the null hypothesis. The usual loss function used for hypothesis testing is called 0/1 loss, where we lose 1 if we make a wrong decision, and lose nothing if we are correct. The loss function combines elements of the tables on testing in (15.7) and on game theory in (20.56):

$L(a, \boldsymbol{\theta})$	Action	
	Accept	Reject
$\boldsymbol{\theta} \in \mathcal{T}_0$	0	1
$\boldsymbol{\theta} \in \mathcal{T}_A$	1	0

$$\tag{22.2}$$

The risk is thus the probability of making an error given $\boldsymbol{\theta}$. Using test functions $\phi : \mathcal{X} \to [0,1]$ as in (21.5), where $\phi(\mathbf{x})$ is the probability of rejecting the null when $\mathbf{x}$ is observed, the risk function is

$$R(\boldsymbol{\theta}\,;\phi) = \begin{cases} E_{\boldsymbol{\theta}}[\phi(\mathbf{X})] & \text{if} \quad \boldsymbol{\theta} \in \mathcal{T}_0 \\ 1 - E_{\boldsymbol{\theta}}[\phi(\mathbf{X})] & \text{if} \quad \boldsymbol{\theta} \in \mathcal{T}_A \end{cases}. \tag{22.3}$$

Note that if $\boldsymbol{\theta} \in \mathcal{T}_A$, the risk is one minus the power.

There are a few different approaches to evaluating tests decision-theoretically, depending on how one deals with the level. The generic approach does not place any restrictions on level, evaluating tests on their power as well as their size function. A more common approach to hypothesis testing is to fix α, and consider only tests ϕ of level α. The question then becomes whether to look at the risk for parameter values in the null, or just worry about the power. For example, suppose $X \sim N(\mu, 1)$ and we test $H_0: \mu \leq 0$ versus $H_A: \mu > 0$, restricting to tests with level $\alpha = 0.05$. If we take risk at the null and alternative into account, then any test that rejects the null when $X \geq c$ for some $c \geq 1.645$ is admissible, since it is the uniformly most powerful test of its size. That is, the test $I[X \geq 1.645]$ is admissible, but so is the test $I[X \geq 1.96]$, which has smaller power but smaller size. If we evaluate only on power, so ignore size except for making sure it is no larger than 0.05, $I[X \geq 1.96]$ is dominated by $I[X \geq 1.645]$; in fact, the latter is the only admissible test. If we restrict to tests with size function exactly equal to α, i.e., $R(\theta; \phi) = \alpha$ for all $\theta \in \mathcal{T}_0$, then the power is the only relevant decider.

The Rao-Blackwell theorem (Theorem 13.8 on page 210) showed that when estimating with squared-error loss, any estimator that is not essentially a function of just the sufficient statistic is inadmissible. For testing, the result is not quite as strong. Suppose $\phi(x)$ is any test, and $\mathbf{T} = \mathbf{t}(\mathbf{X})$ is a sufficient statistic. Then let $e_\phi(\mathbf{t}) = E_\theta[\phi(\mathbf{X}) \,|\, \mathbf{t}(\mathbf{X}) = \mathbf{t}]$. (The conditional expected value does not depend on the parameter by sufficiency.) Note that $0 \leq e_\phi \leq 1$, hence it is also a test function. For each θ, we have

$$E_\theta[e_\phi(\mathbf{T})] = E_\theta[\phi(\mathbf{X})] \implies R(\theta, \phi) = R(\theta, e_\phi). \tag{22.4}$$

That is, any test's risk function can be exactly matched by a test depending on just the sufficient statistic. Thus when analyzing a testing problem, we lose nothing by **reducing by sufficiency**: We look at the same hypotheses, but base the tests on the sufficient statistic $\mathbf{T}$.

The UMP, UMP unbiased, and LMP level α tests we saw in Chapter 21 will be admissible under certain reasonable conditions. See Exercise 22.8.1. In the next section we look at Bayes tests, and conditions under which they are admissible. Section 22.3 looks at necessary conditions for a test to be admissible. Basically, it must be a Bayes test or a certain type of limit of Bayes tests. Section 22.4 considers the special case of compact parameter spaces for the hypotheses, and Section 22.5 contains some cases where tests with convex acceptance regions are admissible. Section 22.6 introduces invariance, which is a method for exploiting symmetries in the model to simplify analysis of test statistics. It is especially useful in multivariate analysis.

We will not say much about minimaxity in hypothesis testing, though it can be useful. Direct minimaxity for typical testing problems is not very interesting since the maximal risk for a level α test is $1 - \alpha$ (if $\alpha < 0.5$, the null and alternative are not separated, and the power function is continuous in θ). See the second graph in Figure 22.1. If we restrict to level α tests, then they all have the same maximal risk, and if we allow all levels, then the minimax tests are the ones with level 0.5. If the alternative is separated from the null, e.g., testing $\theta = 0$ versus $\theta > 1$, then the minimax test will generally be one that is most powerful at the closest point in the alternative, or Bayes wrt a prior concentrated on the set of closest points if there are more than one (as in the hypotheses in (22.30)). More informative is **maximal regret**, where we restrict to level α tests, and define the risk to be the distance between the actual power and the best possible power at each alternative:

$$R(\theta; \phi) = \sup_{\text{level } \alpha \text{ tests } \psi} E_\theta[\psi] - E_\theta[\phi]. \tag{22.5}$$

See van Zwet and Oosterhoff (1967) for some applications.

22.2 Bayes tests

Section 15.4 introduced Bayes tests. Here we give their formal decision-theoretic justification. The prior distribution π over $\mathcal{T}_0 \cup \mathcal{T}_A$ is given in two stages. The marginal probabilities of the hypotheses are $\pi_0 = P[\boldsymbol{\theta} \in \mathcal{T}_0]$ and $\pi_A = P[\boldsymbol{\theta} \in \mathcal{T}_A]$, $\pi_0 + \pi_A = 1$. Then conditionally, $\boldsymbol{\theta}$ given H_0 is true has conditional density $\rho_0(\boldsymbol{\theta})$, and given H_A is true has conditional density $\rho_A(\boldsymbol{\theta})$. If we look at all possible tests, and take into account size and power for the risk, a Bayes test wrt π minimizes

$$R(\pi; \phi) = \pi_A \int_{\mathcal{T}_A} (1 - E_{\boldsymbol{\theta}}[\phi(\mathbf{X})])\rho_A(\boldsymbol{\theta})d\boldsymbol{\theta} + \pi_0 \int_{\mathcal{T}_0} E_{\boldsymbol{\theta}}[\phi(\mathbf{X})]\rho_0(\boldsymbol{\theta})d\boldsymbol{\theta} \qquad (22.6)$$

over ϕ. If $\mathbf{X}$ has pdf $f(\mathbf{x} \mid \boldsymbol{\theta})$, then

$$\begin{aligned}
R(\pi; \phi) &= \pi_A \int_{\mathcal{T}_A} \left(1 - \int_{\mathcal{X}} \phi(\mathbf{x})f(\mathbf{x} \mid \boldsymbol{\theta})d\mathbf{x}\right) \rho_A(\boldsymbol{\theta})d\boldsymbol{\theta} \\
&\quad + \pi_0 \int_{\mathcal{T}_0} \int_{\mathcal{X}} \phi(\mathbf{x})f(\mathbf{x} \mid \boldsymbol{\theta})d\mathbf{x}\rho_0(\boldsymbol{\theta})d\boldsymbol{\theta} \\
&= \int_{\mathcal{X}} \phi(\mathbf{x}) \left(\pi_0 \int_{\mathcal{T}_0} f(\mathbf{x} \mid \boldsymbol{\theta})\rho_0(\boldsymbol{\theta})d\boldsymbol{\theta} - \pi_A \int_{\mathcal{T}_A} f(\mathbf{x} \mid \boldsymbol{\theta})\rho_A(\boldsymbol{\theta})d\boldsymbol{\theta}\right) d\mathbf{x} + \pi_A.
\end{aligned} \tag{22.7}$$

To minimize this Bayes risk, we take $\phi(\mathbf{x})$ to minimize the integrand in the last line. Since ϕ must be in $[0,1]$, we take $\phi(\mathbf{x}) = 0$ if the quantity in the large parentheses is positive, and $\phi(\mathbf{x}) = 1$ if it is negative, yielding a test of the form

$$\phi_\pi(\mathbf{x}) = \begin{cases} 1 & \text{if } B_{A0}(\mathbf{x})\frac{\pi_A}{\pi_0} > 1 \\ \gamma(\mathbf{x}) & \text{if } B_{A0}(\mathbf{x})\frac{\pi_A}{\pi_0} = 1 \\ 0 & \text{if } B_{A0}(\mathbf{x})\frac{\pi_A}{\pi_0} < 1 \end{cases}, \tag{22.8}$$

where B_{A0} is the Bayes factor as in (15.40),

$$B_{A0}(\mathbf{x}) = \frac{\int_{\mathcal{T}_A} f(\mathbf{x} \mid \boldsymbol{\theta})\rho_A(\boldsymbol{\theta})d\boldsymbol{\theta}}{\int_{\mathcal{T}_0} f(\mathbf{x} \mid \boldsymbol{\theta})\rho_0(\boldsymbol{\theta})d\boldsymbol{\theta}}. \tag{22.9}$$

(If $B_{A0}(\mathbf{x})\pi_A/\pi_0$ is $0/0$, take it to be 1.) Thus the Bayes test rejects the null if, under the posterior the null is probably false, and accepts the null if it is probably true.

22.2.1 Admissibility of Bayes tests

Lemma 20.3 (page 349) gives some sufficient conditions for a Bayes procedure ϕ_π to be admissible, which apply here: (a) ϕ_π is admissible among the tests that are Bayes wrt π; (b) ϕ_π is the unique Bayes test (up to equivalence) wrt π; (c) the parameter space is finite or countable, and π places positive probability on each parameter value. Parts (d) and (e) require the risk to be continuous in $\boldsymbol{\theta}$, which is usually not true in hypothesis testing. For example, suppose $X \sim N(\mu, 1)$ and we test $H_0: \mu \leq 0$ versus $H_A: \mu > 0$ using the test that rejects when $X > z_\alpha$. Then the risk at $\mu = 0$ is exactly α, but for μ just a bit larger than zero, the risk is almost $1 - \alpha$. Thus the risk is discontinuous at $\mu = 0$ (unless $\alpha = 1/2$). See Figure 22.1.

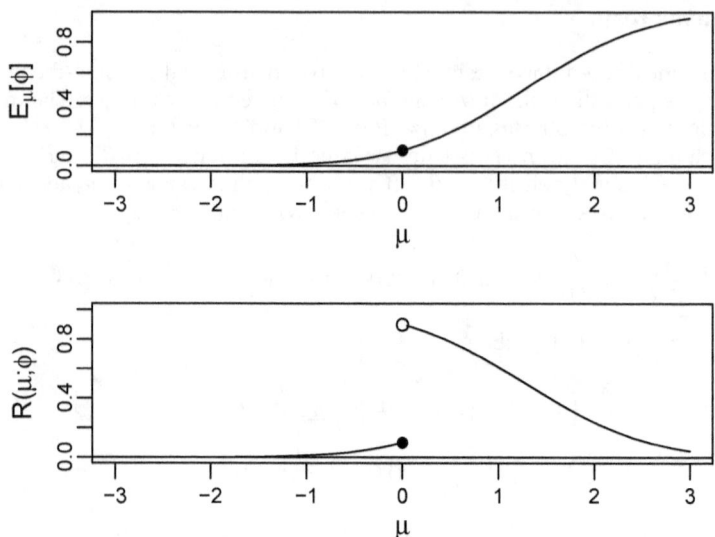

Figure 22.1: Testing $H_0: \mu \leq 0$ versus $H_A: \mu > 0$ based on $X \sim N(\mu, 1)$. The test ϕ rejects when $X > 1.28$, and has level $\alpha = 0.10$. The top graph is the power function, which is continuous. The bottom graph is the risk function, which inverts the power function when $\mu > 0$. It is not continuous at $\mu = 0$.

These parts of the lemma can be extended to hypothesis testing if $E_{\boldsymbol{\theta}}[\phi]$ is continuous in $\boldsymbol{\theta}$ for any test ϕ. We decompose the parameter space into three pieces. Let $\mathcal{T}_*$ be the border between the null and hypothesis spaces, formally, $\mathcal{T}^* = \text{closure}(\mathcal{T}_A) \cap \text{closure}(\mathcal{T}_0)$. It is the set of points $\boldsymbol{\theta}^*$ for which there are points in both the null and alternative spaces arbitrarily close to $\boldsymbol{\theta}^*$. We assume that $\mathcal{T}_0 - \mathcal{T}^*$ and $\mathcal{T}_A - \mathcal{T}^*$ are both open. Then if prior π has $\pi(B) > 0$ for any open set $B \in \mathcal{T}_0 - \mathcal{T}_B$ or $B \in \mathcal{T}_A - \mathcal{T}_B$, the Bayes test ϕ_π wrt π is admissible. The proof is basically the same as in Lemma 20.3, but we also need to note that if a test ϕ is at least as good as ϕ_π, then the two tests have the same risk on the border (at all $\boldsymbol{\theta} \in \mathcal{T}_B$). For example, the test based on the Bayes factor in (15.43) is admissible, as are the tests in Exercises 15.7.6 and 15.7.11.

Return to condition (a), and let $\mathcal{D}_\pi$ be the set of Bayes tests wrt π, so that they all satisfy (22.8) (with probability 1) for some $\gamma(\mathbf{x})$. As in (21.127) of Exercise 21.8.17, the power of any test $\phi \in \mathcal{D}_\pi$ can be written

$$E_{\boldsymbol{\theta}}[\phi] = P_{\boldsymbol{\theta}}[B(\mathbf{X}) > 1] + E_{\boldsymbol{\theta}}[\gamma(\mathbf{X}) \mid B(\mathbf{X}) = 1]P_{\boldsymbol{\theta}}[B(\mathbf{X}) = 1], \quad B(\mathbf{x}) = B_{A0}(\mathbf{x})\frac{\pi_A}{\pi_0}.$$
(22.10)

If $P_{\boldsymbol{\theta}}[B(\mathbf{X}) = 1] = 0$ for all $\boldsymbol{\theta} \in \mathcal{T}_0 \cup \mathcal{T}_A$, then all tests in $\mathcal{D}_\pi$ have the same risk function. They are thus all admissible among the Bayes tests, hence admissible among all tests.

If $P_{\boldsymbol{\theta}}[B(\mathbf{X}) = 1] > 0$ for some $\boldsymbol{\theta}$, then any differences in power are due to the $\gamma(\mathbf{x})$ when $B(\mathbf{x}) = 1$. Consequently, a test is admissible in $\mathcal{D}_\pi$ if and only if it is admissible in the *conditional* testing problem where $\gamma(x)$ is the test, and the distribution under consideration is the conditional distribution of $X \mid B(X) = 1$. For example, if $\{x \mid B(x) = 1\}$ consists of just the one point x_0, the $\gamma(x_0)$ for the conditional problem is a constant, in which case any value $0 \le \gamma(x_0) \le 1$ yields an admissible test. See Exercise 22.8.2.

For another example, consider testing $H_0: \theta = 1$ versus $H_A: \theta > 1$ based on $X \sim \text{Uniform}(0, \theta)$. Let the prior put half the probability on $\theta = 1$ and half on $\theta = 2$. Then

$$B(x) = \begin{cases} \infty & \text{if } x \in [1, 2) \\ 1(= \frac{0}{0}) & \text{if } x \ge 2 \\ \frac{1}{2} & \text{if } x \in (0, 1) \end{cases} . \tag{22.11}$$

Thus any Bayes test ϕ has $\phi(x) = 1$ if $1 \le x < 2$ and $\phi(x) = 0$ if $0 < x < 1$. The γ goes into effect if $x \ge 2$. If $\theta = 1$ then $P_1[B(X) = 1] = 0$, so only power is relevant in comparing Bayes tests on their γ. But $\gamma(x) \equiv 1$ will maximize the conditional power, so the only admissible Bayes test is $\phi(x) = I[x \ge 1]$.

22.2.2 Level α Bayes tests

The Bayes test in (22.8) minimizes the Bayes risk among all tests, with no guarantee about level. An expression for the test that minimizes the Bayes risk among level α tests is not in general easy to find. But if the null is simple and we evaluate the risk only on $\theta \in \mathcal{T}_A$, then the Neyman-Pearson lemma can again be utilized. The hypotheses are now

$$H_0: \theta = \theta_0 \ versus \ H_A: \theta \in \mathcal{T}_A \tag{22.12}$$

for some $\theta_0 \notin \mathcal{T}_A$, and for given $\alpha \in (0, 1)$, we consider just the set of level α tests

$$\mathcal{D}_\alpha = \{\phi \mid E_{\theta_0}[\phi] \le \alpha\}. \tag{22.13}$$

The prior π is the same as above, but since the null has only one point, $\rho_0(\{\theta_0\}) = 1$, the Bayes factor is

$$B_{A0}(x) = \frac{\int_{\mathcal{T}_A} f(x \mid \theta_A) \rho_A(\theta_A) d\theta_A}{f(x \mid \theta_0)}. \tag{22.14}$$

If the Bayes test wrt π in (22.8) for some $\gamma(x)$ has level α, then since it has the best Bayes risk among all tests, it must have the best Bayes risk among level α tests. Suppose its size is larger than α. Consider the test ϕ_α given by

$$\phi_\alpha(x) = \begin{cases} 1 & \text{if } B_{A0}(x) > c_\alpha \\ \gamma_\alpha & \text{if } B_{A0}(x) = c_\alpha \\ 0 & \text{if } B_{A0}(x) < c_\alpha \end{cases} , \tag{22.15}$$

where c_α and γ_α are chosen so that $E_{\theta_0}[\phi_\alpha(X)] = \alpha$. Suppose ϕ is another level α test. It must be that $c_\alpha > \pi_0 / \pi_A$, because otherwise the Bayes test would be level α. Using

(22.7) and (22.14), we can show that the difference in Bayes risks between ϕ and ϕ_α is

$$R(\pi; \phi) - R(\pi; \phi_\alpha) = \int_{\mathcal{X}} (\phi_\alpha(\mathbf{x}) - \phi(\mathbf{x}))(\pi_A B_{A0}(\mathbf{x}) - \pi_0) f(\mathbf{x} \mid \boldsymbol{\theta}_0)) d\mathbf{x}$$

$$= \int_{\mathcal{X}} (\phi_\alpha(\mathbf{x}) - \phi(\mathbf{x}))(\pi_A (B_{A0}(\mathbf{x}) - c_\alpha) + \pi_A c_\alpha - \pi_0) f(\mathbf{x} \mid \boldsymbol{\theta}_0)) d\mathbf{x}$$

$$= \pi_A \int_{\mathcal{X}} (\phi_\alpha(\mathbf{x}) - \phi(\mathbf{x}))(B_{A0}(\mathbf{x}) - c_\alpha) f(\mathbf{x} \mid \boldsymbol{\theta}_0) d\mathbf{x}$$

$$+ (\pi_A c_\alpha - \pi_0)(E_{\boldsymbol{\theta}_0}[\phi_\alpha] - E_{\boldsymbol{\theta}_0}[\phi]). \tag{22.16}$$

In the last line, the integral term is nonnegative by the definition (22.15) and the fact that $\pi_A c_\alpha - \pi_0 > 0$, and $E_{\boldsymbol{\theta}_0}[\phi_\alpha] \geq E_{\boldsymbol{\theta}_0}[\phi]$ because ϕ_α has size α and ϕ has level α. Thus $R(\pi; \phi) \geq R(\pi; \phi_\alpha)$, proving that ϕ_α is Bayes wrt π among level α tests. Exercise 22.8.3 shows that the Bayes test (or any test) is admissible among level α tests if and only if it is level α and admissible among all tests.

22.3   Necessary conditions for admissibility

As in estimation, under reasonable conditions, admissible tests are either Bayes tests or limits of Bayes tests, though not all Bayes tests are admissible. Here we extend the results in Section 20.8 to parameter spaces in (22.1) that are contained in $\mathbb{R}^K$.

We first need to define what we mean by limits of tests, in this case **weak limits**.

Definition 22.1. *Suppose $\phi_1, \phi_2, \ldots$ is a sequence of test functions, and ϕ is another test function. Then ϕ_n **converges weakly** to ϕ, written $\phi_n \to^w \phi$, if*

$$\int_{\mathcal{X}} \phi_n(\mathbf{x}) f(\mathbf{x}) d\mathbf{x} \longrightarrow \int_{\mathcal{X}} \phi(\mathbf{x}) f(\mathbf{x}) d\mathbf{x} \text{ for any } f \text{ such that } \int_{\mathcal{X}} |f(\mathbf{x})| d\mathbf{x} < \infty. \tag{22.17}$$

This definition is apropos for models with pdfs. This convergence is weaker than pointwise, since $\phi_n(\mathbf{x}) \to \phi(\mathbf{x})$ for all $\mathbf{x}$ implies that $\phi_n \to^w \phi$, but we can change ϕ at isolated points without affecting weak convergence. If $\mathcal{X}$ is countable, then the definition replaces the integral with summation, and weak convergence is equivalent to pointwise convergence.

Note that if we do have pdfs, then we can use $f_{\boldsymbol{\theta}}(\mathbf{x})$ for f to show

$$\phi_n \to^w \phi \implies E_{\boldsymbol{\theta}}[\phi_n(\mathbf{X})] \to E_{\boldsymbol{\theta}}[\phi(\mathbf{X})] \text{ for all } \boldsymbol{\theta} \in \mathcal{T}$$

$$\implies R(\boldsymbol{\theta}; \phi_n(\mathbf{X})) \to R(\boldsymbol{\theta}; \phi(\mathbf{X})) \text{ for all } \boldsymbol{\theta} \in \mathcal{T}. \tag{22.18}$$

As for Theorems 20.6 and 20.8 (pages 358 and 360), we need the risk set for any finite collection of $\boldsymbol{\theta}$'s to be closed, convex, and bounded from below. The last requirement is automatic, since all risks are in [0,1]. The first two conditions will hold if the corresponding conditions hold for $\mathcal{D}$:

   1. If $\phi_1, \phi_2 \in \mathcal{D}$ then $\beta \phi_1 + (1 - \beta)\phi_2 \in \mathcal{D}$ for all $\beta \in [0,1]$.
   2. If $\phi_1, \phi_2, \ldots \in \mathcal{D}$ and $\phi_n \to^w \phi$ then $\phi \in \mathcal{D}$. $\tag{22.19}$

These conditions hold, e.g., if $\mathcal{D}$ consists of all tests, or of all level α tests. Here is the main result, a special case of the seminal results in Wald (1950), Section 3.6.

Theorem 22.2. *Suppose $\mathcal{D}$ satisfies the conditions in (22.19), and $E_\theta[\phi]$ is continuous in θ for any test ϕ. Then if $\phi_0 \in \mathcal{D}$ is admissible among the tests in $\mathcal{D}$, there exists a sequence of Bayes tests $\phi_1, \phi_2, \ldots \in \mathcal{D}$ and a test $\phi \in \mathcal{D}$ such that*

$$\phi_n \to^w \phi \text{ and } R(\theta; \phi) = R(\theta; \phi_0) \text{ for all } \theta \in \mathcal{T}. \tag{22.20}$$

Note that the theorem doesn't necessarily guarantee that a particular admissible test is a limit of Bayes tests, but rather that there is a limit of Bayes tests that has the exact same risk. So you will not lose anything if you consider only Bayes tests and their weak limits. We can also require that each π_n in the theorem is concentrated on a finite set of points.

The proof relies on a couple of mathematical results that we won't prove here. First, we need that $\mathcal{T}_0$ and $\mathcal{T}_A$ both have countable dense subsets. For given set $\mathcal{C}$, a countable set $\mathcal{C}^* \subset \mathcal{C}$ is **dense** in $\mathcal{C}$ if for any $x \in \mathcal{C}$, there exists a sequence $x_1, x_2, \ldots \in \mathcal{C}^*$ such that $x_n \to x$. For example, the rational numbers are dense in the reals. As long as the parameter space is contained in $\mathbb{R}^K$, this condition is satisfied.

Second, we need that the set of tests ϕ is **compact** under weak convergence. This condition means that for any sequence $\phi_1, \phi_2, \ldots$ of tests, there exists a subsequence $\phi_{n_1}, \phi_{n_2}, \ldots$ and test ϕ such that

$$\phi_{n_i} \longrightarrow^w \phi \text{ as } i \to \infty. \tag{22.21}$$

See Theorem A.5.1 in the Appendix of Lehmann and Romano (2005).

Proof. (*Theorem 22.2*) Suppose ϕ_0 is admissible, and consider the new risk function

$$R^*(\theta; \phi) = R(\theta; \phi) - R(\theta; \phi_0). \tag{22.22}$$

Let $\mathcal{T}_0^* = \{\theta_{01}, \theta_{02}, \ldots, \}$ and $\mathcal{T}_A^* = \{\theta_{A1}, \theta_{A2}, \ldots, \}$ be countable dense subsets of $\mathcal{T}_0$ and $\mathcal{T}_A$, respectively, and set

$$\mathcal{T}_{0n} = \{\theta_{01}, \ldots, \theta_{0n}\} \text{ and } \mathcal{T}_{An} = \{\theta_{A1}, \ldots, \theta_{An}\}. \tag{22.23}$$

Consider the testing problem

$$H_0 : \theta \in \mathcal{T}_{0n} \text{ versus } H_A : \theta \in \mathcal{T}_{An}. \tag{22.24}$$

The parameter set here is finite, hence we can use Theorem 20.6 on page 358 to show that there exists a test ϕ_n and prior π_n on $\mathcal{T}_{0n} \cup \mathcal{T}_{An}$ such that ϕ_n is Bayes wrt π_n and minimax for R^*. Since ϕ_0 has maximum risk 0 under R^*, ϕ_n can be no worse:

$$R^*(\theta; \phi_n) \leq 0 \text{ for all } \theta \in \mathcal{T}_{0n} \cup \mathcal{T}_{An}. \tag{22.25}$$

Now by the compactness of the set of tests under weak convergence, there exist a subsequence ϕ_{n_i} and test ϕ such that (22.21) holds. Then (22.18) implies that

$$R^*(\theta; \phi_{n_i}) \longrightarrow R^*(\theta; \phi) \text{ for all } \theta \in \mathcal{T}. \tag{22.26}$$

Take any $\theta \in \mathcal{T}_0^* \cup \mathcal{T}_A^*$. Since it is a member of one of the sequences, there is some K such that $\theta \in \mathcal{T}_{0n} \cup \mathcal{T}_{An}$ for all $n \geq K$. Thus by (22.25),

$$R^*(\theta; \phi_{n_i}) \leq 0 \text{ for all } n_i \geq K \Rightarrow R^*(\theta; \phi_{n_i}) \to R^*(\theta; \phi) \leq 0 \text{ for all } \theta \in \mathcal{T}_0^* \cup \mathcal{T}_A^*. \tag{22.27}$$

Exercise 22.8.11 shows that since $E_{\boldsymbol{\theta}}[\phi]$ is continuous in $\boldsymbol{\theta}$ and $\mathcal{T}_0^* \cup \mathcal{T}_A^*$ is dense in $\mathcal{T}_0 \cup \mathcal{T}_A$, we have

$$R^*(\boldsymbol{\theta}\,;\phi) \le 0 \text{ for all } \boldsymbol{\theta} \in \mathcal{T}_0 \cup \mathcal{T}_A, \qquad (22.28)$$

i.e.,

$$R(\boldsymbol{\theta}\,;\phi) \le R(\boldsymbol{\theta}\,;\phi_0) \text{ for all } \boldsymbol{\theta} \in \mathcal{T}_0 \cup \mathcal{T}_A. \qquad (22.29)$$

Thus by the assumed admissibility of ϕ_0, (22.20) holds.   □

If the model is complete as in Definition 19.2 (page 328), then $R(\boldsymbol{\theta}\,;\phi) = R(\boldsymbol{\theta}\,;\phi_0)$ for all $\boldsymbol{\theta}$ means that $P_{\boldsymbol{\theta}}[\phi(\mathbf{X}) = \phi_0(\mathbf{X})] = 1$, so that any admissible test is a weak limit of Bayes tests.

22.4   Compact parameter spaces

The previous section showed that in many cases, all admissible tests must be Bayes or limits of Bayes, but that fact it not easy to apply directly. Here and in the next section, we look at some more explicit characterizations of admissibility.

We first look at testing with compact null and alternatives. That is, both $\mathcal{T}_0$ and $\mathcal{T}_A$ are closed and bounded. This requirement is somewhat artificial, since it means there is a gap between the two spaces. For example, suppose we have a bivariate normal, $\mathbf{X} \sim N(\boldsymbol{\mu}, \mathbf{I}_2)$. The hypotheses

$$H_0: \boldsymbol{\mu} = \mathbf{0} \ \ versus \ \ H_A: 1 \le \|\boldsymbol{\mu}\| \le 2 \qquad (22.30)$$

would fit into the framework. Replacing the alternative with $0 < \|\boldsymbol{\mu}\| \le 2$ or $1 \le \|\boldsymbol{\mu}\| < \infty$ or $\boldsymbol{\mu} \ne \mathbf{0}$ would not fit. Then if the conditions of Theorem 22.2 hold, all admissible tests are Bayes (so we do not have to worry about limits).

Theorem 22.3. *Suppose (22.19) hold for $\mathcal{D}$, and $E_{\boldsymbol{\theta}}[\phi]$ is continuous in $\boldsymbol{\theta}$. Then if ϕ_0 is admissible among the tests in $\mathcal{D}$, it is Bayes.*

The proof uses the following lemmas.

Lemma 22.4. *Let $\phi_1, \phi_2, \dots$ be a sequence of tests, where*

$$\phi_n(\mathbf{x}) = \begin{cases} 1 & if \quad g_n(\mathbf{x}) > 0 \\ \gamma_n(\mathbf{x}) & if \quad g_n(\mathbf{x}) = 0 \\ 0 & if \quad g_n(\mathbf{x}) < 0 \end{cases}$$

for some functions $g_n(\mathbf{x})$. Suppose there exists a function $g(\mathbf{x})$ such that $g_n(\mathbf{x}) \to g(\mathbf{x})$ for each $\mathbf{x} \in \mathcal{X}$, and a test ϕ such that $\phi_n \to^w \phi$. Then (with probability one),

$$\phi(\mathbf{x}) = \begin{cases} 1 & if \quad g(\mathbf{x}) > 0 \\ \gamma(\mathbf{x}) & if \quad g(\mathbf{x}) = 0 \\ 0 & if \quad g(\mathbf{x}) < 0 \end{cases}$$

for some function $\gamma(\mathbf{x})$.

In the lemma, γ is unspecified, so the lemma tells us nothing about ϕ when $g(\mathbf{x}) = 0$.

Proof. Let f be any function with finite integral ($\int |f(x)|dx < \infty$ as in (22.17)). Then the function $f(x)I[g(x) > 0]$ also has finite integral. Hence by Definition 22.1,

$$\int_{\mathcal{X}} f(x)I[g(x) > 0]\phi_n(x)dx \longrightarrow \int_{\mathcal{X}} f(x)I[g(x) > 0]\phi(x)dx. \qquad (22.31)$$

If $g(x) > 0$, then $g_n(x) > 0$ for all sufficiently large n, hence $\phi_n(x) = 1$ for all sufficient large n. Thus $\phi_n(x) \to 1$ if $g(x) > 0$, and

$$\int_{\mathcal{X}} f(x)I[g(x) > 0]\phi_n(x)dx \longrightarrow \int_{\mathcal{X}} f(x)I[g(x) > 0](1)dx. \qquad (22.32)$$

Thus the two limits in (22.31) and (22.32) must be equal for any such f, which means $\phi(x) = 1$ if $h(x) > 0$ with probability one (i.e., $P_{\boldsymbol{\theta}}[\phi(\mathbf{X}) = 1 \,|\, h(\mathbf{X}) > 0] = 1$ for all $\boldsymbol{\theta}$). Similarly, $\phi(x) = 0$ for $g(x) < 0$ with probability one, which completes the proof. $\quad\square$

Weak convergence for probability distributions, $\pi_n \to^w \pi$, is the same as convergence in distribution of the corresponding random variables. The next result from measure theory is analogous to the weak compactness we saw for test functions in (22.21). See Section 5 on Prohorov's Theorem in Billingsley (1999) for a proof.

Lemma 22.5. *Suppose $\pi_1, \pi_2, \ldots$ is a sequence of probability measures on the compact space $\mathcal{T}$. Then there exists a subsequence $\pi_{n_1}, \pi_{n_2}, \ldots$, and probability measure π on $\mathcal{T}$, such that $\pi_{n_i} \to^w \pi$.*

Proof. (Theorem 22.3) Suppose ϕ_0 is admissible. Theorem 22.2 shows that there is a sequence of Bayes tests such that $\phi_n \to^w \phi$, where ϕ and ϕ_0 have the same risk function. Let π_n be the prior for which ϕ_n is Bayes. Decompose π_n into its components $(\rho_{n0}, \rho_{nA}, \pi_{n0}, \pi_{nA})$ as in the beginning of Section 22.2, so that from (22.8),

$$\phi_{\pi_n}(\mathbf{x}) = \begin{cases} 1 & \text{if} \quad B_n(\mathbf{x}) > 1 \\ \gamma_n(\mathbf{x}) & \text{if} \quad B_n(\mathbf{x}) = 1 \\ 0 & \text{if} \quad B_n(\mathbf{x}) < 1 \end{cases}, \quad \text{where} \quad B_n(\mathbf{x}) = \frac{E_{\rho_{nA}}[f(\mathbf{x}\,|\,\boldsymbol{\theta})]\pi_{nA}}{E_{\rho_{n0}}[f(\mathbf{x}\,|\,\boldsymbol{\theta})]\pi_{n0}}. \qquad (22.33)$$

By Lemma 22.5, there exists a subsequence and prior π such that $\pi_{n_i} \to^w \pi$, where the components also converge,

$$\rho_{n_i 0} \to^w \rho_0, \quad \rho_{n_i A} \to^w \rho_A, \quad \pi_{n_i 0} \to \pi_0, \quad \pi_{n_i A} \to \pi_A. \qquad (22.34)$$

If for each $\mathbf{x}$, $f(\mathbf{x}\,|\,\boldsymbol{\theta})$ is bounded and continuous in $\boldsymbol{\theta}$, then the two expected values in (22.33) will converge to the corresponding ones with π, hence the entire ratio converges:

$$B_n(\mathbf{x}) \to \frac{E_{\rho_A}[f(\mathbf{x}\,|\,\boldsymbol{\theta})]\pi_A}{E_{\rho_0}[f(\mathbf{x}\,|\,\boldsymbol{\theta})]\pi_0} \equiv B(\mathbf{x}). \qquad (22.35)$$

Then Lemma 22.4 can be applied to show that

$$\phi_{n_i}(\mathbf{x}) \to^w \phi_{\pi}(\mathbf{x}) = \begin{cases} 1 & \text{if} \quad B(\mathbf{x}) > 1 \\ \gamma(\mathbf{x}) & \text{if} \quad B(\mathbf{x}) = 1 \\ 0 & \text{if} \quad B(\mathbf{x}) < 1 \end{cases} \qquad (22.36)$$

for some $\gamma(\mathbf{x})$. This ϕ_{π} is the correct form to be Bayes wrt π. Above we have $\phi_n \to^w \phi$, hence ϕ and ϕ_{π} must have the same risk function, which is also the same as the original test ϕ_0. That is, ϕ_0 is Bayes wrt π. $\quad\square$

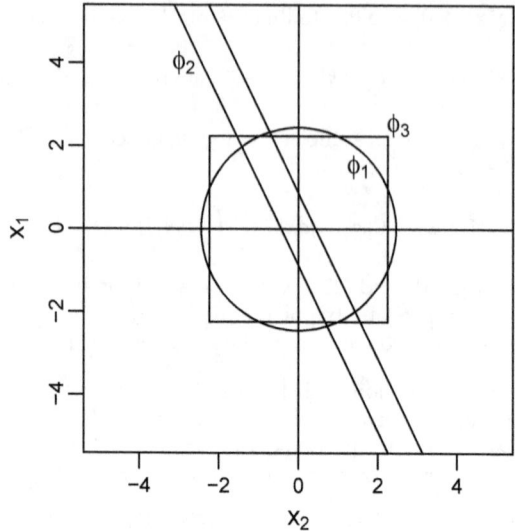

Figure 22.2: Testing $H_0 : \boldsymbol{\mu} = \mathbf{0}$ versus $H_A : 1 \leq \|\boldsymbol{\mu}\|^2 \leq 2$ based on $\mathbf{X} \sim N(\boldsymbol{\mu}, \mathbf{I}_2)$, with level $\alpha = 0.05$. Test ϕ_1 rejects the null when $\|\mathbf{x}\|^2 > \chi^2_{2,\alpha}$, and ϕ_2 rejects when $|2x_1 + x_2| > \sqrt{5} z_{\alpha/2}$. These two tests are Bayes and admissible. Test ϕ_3 rejects the null when $\max\{|x_1|, |x_2|\} > z_{(1-\sqrt{1-\alpha})/2}$. It is not Bayes, hence not admissible.

The theorem will not necessarily work if the parameter spaces are not compact, since there may not be a limit of the π_n's. For example, suppose $\mathcal{T}_A = (0, \infty)$. Then the sequence of Uniform$(0, n)$'s will not have a limit, nor will the sequence of π_n where $\pi_n[\Theta = n] = 1$. The parameter spaces also need to be separated. For example, if the null is $\{0\}$ and the alternative is $(0, 1]$, consider the sequence of priors with $\pi_{n0} = \pi_{nA} = 1/2$, $\rho_0[\Theta = 0] = 1$ and $\rho_n[\Theta = 1/n] = 1$. The limit $\rho_n \to^w \rho$ is $\rho[\Theta = 0] = 1$, same as ρ_0, and *not* a probability on $\mathcal{T}_A$. Now $B(\mathbf{x}) = 1$, and (22.36) has no information about the limit. But see Exercise 22.8.5. Also, Brown and Marden (1989) contains general results on admissibility when the null is simple.

Going back to the bivariate normal problem in (22.30), any admissible test is Bayes. Exercise 22.8.7 shows that the test that rejects the null when $\|\mathbf{x}\|^2 > \chi^2_{2,\alpha}$ is Bayes and admissible, as are any tests that reject the null when $|aX_1 + bX_2| > c$ for some constants a, b, c. However, the test that rejects when $\max\{|x_1|, |x_2|\} > c$ has a square as an acceptance region. It is not admissible, because it can be shown that any Bayes test here has to have "smooth" boundaries, not sharp corners.

22.5 Convex acceptance regions

If all Bayes tests satisfy a certain property, and that property persists through limits, then the property is a necessary condition for a test to be admissible. For example, suppose the set of distributions under consideration is a one-dimensional exponential family distribution. By the discussion around (22.4), we do not lose anything by looking at just tests that are functions of the sufficient statistic. Thus we will assume x itself is the natural sufficient statistic, so that its density is $f(x \mid \theta) = a(x) \exp(\theta x - \psi(\theta))$. We test the two-sided hypotheses $H_0: \theta = \theta_0$ versus $H_A: \theta \neq \theta_0$. A Bayes test wrt π is

$$\phi_\pi(x) = \begin{cases} 1 & \text{if} \quad B_\pi(x) > 1 \\ \gamma_\pi(x) & \text{if} \quad B_\pi(x) = 1 \\ 0 & \text{if} \quad B_\pi(x) < 1 \end{cases}, \tag{22.37}$$

where we can write

$$B_\pi(x) = \frac{\pi_A}{\pi_0} \sum_{i=1}^{K} e^{x(\theta_i - \theta_0) - (\phi(\theta_i) - \phi(\theta_0))} \rho_A(\theta_i), \tag{22.38}$$

at least if π has pmf ρ_A on a finite set of θ's in $\mathcal{T}_A$. Similar to (21.94), $B(x)$ is convex as a function of x, hence the acceptance region of ϕ_π is an interval (possibly infinite or empty):

$$\phi_\pi(x) = \begin{cases} 1 & \text{if} \quad x < a_\pi \text{ or } x > b_\pi \\ \gamma_\pi(x) & \text{if} \quad x = a_\pi \text{ or } x = b_\pi \\ 0 & \text{if} \quad a_\pi < x < b_\pi \end{cases} \tag{22.39}$$

for some a_π and b_π. If ϕ_n is a sequence of such Bayes tests with $\phi_n \to \phi$, then Exercise 22.8.10 shows that ϕ must have the same form (22.39) for some $-\infty \leq a \leq b \leq \infty$, hence any admissible test must have that form.

Now suppose we have a p-dimensional exponential family for $\mathbf{X}$ with parameter space $\mathcal{T}$, and for $\boldsymbol{\theta}_0 \in \mathcal{T}$ test

$$H_0: \boldsymbol{\theta} = \boldsymbol{\theta}_0 \quad versus \quad H_A: \boldsymbol{\theta} \in \mathcal{T} - \{\boldsymbol{\theta}_0\}. \tag{22.40}$$

Then any Bayes test wrt a π whose probability is on a finite set of $\boldsymbol{\theta}_i$'s has the form (22.37) but with

$$B(\mathbf{x}) = \frac{\pi_A}{\pi_0} \sum_{i=1}^{K} e^{\mathbf{x} \cdot (\boldsymbol{\theta}_i - \boldsymbol{\theta}_0) - (\phi(\boldsymbol{\theta}_i) - \phi(\boldsymbol{\theta}_0))} \rho_A(\boldsymbol{\theta}_i). \tag{22.41}$$

Consider the set

$$\mathcal{C} = \{\mathbf{x} \in \mathbb{R}^p \mid B(\mathbf{x}) < 1\}. \tag{22.42}$$

It is convex by the convexity of $B(\mathbf{x})$. Also, it can be shown that the ϕ_π of (22.39) (with $\mathbf{x}$ in place of x) is

$$\phi_\pi(\mathbf{x}) = \begin{cases} 1 & \text{if} \quad \mathbf{x} \notin \text{closure}(\mathcal{C}) \\ \gamma_\pi(\mathbf{x}) & \text{if} \quad \mathbf{x} \in \text{boundary}(\mathcal{C}) \\ 0 & \text{if} \quad \mathbf{x} \in \mathcal{C} \end{cases}. \tag{22.43}$$

(If $\mathcal{C}$ is empty, then let it equal the set with $B(\mathbf{x}) \leq 1$.) Note that if $p = 1$, $\mathcal{C}$ is an interval as in (22.39).

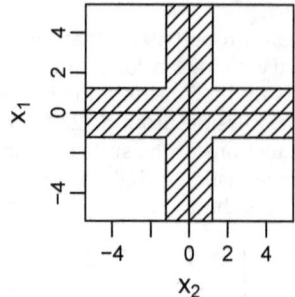

Figure 22.3: The test that rejects the null when $\min\{|x_1|, |x_2|\} > c_\alpha$. The acceptance region is the shaded cross. The test is not admissible.

Suppose ϕ is admissible, so that there exists a sequence of Bayes tests whose weak limit is a test with the same risk function as ϕ. Each Bayes test has the form (22.43), i.e., its acceptance region is a convex set. Birnbaum (1955), with correction and extensions by Matthes and Truax (1967), has shown that a weak limit of such tests also is of that form. If we have completeness, then any admissible test has to have that form. Here is the formal statement of the result.

Theorem 22.6. *Suppose* $\mathbf{X}$ *has a p-dimensional exponential family distribution, where* $\mathbf{X}$ *is the natural sufficient statistic and* $\boldsymbol{\theta}$ *is the natural parameter. Suppose further that the model is complete for* $\mathbf{X}$. *Then a necessary condition for a test* ϕ *to be admissible is that there exists a convex set* C *and function* $\gamma(\mathbf{x})$ *such that*

$$\phi(\mathbf{x}) = \begin{cases} 1 & \text{if} \quad \mathbf{x} \notin \text{closure}(C) \\ \gamma(\mathbf{x}) & \text{if} \quad \mathbf{x} \in \text{boundary}(C) \\ 0 & \text{if} \quad \mathbf{x} \in \text{interior}(C) \end{cases}, \tag{22.44}$$

or at least equals that form with probability 1.

If the distributions have a pdf, then the probability of the boundary of a convex set is zero, hence we can drop the randomization part in (22.44), as well as for Bayes tests in (22.43). The latter fact means that all Bayes tests are essentially unique for their priors, hence admissible.

The three tests in Figure 22.2 all satisfy (22.44), but only ϕ_1 and ϕ_2 are admissible for the particular bivariate normal testing problem in (22.30). Another potentially reasonable test in this case is the one based on the minimum of the absolute x_i's:

$$\phi(x_1, x_2) = \begin{cases} 1 & \text{if} \quad \min\{|x_1|, |x_2|\} > c_\alpha \\ 0 & \text{if} \quad \min\{|x_1|, |x_2|\} \le c_\alpha \end{cases}. \tag{22.45}$$

See Figure 22.3. The test accepts within the shaded cross, which is not a convex set. Thus the test is inadmissible.

22.5.1 Admissible tests

Not all tests with convex acceptance regions are admissible in general. However, if the parameter space is big enough, then they usually are. We continue with the p-dimensional exponential family distribution, with $\mathbf{X}$ as the natural sufficient statistic and $\boldsymbol{\theta}$ the natural parameter. Now suppose that $\mathcal{T} = \mathbb{R}^p$, and we test the hypotheses in (22.40), which are now $H_0: \boldsymbol{\theta} = \boldsymbol{\theta}_0$ versus $H_A: \boldsymbol{\theta} \neq \boldsymbol{\theta}_0$.

For a vector $\mathbf{a} \in \mathbb{R}^p$ and constant c, let $\mathcal{H}_{\mathbf{a},c}$ be the closed half-space defined by

$$\mathcal{H}_{\mathbf{a},c} = \{\mathbf{x} \in \mathbb{R}^p \mid \mathbf{a} \cdot \mathbf{x} \leq c\}. \tag{22.46}$$

In the plane, $\mathcal{H}_{\mathbf{a},c}$ is the set of points on one side of a line, including the line. Any closed half-space is a convex set. It is not hard to show that the test that rejects the null when $\mathbf{x} \in \mathcal{H}_{\mathbf{a},c}^C$ is Bayes (see Exercise 22.8.6), hence likely admissible. (Recall the complement of a set $\mathcal{A}$ is $\mathbb{R}^p - \mathcal{A}$, denoted by $\mathcal{A}^C$.) We can show a stronger result: A test that always rejects outside of the half-space has better power for some parameter values than one that does not so reject. Formally:

Lemma 22.7. *For given* $\mathbf{a}$ *and* c, *suppose for test* ϕ_0 *that* $\phi_0(\mathbf{x}) = 1$ *for all* $\mathbf{x} \in \mathcal{H}_{\mathbf{a},c}^C$, *and for test* ϕ *that* $P_{\boldsymbol{\theta}}[\phi(\mathbf{X}) < 1 \,\&\, \mathbf{X} \in \mathcal{H}_{\mathbf{a},c}^C] > 0$. *Then*

$$e^{\psi(\boldsymbol{\theta}_0 + \eta\mathbf{a}) - \psi(\boldsymbol{\theta}_0) - c\eta} E_{\boldsymbol{\theta}_0 + \eta\mathbf{a}}[\phi_0 - \phi] \longrightarrow \infty \text{ as } \eta \to \infty. \tag{22.47}$$

Thus for some $\boldsymbol{\theta}' \neq \boldsymbol{\theta}_0$,

$$E_{\boldsymbol{\theta}'}[\phi_0] > E_{\boldsymbol{\theta}'}[\phi]. \tag{22.48}$$

The test that rejects the null if and only if $\mathbf{x} \in \mathcal{H}_{\mathbf{a},c}^C$ is then admissible, since it has the smallest size for any test whose rejection region is contained in $\mathcal{H}_{\mathbf{a},c}^C$.

Proof. Write

$$e^{\psi(\boldsymbol{\theta}_0 + \eta\mathbf{a}) - \psi(\boldsymbol{\theta}_0) - c\eta} E_{\boldsymbol{\theta}_0 + \eta\mathbf{a}}[\phi_0 - \phi]$$

$$= e^{\psi(\boldsymbol{\theta}_0 + \eta\mathbf{a}) - \psi(\boldsymbol{\theta}_0) - c\eta} \int_{\mathcal{X}} (\phi_0(\mathbf{x}) - \phi(\mathbf{x})) e^{(\boldsymbol{\theta}_0 + \eta\mathbf{a}) \cdot \mathbf{x} - \psi(\boldsymbol{\theta}_0 + \eta\mathbf{a})} a(\mathbf{x}) d\mathbf{x}$$

$$= \int_{\mathcal{X}} (\phi_0(\mathbf{x}) - \phi(\mathbf{x})) e^{\eta(\mathbf{a} \cdot \mathbf{x} - c)} f(\mathbf{x} \mid \boldsymbol{\theta}_0) d\mathbf{x}$$

$$= \int_{\mathcal{H}_{\mathbf{a},c}} (\phi_0(\mathbf{x}) - \phi(\mathbf{x})) e^{\eta(\mathbf{a} \cdot \mathbf{x} - c)} f(\mathbf{x} \mid \boldsymbol{\theta}_0) d\mathbf{x}$$

$$+ \int_{\mathcal{H}_{\mathbf{a},c}^C} (1 - \phi(\mathbf{x})) e^{\eta(\mathbf{a} \cdot \mathbf{x} - c)} f(\mathbf{x} \mid \boldsymbol{\theta}_0) d\mathbf{x}, \tag{22.49}$$

since $\phi_0(\mathbf{x}) = 1$ for $\mathbf{x} \in \mathcal{H}_{\mathbf{a},c}^C$. For $\eta > 0$, the first integral in the last equality of (22.47) is bounded by ± 1 since $\mathbf{a} \cdot \mathbf{x} \leq c$. The exponential in the second integral goes to infinity as $\eta \to \infty$, and since by assumption $1 - \phi(\mathbf{X}) > 0$ on $\mathcal{H}_{\mathbf{a},c}$ with positive probability (and is never negative), the integral goes to infinity, proving (22.47). Because the constant in front of the expectation in the first expression of (22.49) is positive, we have (22.48). In fact, there exists an η_0 such that

$$E_{\boldsymbol{\theta}_0 + \eta\mathbf{a}}[\phi_0] > E_{\boldsymbol{\theta}_0 + \eta\mathbf{a}}[\phi] \text{ for all } \eta > \eta_0. \tag{22.50}$$

(Compare the result here to Exercise 21.8.24.) $\qquad\square$

If a test rejects the null hypothesis whenever **x** is not in any of a set of half-spaces, then it has better power for some parameter values than any test that doesn't always reject when not in any one of those half-spaces. The connection to tests with convex acceptance regions is based on the fact that a set is closed and convex (other than $\mathbb{R}^p$) if and only if it is an intersection of closed half-spaces. Which is the next lemma, shown in Exercise 22.8.12.

Lemma 22.8. *Suppose the set* $C \subset \mathbb{R}^p$, $C \neq \mathbb{R}^p$, *is closed. Then it is convex if and only if there is a set of vectors* $\mathbf{a} \in \mathbb{R}^p$ *and constants c such that*

$$C = \cap_{\mathbf{a},c} \mathcal{H}_{\mathbf{a},c}. \tag{22.51}$$

Next is the main result of this section, due to Stein (1956a).

Theorem 22.9. *Suppose* C *is a closed convex set. Then the test*

$$\phi_0(\mathbf{x}) = \begin{cases} 1 & \text{if } \mathbf{x} \notin C \\ 0 & \text{if } \mathbf{x} \in C \end{cases} \tag{22.52}$$

is admissible.

Proof. Let ϕ be any test at least as good as ϕ_0, i.e., $R(\boldsymbol{\theta}; \phi) \leq R(\boldsymbol{\theta}; \phi_0)$ for all $\boldsymbol{\theta}$. By Lemma 22.8, there exists a set of half-spaces $\mathcal{H}_{\mathbf{a},c}$ such that (22.51) holds. Thus $\phi_0(\mathbf{x}) = 1$ whenever $\mathbf{x} \notin \mathcal{H}_{\mathbf{a},c}$ for any of those half-spaces. Lemma 22.7 then implies that $\phi_0(\mathbf{x})$ must also be 1 (with probability one) whenever $\mathbf{x} \notin \mathcal{H}_{\mathbf{a},c}$, or else ϕ_0 would have better power somewhere. Thus $\phi(\mathbf{x}) = 1$ whenever $\phi_0(\mathbf{x}) = 1$ (with probability one). Also, $P_{\boldsymbol{\theta}}[\phi(\mathbf{X}) > 0 \,\&\, \phi_0(\mathbf{X}) = 0] = 0$, since otherwise $E_{\boldsymbol{\theta}_0}[\phi] > E_{\boldsymbol{\theta}_0}[\phi_0]$. Thus $P_{\boldsymbol{\theta}}[\phi(\mathbf{X}) = \phi_0(\mathbf{X})] = 1$, hence they have the same risk function, proving that ϕ_0 is admissible.   □

This theorem implies that the three tests in Figure 22.2, which reject the null when $\|\mathbf{x}\|^2 > c$, $|ax_1 + bx_2| > c$, and $\max\{|x_1|, |x_2|\}$, respectively, are admissible for the current hypotheses. Recall that the third one is not admissible for the compact hypotheses in (22.30).

If the distributions of **X** have pdfs, then the boundary of any convex set has probability zero. Thus Theorems 22.6 and 22.9 combine to show that a test is admissible if and only if it is of the form (22.52) with probability one. In the discrete case, it can be tricky, since a test of the form (22.44) may not be admissible if the boundary of C has positive probability.

22.5.2 Monotone acceptance regions

If instead of a general alternative hypothesis, the alternative is one-sided for all θ_i, then it would seem reasonable that a good test would tend to reject for larger values of the components of **X**, but not smaller values. More precisely, suppose the hypotheses are

$$H_0: \boldsymbol{\theta} = \mathbf{0} \ \text{ versus } \ H_A: \boldsymbol{\theta} \in \mathcal{T}_A = \{\boldsymbol{\theta} \in \mathcal{T} \mid \theta_i \geq 0 \text{ for each } i\} - \{\mathbf{0}\}. \tag{22.53}$$

Then the Bayes ratio $B_\pi(\mathbf{x})$ as in (22.38) is nondecreasing in each x_i, since in the exponent we have $\sum x_i \theta_i$ with all $\theta_i \geq 0$. Consequently, if $B_\pi(\mathbf{x}) < 1$, so that the test accepts the null, then $B_\pi(\mathbf{x}) < 1$ for all **y** with $y_i \leq x_i$, $i = 1, \ldots, p$. We can reverse the inequalities as well, so that if we reject at $\mathbf{x}$, we reject at any $\mathbf{y}$ whose components are at least as large as $\mathbf{x}$'s. Assuming continuous random variables, so that the randomized parts of Bayes tests can be ignored, any Bayes test for (22.53) has the form

$$\phi(\mathbf{x}) = \begin{cases} 1 & \text{if } \mathbf{x} \notin \mathcal{A} \\ 0 & \text{if } \mathbf{x} \in \mathcal{A} \end{cases} \tag{22.54}$$

for some nonincreasing convex set $\mathcal{A} \subset \mathbb{R}^p$, where by nonincreasing we mean

$$\mathbf{x} \in \mathcal{A} \implies \mathcal{L}_{\mathbf{x}} \subset \mathcal{A}, \text{ where } \mathcal{L}_{\mathbf{x}} = \{\mathbf{y} \,|\, y_i \le x_i, \ i = 1, \ldots, p\}. \tag{22.55}$$

(Compare $\mathcal{L}_{\mathbf{x}}$ here to $\mathcal{L}_s$ in (20.65).)

Now suppose we have a sequence of Bayes tests ϕ_n, and another test ϕ such that $\phi_n \to \phi$. Eaton (1970) has shown that ϕ has the form (22.54), hence any admissible test must have that form, or be equal to a test of that form with probability one. Furthermore, if the overall parameter space $\mathcal{T}$ is unbounded in such a way that $\mathcal{T}_A = \{\boldsymbol{\theta} \in \mathbb{R}^p \,|\, \theta_i \ge 0 \text{ for each } i\} - \{\mathbf{0}\}$, then an argument similar to that in the proof of Theorem 22.9 shows that all tests of the form (22.54) are admissible.

In the case $\mathbf{X} \sim N(\boldsymbol{\theta}, \mathbf{I}_p)$, the tests in Figure 22.2 are inadmissible for the hypotheses in (22.53) because the acceptance regions are not nonincreasing. Admissible tests include that with rejection region $a_1 x_1 + a_2 x_2 > c$ where $a_1 > 0$ and $a_2 > 0$, and that with rejection region $\min\{x_1, x_2\} > c$. See Exercise 22.8.13 for the likelihood ratio test.

22.6 Invariance

We have seen that in many testing problems, especially multiparameter ones, there is no uniquely best test. Admissibility can help, but there may be a large number of admissible tests, and it can be difficult to decide whether any particular test is admissible. We saw shift equivariance in Sections 19.6 and 19.7, where by restricting consideration to shift equivariant estimators, we could find an optimal estimator in certain models. A similar idea applies in hypothesis testing.

For example, in the Student's t test situation, $X_1, \ldots, X_n$ are iid $N(\mu, \sigma^2)$, and we test $\mu = 0$ with σ^2 unknown. Then the two parameter spaces are

$$\mathcal{T}_0 = \{(0, \sigma^2) \,|\, \sigma^2 > 0\} \text{ and } \mathcal{T}_A = \{(\mu, \sigma^2) \,|\, \mu \ne 0 \text{ and } \sigma^2 > 0\}. \tag{22.56}$$

Changing units of the data shouldn't affect the test. That is, if we reject the null when the data is measured in feet, we should also reject when the data is measured in centimeters. This problem is invariant under multiplication by a constant. That is, let $\mathcal{G} = \{a \in \mathbb{R} \,|\, a \ne 0\}$, the nonzero reals. This is a group under multiplication. The **action** of a group element on the data is to multiply each x_i by the element, which is written

$$a \circ \mathbf{x} = a\mathbf{x} \text{ for } \mathbf{x} \in \mathcal{X} \text{ and } a \in \mathcal{G}. \tag{22.57}$$

For given $a \ne 0$, set $X_i^* = aX_i$. Then the transformed problem has $X_1^*, \ldots, X_n^*$ iid $N(\mu^*, \sigma^{*2})$, where $\mu^* = a\mu^*$ and $\sigma^{*2} = a^2\sigma^2$. The transformed parameter spaces are

$$\mathcal{T}_0^* = \{(0, \sigma^{*2}) \,|\, \sigma^{*2} > 0\} \text{ and } \mathcal{T}_A^* = \{(\mu^*, \sigma^{*2}) \,|\, \mu^* \ne 0 \text{ and } \sigma^{*2} > 0\}. \tag{22.58}$$

Those are the exact same spaces as in (22.56), and the data has the exact same distribution except for asterisks in the notation. That is, these two testing problems are equivalent.

The thinking is that therefore, any test based on $\mathbf{X}$ should have the same outcome as a test based on $\mathbf{X}^*$. Such a test function ϕ is called **invariant** under $\mathcal{G}$, meaning

$$\phi(\mathbf{x}) = \phi(a\mathbf{x}) \quad \text{for all} \quad \mathbf{x} \in \mathcal{X} \text{ and } a \in \mathcal{G}. \tag{22.59}$$

The test which rejects the null when $\overline{X} > c$ is not invariant, nor is the one-sided t test, which rejects when $T = \sqrt{n}\overline{x}/s_* > t_{n-1,\alpha}$. The two-sided t test is invariant:

$$|T^*| = \sqrt{n}\,\frac{|\overline{x}^*|}{s_*^*} = \sqrt{n}\,\frac{|a||\overline{x}|}{|a|s_*} = \sqrt{n}\,\frac{|\overline{x}|}{s_*} = |T|. \tag{22.60}$$

We will later see that the two-sided t test is the uniformly most power invariant level α test.

22.6.1 Formal definition

In Section 17.4 we saw algebraic groups of matrices. Here we generalize slightly to affine groups. Recall affine transformations from Sections 2.2.2 and 2.3, where an affine transformation of a vector $\mathbf{x}$ is $\mathbf{A}\mathbf{x} + \mathbf{b}$ for some matrix $\mathbf{A}$ and vector $\mathbf{b}$. A set $\mathcal{G}$ of affine transformations of dimension p is a subset of $\mathcal{A}_p \times \mathbb{R}^p$, where $\mathcal{A}_p$ is the set of $p \times p$ invertible matrices. For the set $\mathcal{G}$ to be a group, it has to have an operation $\circ$ that combines two elements such that the following properties hold:

1. Closure: $\mathbf{g}_1, \mathbf{g}_2 \in \mathcal{G} \Rightarrow \mathbf{g}_1 \circ \mathbf{g}_2 \in \mathcal{G}$;
2. Associativity: $\mathbf{g}_1, \mathbf{g}_2, \mathbf{g}_3 \in \mathcal{G} \Rightarrow \mathbf{g}_1 \circ (\mathbf{g}_2 \circ \mathbf{g}_3) = (\mathbf{g}_1 \circ \mathbf{g}_2) \circ \mathbf{g}_3$;
3. Identity: There exists $\mathbf{e} \in \mathcal{G}$ such that $\mathbf{g} \in \mathcal{G} \Rightarrow \mathbf{g} \circ \mathbf{e} = \mathbf{e} \circ \mathbf{g} = \mathbf{g}$;
4. Inverse: For each $\mathbf{g} \in \mathcal{G}$ there exists a $\mathbf{g}^{-1} \in \mathcal{G}$ such that $\mathbf{g} \circ \mathbf{g}^{-1} = \mathbf{g}^{-1} \circ \mathbf{g} = \mathbf{e}$.

$$\tag{22.61}$$

Note that we are using the symbol "$\circ$" to represent the action of the group on the sample space as well as the group composition. Which is meant should be clear from the context. For affine transformations, we want to define the composition so that it will conform to taking an affine transformation of an affine transformation. That is, if $(\mathbf{A}_1, \mathbf{b}_1)$ and $(\mathbf{A}_2, \mathbf{b}_2)$ are two affine transformations, then we want the combined transformation $(\mathbf{A}, \mathbf{b}) = (\mathbf{A}_1, \mathbf{b}_1) \circ (\mathbf{A}_2, \mathbf{b}_2)$ to satisfy

$$\mathbf{A}\mathbf{x} + \mathbf{b} = \mathbf{A}_1(\mathbf{A}_2\mathbf{x} + \mathbf{b}_2) + \mathbf{b}_1 \Rightarrow \mathbf{A} = \mathbf{A}_1\mathbf{A}_2 \text{ and } \mathbf{b} = \mathbf{A}_1\mathbf{b}_2 + \mathbf{b}_1$$
$$\Rightarrow (\mathbf{A}_1, \mathbf{b}_1) \circ (\mathbf{A}_2, \mathbf{b}_2) = (\mathbf{A}_1\mathbf{A}_2, \mathbf{A}_1\mathbf{b}_2 + \mathbf{b}_1). \tag{22.62}$$

The groups we consider here then have the composition (22.62). For each, we must also check that the conditions in (22.61) hold.

Let $\mathcal{G}$ be the invariance group, and $\mathbf{g} \circ \mathbf{x}$ be the action on $\mathbf{x}$ for $\mathbf{g} \in \mathcal{G}$. An action has to conform to the group's structure: If $\mathbf{g}_1, \mathbf{g}_2 \in \mathcal{G}$, then $(\mathbf{g}_1 \circ \mathbf{g}_2) \circ \mathbf{x} = \mathbf{g}_1 \circ (\mathbf{g}_2 \circ \mathbf{x})$, and if $\mathbf{e} \in \mathcal{G}$ is the identity element of the group, then $\mathbf{e} \circ \mathbf{x} = \mathbf{x}$. You might try checking these conditions on the action defined in (22.57).

A model is **invariant under** $\mathcal{G}$ if for each $\theta \in \mathcal{T}$ and $\mathbf{g} \in \mathcal{G}$, there exists a parameter value $\theta^* \in \mathcal{T}$ such that

$$\mathbf{X} \sim P_\theta \implies (\mathbf{g} \circ \mathbf{X}) \sim P_{\theta^*}. \tag{22.63}$$

We will denote θ^* by $\mathbf{g} \circ \theta$, the action of the group on the parameter space, though technically the $\circ$'s for the sample space and parameter space may not be the same.

The action in the t test example of (22.56) is $a \circ (\mu, \sigma^2) = (a\mu, a^2\sigma^2)$. The testing problem is invariant if both hypotheses' parameter spaces are invariant, so that for any $\mathbf{g} \in \mathcal{G}$,

$$\boldsymbol{\theta} \in \mathcal{T}_0 \Rightarrow \mathbf{g} \circ \boldsymbol{\theta} \in \mathcal{T}_0 \text{ and } \boldsymbol{\theta} \in \mathcal{T}_A \Rightarrow \mathbf{g} \circ \boldsymbol{\theta} \in \mathcal{T}_A. \tag{22.64}$$

22.6.2 Reducing by invariance

Just below (22.4), we introduced the notion of "reducing by sufficiency," which simplifies the problem by letting us focus on just the sufficient statistics. Similarly, **reducing by invariance** simplifies the problem by letting us focus on just invariant tests. The key is to find the **maximal invariant statistic**, which is an invariant statistic $\mathbf{W} = w(\mathbf{X})$ such that any invariant function of $\mathbf{x}$ is a function of just $w(\mathbf{x})$.

The standard method for showing that a potential such function w is indeed maximal invariant involves two steps:

1. Show that w is invariant: $w(\mathbf{g} \circ \mathbf{x}) = w(\mathbf{x})$ for all $\mathbf{x} \in \mathcal{X}, \mathbf{g} \in \mathcal{G}$;

2. Show that for each $\mathbf{x} \in \mathcal{X}$, there exists a $\mathbf{g}_\mathbf{x} \in \mathcal{G}$ such that $\mathbf{g}_\mathbf{x}\mathbf{x}$ is a function of just $w(\mathbf{x})$.

To illustrate, return to the t test example, where $X_1, \ldots, X_n$ are iid $N(\mu, \sigma^2)$, and we test $\mu = 0$ versus $\mu \neq 0$, so that the parameter spaces are as in (22.56). We could try to find a maximal invariant for $\mathbf{X}$, but instead we first reduce by sufficiency, to $(\overline{X}, S_*^2)$, the sample mean and variance (with $n-1$ in the denominator). The action of the group $\mathcal{G} = \{a \in \mathbb{R} \mid a \neq 0\}$ from (22.57) on the sufficient statistic is

$$a \circ (\overline{x}, s_*^2) = (a\overline{x}, a^2 s_*^2), \tag{22.65}$$

the same as the action on (μ, σ^2). There are a number of equivalent ways to express the maximal invariant (any one-to-one function of a maximal invariant is also maximal invariant). Here we try $w(\overline{x}, s_*^2) = \overline{x}^2/s_*^2$. The two steps:

1. $w(a \circ (\overline{x}, s_*^2)) = w(a\overline{x}, a^2 s_*^2) = (a\overline{x})^2/a^2 s_*^2 = \overline{x}^2/s_*^2 = w(\overline{x}, s_*^2);$ ✔

2. Let $a_{(\overline{x}, s_*^2)} = \text{Sign}(\overline{x})/s_*$. Then $a_{(\overline{x}, s_*^2)} \circ (\overline{x}, s_*^2) = (|\overline{x}|/s_*, 1) = (\sqrt{w(\overline{x}, s_*^2)}, 1)$, a function of just $w(\overline{x}, s_*^2)$. ✔

(If $\overline{x} = 0$, take the sign to be 1, not 0.) Thus $\overline{x}^2/s_*^2$ is a maximal invariant statistic, as is the absolute value of the t statistic in (22.60).

The invariance-reduced problem is based on the random variable $\mathbf{W} = w(\mathbf{X})$, and still tests the same hypotheses. But the parameter can also be simplified (usually), by finding the maximal invariant parameter Δ. It is defined the same as for the statistic, but with $\mathbf{X}$ replaced by $\boldsymbol{\theta}$. In the t test example, the maximal invariant parameter can be taken to be $\Delta = \mu^2/\sigma^2$. The distribution of the maximal invariant statistic depends on $\boldsymbol{\theta}$ only through the maximal invariant parameter. The two parameter spaces for the hypotheses can be expressed through the latter, hence we have

$$H_0 : \Delta \in \mathcal{D}_0 \text{ versus } H_A : \Delta \in \mathcal{D}_A, \text{ based on } \mathbf{W} \sim P_\Delta^*, \tag{22.66}$$

where P_Δ^* is the distribution of $\mathbf{W}$. For the t test, $\mu = 0$ if and only if $\Delta = 0$, hence the hypotheses are simply $H_0 : \Delta = 0$ versus $H_A : \Delta > 0$, with no nuisance parameters.

22.7   UMP invariant tests

Since an invariant test is a function of just the maximal invariant statistic, the risk function of an invariant test in the original testing problem is exactly matched by a test function in the invariance-reduced problem, and vice versa. The implication is that when decision-theoretically evaluating invariant tests, we need to look at only the invariance-reduced problem. Thus a test is uniformly most power invariant level α in the original problem if and only if its corresponding test in the reduced problem is UMP level α in the reduced problem. Similarly for admissibility. In this section we provide some examples where there are UMP invariant tests. Often, especially in multivariate analysis, there is no UMP test even after reducing by invariance.

22.7.1   Multivariate normal mean

Take $\mathbf{X} \sim N(\mu, \mathbf{I}_p)$, and test $H_0 \colon \mu = \mathbf{0}$ versus $H_A \colon \mu \in \mathbb{R}^p - \{\mathbf{0}\}$. This problem is invariant under the group $\mathcal{O}_p$ of $p \times p$ orthogonal matrices. Here the "b" part of the affine transformation is always $\mathbf{0}$, hence we omit it. The action is multiplication, and is the same on $\mathbf{X}$ as on μ: For $\boldsymbol{\Gamma} \in \mathcal{O}_p$,

$$\boldsymbol{\Gamma} \circ \mathbf{X} = \boldsymbol{\Gamma}\mathbf{X} \text{ and } \boldsymbol{\Gamma} \circ \mu = \boldsymbol{\Gamma}\mu. \tag{22.67}$$

The maximal invariant statistic is $w(\mathbf{x}) = \|\mathbf{x}\|^2$, since $\|\boldsymbol{\Gamma}\mathbf{x}\| = \|\mathbf{x}\|$ for orthogonal $\boldsymbol{\Gamma}$, and as in (7.72), if $\mathbf{x} \neq \mathbf{0}$, we can find an orthogonal matrix, $\boldsymbol{\Gamma}_\mathbf{x}$, whose first row is $\mathbf{x}'/\|\mathbf{x}\|$, so that $\boldsymbol{\Gamma}_\mathbf{x}\mathbf{x} = (\|\mathbf{x}\|, \mathbf{0}'_{p-1})'$, a function of just $w(\mathbf{x})$. Also, the maximal invariant parameter is $\Delta = \|\mu\|^2$. The hypotheses become $H_0 \colon \Delta = 0$ versus $H_A \colon \Delta > 0$ again.

From Definition 7.7 on page 113 we see that $W = \|\mathbf{X}\|^2$ has a noncentral chi-squared distribution, $W \sim \chi_p^2(\Delta)$. Exercise 21.8.13 shows that the noncentral chi-square has strict monotone likelihood ratio wrt W and Δ. Thus the UMP level α test for this reduced problem rejects the null when $W > \chi_{p,\alpha}^2$, which is then the UMP invariant level α test for the original problem.

Now for p_1 and p_2 positive integers, $p_1 + p_2 = p$, partition $\mathbf{X}$ and μ as

$$\mathbf{X} = \begin{pmatrix} \mathbf{X}_1 \\ \mathbf{X}_2 \end{pmatrix} \text{ and } \mu = \begin{pmatrix} \mu_1 \\ \mu_2 \end{pmatrix}, \tag{22.68}$$

where $\mathbf{X}_1$ and μ_1 are $p_1 \times 1$, and $\mathbf{X}_2$ and μ_2 are $p_2 \times 1$. Consider testing

$$H_0 \colon \mu_2 = \mathbf{0}, \mu_1 \in \mathbb{R}^{p_1} \text{ versus } H_A \colon \mu_2 \in \mathbb{R}^{p_2} - \{\mathbf{0}\}, \mu_1 \in \mathbb{R}^{p_1}. \tag{22.69}$$

That is, we test just $\mu_2 = \mathbf{0}$ versus $\mu_2 \neq \mathbf{0}$, and μ_1 is a nuisance parameter. The problem is not invariant under $\mathcal{O}_p$ as in (22.67), but we can multiply $\mathbf{X}_2$ by the smaller group $\mathcal{O}_{p_2}$. Also, adding a constant to an element of $\mathbf{X}_1$ adds the constant to an element of μ_1, which respects the two hypotheses. Thus we can take the group to be

$$\mathcal{G} = \left\{ \left(\begin{pmatrix} \mathbf{I}_{p_1} & \mathbf{0} \\ \mathbf{0} & \boldsymbol{\Gamma}_2 \end{pmatrix}, \begin{pmatrix} \mathbf{b}_1 \\ \mathbf{0} \end{pmatrix} \right) \middle| \boldsymbol{\Gamma}_2 \in \mathcal{O}_{p_2}, \mathbf{b}_1 \in \mathbb{R}^{p_1} \right\}. \tag{22.70}$$

Writing group elements more compactly as $\mathbf{g} = (\boldsymbol{\Gamma}_2, \mathbf{b}_1)$, the action is

$$(\mathbf{b}_1, \boldsymbol{\Gamma}_2) \circ \mathbf{X} = \begin{pmatrix} \mathbf{X}_1 + \mathbf{b}_1 \\ \boldsymbol{\Gamma}_2 \mathbf{X}_2 \end{pmatrix}. \tag{22.71}$$

Exercise 22.8.18 shows that the maximal invariant statistic and parameter are, respectively, $W_2 = \|\mathbf{X}_2\|^2$ and $\Delta_2 = \|\boldsymbol{\mu}_2\|^2$. Now $W_2 \sim \chi^2_{p_2}(\Delta_2)$, hence as above the UMP level α invariant test rejects the null when $W_2 > \chi^2_{p_2,\alpha}$.

22.7.2 Two-sided t test

Let $X_1, \ldots, X_n$ be iid $N(\mu, \sigma^2)$, and test $\mu = 0$ versus $\mu \neq 0$ with $\sigma^2 > 0$ as a nuisance parameter, i.e., the parameter spaces are as in (22.56). We saw in Section 22.6.2 that the maximal invariant statistic is $\overline{x}^2/s_*^2$ and the maximal invariant parameter is $\Delta = \mu^2/\sigma^2$. Exercise 7.8.23 defined the noncentral F statistic in (7.132) as $(U/\nu)/(V/\mu)$, where U and V are independent, $U \sim \chi^2_\nu(\Delta)$ (noncentral), and $V \sim \chi^2_\mu$ (central). Here, we know that $\overline{X}$ and S_*^2 are independent,

$$\overline{X} \sim N(\mu, \sigma^2/n) \;\Rightarrow\; \sqrt{n}\,\overline{X}/\sigma \sim N(\sqrt{n}\mu/\sigma, 1) \;\Rightarrow\; n\overline{X}^2/\sigma^2 \sim \chi^2_1(n\Delta), \qquad (22.72)$$

and $(n-1)S_*^2/\sigma^2 \sim \chi^2_{n-1}$. Thus

$$n\frac{\overline{X}^2}{S_*^2} = T^2 \sim F_{1,n-1}(n\Delta), \qquad (22.73)$$

where T is the usual Student's t from (22.60).

The noncentral F has monotone likelihood ratio (see Exercise 21.8.13), hence the UMP invariant level α test rejects the null when $T^2 > F_{1,n-1,\alpha}$. Equivalently, since $t_\nu^2 = F_{1,\nu}$, it rejects when $|T| > t_{n-1,\alpha/2}$, so the two-sided t test is the UMP invariant test.

22.7.3 Linear regression

Section 15.2.1 presents the F test in linear regression. Here, $\mathbf{Y} \sim N(\mathbf{x}\boldsymbol{\beta}, \sigma^2\mathbf{I}_n)$, and we partition $\boldsymbol{\beta}$ as in (22.68), so that $\boldsymbol{\beta}' = (\boldsymbol{\beta}_1', \boldsymbol{\beta}_2')$, where $\boldsymbol{\beta}_1$ is $p_1 \times 1$, $\boldsymbol{\beta}_2$ is $p_2 \times 1$, and $p = p_1 + p_2$. We test whether $\boldsymbol{\beta}_2 = 0$:

$$H_0 : \boldsymbol{\beta}_2 = 0, \; \boldsymbol{\beta}_1 \in \mathbb{R}^{p_1}, \; \sigma^2 > 0 \;\; versus \;\; H_A : \boldsymbol{\beta}_2 \in \mathbb{R}^{p_2} - \{0\}, \; \boldsymbol{\beta}_1 \in \mathbb{R}^{p_1}, \; \sigma^2 > 0. \tag{22.74}$$

We assume that $\mathbf{x}'\mathbf{x}$ is invertible.

The invariance is not obvious, so we take a few preliminary steps. First, we can assume that $\mathbf{x}_1$ and $\mathbf{x}_2$ are orthogonal, i.e., $\mathbf{x}_1'\mathbf{x}_2 = 0$. If not, we can rewrite the model as we did in (16.19). (We'll leave off the asterisks.) Next, we reduce the problem by sufficiency. By Exercise 13.8.19, $(\widehat{\boldsymbol{\beta}}, SS_e)$ is the sufficient statistic, where $\widehat{\boldsymbol{\beta}} = (\mathbf{x}'\mathbf{x})^{-1}\mathbf{x}'\mathbf{Y}$ and $SS_e = \|\mathbf{Y} - \mathbf{x}\widehat{\boldsymbol{\beta}}\|^2$. These two elements are independent, with $SS_e \sim \sigma^2\chi^2_{n-p}$ and

$$\begin{pmatrix} \widehat{\boldsymbol{\beta}}_1 \\ \widehat{\boldsymbol{\beta}}_2 \end{pmatrix} \sim N\left(\begin{pmatrix} \boldsymbol{\beta}_1 \\ \boldsymbol{\beta}_2 \end{pmatrix}, \sigma^2 \begin{pmatrix} (\mathbf{x}_1'\mathbf{x}_1)^{-1} & 0 \\ 0 & (\mathbf{x}_2'\mathbf{x}_2)^{-1} \end{pmatrix} \right). \tag{22.75}$$

See Theorem 12.1 on page 183, where the zeros in the covariance matrix are due to $\mathbf{x}_1'\mathbf{x}_2 = 0$. Now (22.74) is invariant under adding a constant vector to $\widehat{\boldsymbol{\beta}}_1$, and multiplying **Y** by a scalar. But there is also orthogonal invariance that is hidden. Let **B** be the symmetric square root of $\mathbf{x}'_2\mathbf{x}_2$, and set

$$\widehat{\boldsymbol{\beta}}^*_2 = \mathbf{B}\widehat{\boldsymbol{\beta}}_2 \sim N(\boldsymbol{\beta}^*_2, \sigma^2\mathbf{I}_{p_2}), \quad \boldsymbol{\beta}^*_2 = \mathbf{B}\boldsymbol{\beta}_2. \tag{22.76}$$

Now we have reexpressed the data, but have not lost anything since $(\widehat{\boldsymbol{\beta}}_1, \widehat{\boldsymbol{\beta}}_2, SS_e)$ is in one-to-one correspondence with $(\widehat{\boldsymbol{\beta}}_1, \widehat{\boldsymbol{\beta}}^*_2, SS_e)$. The hypotheses are the same as in (22.74) with $\boldsymbol{\beta}^*_2$ in place of $\boldsymbol{\beta}_2$'s. Because the covariance of $\widehat{\boldsymbol{\beta}}^*_2$ is $\sigma^2\mathbf{I}_{p_2}$, we can multiply it by an orthogonal matrix without changing the covariance. Thus the invariance group is similar to that in (22.70), but includes the multiplier a:

$$\mathcal{G} = \left\{ \left(a\begin{pmatrix} \mathbf{I}_{p_1} & \mathbf{0} \\ \mathbf{0} & \boldsymbol{\Gamma}_2 \end{pmatrix}, \begin{pmatrix} \mathbf{b}_1 \\ \mathbf{0} \end{pmatrix} \right) \middle| \; \boldsymbol{\Gamma}_2 \in \mathcal{O}_{p_2}, a \in (0, \infty), \mathbf{b}_1 \in \mathbb{R}^{p_1} \right\}. \tag{22.77}$$

For $(a, \boldsymbol{\Gamma}_2, \mathbf{b}_1) \in \mathcal{G}$, the action is

$$(a, \boldsymbol{\Gamma}_2, \mathbf{b}_1) \circ (\widehat{\boldsymbol{\beta}}_1, \widehat{\boldsymbol{\beta}}^*_2, SS_e) = (a\widehat{\boldsymbol{\beta}}_1 + \mathbf{b}_1, a\boldsymbol{\Gamma}_2\widehat{\boldsymbol{\beta}}^*_2, a^2 SS_e). \tag{22.78}$$

To find the maximal invariant statistic, we basically combine the ideas in Sections 22.7.1 and 22.7.2: Take $a = 1/\sqrt{SS_e}$, $\mathbf{b}_1 = -\widehat{\boldsymbol{\beta}}_1/a$, and $\boldsymbol{\Gamma}_2$ so that $\boldsymbol{\Gamma}_2\widehat{\boldsymbol{\beta}}^*_2 = (\|\widehat{\boldsymbol{\beta}}^*_2\|, \mathbf{0}'_{p-1})'$. Exercise 22.8.19 shows that the maximal invariant statistic is $W = \|\widehat{\boldsymbol{\beta}}^*_2\|^2/SS_e$, or, equivalently, the $F = (n - p)W/p_2$ in (15.24).

The maximal invariant parameter is $\Delta = \|\boldsymbol{\beta}^*_2\|^2/\sigma^2$, and similar to (22.73), $F \sim F_{p_2, n-p}(\Delta)$. Since we are testing $\Delta = 0$ versus $\Delta > 0$, monotone likelihood ratio again proves that the F test is the UMP invariant level α test.

22.8   Exercises

Exercise 22.8.1. Here the null hypothesis is $H_0 : \theta = \theta_0$. Assume that $E_\theta[\phi]$ is continuous in θ for every test ϕ. (a) Take the one-sided alternative $H_A : \theta > \theta_0$. Show that if there is a UMP level α test, then it is admissible. (b) For the one-sided alternative, show that if there is a unique (up to equivalence) LMP level α test, it is admissible. (By "unique up to equivalence" is meant that if ϕ and ψ are both LMP level α, then their risks are equal for all θ.) (c) Now take the two-sided alternative $H_A : \theta \neq \theta_0$. Show that if there is a UMP unbiased level α test, then it is admissible.

Exercise 22.8.2. Suppose we test $H_0 : \theta = 0$ versus $H_A : \theta \neq 0$ based on X, which is a constant x_0 no matter what the value of θ. Show that any test $\phi(x_0)$ (with $0 \leq \phi(x_0) \leq 1$) is admissible.

Exercise 22.8.3. Let $\mathcal{D}_\alpha$ be the set of all level α tests, and $\mathcal{D}$ be the set of all tests. (a) Argue that if ϕ is admissible among all tests, and $\phi \in \mathcal{D}_\alpha$, then ϕ is admissible among tests in $\mathcal{D}_\alpha$. (b) Suppose $\phi \in \mathcal{D}_\alpha$ and is admissible among tests in $\mathcal{D}_\alpha$. Show that it is admissible among all tests $\mathcal{D}$. [Hint: Note that if a test dominates ϕ, it must also be level α.]

Exercise 22.8.4. This exercise is to show the two-sided t test is Bayes and admissible. We have $X_1, \ldots, X_n$ iid $N(\mu, \sigma^2)$, $n \geq 2$, and test $H_0 : \mu = 0, \sigma^2 > 0$ versus $H_A : \mu \neq 0, \sigma^2 > 0$. The prior has been cleverly chosen by Kiefer and Schwartz (1965). We parametrize using $\tau \in \mathbb{R}$, setting $\sigma^2 = 1/(1+\tau^2)$. Under the null, $\mu = 0$, but under the alternative set $\mu = \tau/(1+\tau^2)$. Define the pdfs $\rho_0(\tau)$ and $\rho_A(\tau)$ by

$$\rho_0(\tau) = c_0 \frac{1}{(1+\tau^2)^{n/2}} \quad \text{and} \quad \rho_A(\tau) = c_A \frac{1}{(1+\tau^2)^{n/2}} e^{\frac{1}{2}\frac{n\tau^2}{1+\tau^2}}, \qquad (22.79)$$

where c_0 and c_A are constants so that the pdfs integrate to 1. (a) Show that if τ has the null pdf ρ_0, that $\sqrt{n-1}\,\tau \sim t_{n-1}$, Student's t. Thus $c_0 = \Gamma(n/2)/(\Gamma((n-1)/2)\sqrt{\pi})$. (b) Show that $\int_{-\infty}^{\infty}(\rho_A(\tau)/c_A)d\tau < \infty$, so that ρ_A is a legitimate pdf. [Extra credit: Find c_A explicitly.] [Hint: Make the transformation $u = 1/(1+\tau^2)$. For ρ_A, expand the exponent and find c_A in terms of a confluent hypergeometric function, $_1F_1$, from Exercise 7.8.21.] (c) Let $f_0(\mathbf{x} \mid \tau)$ be the pdf of $\mathbf{X}$ under the null written in terms of τ. Show that

$$\int_{-\infty}^{\infty} f_0(\mathbf{x} \mid \tau)\rho_0(\tau)d\tau = c_0^* e^{-\frac{1}{2}\sum x_i^2} \int_{-\infty}^{\infty} e^{-\frac{1}{2}\tau^2 \sum x_i^2} d\tau = c_0^{**} e^{-\frac{1}{2}\sum x_i^2} \frac{1}{\sqrt{\sum x_i^2}} \qquad (22.80)$$

for some constants c_0^* and c_0^{**}. [Hint: The integrand in the second expression looks like a $N(0, 1/\sum x_i^2)$ pdf for τ, without the constant.] (d) Now let $f_A(\mathbf{x} \mid \tau)$ be the pdf of $\mathbf{X}$ under the alternative, and show that

$$\int_{-\infty}^{\infty} f_A(\mathbf{x} \mid \tau)\rho_A(\tau)d\tau = c_A^* e^{-\frac{1}{2}\sum x_i^2} \int_{-\infty}^{\infty} e^{-\frac{1}{2}\tau^2 \sum x_i^2 + \tau \sum x_i} d\tau$$

$$= c_A^{**} e^{-\frac{1}{2}\sum x_i^2} e^{\frac{1}{2}(\sum x_i)^2/\sum x_i^2} \frac{1}{\sqrt{\sum x_i^2}} \qquad (22.81)$$

for some c_A^* and c_A^{**}. [Hint: Complete the square in the exponent with respect to τ, then note that the integral looks like a $N(\sum x_i/\sum x_i^2, 1/\sum x_i^2)$ pdf.] (e) Show that the Bayes factor $B_{A0}(\mathbf{x})$ is a strictly increasing function of $(\sum x_i)^2/(\sum x_i^2)$, which is a strictly increasing function of T^2, where $T = \sqrt{n}\overline{X}/S_*$, $S_*^2 = \sum(X_i - \overline{X})^2/(n-1)$. Thus the Bayes test is the two-sided t test. (f) Show that this test is admissible.

Exercise 22.8.5. Suppose $\mathbf{X}$ has density $f(\mathbf{x} \mid \theta)$, and consider testing $H_0: \theta = 0$ versus $H_A: \theta > 0$. Assume that the density is differentiable at $\theta = 0$. Define prior π_n to have

$$\pi_n[\Theta = 0] = \frac{n+c}{2n+c}, \quad \pi_n[\Theta = \frac{1}{n}] = \frac{n}{2n+c} \qquad (22.82)$$

for constant c. Take n large enough that both probabilities are in $(0,1)$. (a) Show that a Bayes test (22.33) here has the form

$$\phi_{\pi_n}(\mathbf{x}) = \begin{cases} 1 & \text{if} & \frac{f(\mathbf{x} \mid 1/n)}{f(\mathbf{x} \mid 0)} > 1 + \frac{c}{n} \\ \gamma_n(\mathbf{x}) & \text{if} & \frac{f(\mathbf{x} \mid 1/n)}{f(\mathbf{x} \mid 0)} = 1 + \frac{c}{n} \\ 0 & \text{if} & \frac{f(\mathbf{x} \mid 1/n)}{f(\mathbf{x} \mid 0)} < 1 + \frac{c}{n} \end{cases}. \qquad (22.83)$$

(b) Let ϕ be the limit of the Bayes test, $\phi_{\pi_n} \to^w \phi$. (By (22.21) we know there is such a limit, at least on a subsequence. Assume we are on that subsequence.) Apply Lemma

22.4 with $g_n(\mathbf{x}) = f(\mathbf{x}\,|\,1/n)/f(\mathbf{x}\,|\,0) - 1 - c/n$. What can you say about ϕ? (c) Now rewrite the equations in (22.83) so that

$$g_n(\mathbf{x}) = n\left(\frac{f(\mathbf{x}\,|\,1/n)}{f(\mathbf{x}\,|\,0)} - 1\right) - c. \tag{22.84}$$

Show that as $n \to \infty$,

$$g_n(\mathbf{x}) \longrightarrow l'(0\,;\mathbf{x}) - c = \left.\frac{\partial}{\partial\theta}\log(f_\theta(\mathbf{x}))\right|_{\theta=0} - c. \tag{22.85}$$

What can you say about ϕ now?

Exercise 22.8.6. Let $\mathbf{X}$ have a p-dimensional exponential family distribution, where $\mathbf{X}$ itself is the natural sufficient statistic, and $\boldsymbol{\theta}$ is the natural parameter. We test $H_0: \boldsymbol{\theta} = \mathbf{0}$ versus $H_A: \boldsymbol{\theta} \in \mathcal{T}_A$. (a) For fixed $\mathbf{a} \in \mathcal{T}_A$ and $c \in \mathbb{R}$, show that the test

$$\phi(\mathbf{x}) = \begin{cases} 1 & \text{if} \quad \mathbf{x}\cdot\mathbf{a} > c \\ \gamma(\mathbf{x}) & \text{if} \quad \mathbf{x}\cdot\mathbf{a} = c \\ 0 & \text{if} \quad \mathbf{x}\cdot\mathbf{a} < c \end{cases} \tag{22.86}$$

is Bayes wrt some prior π. [Hint: Take $\pi[\boldsymbol{\theta} = \mathbf{0}] = \pi_0$ and $\pi[\boldsymbol{\theta} = \mathbf{a}] = 1 - \pi_0$.] (b) Show that the test ϕ is Bayes for any $\mathbf{a}$ such that $\mathbf{a} = k\mathbf{b}$ for some $\mathbf{b} \in \mathcal{T}_A$ and $k > 0$.

Exercise 22.8.7. Suppose $\mathbf{X} \sim N(\boldsymbol{\mu}, \mathbf{I}_2)$ and we test (22.30), i.e., $H_0: \boldsymbol{\mu} = \mathbf{0}$ versus $H_A: 1 \le \|\boldsymbol{\mu}\| \le 2$. Let a and b be constants so that $1 \le \|(a,b)\| \le 2$. Define the prior π by

$$\pi[\boldsymbol{\mu} = \mathbf{0}] = \pi_0, \; \pi[\boldsymbol{\mu} = (a,b)] = \pi[\boldsymbol{\mu} = -(a,b)] = \tfrac{1}{2}(1 - \pi_0). \tag{22.87}$$

(a) Show that the Bayes test wrt π can be written

$$\phi_\pi(\mathbf{x}) = \begin{cases} 1 & \text{if} \quad g(\mathbf{x}) > d \\ \gamma(\mathbf{x}) & \text{if} \quad g(\mathbf{x}) = d \\ 0 & \text{if} \quad g(\mathbf{x}) < d \end{cases}, \quad g(\mathbf{x}) = e^{ax_1+bx_2} + e^{-(ax_1+bx_2)}, \tag{22.88}$$

and d is a constant. (b) Why is ϕ_π admissible? (c) Show that there exists a c such that ϕ_π rejects the null when $|ax_1 + bx_2| > c$. [Hint: Letting $u = ax_1 + bx_2$, show that $\exp(u) + \exp(-u)$ is strictly convex and symmetric in u, hence the set where $g(\mathbf{x}) > d$ has the form $|u| > c$.] (d) Now suppose a and b are any constants, not both zero, and c is any positive constant. Show that the test that rejects the null when $|ax_1 + bx_2| > c$ is still Bayes and admissible.

Exercise 22.8.8. Continue with the testing problem in Exercise 22.8.7. For constant $\rho \in [1,2]$, let π be the prior $\pi[\boldsymbol{\mu} = \mathbf{0}] = \pi_0$ (so $\pi[\boldsymbol{\mu} \ne \mathbf{0}] = 1 - \pi_0$), and conditional on H_A being true, has a uniform distribution on the circle $\{\boldsymbol{\mu}\,|\,\|\boldsymbol{\mu}\| = \rho\}$. (a) Similar to what we saw in (1.23), we can represent the prior under the alternative by setting $\boldsymbol{\mu} = \rho(\cos(\Theta), \sin(\Theta))$, where $\Theta \sim \text{Uniform}[0, 2\pi)$. Show that the Bayes factor can then be written

$$B_{A0}(\mathbf{x}) = e^{-\frac{1}{2}\rho^2}\frac{1}{2\pi}\int_0^{2\pi} e^{\rho(x_1\cos(\theta)+x_2\sin(\theta))}\,d\theta. \tag{22.89}$$

(b) Show that with $r = \|x\|$,

$$\frac{1}{2\pi}\int_0^{2\pi} e^{\rho(x_1\cos(\theta)+x_2\sin(\theta))}\,d\theta = \frac{1}{2\pi}\int_0^{2\pi} e^{\rho r\cos(\theta)}\,d\theta$$

$$= \sum_{k=0}^{\infty} c_k \frac{r^{2k}\rho^{2k}}{(2k)!}, \quad \text{where } c_k = \frac{1}{2\pi}\int_0^{2\pi}\cos(\theta)^{2k}\,d\theta.$$

$$(22.90)$$

[Hint: For the first equality, let θ_x be the angle such that $x = r(\cos(\theta_x), \sin(\theta_x))$. Then use the double angle formulas to show that we can replace the exponent with $\rho r\cos(\theta - \theta_x)$. The second expression then follows by changing variables θ to $\theta + \theta_x$. The third expression arises by expanding the e in a Taylor series, and noting that for odd powers l, the the integral of $\cos(\theta)^l$ is zero.] (c) Noting that $c_k > 0$ in (22.90), show that the Bayes factor is strictly increasing in r, and that the Bayes test wrt π has rejection region $r > c$ for some constant c. (d) Is this Bayes test admissible?

Exercise 22.8.9. Let ϕ_n, $n = 1, 2, \ldots$, be a sequence of test functions such that for each n, there exists a constant c_n and function $t(x)$ such that

$$\phi_n(x) = \begin{cases} 1 & \text{if } t(x) > c_n \\ \gamma_n(x) & \text{if } t(x) = c_n \\ 0 & \text{if } t(x) < c_n \end{cases}.$$

$$(22.91)$$

Suppose $\phi_n \to^w \phi$ for test ϕ. We want to show that ϕ has same form. There exists a subsequence of the c_n's and constant $c \in [-\infty, \infty]$ such that $c_n \to c$ on that subsequence. Assume we are on that subsequence. (a) Argue that if $c_n \to \infty$ then $\phi_n(x) \to 0$ pointwise, hence $\phi_n \to^w 0$. What is the weak limit if $c_n \to -\infty$? (b) Now suppose $c_n \to c \in (-\infty, \infty)$. Use Lemma 22.4 to show that (with probability one) $\phi(x) = 1$ if $t(x) > c$ and $\phi(x) = 0$ if $t(x) < c$.

Exercise 22.8.10. Let ϕ_n, $n = 1, 2, \ldots$, be a sequence of test functions such that for each n, there exist constants a_n and b_n, $a_n \le b_n$, such that

$$\phi_n(x) = \begin{cases} 1 & \text{if } t(x) < a_n \text{ or } t(x) > b_n \\ \gamma_n(x) & \text{if } t(x) = a_n \text{ or } t(x) = b_n \\ 0 & \text{if } a_n < t(x) < b_n \end{cases}.$$

$$(22.92)$$

Suppose ϕ is a test with $\phi_n \to^w \phi$. We want to show that ϕ has the form (22.92) or (22.39) for some constants a and b. (a) Let ϕ_{1n} be as in (22.91) with $c_n = b_n$, and ϕ_{2n} similar but with $c_n = -a_n$ and $t(x) = -t(x)$, so that $\phi_n = \phi_{1n} + \phi_{2n}$. Show that if $\phi_{1n} \to^w \phi_1$ and $\phi_{2n} \to^w \phi_2$, then $\phi_n \to^w \phi_1 + \phi_2$. (b) Find the forms of the tests ϕ_1 and ϕ_2 in part (a), and show that $\phi = \phi_1 + \phi_2$ (with probability one) and

$$\phi(x) = \begin{cases} 1 & \text{if } t(x) < a \text{ or } t(x) > b \\ \gamma(x) & \text{if } t(x) = a \text{ or } t(x) = b \\ 0 & \text{if } a < t(x) < b \end{cases}$$

$$(22.93)$$

for some a and b (one or both possibly infinite).

Exercise 22.8.11. This exercise verifies (22.27) and (22.28). Let $\{\theta_1, \theta_2, \ldots\}$ be a countable dense subset of $\mathcal{T}$, and $g(\theta)$ be a continuous function. (a) Show that if $g(\theta_i) = 0$ for all $i = 1, 2, \ldots$, then $g(\theta) = 0$ for all $\theta \in \mathcal{T}$. (b) Show that if $g(\theta_i) \le 0$ for all $i = 1, 2, \ldots$, then $g(\theta) \le 0$ for all $\theta \in \mathcal{T}$.

Exercise 22.8.12. This exercise is to prove Lemma 22.8. Here, $\mathcal{C} \subset \mathbb{R}^p$ is a closed set, and the goal is to show that $\mathcal{C}$ is convex if and only if it can be written as $\mathcal{C} = \cap_{\mathbf{a},c} \mathcal{H}_{\mathbf{a},c}$ for some set of vectors $\mathbf{a}$ and constants c, where $\mathcal{H}_{\mathbf{a},c} = \{\mathbf{x} \in \mathbb{R}^p \,|\, \mathbf{a} \cdot \mathbf{x} \le c\}$ as in (22.46). (a) Show that the intersection of convex sets is convex. Thus $\cap_{\mathbf{a},c} \mathcal{H}_{\mathbf{a},c}$ is convex, since each halfspace is convex, which proves the "if" part of the lemma. (b) Now suppose $\mathcal{C}$ is convex. Let $\mathcal{C}^*$ be the intersection of all halfspaces that contain $\mathcal{C}$:

$$\mathcal{C}^* = \cap_{\{\mathbf{a},c \,|\, \mathcal{C} \subset \mathcal{H}_{\mathbf{a},c}\}} \mathcal{H}_{\mathbf{a},c}. \tag{22.94}$$

(i) Show that $\mathbf{z} \in \mathcal{C}$ implies that $\mathbf{z} \in \mathcal{C}^*$. (ii) Suppose $\mathbf{z} \notin \mathcal{C}$. Then by (20.104) in Exercise 20.9.28, there exists a non-zero vector $\boldsymbol{\gamma}$ such that $\boldsymbol{\gamma} \cdot \mathbf{x} < \boldsymbol{\gamma} \cdot \mathbf{z}$ for all $\mathbf{x} \in \mathcal{C}$. Show that there exists a c such that $\boldsymbol{\gamma} \cdot \mathbf{x} \le c < \boldsymbol{\gamma} \cdot \mathbf{z}$ for all $\mathbf{x} \in \mathcal{C}$, hence $\mathbf{z} \notin \mathcal{H}_{\boldsymbol{\gamma},c}$ but $\mathcal{C} \subset \mathcal{H}_{\boldsymbol{\gamma},c}$. Argue that consequently $\mathbf{z} \notin \mathcal{C}^*$. (iii) Does $\mathcal{C}^* = \mathcal{C}$?

Exercise 22.8.13. Suppose $\mathbf{X} \sim N(\boldsymbol{\mu}, \mathbf{I}_p)$, and we test $H_0 : \boldsymbol{\mu} = \mathbf{0}$ versus the multivariate one-sided alternative $H_A : \boldsymbol{\mu} \in \{\boldsymbol{\mu} \in \mathbb{R}^p \,|\, \mu_i \ge 0 \text{ for each } i\} - \{\mathbf{0}\}$. (a) Show that under the alternative, the MLE of $\boldsymbol{\mu}$ is given by $\widehat{\mu}_{Ai} = \max\{0, x_i\}$ for each i, and that the likelihood ratio statistic is $-2\log(LR) = \sum \max\{0, x_i\}^2$. (b) For $p = 2$, sketch the acceptance region of the likelihood ratio test, $\{\mathbf{x} \,|\, -2\log(LR) \le c\}$, for fixed $c > 0$. Is the acceptance region convex? Is it nonincreasing? Is the test admissible? [Extra credit: Find the c so that the level of the test in part (b) is 0.05.]

Exercise 22.8.14. Continue with the multivariate one-sided normal testing problem in Exercise 22.8.13. Exercise 15.7.9 presented several methods for combining independent p-values. This exercise determines their admissibility in the normal situation. Here, the ith p-value is $U_i = 1 - \Phi(X_i)$, where Φ is the $N(0,1)$ distribution function. (a) Fisher's test rejects when $T_P(\mathbf{U}) = -2\sum \log(U_i) > c$. Show that T_P as a function of $\mathbf{x}$ is convex and increasing in each component, hence the test is admissible. [Hint: It is enough to show that $-\log(1 - \Phi(x_i))$ is convex and increasing in x_i. Show that

$$-\frac{d}{dx_i} \log(1 - \Phi(x_i)) = \left(\int_{x_i}^{\infty} e^{-\frac{1}{2}(y^2 - x_i^2)} dy \right)^{-1} = \left(\int_{0}^{\infty} e^{-\frac{1}{2}u(u + 2x_i)} du \right)^{-1}, \tag{22.95}$$

where $u = y - x_i$. Argue that that final expression is positive and increasing in x_i.] (b) Tippett's test rejects the null when $\min\{U_i\} < c$. Sketch the acceptance region in the $\mathbf{x}$-space for this test when $p = 2$. Argue that the test is admissible. (c) The maximum test rejects when $\max\{U_i\} < c$. Sketch the acceptance region in the $\mathbf{x}$-space for this test when $p = 2$. Argue that the test is inadmissible. (d) The Edgington test rejects the null when $\sum U_i < c$. Take $p = 2$ and $0 < c < 0.5$. Sketch the acceptance region in the $\mathbf{x}$-space. Show that the boundary of the acceptance region is asymptotic to the lines $x_1 = \Phi^{-1}(1 - c)$ and $x_2 = \Phi^{-1}(1 - c)$. Is the acceptance region convex in $\mathbf{x}$? Is the test admissible? (e) The Liptak-Stouffer test rejects the null when $-\sum \Phi^{-1}(U_i) > c$. Show that in this case, the test is equivalent to rejecting when $\sum X_i > c$. Is it admissible?

Exercise 22.8.15. This exercise finds the UMP invariant test for the two-sample normal mean test. That is, suppose $X_1, \ldots, X_n$ are iid $N(\mu_x, \sigma^2)$ and $Y_1, \ldots, Y_m$ are iid $N(\mu_y, \sigma^2)$, and the X_i's are independent of the Y_i's. Note that the variances are equal. We test

$$H_0 : \mu_x = \mu_y, \ \sigma^2 > 0 \ \textit{versus} \ H_A : \mu_x \ne \mu_y, \ \sigma^2 > 0. \tag{22.96}$$

Consider the affine invariance group $\mathcal{G} = \{(a,b) \mid a \in \mathbb{R} - \{0\}, b \in \mathbb{R}\}$ with action

$$(a,b) \circ (X_1, \ldots, X_n, Y_1, \ldots, Y_m) = a(X_1, \ldots, X_n, Y_1, \ldots, Y_m) + (b, b, \ldots, b). \quad (22.97)$$

(a) Show that the testing problem is invariant under $\mathcal{G}$. What is the action of $\mathcal{G}$ on the parameter (μ_x, μ_y, σ^2)? (b) Show that the sufficient statistic is $(\overline{X}, \overline{Y}, U)$, where $U = \sum(X_i - \overline{X})^2 + \sum(Y_i - \overline{Y})^2$. (c) Now reduce the problem by sufficiency to the statistics in part (b). What is the action of the group on the sufficient statistic? (d) Show that the maximal invariant statistic can be taken to be $|\overline{X} - \overline{Y}|/\sqrt{U}$, or, equivalently, the square of the two-sample t statistic:

$$T^2 = \frac{(\overline{X} - \overline{Y})^2}{\left(\frac{1}{n} + \frac{1}{m}\right) S_P^2}, \quad \text{where } S_P^2 = \frac{U}{n + m - 2}. \quad (22.98)$$

(e) Show that $T^2 \sim F_{1, n+m-2}(\Delta)$, the noncentral F. What is the noncentrality parameter Δ? Is it the maximal invariant parameter? (f) Is the test that rejects the null when $T^2 > F_{1, n+m-2, \alpha}$ the UMP invariant level α test? Why or why not?

Exercise 22.8.16. This exercise follows on Exercise 22.8.15, but does not assume equal variances. Thus $X_1, \ldots, X_n$ are iid $N(\mu_x, \sigma_x^2)$ and $Y_1, \ldots, Y_m$ are iid $N(\mu_y, \sigma_y^2)$, the X_i's are independent of the Y_i's, and we test

$$H_0: \mu_x = \mu_y, \ \sigma_x^2 > 0, \ \sigma_y^2 > 0 \ \text{versus} \ H_A: \mu_x \neq \mu_y, \ \sigma_x^2 > 0, \ \sigma_y^2 > 0. \quad (22.99)$$

Use the same affine invariance group $\mathcal{G} = \{(a,b) \mid a \in \mathbb{R} - \{0\}, b \in \mathbb{R}\}$ and action (22.97). (a) Show that the testing problem is invariant under $\mathcal{G}$. What is the action of $\mathcal{G}$ on the parameter $(\mu_x, \mu_y, \sigma_x^2, \sigma_y^2)$? (b) Show that the sufficient statistic is $(\overline{X}, \overline{Y}, S_x^2, S_y^2)$, where $S_x^2 = \sum(X_i - \overline{X})^2/(n-1)$ and $S_y^2 = \sum(Y_i - \overline{Y})^2/(m-1)$. (c) What is the action of the group on the sufficient statistic? (d) Find a two-dimensional maximal invariant statistic and maximal invariant parameter. Does it seem reasonable that there is no UMP invariant test?

Exercise 22.8.17. Consider the testing problem at the beginning of Section 22.7.1, but with unknown variance. Thus $X \sim N(\mu, \sigma^2 I_p)$, and we test

$$H_0: \mu = 0, \ \sigma^2 > 0 \ \text{versus} \ H_A: \mu \in \mathbb{R}^p - \{0\}, \ \sigma^2 > 0. \quad (22.100)$$

Take the invariance group to be $\mathcal{G} = \{a\Gamma \mid a \in (0, \infty), \Gamma \in \mathcal{O}_p\}$. The action is $a\Gamma \circ X = a\Gamma X$. (a) Show that the testing problem is invariant under the group. (b) Show that the maximal invariant statistic can be taken to be the constant "1" (or any constant). [Hint: Take Γ_x so that $\Gamma_x x = (\|x\|, 0, \ldots, 0)'$ to start, then choose an a.] (c) Show that the UMP level α test is just the constant α. Is that a very useful test? (d) Show that the usual level α two-sided t test (which is not invariant) has better power than the UMP invariant level α test.

Exercise 22.8.18. Here $X \sim N(\mu, I_p)$, where we partition $X = (X_1', X_2')'$ and $\mu = (\mu_1', \mu_2')'$ with X_1 and μ_1 being $p_1 \times 1$ and X_2 and μ_2 being $p_2 \times 1$, as in (22.68). We test $\mu_2 = 0$ versus $\mu_2 \neq 0$ as in (22.69). Use the invariance group $\mathcal{G}$ in (22.70), so that $(b_1, \Gamma_2) \circ X = ((X_1 + b_1)', (\Gamma_2 X_2)')'$. (a) Find the action on the parameter space, $(b_1, \Gamma_2) \circ \mu$. (b) Let $b_{1x} = -x_1$. Find Γ_{2x} so that $(b_{1x}, \Gamma_{2x}) \circ x = (0_{p_1}', \|x_2\|, 0_{p_2-1}')'$. (c) Show that $\|X_2\|^2$ is the maximal invariant statistic and $\|\mu_2\|^2$ is the maximal invariant parameter.

Exercise 22.8.19. This exercise uses the linear regression testing problem in Section 22.7.3. The action is $(a, \Gamma_2, \mathbf{b}_1) \circ (a\widehat{\boldsymbol{\beta}}_1 + \mathbf{b}_1, a\Gamma_2\widehat{\boldsymbol{\beta}}_2^*, a^2 SS_e)$ as in (22.78). (a) Using $a = 1/\sqrt{SS_e}$, $\mathbf{b}_1 = -\widehat{\boldsymbol{\beta}}_1/a$, and Γ_2 so that $\Gamma_2\widehat{\boldsymbol{\beta}}_2^* = (\|\widehat{\boldsymbol{\beta}}_2^*\|, \mathbf{0}_{p-1}')'$, show that the maximal invariant statistic is $\|\widehat{\boldsymbol{\beta}}_2^*\|^2/SS_e$. (b) Show that that maximal invariant statistic is a one-to-one function of the F statistic in (15.24). [Hint: From (22.76), $\widehat{\boldsymbol{\beta}}_2^* = \mathbf{B}\widehat{\boldsymbol{\beta}}_2$, where $\mathbf{BB} = (\mathbf{x}_2'\mathbf{x}_2)^{-1}$, and note that in the current model, $\mathbf{C}_{22} = (\mathbf{x}_2'\mathbf{x}_2)^{-1}$.]

Bibliography

Agresti, A. (2013). *Categorical Data Analysis*. Wiley, third edition.

Akaike, H. (1974). A new look at the statistical model identification. *IEEE Transactions on Automatic Control*, 19:716–723.

Anscombe, F. J. (1948). The transformation of Poisson, binomial, and negative-binomial data. *Biometrika*, 35(3/4):246–254.

Appleton, D. R., French, J. M., and Vanderpump, M. P. J. (1996). Ignoring a covariate: An example of Simpson's paradox. *The American Statistician*, 50(4):340–341.

Arnold, J. (1981). Statistics of natural populations. I: Estimating an allele probability in cryptic fathers with a fixed number of offspring. *Biometrics*, 37(3):495–504.

Bahadur, R. R. (1964). On Fisher's bound for asymptotic variances. *The Annals of Mathematical Statistics*, 35(4):1545–1552.

Bassett, G. and Koenker, R. W. (1978). Asymptotic theory of least absolute error regression. *Journal of the American Statistical Association*, 73(363):618–622.

Berger, J. O. (1993). *Statistical Decision Theory and Bayesian Analysis*. Springer, New York, second edition.

Berger, J. O. and Bayarri, M. J. (2012). Lectures on model uncertainty and multiplicity. *CBMS Regional Conference in the Mathematical Sciences*. https://cbms-mum.soe.ucsc.edu/Material.html.

Berger, J. O., Ghosh, J. K., and Mukhopadhyay, N. (2003). Approximations and consistency of Bayes factors as model dimension grows. *Journal of Statistical Planning and Inference*, 112(1-2):241–258.

Berger, J. O. and Sellke, T. (1987). Testing a point null hypothesis: The irreconcilability of P values and evidence. *Journal of the American Statistical Association*, 82:112–122. With discussion.

Berger, J. O. and Wolpert, R. L. (1988). *The Likelihood Principle*. Institute of Mathematical Statistics, Hayward, CA, second edition.

Bickel, P. J. and Doksum, K. A. (2007). *Mathematical Statistics: Basic Ideas and Selected Topics, Volume I*. Pearson, second edition.

Billingsley, P. (1995). *Probability and Measure*. Wiley, New York, third edition.

Billingsley, P. (1999). *Convergence of Probability Measures*. Wiley, New York, second edition.

Birnbaum, A. (1955). Characterizations of complete classes of tests of some multi-parametric hypotheses, with applications to likelihood ratio tests. *The Annals of Mathematical Statistics*, 26(1):21–36.

Blyth, C. R. (1951). On minimax statistical decision procedures and their admissibility. *The Annals of Mathematical Statistics*, 22(1):22–42.

Box, G. E. P. and Muller, M. E. (1958). A note on the generation of random normal deviates. *Annals of Mathematical Statistics*, 29(2):610–611.

Box, J. F. (1978). *R. A. Fisher: The Life of a Scientist*. Wiley, New York.

Bradley, R. A. and Terry, M. E. (1952). Rank analysis of incomplete block designs: I. The method of paired comparisons. *Biometrika*, 39(3/4):324–345.

Brown, L. D. (1971). Admissible estimators, recurrent diffusions, and insoluble boundary value problems. *The Annals of Mathematical Statistics*, 42(3):855–903.

Brown, L. D. (1981). A complete class theorem for statistical problems with finite sample spaces. *The Annals of Statistics*, 9(6):1289–1300.

Brown, L. D., Cai, T. T., and DasGupta, A. (2001). Interval estimation for a binomial proportion. *Statistical Science*, 16(2):101–133. With discussion.

Brown, L. D. and Hwang, J. T. (1982). A unified admissibility proof. In Gupta, S. S. and Berger, J. O., editors, *Statistical Decision Theory and Related Topics III*, volume 1, pages 205–230. Academic Press, New York.

Brown, L. D. and Marden, J. I. (1989). Complete class results for hypothesis testing problems with simple null hypotheses. *The Annals of Statistics*, 17:209–235.

Burnham, K. P. and Anderson, D. R. (2003). *Model Selection and Multimodel Inference: A Practical Information-Theoretic Approach*. Springer-Verlag, New York, second edition.

Casella, G. and Berger, R. (2002). *Statistical Inference*. Thomson Learning, second edition.

Cramér, H. (1999). *Mathematical Methods of Statistics*. Princeton University Press. Originally published in 1946.

Diaconis, P. and Ylvisaker, D. (1979). Conjugate priors for exponential families. *The Annals of Statistics*, 7(2):269–281.

Duncan, O. D. and Brody, C. (1982). Analyzing n rankings of three items. In Hauser, R. M., Mechanic, D., Haller, A. O., and Hauser, T. S., editors, *Social Structure and Behavior*, pages 269–310. Academic Press, New York.

Durrett, R. (2010). *Probability: Theory and Examples*. Cambridge University Press, fourth edition.

Eaton, M. L. (1970). A complete class theorem for multidimensional one-sided alternatives. *The Annals of Mathematical Statistics*, 41(6):1884–1888.

Efron, B., Hastie, T., Johnstone, I., and Tibshirani, R. (2004). Least angle regression. *The Annals of Statistics*, 32(2):407–499.

Fahrmeir, L. and Kaufmann, H. (1985). Consistency and asymptotic normality of the maximum likelihood estimator in generalized linear models. *The Annals of Statistics*, 13(1):342–368.

Feller, W. (1968). *An Introduction to Probability Theory and its Applications, Volume I.* Wiley, New York, third edition.

Ferguson, T. S. (1967). *Mathematical Statistics: A Decision Theoretic Approach.* Academic Press, New York.

Fieller, E. C. (1932). The distribution of the index in a normal bivariate population. *Biometrika*, 24(3/4):428–440.

Fienberg, S. (1971). Randomization and social affairs: The 1970 draft lottery. *Science*, 171:255–261.

Fink, D. (1997). A compendium of conjugate priors. Technical report, Montana State University, http://www.johndcook.com/CompendiumOfConjugatePriors.pdf.

Fisher, R. A. (1935). *Design of Experiments.* Oliver and Boyd, London. There are many editions. This is the first.

Fraser, D. A. S. (1957). *Nonparametric Methods in Statistics.* Wiley, New York.

Gibbons, J. D. and Chakraborti, S. (2011). *Nonparametric Statistical Inference.* CRC Press, Boca Raton, Florida, fifth edition.

Hastie, T. and Efron, B. (2013). lars: Least angle regression, lasso and forward stagewise. https://cran.r-project.org/package=lars.

Hastie, T., Tibshirani, R., and Friedman, J. H. (2009). *The Elements of Statistical Learning: Data Mining, Inference, and Prediction.* Springer-Verlag, second edition.

Henson, C., Rogers, C., and Reynolds, N. (1996). Always Coca-Cola. Technical report, University Laboratory High School, Urbana, Illinois.

Hoeffding, W. (1952). The large-sample power of tests based on permutations of observations. *The Annals of Mathematical Statistics*, 23(2):169–192.

Hoerl, A. E. and Kennard, R. W. (1970). Ridge regression: Biased estimation for nonorthogonal problems. *Technometrics*, 12(1):55–67.

Hogg, R. V., McKean, J. W., and Craig, A. T. (2013). *Introduction to Mathematical Statistics.* Pearson, seventh edition.

Huber, P. J. and Ronchetti, E. M. (2011). *Robust Statistics.* Wiley, New York, second edition.

Hurvich, C. M. and Tsai, C.-L. (1989). Regression and time series model selection in small samples. *Biometrika*, 76(2):297–307.

James, W. and Stein, C. (1961). Estimation with quadratic loss. In *Proceedings of the Fourth Berkeley Symposium on Mathematical Statistics and Probability, Volume 1: Contributions to the Theory of Statistics*, pages 361–379. University of California Press, Berkeley.

Jeffreys, H. (1961). *Theory of Probability*. Oxford University Press, Oxford, third edition.

Johnson, B. M. (1971). On the admissible estimators for certain fixed sample binomial problems. *The Annals of Mathematical Statistics*, 42(5):1579–1587.

Jonckheere, A. R. (1954). A distribution-free k-sample test against ordered alternatives. *Biometrika*, 41(1/2):133–145.

Jung, K., Shavitt, S., Viswanathan, M., and Hilbe, J. M. (2014). Female hurricanes are deadlier than male hurricanes. *Proceedings of the National Academy of Sciences*, 111(24):8782–8787.

Kass, R. E. and Raftery, A. E. (1995). Bayes factors. *Journal of the American Statistical Association*, 90(430):773–795.

Kass, R. E. and Wasserman, L. (1995). A reference Bayesian test for nested hypotheses and its relationship to the Schwarz criterion. *Journal of the American Statistical Association*, 90(431):928–934.

Kass, R. E. and Wasserman, L. (1996). The selection of prior distributions by formal rules. *Journal of the American Statistical Association*, 91(435):1343–1370.

Kendall, M. G. and Gibbons, J. D. (1990). *Rank Correlation Methods*. E. Arnold, London, fifth edition.

Kiefer, J. and Schwartz, R. (1965). Admissible Bayes character of T^2-, R^2-, and other fully invariant tests for classical multivariate normal problems (corr: V43 p1742). *The Annals of Mathematical Statistics*, 36(3):747–770.

Knight, K. (1999). *Mathematical Statistics*. CRC Press, Boca Raton, Florida.

Koenker, R. W. and Bassett, G. (1978). Regression quantiles. *Econometrica*, 46(1):33–50.

Koenker, R. W., Portnoy, S., Ng, P. T., Zeileis, A., Grosjean, P., and Ripley, B. D. (2015). quantreg: Quantile regression. https://cran.r-project.org/package=quantreg.

Kyung, M., Gill, J., Ghosh, M., and Casella, G. (2010). Penalized regression, standard errors, and Bayesian lassos. *Bayesian Analysis*, 5(2):369–411.

Lamport, L. (1994). *LaTeX: A Document Preparation System*. Addison-Wesley, second edition.

Lazarsfeld, P. F., Berelson, B., and Gaudet, H. (1968). *The People's Choice: How the Voter Makes up his Mind in a Presidential Campaign*. Columbia University Press, New York, third edition.

Lehmann, E. L. (1991). *Theory of Point Estimation*. Springer, New York, second edition.

Lehmann, E. L. (2004). *Elements of Large-Sample Theory*. Springer, New York.

Lehmann, E. L. and Casella, G. (2003). *Theory of Point Estimation*. Springer, New York, second edition.

Lehmann, E. L. and Romano, J. P. (2005). *Testing Statistical Hypotheses*. Springer, New York, third edition.

Li, F., Harmer, P., Fisher, K. J., McAuley, E., Chaumeton, N., Eckstrom, E., and Wilson, N. L. (2005). Tai chi and fall reductions in older adults: A randomized controlled trial. *The Journals of Gerontology: Series A*, 60(2):187–194.

Lumley, T. (2009). leaps: Regression subset selection. Uses Fortran code by Alan Miller. https://cran.r-project.org/package=leaps.

Madsen, L. and Wilson, P. R. (2015). memoir — Typeset fiction, nonfiction and mathematical books. https://www.ctan.org/pkg/memoir.

Mallows, C. L. (1973). Some comments on C_p. *Technometrics*, 15(4):661–675.

Mann, H. B. and Whitney, D. R. (1947). On a test of whether one of two random variables is stochastically larger than the other. *The Annals of Mathematical Statistics*, 18(1):50–60.

Matthes, T. K. and Truax, D. R. (1967). Tests of composite hypotheses for the multivariate exponential family. *The Annals of Mathematical Statistics*, 38(3):681–697.

Mendenhall, W. M., Million, R. R., Sharkey, D. E., and Cassisi, N. J. (1984). Stage T3 squamous cell carcinoma of the glottic larynx treated with surgery and/or radiation therapy. *International Journal of Radiation Oncology·Biology·Physics*, 10(3):357–363.

Pitman, E. J. G. (1939). The estimation of the location and scale parameters of a continuous population of any given form. *Biometrika*, 30(3/4):391–421.

Reeds, J. A. (1985). Asymptotic number of roots of Cauchy location likelihood equations. *The Annals of Statistics*, 13(2):775–784.

Sacks, J. (1963). Generalized Bayes solutions in estimation problems. *The Annals of Mathematical Statistics*, 34(3):751–768.

Schwarz, G. (1978). Estimating the dimension of a model. *The Annals of Statistics*, 6(2):461–464.

Sellke, T., Bayarri, M. J., and Berger, J. O. (2001). Calibration of p values for testing precise null hypotheses. *The American Statistician*, 55(1):62–71.

Sen, P. K. (1968). Estimates of the regression coefficient based on Kendall's tau. *Journal of the American Statistical Association*, 63(324):1379–1389.

Serfling, R. J. (1980). *Approximation Theorems of Mathematical Statistics* Wiley, New York.

Spiegelhalter, D. J., Best, N. G., Carlin, B. P., and van der Linde, A. (2002). Bayesian measures of model complexity and fit. *Journal of the Royal Statistical Society: Series B (Statistical Methodology)*, 64(4):583–616.

Stein, C. (1956a). The admissibility of Hotelling's T^2-test. *The Annals of Mathematical Statistics*, 27:616–623.

Stein, C. (1956b). Inadmissibility of the usual estimator for the mean of a multivariate normal distribution. In *Proceedings of the Third Berkeley Symposium on Mathematical Statistics and Probability, Volume 1: Contributions to the Theory of Statistics*, pages 197–206. University of California Press, Berkeley.

Stichler, R. D., Richey, G. G., and Mandel, J. (1953). Measurement of treadwear of commercial tires. *Rubber Age*, 73(2).

Strawderman, W. E. and Cohen, A. (1971). Admissibility of estimators of the mean vector of a multivariate normal distribution with quadratic loss. *The Annals of Mathematical Statistics*, 42(1):270–296.

Student (1908). The probable error of a mean. *Biometrika*, 6(1):1–25.

Terpstra, T. J. (1952). The asymptotic normality and consistency of Kendall's test against trend, when ties are present in one ranking. *Indagationes Mathematicae (Proceedings)*, 55:327–333.

Theil, H. (1950). A rank-invariant method of linear and polynomial regression analysis I. *Indagationes Mathematicae (Proceedings)*, 53:386–392.

Tukey, J. W. (1977). *Exploratory Data Analysis*. Addison-Wesley, Reading, Massachusetts.

van Zwet, W. R. and Oosterhoff, J. (1967). On the combination of independent test statistics. *The Annals of Mathematical Statistics*, 38(3):659–680.

Venables, W. N. and Ripley, B. D. (2002). *Modern Applied Statistics with S*. Springer, New York, fourth edition.

von Neumann, J. and Morgenstern, O. (1944). *Theory of Games and Economic Behavior*. Princeton University Press, Princeton, New Jersey.

Wald, A. (1950). *Statistical Decision Functions*. Wiley, New York.

Wijsman, R. A. (1973). On the attainment of the Cramér-Rao lower bound. *The Annals of Statistics*, 1(3):538–542.

Wilcoxon, F. (1945). Individual comparisons by ranking methods. *Biometrics Bulletin*, 1(6):80–83.

Zelazo, P. R., Zelazo, N. A., and Kolb, S. (1972). "Walking" in the newborn. *Science*, 176(4032):314–315.

Author Index

Wolpert, R. L. 200

Ylvisaker, D. 170

Zeileis, A. 192

Zelazo, N. A. 286

Zelazo, P. R. 286

Subject Index

The italic page numbers are references to Exercises.